Network and Discrete Location

**WILEY-INTERSCIENCE
SERIES IN DISCRETE MATHEMATICS AND OPTIMIZATION**

ADVISORY EDITORS

RONALD L. GRAHAM
AT & T Bell Laboratories, Murray Hill, New Jersey, U.S.A.

JAN KAREL LENSTRA
*Centre for Mathematics and Computer Science,
Amsterdam, The Netherlands
Erasmus University, Rotterdam, The Netherlands*

ROBERT E. TARJAN
*Princeton University, New Jersey, and
NEC Research Institute, Princeton, New Jersey, U.S.A.*

A complete list of titles in this series appears at the end of this volume

Network and Discrete Location

Models, Algorithms, and Applications

MARK S. DASKIN
Northwestern University

A Wiley-Interscience Publication
JOHN WILEY & SONS, INC.
New York • Chichester • Brisbane • Toronto • Singapore

This text is printed on acid-free paper.

Copyright © 1995 by John Wiley & Sons, Inc.

All rights reserved. Published simultaneously in Canada.

No part of this publication may be reproduced, stored in a retrieval sys
in any form or by any means, electronic, mechanical, photocopying, re
or otherwise, except as permitted under Sections 107 or 108 of the 19
Copyright Act, without either the prior written permission of the Publi
authorization through payment of the appropriate per-copy fee to the C
Clearance Center, 222 Rosewood Drive, Danvers, MA 01923, (978) 7:
(978) 750-4470. Requests to the Publisher for permission should be a
Permissions Department, John Wiley & Sons, Inc., 111 River Street, F
(201) 748-6011, fax (201) 748-6008.

To order books or for customer service please, call 1(800)-CALL-WIL

Library of Congress Cataloging in Publication Data:
Daskin, Mark S. 1952-
  Network and discrete location: models, algorithms, and
 applications / Mark S. Daskin.
    p. cm.
  Includes bibliographical references.
  ISBN 0-471-01897-X
  1. Industrial location—Mathematical models. I. Title.
  T57.6.D373 1995
  658.2'1'01156—dc20   94-24488
                                                        CIP

To

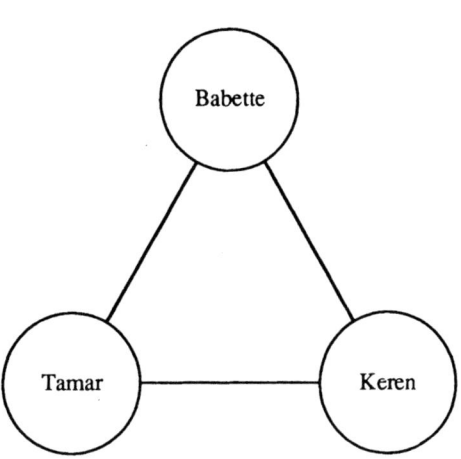

the absolute centers of my life.

Contents

Preface		xi
1 Introduction to Location Theory and Models		**1**
1.1.	Introduction	1
1.2	Key Questions Addressed by Location Models	3
1.3	Example Problem Descriptions	4
1.4	A Taxonomy of Location Problems and Models	10
1.5	Summary	18
	Exercises	19
2 Review of Linear Programming Problem		**20**
2.1	Introduction	20
2.2	The Canonical Form of a Linear Programming Problem	20
2.3	Constructing the Dual of an LP Problem	23
2.4	Complementary Slackness and the Relationships Between the Primal and the Dual Linear Programming Problems	25
2.5	The Transportation Problem	31
2.6	The Shortest Path Problem	41
2.7	The Out-of-Kilter Flow Algorithm	53
2.8	Summary	64
	Exercises	64
3 An Overview of Complexity Analysis		**80**
3.1	Introduction	80
3.2	Basic Concepts and Notation	81
3.3	Example Computation of an Algorithm's Complexity	84
3.4	The Classes P and NP (and NP-Hard and NP-Complete)	85
3.5	Summary	89
	Exercises	90

4 Covering Problems — 92

4.1	Introduction and the Notion of Coverage	92
4.2	The Set Covering Model	92
4.3	Applications of the Set Covering Model	105
4.4	Variants of the Set Covering Location Model	107
4.5	The Maximum Covering Location Model	110
4.6	The Maximum Expected Covering Location Model	130
4.7	Summary	134
	Exercises	135

5 Center Problems — 154

5.1	Introduction	154
5.2	Vertex P-Center Formulation	160
5.3	The Absolute 1- and 2-Center Problems on a Tree	162
5.4	The Unweighted Vertex P-Center Problem on a General Graph	173
5.5	The Unweighted Absolute P-Center Problem on a General Graph	176
5.6	Summary	191
	Exercises	191

6 Median Problems — 198

6.1	Introduction	198
6.2	Formulation and Properties	200
6.3	1-Median Problem on a Tree	203
6.4	Heuristic Algorithms for the P-Median Problem	208
6.5	An Optimization-Based Lagrangian Algorithm for the P-Median Problem	221
6.6	Computational Results Using the Heuristic Algorithms and the Lagrangian Relaxation Algorithm	232
6.7	Summary	236
	Exercises	238

7 Fixed Charge Facility Location Problems — 247

7.1	Introduction	247
7.2	Uncapacitated Fixed Charge Facility Location Problems	250
7.3	Capacitated Fixed Charge Facility Location Problems	275
7.4	Summary	302
	Exercises	303

Contents

8 Extensions of Location Models — 309

- 8.1 Introduction — 309
- 8.2 Multiobjective Problems — 309
- 8.3 Hierarchical Facility Location Problems — 317
- 8.4 Models of Interacting Facilities — 328
- 8.5 Multiproduct Flows and Production/Distribution Systems — 333
- 8.6 Location/Routing Problems — 339
- 8.7 Hub Location Problems — 349
- 8.8 Dispersion Models and Models for the Location of Undesirable Facilities — 363
- 8.9 Summary — 373
- Exercises — 374

9 Location Modeling in Perspective — 383

- 9.1 Introduction — 383
- 9.2 The Planning Process for Facility Location — 383
- 9.3 Summary — 398
- Exercises — 399

Appendices

- A SITATION Operations Guide — 401
- B NET-SPEC Operations Guide — 446
- C MOD-DIST Operations Guide — 450
- D COLORSET Operations Guide — 459
- E MENU-OKF Operations Guide — 463
- F PRINTER.CNS File Description — 474
- G Longitudes, Latitudes, Demands, and Fixed Costs for CITY1990.GRT: An 88-Node Problem Defined on the Continental United States — 476
- H Longitudes, Latitudes, Demands, and Fixed Costs for SORTCAP.GRT: A 49-Node Problem Defined on the Continental United States — 480

References — 483

Author Index — 491

Subject Index — 494

Preface

IMPORTANCE OF LOCATION MODELING

Almost every private and public sector enterprise that we can think of has been faced with the problem of locating facilities at one time or another in its history. Industrial firms must determine locations for fabrication and assembly plants as well as warehouses. Retail outlets must locate stores. Government agencies must locate offices and other public services including schools, hospitals, fire stations, ambulance bases, vehicle inspection stations, and landfills.

In every case, the ability of a firm to produce and market its products effectively or of an agency to deliver high-quality services is dependent in part on the location of the firm's or agency's facilities in relation to other facilities and to its customers. The focus of this book is on the development and solution of mathematical facility location models that can assist public and private sector decision makers.

GOALS OF THE TEXT

This text attempts to realize a number of goals. First, the book attempts to introduce the reader to a number of classical facility location models on which other more complicated and realistic models are based. Second, the book tries to assist the reader in developing his or her own modeling skills. Toward that end, many of the exercises at the end of the chapters ask the reader to formulate problems or to extend traditional formulations. Third, the text introduces the reader to a number of key methodologies that are used in solving facility location problems and the related problem of allocating demands to facilities. These methodologies include: linear programming, selected graph-theoretic algorithms, heuristic algorithms, Lagrangian relaxation, branch and bound, dual ascent algorithms, and Bender's decomposition. When it comes to methodologies, the goal is to teach the reader how the basic approach works, when it is useful, what information is provided by the approach, and how good the results are likely to be when the approach is employed. It is my hope that the modeling and methodological skills that the reader develops in the course of using this text will be transferable to problem contexts above and beyond those that arise in locating facilities. Fourth, the text attempts to introduce students to selected applications both through the text itself and through the exercises at the end of the chapters. Finally, the book provides students with software capable of solving most of the basic location problems on moderate-sized networks.

REQUIRED BACKGROUND

The background required for most of this book is quite minimal. Students should be familiar with *elementary* notions of linear algebra including the use of summation signs, subscripted variables, and other basic notation. In addition, some exposure to linear programming is desirable though not absolutely necessary as Chapter 2 reviews the essential concepts of linear programming that are used later in the text. To use the software, students should be familiar with the basic concepts and techniques of using an IBM-compatible computer. Specifically, students should know how to set up directories, copy and rename files, and start programs.

OUTLINE OF THE TEXT

Chapter 1 introduces the reader to facility location problems. A taxonomy of facility location problems and models is presented. Using this taxonomy, we distinguish between problems and models that are covered in the text and those that are outside the scope of the book. Specifically, within the broad family of facility location models, the text addresses problems that can be posed as network or discrete location problems. In fact, many real-world problems are of this sort. Specifically, decision makers must select sites from some finite set of candidate locations (as in discrete location problems) or they must locate facilities on a network (e.g., a highway network). Problems in which facilities can be located anywhere on the plane are outside the scope of the text.

Virtually all of the problems addressed in the text are formulated as integer linear programming problems. As such, linear programming is a key methodological tool used in solving the problems of interest. In addition, specialized linear programming problems are needed to obtain inputs to location problems or to solve allocation problems that arise once facility locations are known. For example, the shortest path distances between candidate locations and customers or demands are needed as inputs into many location models. Also, once facility locations are known, a special linear programming model known as the transportation problem is often used to assign customers to facilities. Chapter 2 provides an overview of linear programming in general. It also covers shortest path problems, the

Preface

transportation problem, and the out-of-kilter flow algorithm—a general-purpose algorithm for solving linear programming problems with network structure.

Chapter 3 introduces the reader to complexity theory. Using concepts presented in this chapter, we can distinguish between problems that can be optimally solved in an efficient manner and those for which no efficient, provably optimal, algorithm exists. For those problems for which an efficient, provably optimal, algorithm does not exist—and this includes most of the problems of interest in network and discrete location modeling—we are justified in using a heuristic or approximate algorithm. In addition, using notions introduced in this chapter, we can discuss the (worst case) running time of algorithms in a manner that is (with the exception of parallel processor machines) independent of the computer on which the algorithms are executed.

Chapters 4 through 7 introduce four classical facility location problems: covering, center, median, and fixed charge location problems. In many cases, the quality of service depends on whether or not the nearest facility is within some distance or time standard of the customer requesting the service. For example, one well-known pizza firm promises to deliver a pizza within 30 minutes. To realize this sort of promise, the firm must be sure that it has enough stores to reach all customers within 30 minutes. Many express delivery services have similar service guarantees for either pickups or deliveries. This leads to the notion of facilities being able to cover demands. Chapter 4 presents location covering models. The chapter begins with the set covering model which finds the locations of the minimum number of facilities needed to ensure a given level of service. Often this number is excessively large. This leads to the maximum covering location model which finds the locations of a given number of facilities to maximize the number of demands that can be served within the service standard. When we cannot serve all customers within the service standard with a reasonable number of facilities, another alternative is to relax the service standard. Center problems find the locations of a given number of facilities so that all customers are served within as tight a service standard as possible. In other words, they find the most stringent service standard that allows all customers to be served using a given number of facilities. Chapter 5 deals with center problems.

Covering and center problems focus on the worst case service. In many contexts, it is also important to examine the average service time or distance. The operating costs of many systems are strongly influenced by the average distance between facilities and customers. For example, in delivery systems, the trucking costs increase with the number of vehicle miles that must be traversed. Median problems, which are the focus of Chapter 6, find the locations of a given number of facilities to minimize the average distance between customers and the nearest facility.

When the number of facilities is an input to the model, we are implicitly assuming that the cost of locating at each of the candidate sites is approximately equal.[1] In that case, the number of facilities is a good proxy for the facility location costs. In many situations, however, the location costs may differ significantly across the candidate locations. In such cases, we will want to minimize both the operating costs (as is done using median models) and the fixed facility location costs. Chapter 7 deals with fixed charge location problems. These models generalize many of the earlier models. In addition, Chapter 7 introduces models in which facilities can serve only a limited number of demands.

Chapter 8 introduces a number of extensions to the basic models discussed in Chapters 4 through 7. Facility location problems are inherently strategic in nature. As such, multiple objectives must be considered in location analyses. Therefore, Chapter 8 begins with a discussion of multiobjective problems. In many cases, the facilities being located interact with each other. This leads to a discussion of hierarchical facility location models and, more generally, of models of interacting facilities. Problems with multiple products represent an important class of facility location problems. Within this class, production and distribution problems are of considerable importance. Such problems are covered in Section 8.5. Distribution of goods often occurs using routes with multiple customers. Chapter 8 also introduces location/routing models. In other cases, logistics systems operate using a hub-and-spoke system. The networks associated with many airlines also exhibit this fundamental structure. Hub location models are also discussed in this chapter. Dispersion models and location models for obnoxious facilities are the final topic covered in Chapter 8. Such models are useful in cases in which the facilities being located pose a risk of physical, economic, or psychological harm to either people or other facilities. Specifically, dispersion models might be used in locating missile silos or franchise stores (which should be located to minimize the degree of competition between stores of the same franchise), while obnoxious facility location models could be used to site landfills, hazardous waste repositories, or nuclear power plants. In all such cases, one objective is to maximize distances rather than to minimize distances.

As important as mathematical models are in assisting decision makers in locating facilities, they are only one part of a broader planning process. Chapter 9 presents one paradigm of the overall planning process which includes four major steps: (i) problem definition, (ii) analysis, (iii) communication and decision, and (iv) implementation.

Finally, Appendices A through E are operations guides for the five major programs provided at *http://sitemaker.umich.edu/msdaskin/software*. Appendix F

---

[1] In public sector problems, we often take the number of facilities as given since the costs associated with locating and operating the facilities are frequently borne by one group while another group frequently derives the benefits from the facilities being located. In these cases, combining the costs and benefits into one performance measure may be inappropriate; it may be preferable to solve the problem using a range of values for the number of facilities to locate.

documents a file that the user can modify to allow the programs to drive different printers. Appendices G and H summarize two sample data sets that are also included on the following website: *http://sitemaker.umich.edu/msdaskin/software*.

ACKNOWLEDGMENTS

No project of this magnitude could be accomplished without the help, support, and encouragement of many other individuals. I am indebted to many people including:

- My parents who have always been there when I needed them and who tried, over the years, to instill within me a love of mathematics, science, and engineering;
- My brother who has so often helped me maintain a sense of humor by reminding me not to take myself too seriously;
- Richard de Neufville who introduced me to systems analysis and optimization and who later suffered as my dissertation advisor;
- David Marks from whom I took my first graduate course on optimization and who is responsible, in many ways, for my having majored in civil engineering;
- Amedeo Odoni who has been an important professional mentor for the last 20 years;
- David Eaton with whom I worked on my first applied location problem and from whom I learned much of what I know about real problems;
- Joe Schofer and Frank Koppelman for their many years of guidance and support. Special thanks are due to Joe for his advice on parts of Chapter 9;
- Karen Donohue, Yinyi Xie, Susan Lash, and Dawn Barnes-Shuster who served as teaching assistants in IE C28, the course which motivated this project;
- The many students of IE C28 whose questions, comments, and ideas helped clarify my own thinking on numerous occasions;
- Art Hurter, Wally Hopp, and Phil Jones for their encouragement during the course of this project. Special thanks are due to Wally for having read early versions of the book as carefully as he did;
- John Ivan, Rapee Vattanakul, Mirali Sharifi-Takieh, and Chandra Bhat for their detailed comments on sections of the book;
- Michael Kuby whose careful reading of the entire text and thoughtful comments and suggestions significantly improved the original draft; and,
- Kate Roach and Maria Allegra of John Wiley & Sons, Inc., whose encouragement and support of this project were clearly critical.

Finally, I am most deeply indebted to my wife, Babette, and my daughters, Tamar and Keren, for their support, encouragement, love, and patience throughout this project. They spent many evenings and weekends staring at my back as I typed text, figures, equations, and code into one or more of my computers. Without them, this book would never have been finished. I hope that we can reclaim our time together now that this book is completed.

MARK S. DASKIN

1

Introduction to Location Theory and Models

1.1 INTRODUCTION

If you ask what to look for in buying a house, any realtor will tell you that there are three things that are important: location, location, and location. The theory behind this answer is that the community in which you elect to live and the location within that community are likely to affect your quality of life at least as much as the amenities within your house. For example, if you live within walking distance of the local elementary school, your children will not need to be bused to school. If you live near a community center, you may be able to avoid involvement in car pools taking children to and from activities. If your house is too close to a factory, noise, traffic, and pollution from the factory may degrade your quality of life.

Location decisions also arise in a variety of public and private sector problems. For example, state governments need to determine locations for bases for emergency highway patrol vehicles. Similarly, local governments must locate fire stations and ambulances. In all three of these cases, poor locations can increase the likelihood of property damage and/or loss of life. In the private sector, industry must locate offices, production and assembly plants, distribution centers, and retail outlets. Poor location decisions in this environment lead to increased costs and decreased competitiveness.

In short, the success or failure of both private and public sector facilities depends in part on the locations chosen for those facilities. This book presents methods for finding desirable or optimal facility locations.

We should emphasize from the beginning that the word "optimal" is used in a *mathematical* sense. That is, we will define quantifiable objectives that depend on the locations of the facilities. We will then identify algorithms (rigorous procedures) for finding optimal or at least good facility locations.

Two factors limit the broader optimality of the sites suggested by the optimization models discussed in this text. First, in many cases, nonquantifiable objectives and concerns will influence siting decisions to a great extent. Often, the qualitative factors that influence siting decisions are critically important. Thus, to the extent that the procedures discussed in this text ignore qualitative concerns and factors, the sites identified by the mathematical algorithms are *optimal* only in a narrow sense of the word. Second, the performance of a system is affected by many factors of which location is only one. For example, the ability of an ambulance service to save lives (an objective that many would attribute to such a service) depends not only on the proximity of the ambulance bases to the calls for service (which can be measured and optimized to some extent) but also on such factors as: the training and skill of the paramedics, the public's knowledge of emergency medical procedures and when it is appropriate to call for an ambulance, the existence of a 911 emergency line, and the protocols and technologies employed by the paramedics.

In the face of (1) exogenous qualitative concerns that influence siting decisions and (2) nonlocation factors that affect the performance of facilities, one might legitimately ask, "Why bother developing mathematical location models?" There are a number of answers to this question. First, while location is not the only factor influencing the success or failure of an enterprise, it is critical in many cases. Poorly sited ambulances will lead to an increased average response time with the associated increase in mortality, that is, more deaths. Second, while exogenous qualitative factors will influence siting decisions, mathematical models allow us to quantify the degradation in the quantifiable objectives that comes from recognizing the qualitative concerns. Thus, if it is important to locate an ambulance in one district for political reasons, the increase in average response time (or maximum response time) resulting from the imposition of this political constraint can be quantified. Third, the modeling process (identifying objectives and constraints and collecting data) often improves the decisions that are made even if the models are never run. Fourth, there are nonlocation problems in which models identical to those discussed in later chapters arise. For example, the problem of selecting tools in a flexible manufacturing context is mathematically similar to that of locating ambulances in a city (Daskin, Jones, and Lowe, 1990).

Section 1.2 outlines a number of key questions that are addressed in location models. Section 1.3 extends the discussion of ambulance location problems and introduces some of the terminology used in location modeling. In this section we also describe qualitatively another location problem—that of locating landfill sites for solid wastes. Section 1.4 outlines a taxonomy of location models that facilitates identification and classification of different types of location problems. Finally, Section 1.5 summarizes the chapter.

1.2 KEY QUESTIONS ADDRESSED BY LOCATION MODELS

Mathematical location models are designed to address a number of questions including:

a. How many facilities should be sited?
b. Where should each facility be located?
c. How large should each facility be?
d. How should demand for the facilities' services be allocated to the facilities?

The answers to these questions depend intimately on the context in which the location problem is being solved and on the objectives underlying the location problem. In some cases, such as ambulance siting problems, we will want to locate the facilities as near as possible to the demand sites. In locating radioactive waste repositories, we will want to be in a geologically stable region and would like to be as far as possible from major population centers.

The number of facilities to be located as well as the size of the individual facilities is often a function of the service/cost tradeoffs. In many cases, the quality of service improves as the number of facilities located increases, but the cost of providing the service also increases. For example, having more ambulances is generally preferable to having fewer ambulances since the likelihood of having an available ambulance near a call for service increases with the number of vehicles deployed. In addition, there are no significant economies of scale in ambulance operations; that is, having a single site with multiple ambulances is not much cheaper than having the same number of ambulances at multiple locations. Thus, having a large number of single vehicle ambulance bases is likely to be preferable to having a small number of multivehicle bases. At some point, however, the quality of medical care provided by the paramedics *degrades* as more ambulances are added to the system. The reason for this is that there are not enough demands requiring a variety of medical skills to maintain the paramedics' training. In many manufacturing contexts, there are significant economies of scale, which drive the location decisions toward having a smaller number of large facilities.

Facility location models are also concerned with the allocation of demands to facilities. In some cases, it is important that demands at a site not be split between facilities. For example, in some retailing operations, a retail store must be supplied by a single warehouse. For administrative reasons, the store's supply cannot be split between different warehouses. In other cases, such as ambulance services, demands can be served by any available facility. Facility location models must reflect these different demand allocation policies and must then allocate demands (or fractions of the total demand in a region) to different facilities. In many cases, demands will be allocated to the nearest (available) facility; in other cases, doing so may not be optimal.

1.3 EXAMPLE PROBLEM DESCRIPTIONS

In this section we outline a number of different facility location contexts and qualitatively define some of the classical location problems.

1.3.1 Ambulance Location

As indicated above, poor ambulance locations can cost lives! To illustrate this point, a commonly cited statistic is that if a person's brain is denied oxygen for more than 4 minutes (as a result of a stroke or heart attack, for example), the likelihood of the individual surviving to lead a normal life drops below 50 percent. This suggests that we would like to locate ambulances so that the maximum response time is well under 4 minutes. Thus, one objective might be to *minimize the number of ambulances needed so that all demand nodes are within a given number of minutes (the service standard) of the nearest ambulance*. Such a model formulation is known as a *set covering model*. Demands are said to be *covered* if the nearest ambulance is located not more than X minutes away, where X is the service standard used in the model (e.g., 4 minutes). Set covering models have been used by a number of authors in locating ambulances and other emergency service vehicles (e.g., Toregas, Swain, ReVelle, and Bergman, 1971; Walker, 1974; Plane and Hendrick, 1977; Daskin and Stern, 1981; Jarvis, Stevenson, and Willemain, 1975).

One of the common problems associated with the set covering model is that the solution is likely to call for locating more vehicles than the community can afford. If we deployed one less vehicle and relocated the remaining vehicles to maximize the number of demands that can be served within the given service standard (e.g., 4 minutes), the fraction of the demands that would not be serviceable within the service standard would generally be far less than $1/N$, where N is the number of ambulances called for by the set covering model. In other words, the last few ambulances add relatively little to the fraction of demands that can be served within the service standard but add significantly to the cost of the ambulance service. This suggests an alternate objective: *maximize the number of demands that can be covered within a specified service standard using a given number of vehicles*. Such a model is known as a *maximum covering model*. In practice, the fleet size that is input into such a model is often varied from 1 up to the number required for full coverage as indicated by the set covering model. This allows us to trace out the tradeoff between additional vehicles and coverage. Such a curve is shown in Figure 1.1. In this example, 8 vehicles are needed to cover all demands. In other words, the solution to the set covering model for this problem is 8 vehicles. The maximum covering model would then be solved for

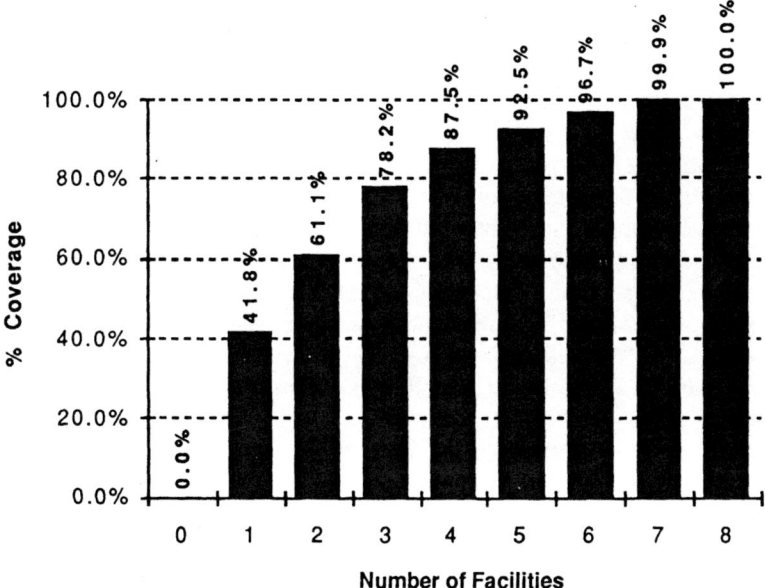

Figure 1.1. Typical tradeoff in maximum covering model.[1]

1 through 7 vehicles in the system. Notice that the incremental coverage *decreases* as additional vehicles are added to the system. Maximum covering models and their variants have also been used in analyzing ambulance systems and related emergency services (e.g., Daskin, 1982; Daskin, 1983; Eaton et al., 1985; Church and ReVelle, 1974; Belardo et al., 1984).

In some cases, logical choices for the service standard might not be readily available. The choice of 4 minutes in the discussion of the covering models was predicated on the observation that irreversible brain damage is likely to occur if the brain is denied oxygen for more than 4 minutes. However, this does not necessarily imply that 4 minutes is the appropriate service standard. A shorter service standard could be justified by the observation that the clock for brain damage beings at the onset of the medical incident (the stroke or heart attack that denies the brain oxygen), while the clock for the response time begins only once the vehicle begins to roll out of its base. There is often a long time (several minutes) between when a medical incident arises and when a vehicle begins traveling to the scene. This additional time is consumed by the time required to recognize the need for an ambulance, the time to notify the dispatcher of the need, and the time required by the dispatcher to assign the call to a vehicle and to notify the vehicle's crew of the call. On the other hand, a longer service standard may be dictated by budgetary

---

[1]The tradeoff curve shown is for the 88-node, CITY1990. GRT, data set described in Appendix G with a coverage distance of 410 miles. The results shown are optimal values.

considerations. Requiring all demands to be served within 4 minutes may be too costly. It would be cheaper to have all demands served within 5 minutes or some longer time period. This suggests yet another model and another objective function: *minimize the maximum response time (the time between a demand site and the nearest ambulance) using a given number (P) of vehicles.* Such a model is referred to as the *P-center problem*.

Covering and center problems focus on the worst case behavior of the system, for example, the maximum response time. In practice, there is often a tradeoff between minimizing the maximum response time and minimizing the *average* response time. This suggests yet a fourth model and objective that might be used in locating ambulances: *minimize the average response time (the time between a demand site and the nearest ambulance) using a given number (P) of vehicles.* This model is called the *P-median problem* (Hakimi, 1964; Hakimi, 1965).

While we have outlined a number of different objectives that might be used in locating ambulances, it is important that we realize explicitly some of the factors that have been ignored in this discussion. The importance of doing so was enunciated particularly well by Jacobsen (1990, p. 205) who pointed out that formulating a problem incorrectly (e.g., failing to account for important problem factors) is likely to be far more important than whether or not you obtain an optimal or suboptimal solution to a particular problem formulation. Thus, while the focus of this book is generally on finding optimal solutions (or near optimal solutions) to specific mathematical statements of facility location problems, we must always ask whether the model being solved adequately represents the real problem being analyzed.

In the context of ambulance location, at least three facets of the real-world problem have been ignored in the discussion above. First, the models outlined above ignore the stochastic (or random) nature of demands and the fact that the nearest vehicle might not be available when called upon to serve a demand. A variety of approaches have been adopted to address this problem including: extending the deterministic models outlined above (Aly and White, 1978; Weaver and Church, 1983b; Weaver and Church, 1984; Daskin, 1982; Daskin, 1983); incorporating queuing theory into location models (Larson, 1974; Larson, 1975; Fitzsimmons, 1973); and simulation approaches (Swoveland et al., 1973). Once the inputs to the model are recognized as being random variables, the outputs are likely to be random variables as well. Thus, we might no longer be interested only in the *average* response time (as in the *P*-median model) but also in the distribution of response times. Also, just as the demands are stochastic, so are the travel times. Models with stochastic travel times have also been developed (Weaver and Church, 1983a; Mirchandani and Odoni, 1979; Daskin and Haghani, 1984; Daskin, 1987).

Second, there is a need to balance the workload of the different vehicles. This stems from the needs (1) to preserve morale among the emergency medical service employees and (2) to maintain the skill level of all paramedics at some minimal level by ensuring that they are all exposed to a minimum number of medical emergencies of differing types.

As in all situations, we must ask whether facility location is really the correct problem. The quality of medical care delivered to the public and the likelihood of people surviving major medical incidents (e.g., auto crashes, assaults with deadly weapons, heart attacks and strokes) depend on many factors in addition to the location of ambulances in the community. The installation of a 911 emergency phone line can reduce the time needed to contact an ambulance dispatcher. This reduction in time may be greater (as compared with its cost) than that achievable by any relocation of ambulances. Improving the quality of hospital emergency room care may also go a long way toward reducing fatalities. It may be more cost effective to spend public funds on such improvements than it would be on relocating vehicles or adding ambulances. Instituting a community-wide CPR (cardiopulmonary resuscitation) education program might also be a cost-effective way of saving lives.[2]

The analysis assumed that all calls are equally important. In fact, this is not the case. Calls are often differentiated into critical (life-threatening) and noncritical calls. Also, some patients can be cared for at the scene of the incident, while others require transport to the hospital. In addition, the models outlined above fail to recognize the temporal variation in the overall intensity of calls (typically Friday night is the busiest time of the week) and the temporal variation in the spatial distribution of calls (more incidents will be reported from business districts during working hours than during the early morning hours). This temporal variation in demand suggests that having fixed sites may not be optimal; using relocatable ambulances may be preferable (Carson and Batta, 1990).

Once we distinguish between the severity of different demands for ambulance services, we recognize it may be advantageous to institute a multitiered system in which paramedics with differing levels of training are deployed (along with vehicles with correspondingly different levels of equipment). It may not be cost effective to have all paramedics trained at the highest level and to have all vehicles capable of responding to all types of medical emergencies. We may be able to deploy more vehicles and paramedics by (i) disaggregating calls based on their severity and (ii) allowing response units and personnel to be specialized for certain types of calls. In a multitiered system, dispatching rules become even more complicated. Not only might it not be advantageous to dispatch the nearest available vehicle (since doing so

---

[2] The notion of asking whether location solutions are the best way to attack a problem extends well beyond emergency services. For example, in considering problems of energy management, one solution might be to install additional power generation facilities. Another solution might be to manage the peak demand for power better.

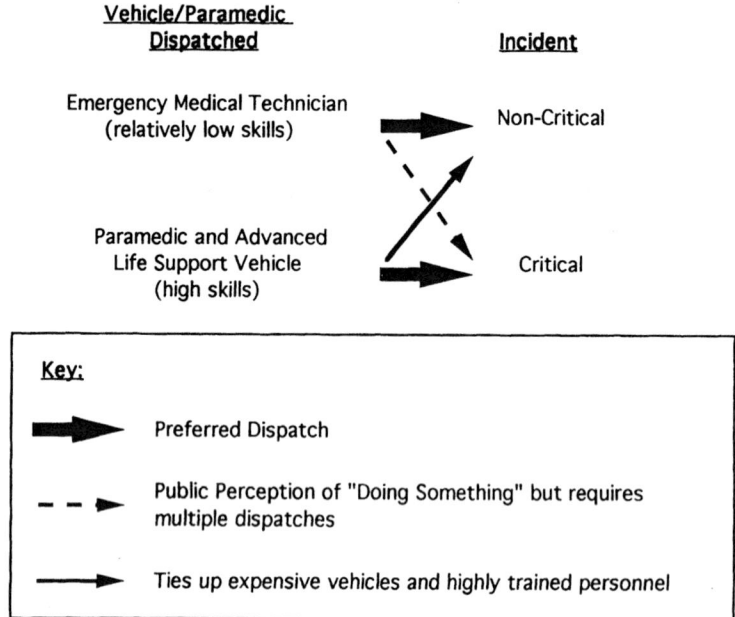

Figure 1.2. Dispatching options in multitiered system.

might leave large portions of the service area uncovered), but we must now decide which type of vehicle and crew to dispatch to each event. The possibilities are shown schematically in Figure 1.2. If possible, we should dispatch an EMT (emergency medical technician) to a noncritical call and a paramedic in an advanced life support (ALS) vehicle to a critical incident. However, we might also want to dispatch an EMT to a critical call if the vehicle is likely to get to the scene before the ALS vehicle. Doing so would give the public the impression that something is being done about the emergency. This policy, however, ties up extra resources since two vehicles would then be dispatched under these conditions. Similarly, if an ALS vehicle is very near a noncritical incident, we might elect to dispatch the ALS vehicle. This policy has the advantage of getting medical assistance to the scene quickly, but has the disadvantage of tying up an expensive ALS vehicle and highly trained crew which might be needed elsewhere while they are busy serving the noncritical incident.

Finally, we note that the models we have briefly outlined, in particular the set covering model, the maximum covering model, and the P-center model, have all been used extensively in a broad range of public sector facility location problems including the location of libraries, schools, clinics, hospitals, and bus stops.

1.3.2 Siting Landfills for Hazardous Wastes

We turn briefly to another problem, that of locating landfills for disposal of hazardous wastes. First, any such site must be deemed geologically stable and suitable. Given this condition, the choice between sites might be dictated by a number of objectives. First, we would like to be as close as possible to the waste generation sites to minimize the transport costs as well as the exposure of the public to the hazardous wastes while they are en route to the disposal site. Minimizing the average (or total) shipping distance over some period of time results in a P-median formulation. However, many of the waste generation sites may be close to heavily populated regions. In this case, we would like the disposal sites to be far from populated areas. This suggests the use of a *maxisum or maxian model* in which we attempt *to locate a given number (P) of facilities to maximize the (population weighted) distance between population centers and the nearest sites* (Church and Garfinkel, 1978; Minieka, 1983). Clearly, this objective and the P-median objective conflict. The presence of conflicting objectives is common in facility location problems.

Both models take the number of facilities as given. In practice, we need to balance the initial capital investment costs against the ongoing operating costs. Thus, we might like to minimize the sum of the fixed site preparation costs and the discounted present cost of the stream of operating costs (on-site operating costs and transport costs, for example). This leads to what is known as a *fixed charge facility location problem*.

Finally, we would like to reduce the inequities across communities. No community wants to be the dumping site for the rest of the state or the rest of the country. Thus, we might like to spread the risk or disbenefit around to the extent that it is possible to do so (e.g., Ratick and White, 1988; Erkut and Neuman, 1992; Wyman and Kuby, 1994). This has recently become an issue not only in the location of disposal sites, but also in the routing of materials from generation sites to disposal facilities (Lindner-Dutton, Batta, and Karwan, 1991; ReVelle, Cohon, and Shobrys, 1991; List and Mirchandani, 1991; List et al., 1991).

1.3.3 Summary

In summary, modeling location problems requires an understanding of the real-world operations that are to be reflected in the model. Models need not reflect *every* aspect of the real-world operations. In fact, parsimonious models are generally better than complex inscrutable models. The ability to know what must be incorporated into a model and what can safely be treated as exogenous is both an art and a science. As illustrated above, location problems often involve multiple conflicting objectives. The purpose of modeling is to identify the tradeoffs between the objectives while capturing as much of the richness of the real-world problem as is necessary to ensure the credibility of the modeler and model itself. Finally, we must always ask whether improving facility locations is the most cost-effective way of improving the system under study.

1.4 A TAXONOMY OF LOCATION PROBLEMS AND MODELS[3]

Location problems and models may be classified in a number of ways. The classification may be based on the topography that is used (e.g., planar problems versus discrete location problems, problems on trees versus those on general graphs, and problems using different distance metrics), or the number of facilities to be located. Problems may also be classified based on the nature of the inputs (e.g., whether they are static or dynamic, known with certainty or only known in a probabilistic sense). Models may further be classified based on whether single or multiple products or demands must be accommodated by the facilities being located, whether there is one objective or multiple objectives, whether the beneficiaries and investors are the same or different actors, whether the facilities are of unlimited capacity or are capacitated, as well as a variety of other classification criteria. This section develops a classification scheme for location models and uses that scheme to help identify those problems that will be the primary focus of this text.

1.4.1 Planar versus Network versus Discrete Location Models

One of the key differences between location models is in the way in which demands and candidate facility locations are represented. In *planar location models*, demands occur anywhere on a plane. We often represent demands using a spatially distributed probability distribution [which gives the likelihood of demands arising at any given (X, Y) coordinate]. In such problems, facilities may be located anywhere on the plane. This modeling approach is to be contrasted with *network location models*, in which demands and travel between demand sites and facilities are assumed to occur only on a network or graph composed of nodes and links. Often, we assume that demands occur only at the nodes of the network, though some network location models have permitted demands to be generated anywhere on the links of the network. In network location models, facilities can be located only on the nodes or links of the network. One of the key questions we will be interested in considering is: when is location only on the nodes of the network optimal? The presence of an underlying network often facilitates the development of solution algorithms. *Discrete location models* allow for the use of arbitrary distances between nodes. As such, the structure of the underlying network is lost.

[3]Similar taxonomies have been developed by Brandeau and Chiu (1989) and Krarup and Pruzan (1990).

However, by removing the restriction that the distances between nodes be obtained from an underlying network, the more general class of discrete location models allows a broader range of problems to be modeled. Discrete location problems are generally formulated as mixed integer programming problems as discussed below. For a further discussion of the differences between these three types of models, the reader is referred to Chhajed, Francis, and Lowe (1993).

The focus of this book is on network and discrete location models. Handler and Mirchandani (1979) and Mirchandani and Francis (1990) provide excellent overviews of network location models, while planar location models are discussed in Hurter and Martinich (1989) and Love, Morris, and Wesolowsky (1988).

1.4.2 Tree Problems versus General Graph Problems

Within the class of network location models, we often distinguish between problems that arise on *trees* and those that must be formulated on a more general (fully connected) *graph*. Figure 1.3 illustrates a number of different trees and general networks.

A tree is a network in which there is at most one path from any node to any other node. In other words, a tree is an acyclic graph or a graph with no cycles. In general, we will focus our attention on spanning trees (trees in which there is exactly one path between any node and any other node). If such a tree has N nodes, it will have $N - 1$ links.

Our interest in trees as opposed to more general graphs results from two considerations. First, many real-life problems can be represented quite well as trees. For example, the links depicting major highways within a region often form a tree as long as we ignore the cycles formed by beltways surrounding major urban areas. Also, major parts of power transmission and telecommunication networks—particularly the portions used for the delivery of local services—are essentially trees. Second, given a verbal or mathematical statement of a location problem, it is often the case that we can solve the problem easily on a tree while solving it on a more general network is exceptionally difficult. In Chapter 3 we formalize the notions of easy and difficult problems using complexity theory.

1.4.3 Distance Metrics

Location models are also often characterized by the distance metric (the method of measuring distances) that is used. For network location models, we will generally use the shortest distance between any pair of points using links in the network. In Chapter 2 we discuss algorithms for finding the shortest paths between points in a network. In planar location problems, one

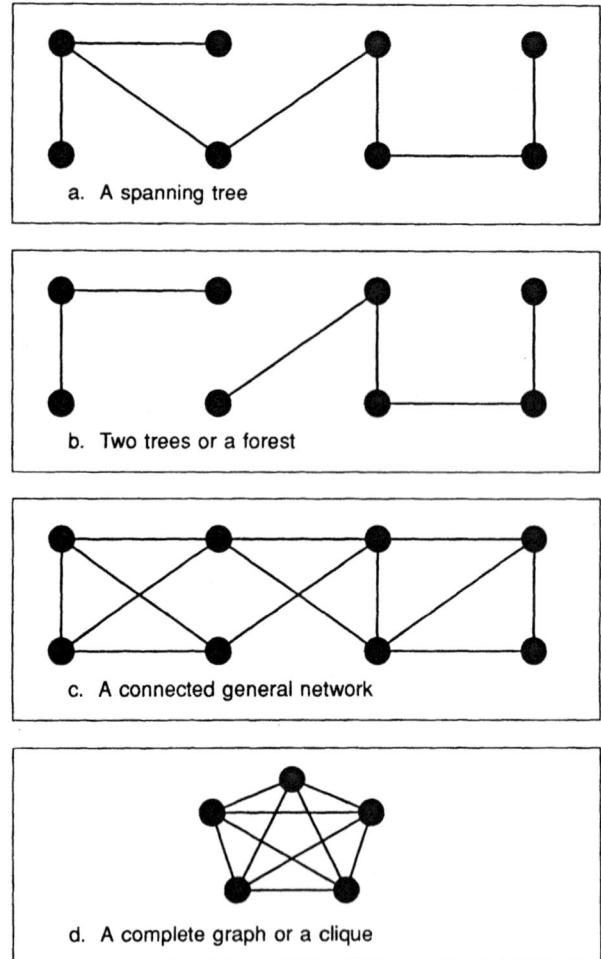

Figure 1.3. Example trees and graphs.

of three distance metrics is typically employed:

a. Manhattan or right-angle distance metric

$$d[(x_i, y_i); (x_j, y_j)] = |x_i - x_j| + |y_i - y_j|$$

b. Euclidean or straight-line distance metric

$$d[(x_i, y_i); (x_j, y_j)] = \{(x_i - x_j)^2 + (y_i - y_j)^2\}^{0.5}$$

c. l_p distance metric

$$d[(x_i, y_i);(x_j, x_j)] = \left\{ \left(|x_i - x_j|\right)^p + \left(|y_i - y_j|\right)^p \right\}^{1/p}$$

where $d[(x_i, y_i);(x_j, y_j)]$ is the distance between the ith and jth points and (x_i, y_i) gives the coordinates of the ith point. A bit of thought will show that the l_2 metric (the l_p metric when $p = 2$) is the same as the Euclidean distance between two points and that the l_1 metric is equivalent to the Manhattan or right-angle distance. What is the l_∞ metric? This is left as an exercise for the reader.

1.4.4 Number of Facilities to Locate

Another way of characterizing facility location problems is by the number of facilities to be located. In some problems (e.g., the P-median, P-center, and maximum covering problems), the number of facilities to locate is *exogenously* specified. In other cases (e.g., the set covering problem and the fixed charge facility location problem), the number of facilities is *endogenous* to the problem and is a model output. For those problem statements in which the number of facilities to locate is exogenously specified, we also distinguish between *single-facility* location problems and those in which *multiple facilities* are to be sited. Often, single-facility location problems are dramatically easier than are their multifacility counterparts.

1.4.5 Static versus Dynamic Location Problems

Most of the location models that we will consider will be *static* problems. In static models, the inputs do not depend on time; typically, we will use a single "representative" set of inputs and solve the problem for a single "representative" period.

As noted above in the discussion of ambulance systems, inputs are rarely static. Thus, while most location *models* are static, most location *problems* are *dynamic* in that the inputs (and consequently the outputs as well) depend on time. Inputs that may depend on time include demands, costs, and available and preexisting candidate facility locations. In dynamic problems, the models must explicitly include multiple periods of time. Different periods might allow us (i) to capture hourly differences in the mean number of demands for service, (ii) to reflect differences between the spatial patterns of demands on weekdays and weekends, or (iii) to account for increases in demands or costs over a period of years.

In dynamic problems, we are concerned not only with the question of *where* to locate facilities, but also with the question of *when* to invest in new facilities or to close existing facilities. In some models of dynamic location problems, once a facility is opened it is assumed to be available for all future periods. In other models, facilities may be opened, closed, or moved throughout the planning horizon (Ballou, 1968; Sweeney and Tatham, 1976; Van Roy and Erlenkotter, 1982; Wesolowsky, 1973; Wesolowsky and Truscott, 1975).

While most researchers and planners have a good idea of what is meant by a static location model, there is considerably less agreement about what is meant by a dynamic location model. One approach might be to identify a single set of locations that perform well with respect to a number of spatially different demand patterns that occur at different times. Such a problem statement might arise in locating fire stations that need to respond well to demands during working hours as well as on weekends. This approach might also be appropriate in locating facilities to serve demands that vary in a cyclic manner (e.g., Osleeb and Ratick, 1990). A second approach to the dynamic location problem would be that of identifying the optimal evolution of facility locations over time. Such a model would be appropriate for a firm that needs to locate warehouses to supply its customers and that plans to expand from a set of regional retail outlets to a national chain. In some cases, it is best to find an optimal first period decision as opposed to a plan for all future time periods (Daskin, Hopp, and Medina, 1992). Finally, an alternate definition of the dynamic location problem would be that of positioning vehicles in real time to respond to minute-by-minute changes in the fleet of available (nonbusy) vehicles. This problem in particular has been analyzed by Kolesar and Walker (1974) using set covering models.[4]

1.4.6 Deterministic versus Probabilistic Models

Just as the inputs to models may be either static or dynamic, so too the inputs may be deterministic (certain) or probabilistic (subject to uncertainty). In dealing with location problems over time, many of the inputs are likely to be uncertain. For example, future calls for ambulance services are not known with certainty. Instead they must be predicted and, as such, are subject to uncertainty. This book focuses on deterministic models, though in some cases we can readily generalize the algorithms or model formulations to include some probabilistic components. Louveaux (1993) reviews stochastic location models.

1.4.7 Single- versus Multiple-Product Models

The models outlined above have all implicitly assumed that we are dealing with a single homogenous product or service and that all demands are identical. Most location models make this assumption. However, in practice, it is often important to distinguish between different products or services all

---

[4]Ratick et al. (1987) review dynamic location models. They distinguish between models in which facilities remain in the siting plan once they are opened and models which allow facilities to be opened and closed throughout the planning horizon.

of which will be served by the same set of facilities. For example, it may be important to distinguish between critical and noncritical calls.

In some cases, products are distinguished by having different origins and destinations. For example, a single set of transshipment facilities may be used by an automobile manufacturer in shipping finished vehicles from assembly plants to dealers. At such transshipment points, vehicles are offloaded from railcars and loaded onto trucks for final delivery to customers (typically, dealers). Each assembly plant/customer combination would represent a different product. In other words, we would need to distinguish between Cadillac Sevilles going from a Cadillac assembly plant to a Cadillac dealer in San Diego and Chevrolet Corvettes going from a Chevrolet plant to a dealer in Los Angeles, even though both vehicles might use the same transshipment point in southern California.

1.4.8 Private versus Public Sector Problems

In private sector problems, the investment costs and benefits are typically measured in monetary units. Furthermore, the costs and benefits are generally incident on the same actors: the firm, its management, and its investors all of whom share common objectives and goals. All this makes cost/benefit analysis relatively easy.

In public sector location problems, many nonmonetary cost and benefits must also be considered. For example, in locating hazardous waste repositories there are a number of environmental costs which may be difficult to translate into monetary units. In locating emergency services, the dollar value of the lives saved as a result of shorter travel times may be exceedingly difficult to assess. In siting public schools, the benefits may be measured in terms of the number of students who graduate from high school. In public sector problems, not only are costs and benefits often incommensurable, but there are often multiple benefit measures (as discussed in Section 1.4.9). In addition, while the costs of public sector projects may be borne by the public at large, the benefits are often concentrated on fewer people. Thus, investment in public schools directly benefits school-aged children and their parents. Such investments do not directly benefit other members of society such as the elderly. Finally, public sector investments are often complicated by the political process in which beneficiaries of one investment may agree to support projects from which they do not directly benefit. Thus, groups representing the elderly may agree to support additional funding of public schools provided other groups support enhanced health care legislation.[5]

---

[5]ReVelle, Marks, and Liebman (1970) were among the first to distinguish between public and private sector location problems. Ghosh and Harche (1993) provide a recent review of location models used in private sector decision making.

1.4.9 Single-versus Multiple-Objective Problems and Models

Most models capture a single objective; however, most problems are inherently multiobjective in nature. Since one of the purposes of location modeling is to help identify tradeoffs, single-objective models must often be run with a range of input parameters (e.g., running a P-median problem with a number of values of P to trace the tradeoff between average distance to a facility and the number of facilities sited). Alternatively, multiple models need to be employed (e.g., Eaton and Daskin, 1980).

1.4.10 Elastic versus Inelastic Demand

Most models treat demand as given and independent of the level of service. In fact, demand in almost all cases depends on the level of service provided. This, in turn, depends on the facility locations and the types and sizes of facilities used. In some cases, demand is likely to be relatively inelastic (independent of the level of service). For example, if someone needs an ambulance, he or she is unlikely to inquire about the cost. An individual is also unlikely to bother to find out the expected arrival time of the vehicle and to identify alternative means of getting to the hospital if the expected response time is too long. On the other hand, consumers' choices of where to shop depend critically on the amenities within the shopping center, the location of the center, and the number and variety of stores in the shopping center. Despite the fact that demand in most real-world location problems exhibits some degree of elasticity with respect to service (which depends in part on the location decisions), we will generally treat demand as inelastic. Recent work by Perl and Ho (1990) has examined some of the implications of elastic demand on public facility location models. Kuby (1989) formulates a model that maximizes the number of firms that can coexist in a market. His model also incorporates elastic demand.

1.4.11 Capacitated versus Uncapacitated Facilities

Many facility location models (e.g., standard set covering, maximum covering, P-median, and P-center models) treat facilities as having unlimited capacity. Other models impose explicit capacity limits on facilities. In still other cases, the size of a facility is a model output.

1.4.12 Nearest Facility versus General Demand Allocation Models

As discussed above, the allocation of demand to facilities is a critical issue in location modeling. Often, demands are assigned to the nearest facility provided that facility has the capacity to serve the demand. In capacitated problems, this may result in the need to split the demand at a site between several facilities. If this is not permissible in a particular problem setting, explicit constraints must be included in the model (typically in the form of integer variables) to force all of the demand at a particular location to be assigned to a single facility. This may result in some demands being assigned to facilities other than the closest site. In still other cases, models must recognize that a fraction of the demand at a site will be served by the nearest facility and the remainder of the demand will be served by more remote facilities when the nearest facility is busy.

1.4.13 Hierarchical versus Single-Level Models

In many systems, a hierarchy of facilities exists with flows between the facilities that are being located. For example, in a national health care system, rural health centers are likely to refer patients to clinics which, in turn, may refer patients to community hospitals. In some such systems, services provided at the lower level (e.g., the rural health center) are offered at higher levels; in other cases, these services are not replicated. Narula (1986) refers to these as *successively inclusive* and *successively exclusive facility hierarchies*, respectively. Also, in some systems, patients may elect to go to the facility of their choice; in others, they must begin service at the lowest level facility in the hierarchy and be referred up from there. In such hierarchical location problems, the locations of the different facilities interact significantly through the flows between the facilities. Facility interactions also arise in many facility layout problems (Francis, McGinnis, and White, 1992).

1.4.14 Desirable versus Undesirable Facilities

In most location problems, we are interested in locating desirable facilities. In other words, value increases, in some sense, the closer the facilities are to the people or goods being served. Ambulances, fire stations, schools, hospitals, post offices, warehouses, and production plants are all considered desirable facilities in this sense.[6] Some facilities, however, are considered undesirable in the sense that most people want them located as far away as possible. Typically, such facilities are either noxious—posing a health or welfare hazard to people—or obnoxious—posing a threat to people's lifestyles—facilities (Erkut and Neuman, 1989). Hazardous waste sites, landfills,

---

[6]While these facilities are considered desirable in a general sense, it is clear that many people might not want to buy a house immediately adjacent to a fire station, for example, since the disruption associated with the fire engines responding to calls for service may outweigh the benefit of being near the station. Nevertheless, it is generally better to be near a fire station than to be far from a station. Similar issues might arise in the location of other generally desirable facilities.

incinerators, missile silos, and prisons generally fall into this category. In the location of undesirable facilities, it is often useful to distinguish between cases in which we are only concerned with the distance between facilities, as might be the case in locating nuclear missile silos, and those in which we are concerned with the distance between the facilities being located and population centers, as might be the case in locating landfills. In almost all practical location contexts involving the location of undesirable facilities of any kind, multiple conflicting objectives are likely to come into play. Thus, while we would like landfills to be located far from population centers, we also want to minimize the costs of transporting material from the waste generation sites to the landfill, as discussed above in Section 1.3.2. Unfortunately, much of the waste that is deposited in landfills is generated in highly populated areas. Thus, in locating landfills, the tradeoff between minimizing transportation costs and minimizing the number of people affected by the landfills needs to be identified.

1.5 SUMMARY

In this chapter we have identified the key questions answered by facility location models. We have qualitatively introduced a number of classical facility location models through example problems. Finally, we have outlined a taxonomy of location models and problems. In the course of this discussion, we identified those areas that will be the primary focus of the remainder of this text. In particular, the text will focus on network and discrete location problems, ignoring planar or continuous location problems and models.

Most network and discrete location problems of interest to us can be formulated as *linear programming problems* in which some of the variables are constrained to take on only integer values. Such problems are called *integer linear programming problems*. An understanding of linear programming is essential to the formulation and solution of many facility location problems. In addition, certain pure linear programming problems must be solved before most facility location problems can be attacked. For example, the problem of finding the shortest path from a facility to a demand node can be formulated as a linear programming problem. Often, shortest path distances are needed as inputs to facility location problems. Finally, once the facility locations are known, the problem of assigning demand nodes to facilities, particularly when the facilities have limited service capacities, can often be cast as another linear programming problem called the *transportation problem*. Chapter 2 reviews linear programming in general as well as a number of special linear programming problems that are intimately linked to facility location problems including the transportation problem and the shortest path problem.

EXERCISES

1.1 The l_p distance metric was defined as follows:

$$l_p = d[(x_i, y_i);(x_j, y_j)] = \left\{\left(|x_i - x_j|\right)^p + \left(|y_i - y_j|\right)^p\right\}^{1/p}$$

If we let $l_\infty = \lim_{p \to \infty}\{l_p\}$, what is l_∞ equal to?

Note: This distance metric is used in a number of industrial contexts. For example, it can be used to compute the time that it takes for an automated picker to move from one location to another in a warehouse when movements in both the X and Y directions can occur simultaneously, but the time to move between locations is governed by the larger of the two distances. Its three-dimensional extension has similar applications in robotics.

1.2 Use the real estate listings in your local newspaper to identify at least four or more houses in your city that are comparable in terms of the number of bedrooms and the number of bathrooms.

(a) What are the asking prices of the houses?

(b) What is the ratio of the largest asking price to the smallest asking price?

(c) What location factors might account for the differences in prices between the homes?

(d) What nonlocation factors might account for the price differences?

1.3 Identify at least two different objectives that public officials might have in locating new prisons.

1.4 With the ever-growing concerns about the environment, vehicle emission inspection policies are coming under increasing review.

(a) Discuss at least two different objectives that state officials would have in determining the locations of vehicle emission testing stations.

(b) Discuss nonlocational strategies that might be employed to increase public cooperation with emission testing laws.

(c) Discuss how the problem of locating vehicle emission testing stations fits into the location problem taxonomy outlined in Section 1.4.

2

Review of Linear Programming

2.1 INTRODUCTION

All the location problems outlined in Section 1.3 can be formulated as integer linear programming problems. This will be true of many of the problems that we will formulate and discuss later in the text. As such, one (perhaps naive) way to solve these problems is to relax the integer requirement on the decision variables, solve the resulting linear programming problem, and then employ a branch-and-bound strategy to force integrality of the variables that must be integer.[1] While we may often be able to do better than this, linear programming theory provides a key underpinning for much of what we intend to do. Therefore, this chapter reviews linear programming theory. In particular, we focus on the relationship between the primal and dual problems and the complementary slackness conditions. We also review two special linear programming problems which arise frequently in location analyses: the shortest path problem and the transportation problem. More details on all of these topics may be found in any of the introductory operations research texts listed in the References (Ecker and Kupferschmid, 1988; Hillier and Lieberman, 1986; Wagner, 1969).

2.2 THE CANONICAL FORM OF A LINEAR PROGRAMMING PROBLEM

In formulating the canonical form of a linear programming (LP) problem, we will use the following notation:

Inputs c_j, a_{ij}, b_i (These are constants that are given in the problem definition and statement.)

---

[1] Branch and bound is discussed in Chapter 4.

The Canonical Form of A Linear Programming Problem

Decision Variables X_j (These are the quantities that we are trying to find.)

Surplus Variables S_i (These variables are used to convert inequality constraints into equality constraints in one of the forms of the problem that we will discuss.)

With this notation, we can define the *primal linear programming problem in canonical form* as[2]

MINIMIZE	$\sum_j c_j X_j$	(2.1a)
SUBJECT TO:	$\sum_j a_{ij} X_j \geq b_i \quad \forall\, i$	(2.1b)
	$X_j \geq 0 \quad \forall\, j$	(2.1c)

We sometimes also want to discuss the same problem in standard form. The *standard form of the primal linear programming problem* is obtained from the canonical form by introducing surplus variables, S_i, into each of the constraints. Surplus variables facilitate the discussion below of the complementary slackness conditions. The standard form is

MINIMIZE	$\sum_j c_j X_j$	(2.2a)
SUBJECT TO:	$\sum_j a_{ij} X_j - S_i = b_i \quad \forall\, i$	(2.2b)
	$X_j \geq 0 \quad \forall\, j$	(2.2c)
	$S_i \geq 0 \quad \forall\, i$	(2.2d)

Before proceeding with a discussion of the properties of linear programming problems, it is useful to show that any linear programming problem can be placed in canonical form (i.e., a minimization problem subject to greater than or equal to constraints and nonnegative decision variables). If we could not do so, we would need to develop a separate theory of linear programming for each different case (e.g., a separate theory for maximization problems and a separate theory for minimization problems). Fortunately, we can convert any problem into this form. Therefore, we will need to outline only one theory!

Many optimization problems are naturally stated as *maximization* problems. A maximization problem may be converted into a minimization problem by simply multiplying the objective function by -1 and minimizing the resulting function.

---

[2] The notation $\sum_j$ means that we sum the quantity that follows over all values of the subscript or index j. The notation $\forall\, i$ means that the preceding expression (or constraint) applies to all values of the index i.

In many cases, constraints are most naturally expressed as *less than or equal to* constraints. Such a constraint can be converted into a canonical form constraint (a greater than or equal to constraint) by multiplying the entire constraint by -1 and changing the sense of the inequality in the process of multiplying by -1.

In some cases, constraints are expressed as *equality* constraints. An equality constraint can be converted into canonical form by replacing the single constraint by two constraints: one less than or equal to constraint and one greater than or equal to constraint. That is, the constraint

$$\sum_j a_{ij} X_j = b_i \qquad (2.3a)$$

is equivalent to the pair of constraints:

$$\sum_j a_{ij} X_j \le b_i \qquad (2.3b)$$

and

$$\sum_j a_{ij} X_j \ge b_i \qquad (2.3c)$$

Constraint (2.3c) is identical to constraint (2.1b) of the canonical form. Constraint (2.3b) simply needs to be multiplied by -1 to obtain the canonical form.

Finally, some problems have variables that are unrestricted in sign. In this case, we replace such a variable, X_j, by the difference between a pair of variables, $X_j^+ - X_j^-$, and restrict each of the new variables, X_j^+ and X_j^-, to be nonnegative. In other words, a problem of the following form:

MINIMIZE $\qquad\qquad 5X_1 - 2X_2 \qquad\qquad (2.4a)$

SUBJECT TO: $\qquad 6X_1 + X_2 \ge 10 \qquad\qquad (2.4b)$

$\qquad\qquad\qquad\quad X_1 - X_2 \ge -17 \qquad\qquad (2.4c)$

$\qquad\qquad\qquad\quad X_1$ and X_2 unrestricted $\qquad (2.4d)$

may be reformulated as follows:

MINIMIZE $\qquad 5(X_1^+ - X_1^-) - 2(X_2^+ - X_2^-) \qquad (2.5a)$

SUBJECT TO: $\quad 6(X_1^+ - X_1^-) + (X_2^+ - X_2^-) \ge 10 \qquad (2.5b)$

$\qquad\qquad\qquad (X_1^+ - X_1^-) - (X_2^+ - X_2^-) \ge -17 \qquad (2.5c)$

$\qquad\qquad\qquad X_1^+ \ge 0,\ X_1^- \ge 0,\ X_2^+ \ge 0,\ X_2^- \ge 0 \qquad (2.5d)$

Constructing the Dual of an LP Problem

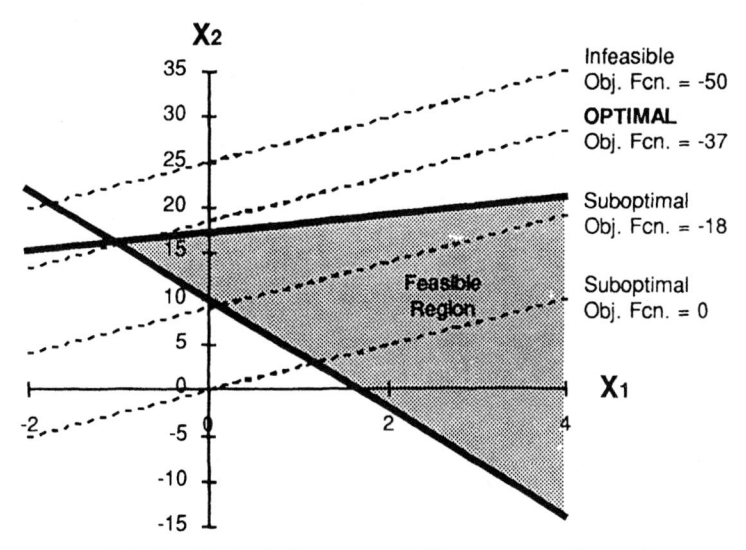

Figure 2.1. Graphical solution to example linear programming problem.

The solution to the original problem is $X_1 = -1$; $X_2 = 16$; and the objective function is equal to -37. For the reformulated problem, we have $X_1^+ = 0$; $X_1^- = 1$; $X_2^+ = 16$; and $X_2^- = 0$. Again, the objective function is equal to -37. Figure 2.1 plots the feasible region for this problem as well as four contours of the objective function. The feasible region is the triangular-shaped area below constraint 2.4c and above constraint 2.4b. The optimal solution is at the intersection of these two constraints $(-1, 16)$.

Having established that any linear programming problem can be converted into a problem in canonical form (or for that matter in standard form), we can now proceed with a discussion of the basic elements of the theory of linear programming.

2.3 CONSTRUCTING THE DUAL OF AN LP PROBLEM

Associated with every *primal* linear programming problem is another linear programming problem known as the *dual linear programming problem*. The dual problem is important for a number of reasons. First, in many cases, it is easier to solve the dual problem than it is to solve the primal problem. Second, the dual variables give us information about how the objective function will change as a result of small changes in the constraint values. Third, duality theory is important in developing solution algorithms for more complicated problems with embedded linear programming models.

To formulate the dual problem of the primal linear programming problems shown above, we introduce the following additional notation:

Dual Decision Variables U_i

Slack Variables T_j

With this additional notation, the canonical form of the dual linear programming problem becomes

MAXIMIZE $$\sum_i b_i U_i \qquad (2.6a)$$

SUBJECT TO: $$\sum_i a_{ij} U_i \le c_j \quad \forall j \qquad (2.6b)$$

$$U_i \ge 0 \quad \forall i \qquad (2.6c)$$

We can also formulate the standard form dual linear programming problem as

MAXIMIZE $$\sum_i b_i U_i \qquad (2.7a)$$

SUBJECT TO: $$\sum_i a_{ij} U_i + T_j = c_j \quad \forall j \qquad (2.7b)$$

$$U_i \ge 0 \quad \forall i \qquad (2.7c)$$

$$T_j \ge 0 \quad \forall j \qquad (2.7d)$$

Note that in going from the primal problem to the dual, we do the following:

a. Introduce a dual variable, U_i, associated with each of the primal constraints. That is, there is one dual variable for each primal constraint.
b. Convert the problem to a maximization problem from a minimization problem.
c. Make the right-hand sides of the primal constraints (the b_i terms) the coefficients of the U_i terms in the dual objective function.
d. Make the coefficients in the primal objective function into the right-hand sides of the dual constraints.
e. Transpose the coefficient matrix of the constraints. (In other words, we sum over i instead of summing over j.)
f. Change the sense of the inequality constraints (so that they are now less than or equal to constraints).
g. Constrain the dual decision variables, U_i, to be nonnegative.

To illustrate this procedure numerically, the reader is referred to the primal/dual pair in (2.14a)–(2.14e)/(2.15a)–(2.15d) below. What do you think might happen in the dual if there is an equality constraint in the primal? The answer to this question is left as an exercise for the reader.

2.4 COMPLEMENTARY SLACKNESS AND THE RELATIONSHIPS BETWEEN THE PRIMAL AND THE DUAL LINEAR PROGRAMMING PROBLEMS

Our interest in formulating the dual of a linear programming problem arises from (i) the tremendous insight that can be gained into the problem by studying the relationship between the primal and dual formulations and (ii) the fact that many solution algorithms for linear programming problems key off of these relationships. Our primary focus will be on the relationships between these two problem formulations.

First, we need to consider the issues of feasibility and boundedness of the objective function. A linear programming problem is *feasible* if there is at least one set of values of the decision variables that satisfy all of the constraints. A linear programming problem is said to have an *unbounded* objective function if the problem is feasible and if we can make the value of the objective function of the primal (minimization) problem as small as we desire. If the dual is unbounded, we can make the value of the dual objective function as large as desired. With these definitions, we can state the following properties:

1a. If the primal problem is infeasible, then either (i) the dual problem is infeasible as well or (ii) the dual problem is unbounded.

1b. If the primal is unbounded, then the dual is infeasible.

To illustrate these properties, consider the following three primal/dual pairs:

Problem 1—Primal

MINIMIZE	$2X - 3Y$	(2.8a)
SUBJECT TO:	$X - Y \geq 1$	(2.8b)
	$-X + Y \geq 1$	(2.8c)
	$X \geq 0; Y \geq 0$	(2.8d)

Problem 1—Dual

MAXIMIZE	$U + V$	(2.9a)
SUBJECT TO:	$U - V \leq 2$	(2.9b)
	$-U + V \leq -3$	(2.9c)
	$U \geq 0; V \geq 0$	(2.9d)

In Problem 1 the primal is infeasible since the quantity $X - Y$ cannot simultaneously be greater than or equal to 1 [as required by constraint (2.8b)] and less than or equal to -1 [as required by constraint (2.8c)]. Similarly, the dual is infeasible since the quantity $U - V$ cannot simultaneously be less than or equal to 2 [as required by constraint (2.9b)] and greater than or equal to 3 [as required by constraint (2.9c)]. This illustrates property 1a (i) above. Next consider the following primal/dual pair:

Problem 2—Primal

MINIMIZE	X	(2.10a)
SUBJECT TO:	$X \geq 4$	(2.10b)
	$-X \geq -2$	(2.10c)
	$X \geq 0$	(2.10d)

Problem 2—Dual

MAXIMIZE	$4U - 2V$	(2.11a)
SUBJECT TO:	$U - V \leq 1$	(2.11b)
	$U \geq 0, V \geq 0$	(2.11c)

In this problem, the primal is again infeasible since X cannot simultaneously be greater than or equal to 4 [as required by constraint (2.10b)] and less than or equal to 2 [as required by constraint (2.10c)]. The dual constraint will be satisfied by $U = 1$ and $V = 0$, yielding a dual objective function of 2. However, we can then increase both U and V by one unit, retain dual feasibility, and increase the dual objective function by two units. We can continue this process indefinitely. Thus, the dual problem is unbounded. This illustrates property 1a (ii) above. Finally, consider the following pair:

Problem 3—Primal

MINIMIZE	$2X - 4Y$	(2.12a)
SUBJECT TO:	$X - Y \geq 1$	(2.12b)
	$X \geq 0, Y \geq 0$	(2.12c)

Problem 3—Dual

MAXIMIZE	U	(2.13a)
SUBJECT TO:	$U \leq 2$	(2.13b)
	$-U \leq -4$	(2.13c)
	$U \geq 0$	(2.13d)

In this problem, the dual is infeasible since U cannot simultaneously be less than or equal to 2 [as required by (2.13b)] and greater than or equal to 4 [as required by (2.13c)]. The primal is feasible, however, since $X = 1$, $Y = 0$ satisfies the constraints and yields a primal objective function of 2. However, we can simultaneously increase both X and Y by one unit, retain primal feasibility, and reduce the primal objective function by two units. We can continue this process indefinitely. Thus, the primal is unbounded. This illustrates property 1b above.

Problems of infeasibility and unboundedness generally arise in one of four scenarios. First, the original problem may be misformulated mathematically. Second, the computer program that generates the input file for the linear programming problem solver (often a separate computer problem) incorrectly translated the correct problem formulation into a form that is understandable by the LP solver. In very large linear programming problems, this is not uncommon. Third, infeasibility or unboundedness may be the result of data errors. This, too, is common. Fourth, infeasibility and unboundedness also arise when linear programming problems are used in branch and bound (discussed below). In those cases, we often begin with a well-formulated integer programming problem. We attempt to solve the problem by relaxing (ignoring) the integrality constraints and solving the resulting linear programming problem. If the solution to the linear programming relaxation does not satisfy all of the original integer constraints, we introduce additional linear constraints. The introduction of these additional constraints may result in problems that are infeasible.

Generally, however, our primary interest lies in cases in which both the primal and dual linear programming problems are feasible and have bounded (finite) objective functions. Under those conditions, the following conditions hold:

2a. $\sum_j c_j X_j \geq \sum_i b_i U_i$, for any *feasible* (though not necessarily optimal) values of X_j and U_i. This is known as the *weak duality* condition.

2b. $\sum_j c_j X_j^* = \sum_i b_i U_i^*$, where X_j^* and U_i^* are the *optimal* primal and dual solutions. This is known as the *strong duality* condition.

2c. At the optimum, the *complementary slackness* conditions hold, namely: (i) $S_i^* U_i^* = 0$, for all values of i, and (ii) $T_j^* X_j^* = 0$, for all values of j, where X_j^* and S_i^* are the optimal primal decision variables and surplus variables, respectively, and U_i^* and T_j^* are the optimal dual decision variables and slack variables, respectively.

2d. At least one optimal solution is at an extreme point (corner) of the feasible region.

2e. The optimal dual variables give the rate of change of the objective function with respect to changes in the right-hand side of the associated primal constraints. [We note in passing that care must be taken in this interpretation as observed by Rubin and Wagner (1990).]

Conditions 2c are of primary importance. They state that if there is slack in a primal constraint ($S_i^* > 0$), then the associated dual variable must be 0 ($U_i^* = 0$). Similarly, if a dual variable is nonnegative ($U_i^* > 0$), then the slack variable in the associated primal constraint must be 0 ($S_i^* = 0$). Similar relationships hold between the optimal primal decision variables (X_j^*) and the optimal dual slack variables (T_j^*).

Finally, we note two other characteristics of the relationship between the primal and dual linear programming problems:

3. If a solution (X_j, S_i, U_i, T_j) satisfies primal feasibility, dual feasibility, and the complementary slackness conditions, then the solution is optimal.
4. The dual of the dual is the primal.

Condition 3 is often used in developing solution algorithms for linear programming problems. One or more of the properties is preserved throughout the solution algorithm and the algorithm drives toward achieving the unsatisfied properties.

We have been able to gain considerable insight into linear programming solutions by examining the dual linear programming problem. This suggests that we could gain even more insight by formulating the dual of the dual problem. Condition 4 is the bad news: doing so will only bring us back to the original primal problem; we will not be able to gain any further insight into the problem by formulating the dual of the dual.

To illustrate these properties, we consider the following problem:

Problem 4—Primal

MINIMIZE

$$5X_1 + 10X_2 \tag{2.14a}$$

SUBJECT TO:

$$X_1 + X_2 \geq 100 \tag{2.14b}$$

$$X_1 \leq 3X_2 \quad \text{or equivalently} \quad -X_1 + 3X_2 \geq 0 \tag{2.14c}$$

$$X_2 \leq 2X_1 \quad \text{or equivalently} \quad 2X_1 - X_2 \geq 0 \tag{2.14d}$$

$$X_1 \geq 0, X_2 \geq 0 \tag{2.14e}$$

Problem 4 might represent the problem of trying to minimize the cost of producing some item using two different processes: process 1 and process 2. Producing one unit of the item using process 1 costs $5, while producing one unit of the item using process 2 costs $10. Constraint (2.14b) stipulates that we must produce at least 100 units. Constraints (2.14c) and (2.14d) state that we do not want to put all our eggs in one basket. We do not want the amount

Primal and Dual Linear Programming Problems

produced using process 1 to exceed three times the amount produced using process 2 [constraint (2.14c)] and we do not want the amount produced using process 2 to exceed twice the amount produced using process 1 [constraint (2.14d)]. Such constraints may be imposed so that the firm involved retains some expertise in the use of both processes. Finally, constraints (2.14e) are standard nonnegativity constraints. The dual of this problem is

Problem 4—Dual

MAXIMIZE	$100U_1$	(2.15a)
SUBJECT TO:	$U_1 - U_2 + 2U_3 \leq 5$	(2.15b)
	$U_1 + 3U_2 - U_3 \leq 10$	(2.15c)
	$U_1 \geq 0, U_2 \geq 0, U_3 \geq 0$	(2.15d)

In this case, the primal problem (which has only two decision variables) may be solved graphically. The solution is shown in Figure 2.2.

The solution is

$X_1 = 75$(Graphically) $T_1 = 0$(Complementary slackness)

$X_2 = 25$(Graphically) $T_2 = 0$(Complementary slackness)

$S_1 = 0$[Constraint (2.14b)] $U_1 = 6.25$(by property 2b)

$S_2 = 0$[Constraint(2.14c)] $U_2 = 1.25$[constraint (2.15b)

                                             after all other values are determined]

$S_3 = 125$[Constraint (2.14 d)] $U_3 = 0$[by property 2c (i)]

and the objective function value is 625.

The interpretation of the dual variables is now readily apparent. If we change the right-hand side of constraint (2.14b) to 101 (by 1 unit), the objective function will increase to 631.25 (by 6.25 units, the exact value of U_1). The new primal decision variables would be $X_1 = 75.75$ and $X_2 = 25.25$. The dual variables do not change. Property 2b remains true, despite the change in the value of the objective function (and the lack of change in the value of the dual variables) because the dual objective function changes from $100U_1$ to $101U_1$. Similarly, if the right-hand side of constraint (2.14c) were to change from 0 to 1, the objective function would change to 626.25 (an increase of 1.25 units, the exact value of U_2). The new decision variables would be $X_1 = 74.75$ and $X_2 = 25.25$. Again, the dual variable values do not change. How does the dual objective function change this time so that property 2b remains true? Finally, we note that if we change the right-hand side of constraint (2.14d) by one unit, the optimal solution will not change

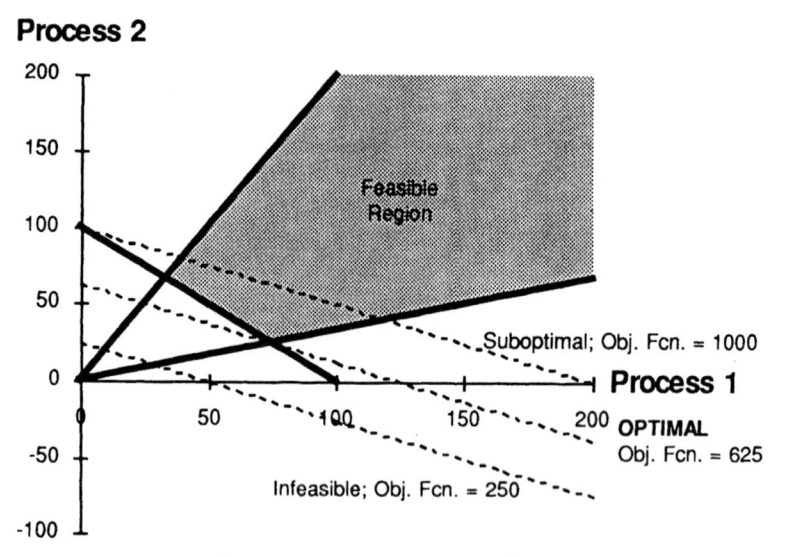

Figure 2.2. Graphical solution of problem 4 [(2.14a)–(2.14e)].

since constraint (2.14d) is not binding. This is further verified by the fact that $U_3 = 0$.

Linear programming models are generally solved using one of two methods: the simplex algorithm or interior point algorithms. The simplex algorithm moves from one extreme point of the feasible region to an adjacent extreme point until a move to any one of the adjacent extreme points would not improve the objective function. At that point, the algorithm stops with an optimal solution. Interior point algorithms solve the problem by moving from one feasible solution within the feasible region to another, moving toward the optimal solution. Interior point algorithms have a number of theoretical advantages over the simplex algorithm. In practical applications, however, the two approaches appear to be competitive. There are some problems for which interior point algorithms are superior and there are other problems for which the simplex algorithm is better. For this reason, commercial linear programming packages are beginning to include both algorithms. Development of these two approaches remains an exciting and active research area.

For problems in which some of the variables must be integer valued, including most location problems, we often need to solve a series of intimately related linear programming problems. For such problems, the simplex algorithm is currently superior to interior point algorithms because the simplex algorithm can be "restarted" with a solution that is very near the optimal solution for closely related problems; interior point algorithms must be restarted from scratch each time a new problem is solved.

In this text we will not go into additional detail about the workings of either of these algorithms. The reader interested in more detail about the

The Transportation Problem

simplex algorithm can consult any one of a large number of basic operations research texts including: Ecker and Kupferschmid (1988), Hillier and Lieberman (1986), Wagner (1969), and Zionts (1974). For an introduction to interior point algorithms, the reader should consult Hooker (1986) or Marsten et al. (1990).

2.5 THE TRANSPORTATION PROBLEM

There are a number of linear programming problems that are of special interest. The first such problem that we discuss is the *transportation problem*. The transportation problem is the following: given a set of suppliers each with supply S_i, a set of demand points each with demand D_j, and the unit cost c_{ij} of shipping from supply point i to demand point j, find the shipment pattern (shipments from the supply points to the demand points) such that all demands are satisfied and the total cost is minimized.[3] This problem may be formulated mathematically using the following notation:

Inputs

S_i = the supply at node i
D_j = the demand at node j
c_{ij} = the unit cost of shipping from supply node i to demand node j

Decision Variables

X_{ij} = the amount shipped from supply node i to demand node j

With these inputs and decision variables, the transportation problem may be formulated as follows:

Transportation Problem—Primal

MINIMIZE	$\sum_i \sum_j c_{ij} X_{ij}$	(2.16a)
SUBJECT TO:	$\sum_j X_{ij} \leq S_i \quad \forall\, i$	(2.16b)
	$\sum_i X_{ij} \geq D_j \quad \forall\, j$	(2.16c)
	$X_{ij} \geq 0 \quad \forall\, i,j$	(2.16d)

---

[3]The transportation problem is important in location modeling for the following reason. Suppose we are asked to locate facilities with known capacities, S_i, if a facility is located at candidate site i, to serve demand sites with known demands, D_j, to minimize the combined facility location and transport costs. For any given siting plan, the problem boils down to one of allocating demands to facilities to minimize the transport costs. This, however, is simply the transportation problem.

The objective function (2.16a) minimizes the total cost. Constraints (2.16b) stipulate that the total amount shipped from each supply point i must be less than or equal to the total supply at node i. Similarly, constraints (2.16c) ensure that each demand point receives at least as much as is required at that point. Finally, constraints (2.16d) are the nonnegativity constraints. The dual of the transportation problem is

Transportation Problem—Dual

MAXIMIZE $$\sum_j D_j W_j - \sum_i S_i V_i \qquad (2.17a)$$

SUBJECT TO: $$W_j - V_i \leq c_{ij} \qquad \forall\, i, j \qquad (2.17b)$$

$$W_j \geq 0 \qquad \forall\, j \qquad (2.17c)$$

$$V_i \geq 0 \qquad \forall\, i \qquad (2.17d)$$

where V_i is the dual variable associated with the ith supply constraint (2.16b) and W_j is the dual variable associated with the jth demand constraint (2.16c).

We will assume that the total supply equals the total demand (i.e., $\sum_i S_i = \sum_j D_j$). If this is not the case for the inputs that apply to a particular problem, we can readily convert the problem to one in which this assumption holds. For example, if supply exceeds demand (i.e., $\sum_i S_i > \sum_j D_j$), we can add a dummy demand node (node $J + 1$) with demand, $D_{J+1} = \sum_i S_i - \sum_j D_j$. All costs from a supply node to the new dummy demand node will be set equal to 0 (i.e., $c_{i, J+1} = 0$, $\forall\, i$). Flows from a supply node to the dummy demand node indicate the amount by which the supply at that node is not fully utilized.

Similarly, if demand exceeds supply, we can create a dummy supply node (node $I + 1$) with supply equal to $S_{I+1} = \sum_j D_j - \sum_i S_i$. As before, all costs from the dummy supply node to each real demand node will be set equal to 0 (i.e., $C_{I+1, j} = 0$, $\forall\, j$). In this case, flows from the dummy supply node to a real demand node indicate the amount by which we will *undersatisfy* the demand at that node. Thus, the assumption that demand equals supply is not overly restrictive.

With this assumption, we eliminate constraints (2.17c) and (2.17d) and allow W_j and V_i to be unrestricted in sign. We do so because we know that constraints (2.16b) and (2.16c) will be satisfied by strict equality at any feasible solution. Hence, these inequalities may be replaced by equality constraints (causing the associated dual variables to be unrestricted in sign). This will facilitate our finding an optimal solution to the problem. Note, however, that if we prefer to have nonnegative dual variables (due for example to the interpretation we will impose on the dual variables), we can always add an appropriate constant δ to all of the dual variables. Since supply equals demand, the value of the dual objective function will not

The Transportation Problem

change. Furthermore, since only the difference between dual variables matters in the dual constraints, adding a constant to all dual variables will not affect dual feasibility. Thus, any dual feasible solution in which some variables are negative can be converted to an equivalent dual solution in which all dual variables are nonnegative by adding $\delta = |$ most negative dual variable $|$ to all the dual variables.

We are particularly interested in the transportation problem in a location context for two reasons. First, the transportation problem is often a subproblem of larger problems we want to solve. Second, we can readily extend this formulation to that of a location problem by incorporating a fixed cost f_i associated with all supply nodes (now considered candidate locations) and a new decision variable Y_i which will take on a value of 1 if we locate at candidate site i and 0 if not. The exact formulation of this problem is left as an exercise to the reader.

Finally, we note that at most $I + J - 1$ of the flow variables X_{ij} will be nonnegative at optimality where I is the total number of supply nodes (including any dummy supply node) and J is the total number of demand nodes (including any dummy demand node). This is so because only $I + J - 1$ of these constraints are linearly independent under the assumption that supply equals demand.

We now turn to the complementary slackness conditions for this problem. They are

$$\left[S_i - \sum_j X_{ij} \right] V_i = 0 \quad \forall \, i \qquad (2.18a)$$

$$\left[\sum_i X_{ij} - D_j \right] W_j = 0 \quad \forall \, j \qquad (2.18b)$$

$$\left[c_{ij} + V_i - W_j \right] X_{ij} = 0 \quad \forall \, i, j \qquad (2.18c)$$

Conditions (2.18a) are always satisfied for any feasible flows X_{ij} since the sum over all demand nodes of the flows leaving a supply node must equal the supply at that node (since supply equals demand). Similarly, conditions (2.18b) are always satisfied for any feasible flows X_{ij}. Finally, since $X_{ij} = 0$ for at least $IJ - (I + J - 1)$ of the possible flows, conditions (2.18c) are satisfied *in most cases* for a feasible and basic set of flows and any set of dual variables. Clearly, if we have a basic feasible solution[4] and if we can find a set

---

[4]A *basic* solution to the primal problem is one in which the number of nonzero primal decision variables is (less than or) equal to the number of linearly independent primal constraints, not counting the nonnegativity constraints. The reader who is unfamiliar with the notion of linearly independent equations is referred to any standard linear algebra text (e.g., Noble, 1969). A basic solution to the dual problem is defined in a similar manner in terms of the dual constraints. A *basic feasible* solution to the primal problem is one which is both basic and satisfies all of the primal constraints.

of dual variables which satisfy condition (2.18c) as well as the dual feasibility condition (2.17b), we will have an optimal solution. Finding such a set of dual variables may necessitate our changing the basic primal feasible solution (the flow variables X_{ij}).

The dual variables may be thought of as the prices or value of the commodity being shipped at the suppliers (V_i) and at the demanders or consumers (W_j). The dual constraint (2.17b) states that the value of the commodity at destination j must be less than or equal to the value at each supply point i plus the transport cost between i and j. If this were not so, that is, if the value at j were more than the value at i plus the transport costs, then more firms would enter the market, thereby driving down the price at j (or increasing the value of the commodity at supply point i). Complementary slackness condition (2.18c) states that if these two values (W_j—the value at j—and $V_i + c_{ij}$—the value at i plus the transport costs) are not equal, then there can be no flow between i and j. Recall that if they are not equal, it implies that the value at the supplier i plus the cost of transport between i and j exceeds the value at the destination j by constraint (2.17b); thus, there is no benefit in shipping from i to j. If they are equal, there can be flow.

With this background, we can now state an algorithm for solving the transportation problem. Like many such algorithms, the approach we adopt is one of first finding a basic feasible solution to the primal problem. We do this by using a method known as the *northwest corner method*. We then determine dual variables such that condition (2.18c) is satisfied everywhere. Then we check to see if constraint (2.17b) is satisfied. If it is, we have an optimal solution. If it is not, we have identified a candidate flow (a supply node and a demand node) which is currently 0 but whose increase will result in a decrease in the primal objective function. We then increase this flow and modify all other flows appropriately so that we retain a basic feasible solution, recompute the dual variables so that (2.18c) is satisfied, and again check constraint (2.17b). This process continues until constraint (2.17b) is satisfied. At that point, we have a solution that satisfies primal feasibility, dual feasibility, and complementary slackness. Hence, the solution is optimal. Formally, the algorithm may be described as follows:

Algorithm for Solving the Transportation Problem

Step 1: Find an initial primal feasible (basic) flow using the *northwest corner method*.

    1.1. Set $i = 1, j = 1$ (indices)

$$\hat{S}_i = S_i \quad \forall i \quad (\hat{S}_i = \text{temporary supply})$$
$$\hat{D}_j = D_j \quad \forall j \quad (\hat{D}_j = \text{temporary demand})$$
$$X_{ij} = 0 \quad \forall i, j$$

The Transportation Problem

    1.2. Assign $f = \min(\hat{S}_i, \hat{D}_j)$ flow units to origin i and destination j (i.e., set $X_{ij} = f$). If $f = \hat{S}_i$, increment i and set $\hat{D}_j = \hat{D}_j - f$ (using a FORTRAN-like assignment); otherwise, increment j and set $\hat{S}_i = \hat{S}_i - f$.

    1.3. Are all demands satisfied? If not, go to step 1.2; if yes, stop, we have a basic feasible solution to the primal.

Step 2: Compute dual variables that satisfy condition (2.18c).

    2.1. Let $V_i = 0$. (Since we have $I + J$ dual variables and only $I + J - 1$ equations, we can pick any single dual variable and set it to 0 arbitrarily.)

    2.2. For the $I + J - 1$ cases in which X_{ij} is a basic variable, solve for V_i and W_j to satisfy the complementary slackness conditions (2.18c). That is, choose values of V_i and W_j to satisfy $c_{ij} + V_i - W_j = 0$ if X_{ij} is a basic variable. (Note that this may be done quite easily. There is no need to resort to messy ways of solving simultaneous linear equations. These equations may be solved *by substitution*.)

Step 3: At this point we satisfy (i) primal feasibility, (ii) complementary slackness, and (iii) dual feasibility for those i, j pairs such that X_{ij} is a basic variable. We need only check dual feasibility for the i, j pairs for which X_{ij} is not a basic variable. If $c_{ij} + V_i - W_j \geq 0$ for all i, j, then stop. The solution is *optimal*. If not, select one i, j pair for which this is not true and go to step 4.

Step 4: Revise the primal solution as follows. For the i, j pair found in step 3 for which $c_{ij} + V_i - W_j < 0$, find the minimum amount by which the flow may be increased while maintaining a basic feasible primal solution. Note that some flow will go to 0. This is done by finding a flow "augmenting" path from j to i. [A variety of network data structures can be used to facilitate this search since the solution to a transportation problem always forms a tree in graph theory terminology. The interested reader is referred to such texts as Aho, Hopcroft, and Ullman (1983), Ahuja, Magnanti, and Orlin (1993), Jensen and Barnes (1980), and Kennington and Helgason (1980) for a discussion of these techniques.]

    Using the tableau format shown below, find a sequence of cells, beginning with the cell in which flow is to be added and then using *only* cells containing basic variables, moving either horizontally or vertically, that returns to the cell whose flow is to increase. In every other cell, the flow either will increase or decrease. The cell whose flow will decrease and with the smallest flow dictates the amount by which the flow around the cycle can change.

    Change the flows by this amount around the flow augmenting cycle and set the flow X_{ij} equal to this amount. Note that this preserves a basic primal feasible solution. Go to step 2.

Review of Linear Programming

				Supply	V_i
[5] 70	[9] 50	[6] 20	[4]	140	
[7]	[2]	[8] 105	[11] 25	130	
[3]	[1]	[10]	[14] 110	110	
Demand: 70	50	125	135	Obj. Fcn. = 3575	

W_j

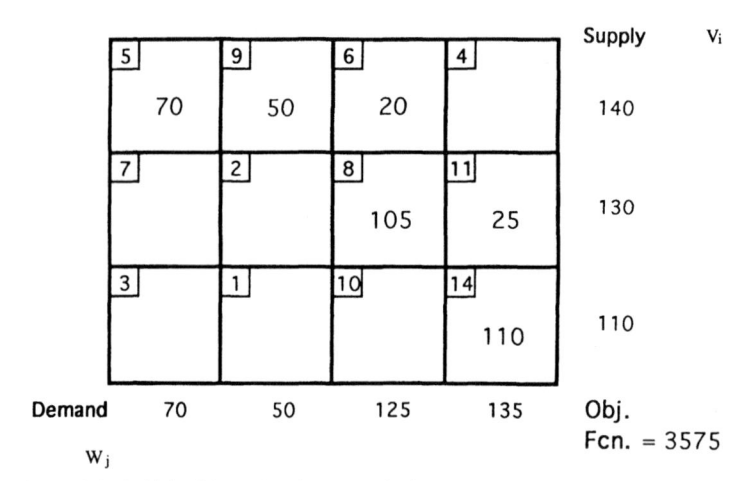

Figure 2.3. Initial tableau showing the solution from the northwest corner method.

We will illustrate this algorithm using the problem shown in Figure 2.3. Numbers in the top left-hand corners are the unit costs. Numbers in the middle of the cells represent the flows X_{ij}. Circled numbers in each cell represent the value of $c_{ij} + V_i - W_j$. The following tableaus (tables) show the results after each step of the algorithm. We begin with the results after the application of the northwest corner method.

Having found the initial solution, we now compute the dual variables so that the complementary slackness condition (2.18c) is satisfied for all cells in the tableau containing basic (primal) variables. The results are shown in Figure 2.4.

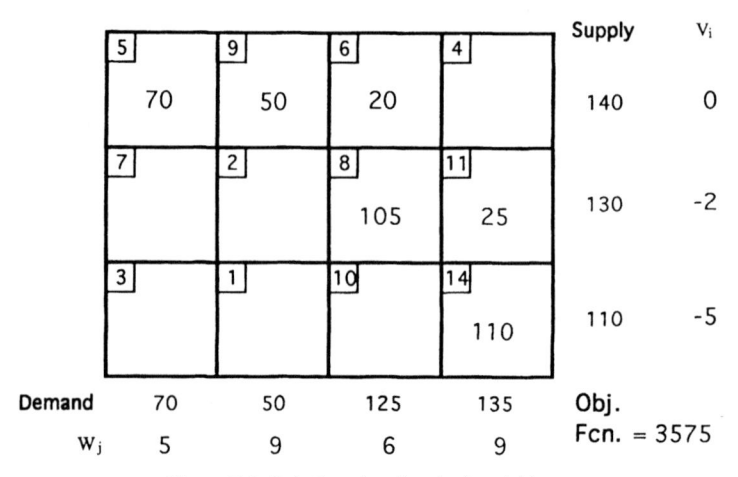

				Supply	V_i
[5] 70	[9] 50	[6] 20	[4]	140	0
[7]	[2]	[8] 105	[11] 25	130	-2
[3]	[1]	[10]	[14] 110	110	-5
Demand: 70	50	125	135	Obj. Fcn. = 3575	
W_j: 5	9	6	9		

Figure 2.4. Solution showing dual variables.

The Transportation Problem

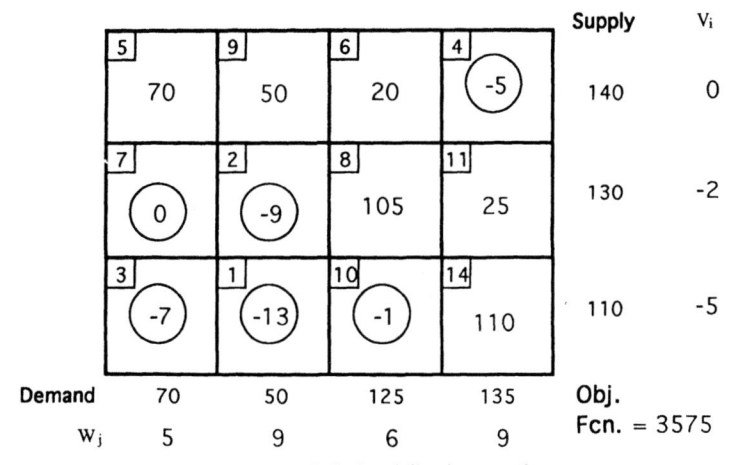

				Supply	V_i
5 70	**9** 50	**6** 20	**4** (-5)	140	0
7 (0)	**2** (-9)	**8** 105	**11** 25	130	-2
3 (-7)	**1** (-13)	**10** (-1)	**14** 110	110	-5
Demand 70	50	125	135	Obj. Fcn. = 3575	
w_j 5	9	6	9		

Figure 2.5. Solution following step 3.

We now compute the values of the slack variable in the dual constraint (2.17b). These values are shown in Figure 2.5 as circled numbers. Recall that all such values must be nonnegative in an optimal solution; negative values indicate dual infeasibility and also indicate cells in which an increase in flow will result in a decrease in total cost.

Having computed the slack variables in constraint (2.17b), we can identify cells in which an increase in flow will decrease the total cost. Any cell with a negative value can be used. In Figure 2.6 we have chosen cell (3, 2)—the cell with the most negative value. Using this cell will decrease the objective function at the fastest rate, but may not decrease the objective function as

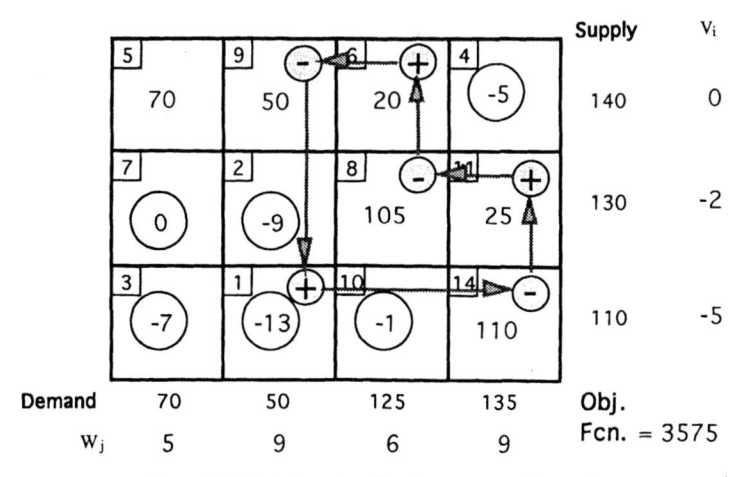

Figure 2.6. Solution showing flow augmenting path.

much as using some other cell might decrease the objective function. The reader is encouraged to consider why this is so. Having identified this cell as the one in which we want to increase the flow, we must now determine the amount by which the flow here can be changed. We must at all times satisfy primal feasibility (the supply and demand constraints). Thus, we find a flow augmenting path.

Note that the net change in cost around the flow augmenting path is -13, exactly equal to the value computed in cell (3, 2) in Figure 2.6. This is not a coincidence. The circled values indicate the amount by which the cost would change if we added one unit of flow to the corresponding cell while changing all other values appropriately to satisfy the supply and demand conditions. To see this, we sum the costs around the cycle, beginning with cell (3, 2):

Cell	Cost/Unit Flow
(3, 2)	1
(3, 4)	-14
(2, 4)	11
(2, 3)	-8
(1, 3)	6
(1, 2)	-9
Net change	-13

Also note that of the three cells in which flow will be decreased [cells (3, 4), (2, 3), and (1, 2)], cell (1, 2) has the smallest flow. Thus, we can change the flow around this path by 50 units. After doing so, the flow in cell (1, 2) will be 0 and this flow variable will no longer be a basic variable. The solution after this change is shown in Figure 2.7.

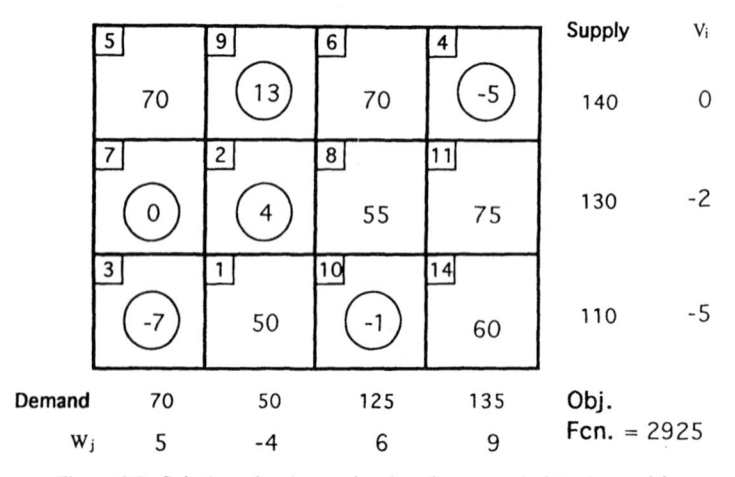

Figure 2.7. Solution after increasing flow from supply 3 to demand 2.

The Transportation Problem

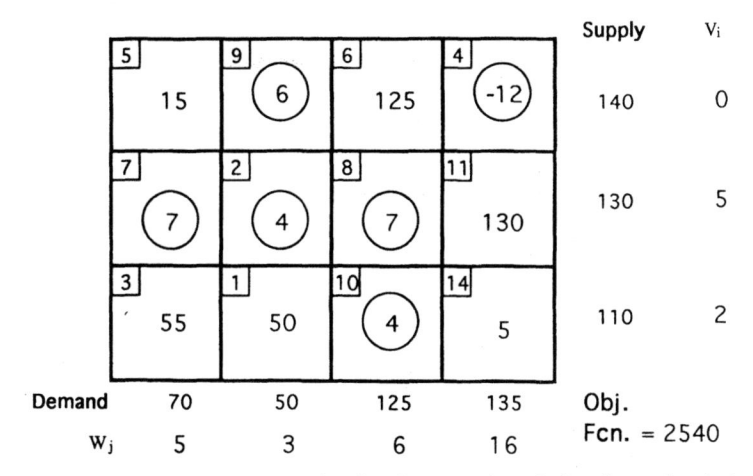

Figure 2.8. Solution after increasing flow from supply node 3 to demand node 1.

Note that the objective function decreased by 650 units [equal to -13 in cell (3, 2) from the previous tableau times 50 units of allowable flow increase]. Also, note that the flow from 1 to 2 decreased to 0; hence, this flow is no longer basic. Finally, note that the solution is still a basic feasible solution. We will next increase the flow from supply 3 to demand 1. The reader is encouraged to identify the flow augmenting path used in effecting this flow change. The results of this flow increase are shown in Figure 2.8.

After several additional iterations, we obtain the solution to the problem shown in Figure 2.9. The details of these steps are left as an exercise for the reader.

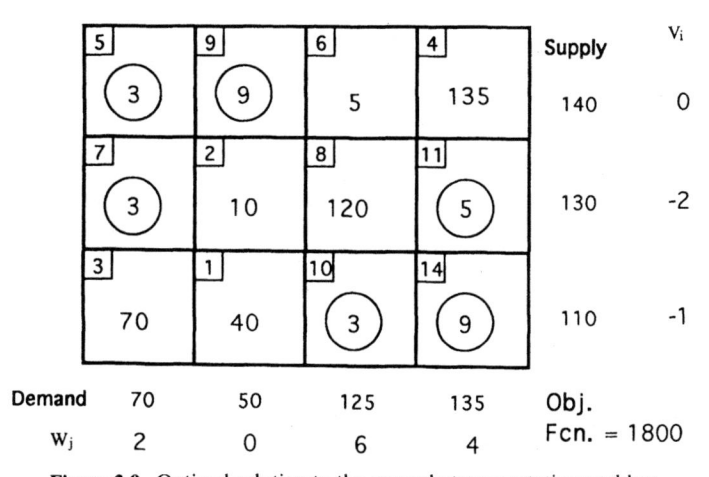

Figure 2.9. Optimal solution to the example transportation problem.

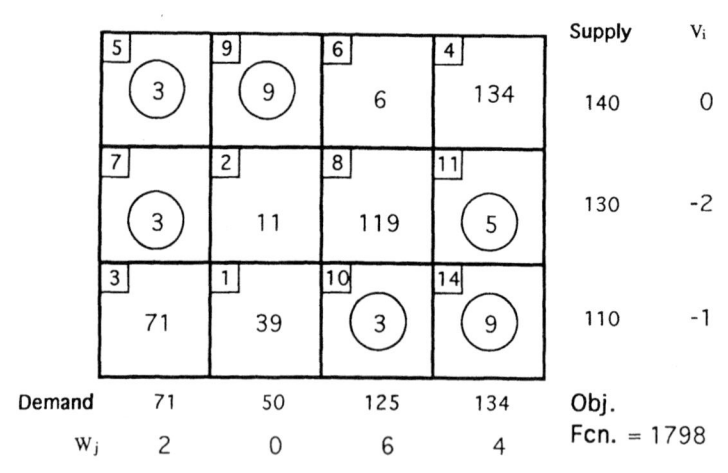

Figure 2.10. Solution after increasing demand at node 1 and decreasing demand at node 4 by 1 unit each.

We can use the optimal solution (and the dual variables from the optimal dual problem) to conduct sensitivity analyses. For example, if we increase the demand at some node m by one unit and decrease the demand at some other node n by one unit, the change in the objective function will be $W_m - W_n$. To see this, suppose we increase the demand at demand node 1 to 71 and decrease the demand to demand node 4 by one unit each. The dual variables suggest that the total cost will *decrease* by two units ($2 - 4 = -2$). The new optimal solution is given in Figure 2.10 (and the new total cost is, in fact, 1798, or two units less than the cost of the optimal solution shown in Figure 2.9).

Similarly, we can increase the supply at some node p by one unit and simultaneously decrease the supply at some other node q by one unit. The objective function value will change by $V_q - V_p$. (Note that, in this case, the dual variable with the index of the supply node at which supply increases has a negative sign.) Finally, if we simultaneously increase supply at some node p and demand at some node m by one unit each, the objective function will change by $W_m - V_p$ cost units. For example, if we increase demand at demand node 3 and increase supply at supply node 3 by one unit each, the cost will increase by seven units [$6 - (-1) = 7$]. The solution, after making this change to the solution of Figure 2.9, is shown in Figure 2.11.

Note that the dual variables give the rate of change in the objective function for *small* changes in the flows. For large changes, the basis is likely to change and the dual variables can no longer be used to predict the cost changes. The reader is also referred to the paper by Rubin and Wagner (1990) for a discussion of some of the other pitfalls associated with naive

The Shortest Path Problem

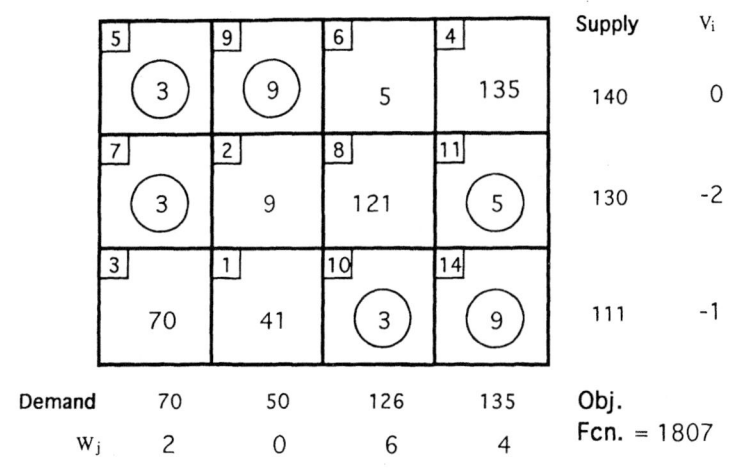

	5	9	6	4	Supply	V_i
	③	⑨	5	135	140	0
	7	2	8	11		
	③	9	121	⑤	130	-2
	3	1	10	14		
	70	41	③	⑨	111	-1
Demand	70	50	126	135	Obj.	
W_j	2	0	6	4	Fcn. = 1807	

Figure 2.11. Solution after increasing demand at node 3 and increasing supply at node 3 by 1 unit each.

interpretation of the dual variables in the context of the transportation problem.

2.6 THE SHORTEST PATH PROBLEM

The second special linear programming problem that we discuss is the *shortest path problem* which may be stated as follows: given a network with costs associated with each of the links, find the shortest path from a specified node (s) to another node (t). This problem arises in location problems for a number of reasons, not the least of which is that for most network location problems we need to know the shortest path between demand nodes and candidate locations.

The shortest path problem may be formulated as a linear programming problem using the following notation:

Inputs

$\quad c_{ij}$ = unit cost of traversing link (i, j)

Decision Variables

$$X_{ij} = \begin{cases} 1 & \text{if link } (i, j) \text{ is on the shortest path from } s \text{ to } t \\ 0 & \text{if not} \end{cases}$$

With these definitions, the primal may be formulated as follows:

Shortest Path Problem—Primal

MINIMIZE
$$\sum_i \sum_j c_{ij} X_{ij} \qquad (2.19a)$$

SUBJECT TO:

$$\sum_j X_{ji} - \sum_k X_{ik} = \begin{cases} -1 & \text{if node } i = s \\ 0 & \text{otherwise} \\ 1 & \text{if node } i = t \end{cases} \qquad (2.19b)$$

$$X_{ij} = 0, 1 \qquad \forall i, j \quad (2.19c)$$

The objective function (2.19a) minimizes the total cost of travel for the selected links. Constraint (2.19b) stipulates that the flow into node i ($\sum_j X_{ji}$) minus the flow out of node i ($\sum_k X_{ik}$) must be: -1 if node i is node s; 1 if node i is node t; and 0 otherwise. Constraint (2.19c) stipulates that each link appear at most once on the path from s to t.

The dual of this linear programming problem is

Shortest Path Problem—Dual

MAXIMIZE $\qquad V_t - V_s \qquad (2.20a)$

SUBJECT TO: $\qquad V_j - V_i \leq c_{ij} \quad \forall i, j \qquad (2.20b)$

$\qquad\qquad\qquad V_i \geq 0 \qquad \forall i \qquad (2.20c)$

It is important to note that technically this problem, as formulated above, is not a linear programming problem. The reason for this is that the decision variables, X_{ij}, are integer (in fact, binary) variables and not real variables. Thus, they are restricted to take on values of either 0 or 1 and cannot take on fractional values as is the case in other linear programming problems. However, the constraint matrix is totally unimodular (see Nemhauser and Wolsey, 1988). This property of the constraint matrix guarantees that if we solve this problem as a linear programming problem, the result will be all-integer decision variables. Thus, we can redefine the decision variables to be nonnegative variables ($X_{ij} \geq 0$) and solve the resulting problem as a linear programming problem.

The Shortest Path Problem

Note that if some link costs are negative, the solution to either the integer or linear programming formulations might contain negative cycles. A negative cycle is sequence of links which form a cycle (i.e., the sequence begins at some node p goes to a number of other nodes and returns to node p) and whose total cost of traversal is negative. In what follows, we assume that all link costs are nonnegative.

In solving the shortest path problem, using a standard linear programming package or algorithm would be terribly inefficient. There are far more effective means of solving such a problem. We outline one such approach below. This approach is known as *Dijkstra's algorithm*.

The solution to any shortest path problem should contain two pieces of information: the cost of the shortest path between the origin s and the destination t and the actual sequence of links (or nodes) that are traversed in going from s to t. Thus, we need to keep track of both the cost of getting to each intermediate node and the node from which we came (so that we can trace the best path from s to t). A *node label* will therefore be composed of two parts: (1) the shortest known cost of getting from s to the node and (2) the node from which we came in getting to the node in question. The notation we will use for a node label is

[shortest known cost of getting to the node, predecessor node]

The best known cost of getting to node j is V_j, the dual variable associated with node j. Note that the interpretation of the dual constraint (2.20b) is similar to that of the dual constraint for the transportation problem (2.17b). Thus, the node label will actually be

$$[V_j, P_j]$$

where P_j is the predecessor node of node j on the shortest path from node s to node j.

Node labels may be either temporary or permanent. We will associate a permanent label with a node when we know that the best currently known cost of getting to the node is also the absolute best we will ever be able to attain. Nodes with permanent labels will be called *scanned*; those with temporary labels will be called *unscanned*. The algorithm may now be stated as follows for the case in which we want to find the shortest path from node s to all other nodes. Generally, the computational cost of finding the shortest path from s to all other nodes is not much more than that of finding the shortest path from s to some specific node t. Furthermore, we often need the shortest paths from s to all other nodes and much of the work involved in finding the shortest path from s to t would have to be repeated if we later wanted the shortest path from s to some other node u.

Algorithm for Finding the Shortest Path from s to All Other Nodes When All Link Costs Are Nonnegative

Step 1: Initialization
  a. Label node s [0, –] (i.e., set $V_s = 0$).
  b. Label all other nodes [∞, –] (i.e., set $V_j = \infty$ for all nodes j other than node s).
  c. Set node s as scanned.

Step 2: Label Updates
  a. Call the last scanned node, node m.
  b. For all links (m, j) such that node j is not scanned, compute
   b.1. $T_j = V_m + c_{mj}$.
   b.2. If $T_j < V_j$, relabel node j with $[T_j, m]$.

Step 3: Scan a Node
  a. Find the unscanned node with the smallest label V_j. In the event of a tie, arbitrarily choose one of the nodes with the smallest label V_j.
  b. Scan this node.

Step 4: Termination Check
  a. Are all nodes scanned?
   a.1. YES—stop.
   a.2. NO—go to step 2.

We illustrate the use of this algorithm on the network shown in Figure 2.12.

Suppose we want to find the shortest path from node A to all other nodes. We begin by labeling node A [0, –] and all other nodes [∞, –]. Node A is then called scanned. This results in the network shown in Figure 2.13 (in which scanned nodes are shaded).

We now update the node labels on all nodes that can be reached directly from node A (the last node to be labeled permanently). This results in the node labels shown in Figure 2.14.

We can now label node F permanently and proceed from there. This results in the network shown in Figure 2.15 in which we show links that are

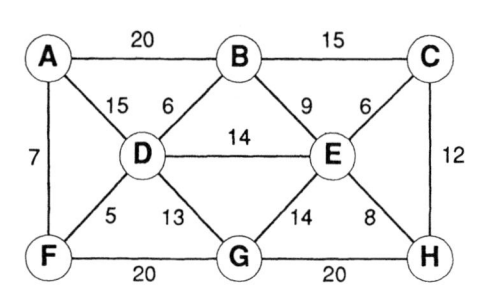

Figure 2.12. Example network for shortest path problems.

The Shortest Path Problem

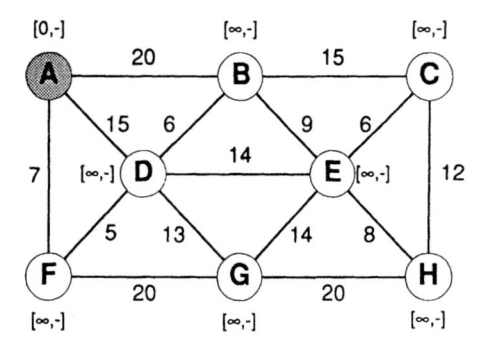

Figure 2.13. Example problem with initial node labels and node A scanned.

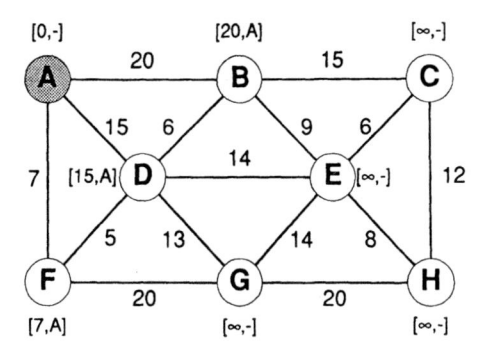

Figure 2.14. Network with nodes labeled from node A.

on the emerging shortest path tree using heavier lines. In this network, node F is now scanned and we have updated some of the node labels. Note that the label on node D has changed from $[15, A]$ to $[12, F]$ to indicate that the best currently known path from node A (the root node of the tree) to node D has a cost of 12 and goes through node F immediately prior to getting to node D.

Figures 2.16 through 2.21 show the resulting network after each remaining iteration of the algorithm.

The reader should notice how the temporary node labels changed with each iteration. Also, note that at each iteration a new node is labeled

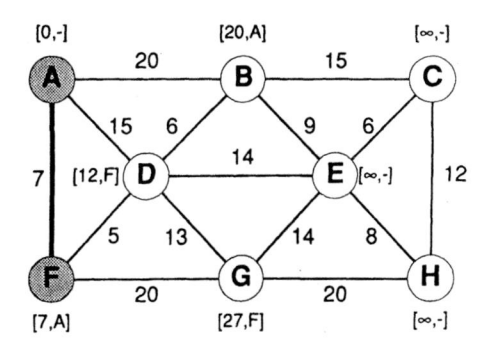

Figure 2.15. Network after node F is labeled permanently.

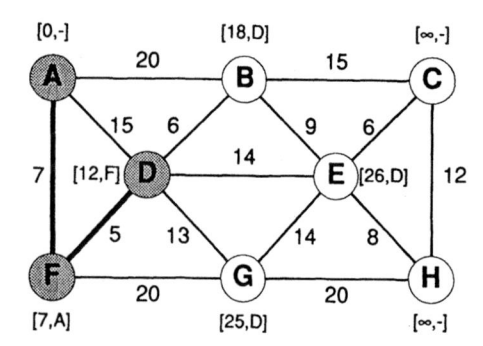

Figure 2.16. Network after node D is labeled permanently.

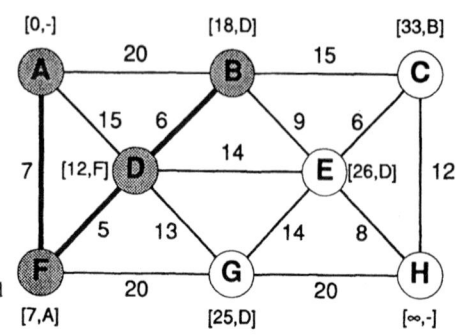

Figure 2.17. Network after node B is labeled permanently.

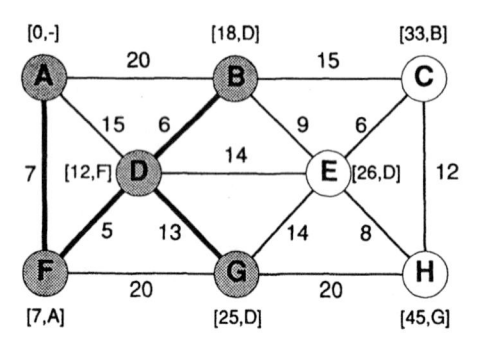

Figure 2.18. Network after node G is labeled permanently.

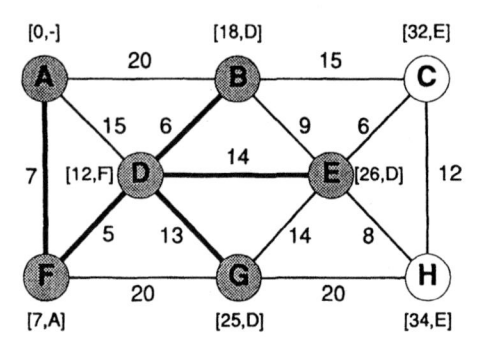

Figure 2.19. Network after node E is labeled permanently.

The Shortest Path Problem

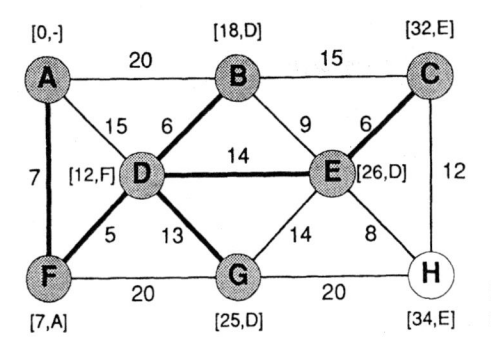

Figure 2.20. Network after node C is labeled permanently.

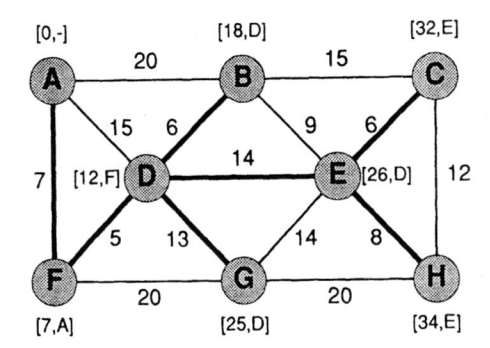

Figure 2.21. Network after node H is labeled permanently.

permanently. Finally, note that the network composed only of links on the shortest paths from node A to all other nodes is a *spanning tree* (i.e., a graph in which all nodes are connected and in which there is exactly one way to go from any node to any other node). This will be true of any such set of links; that is, they constitute a spanning tree rooted at the node from which we are finding the shortest paths.

In some cases, we would like to compute the shortest paths from all nodes to all other nodes at the same time. The algorithm given above may be used repeatedly (beginning with each origin node) to solve this problem. Alternatively, we can use Floyd's (1962) matrix approach outlined below. This approach updates two matrices simultaneously. The first matrix $\mathbf{V} = [V_{ij}]$ gives, at the end of the algorithm, the shortest path distances. At any stage m of the algorithm, the matrix $\mathbf{V}^m = [V_{ij}^m]$ gives the shortest path between each pair of nodes using only the first m modes. Similarly, the matrix $\mathbf{P} = [P_{ij}]$ gives, at the end of the algorithm, the predecessor nodes on the shortest paths. The interpretation of the matrix $\mathbf{P}^m = [P_{ij}^m]$ at the mth stage of the algorithm is similar to that of the matrix $\mathbf{V}^m$. Note that this approach does not restrict us to considering only cases with nonnegative costs, c_{ij}. The approach will detect a negative cycle if one exists (in step 2). Further details of this approach may be found in Floyd (1962) who originally proposed this approach or in Lawler (1976).

Floyd's Matrix Approach to Solving for the Shortest Paths from All Origins to All Destinations

Step 1: Initialization

    a. Set

$$V_{ij}^0 = \begin{cases} 0 & \text{if } i = j \\ c_{ij} & \text{if a direct link between nodes } i \text{ and } j \text{ exists.} \\ \infty & \text{otherwise} \end{cases}$$

    b. Set $P_{ij}^0 = i$ if a direct link exists between i and j; and "−" otherwise.

    c. Set $m = 0$ (an index).

Step 2: Check for Negative Cycles

    a. Are any $V_{ij}^m < 0$? If so, stop, a negative cycle has been detected.

    b. Increment m by 1.

    c. If $m \leq n$ (where n is the number of nodes in the graph), go to step 3; otherwise, stop, the algorithm is finished.

Step 3: Matrix Updates

    a. If $V_{im}^m + V_{mj}^m < V_{ij}^m$, set $V_{ij}^{m+1} = V_{im}^m + V_{mj}^m$ and set $P_{ij}^{m+1} = P_{mj}^m$; otherwise, set $V_{ij}^{m+1} = V_{ij}^m$ and $P_{ij}^{m+1} = P_{ij}^m$.

    b. Repeat step 3a for all cells (i, j) where $i \neq m$ and/or $j \neq m$.

This approach is illustratled using the network shown in Figure 2.12. The initial matrices are

$$\mathbf{V}^0$$

$$\begin{bmatrix} 0 & 20 & \infty & 15 & \infty & 7 & \infty & \infty \\ 20 & 0 & 15 & 6 & 9 & \infty & \infty & \infty \\ \infty & 15 & 0 & \infty & 6 & \infty & \infty & 12 \\ 15 & 6 & \infty & 0 & 14 & 5 & 13 & \infty \\ \infty & 9 & 6 & 14 & 0 & \infty & 14 & 8 \\ 7 & \infty & \infty & 5 & \infty & 0 & 20 & \infty \\ \infty & \infty & \infty & 13 & 14 & 20 & 0 & 20 \\ \infty & \infty & 12 & \infty & 8 & \infty & 20 & 0 \end{bmatrix}$$

$$\mathbf{P}^0$$

$$\begin{bmatrix} - & A & - & A & - & A & - & - \\ B & - & B & B & B & - & - & - \\ - & C & - & - & C & - & - & C \\ D & D & - & - & D & D & D & - \\ - & E & E & E & - & - & E & E \\ F & - & - & F & - & - & F & - \\ - & - & - & G & G & G & - & G \\ - & - & H & - & H & - & H & - \end{bmatrix}$$

After one iteration we have the following matrices. Note that the first row and column of the **V** matrix are shown in *italics* to indicate that we have operated off of that row and column (i.e., $m = 1$). Also, in both matrices we have enclosed the cells that changed within a box.

The Shortest Path Problem

$$\mathbf{V}^1 = \begin{bmatrix} 0 & 20 & \infty & 15 & \infty & 7 & \infty & \infty \\ 20 & 0 & 15 & 6 & 9 & \boxed{27} & \infty & \infty \\ \infty & 15 & 0 & \infty & 6 & \infty & \infty & 12 \\ 15 & 6 & \infty & 0 & 14 & 5 & 13 & \infty \\ \infty & 9 & 6 & 14 & 0 & \infty & 14 & 8 \\ 7 & \boxed{27} & \infty & 5 & \infty & 0 & 20 & \infty \\ \infty & \infty & \infty & 13 & 14 & 20 & 0 & 20 \\ \infty & \infty & 12 & \infty & 8 & \infty & 20 & 0 \end{bmatrix}$$

$$\mathbf{P}^1 = \begin{bmatrix} - & A & - & A & - & A & - & - \\ B & - & B & B & B & \boxed{A} & - & - \\ - & C & - & - & C & - & - & C \\ D & D & - & - & D & D & D & - \\ - & E & E & E & - & - & E & E \\ F & \boxed{A} & - & F & - & - & F & - \\ - & - & - & G & G & G & - & G \\ - & - & H & - & H & - & H & - \end{bmatrix}$$

After the second iteration we obtain the following matrices:

$$\mathbf{V}^2 = \begin{bmatrix} 0 & 20 & \boxed{35} & 15 & \boxed{29} & 7 & \infty & \infty \\ 20 & 0 & 15 & 6 & 9 & 27 & \infty & \infty \\ \boxed{35} & 15 & 0 & \boxed{21} & 6 & \boxed{42} & \infty & 12 \\ 15 & 6 & \boxed{21} & 0 & 14 & 5 & 13 & \infty \\ \boxed{29} & 9 & 6 & 14 & 0 & \infty & 14 & 8 \\ 7 & 27 & \boxed{42} & 5 & \infty & 0 & 20 & \infty \\ \infty & \infty & \infty & 13 & 14 & 20 & 0 & 20 \\ \infty & \infty & 12 & \infty & 8 & \infty & 20 & 0 \end{bmatrix}$$

$$\mathbf{P}^2 = \begin{bmatrix} - & A & \boxed{B} & A & \boxed{B} & A & - & - \\ B & - & B & B & B & A & - & - \\ \boxed{B} & C & - & \boxed{B} & C & \boxed{A} & - & C \\ D & D & \boxed{B} & - & D & D & D & - \\ \boxed{B} & E & E & E & - & - & E & E \\ F & A & \boxed{B} & F & - & - & F & - \\ - & - & - & G & G & G & - & G \\ - & - & H & - & H & - & H & - \end{bmatrix}$$

Review of Linear Programming

After the third iteration we obtain the following matrices:

$$\mathbf{V}^3 = \begin{bmatrix} 0 & 20 & 35 & 15 & 29 & 7 & \infty & \boxed{47} \\ 20 & 0 & 15 & 6 & 9 & 27 & \infty & \boxed{27} \\ 35 & 15 & 0 & 21 & 6 & 42 & \infty & 12 \\ 15 & 6 & 21 & 0 & 14 & 5 & 13 & \boxed{33} \\ 29 & 9 & 6 & 14 & 0 & \infty & 14 & 8 \\ 7 & 27 & 42 & 5 & \infty & 0 & 20 & \boxed{54} \\ \infty & \infty & \infty & 13 & 14 & 20 & 0 & 20 \\ \boxed{47} & \boxed{27} & 12 & \boxed{33} & 8 & \boxed{54} & 20 & 0 \end{bmatrix}$$

$$\mathbf{P}^3 = \begin{bmatrix} - & A & B & A & B & A & - & \boxed{C} \\ B & - & B & B & B & A & - & \boxed{C} \\ B & C & - & B & C & A & - & C \\ D & D & B & - & D & D & D & \boxed{C} \\ B & E & E & E & - & - & E & E \\ F & A & B & F & - & - & F & \boxed{C} \\ - & - & - & G & G & G & - & G \\ \boxed{B} & \boxed{C} & H & \boxed{B} & H & \boxed{A} & H & - \end{bmatrix}$$

After the fourth iteration we obtain the following matrices:

$$\mathbf{V}^4 = \begin{bmatrix} 0 & 20 & 35 & 15 & 29 & 7 & \boxed{28} & 47 \\ 20 & 0 & 15 & 6 & 9 & \boxed{11} & \boxed{19} & 27 \\ 35 & 15 & 0 & 21 & 6 & \boxed{26} & \boxed{34} & 12 \\ 15 & 6 & 21 & 0 & 14 & 5 & 13 & 33 \\ 29 & 9 & 6 & 14 & 0 & \boxed{19} & 14 & 8 \\ 7 & \boxed{11} & \boxed{26} & 5 & \boxed{19} & 0 & \boxed{18} & \boxed{38} \\ \boxed{28} & \boxed{19} & \boxed{34} & 13 & 14 & \boxed{18} & 0 & 20 \\ 47 & 27 & 12 & 33 & 8 & \boxed{38} & 20 & 0 \end{bmatrix}$$

The Shortest Path Problem

$$\mathbf{P}^4$$

$$\begin{bmatrix}
- & A & B & A & B & A & \boxed{D} & C \\
B & - & B & B & B & \boxed{D} & \boxed{D} & C \\
B & C & - & B & C & \boxed{D} & \boxed{D} & C \\
D & D & B & - & D & D & D & C \\
B & E & E & E & - & \boxed{D} & E & E \\
F & \boxed{D} & B & F & \boxed{D} & - & \boxed{D} & \boxed{D} \\
\boxed{D} & \boxed{D} & \boxed{B} & G & G & \boxed{D} & - & G \\
B & C & H & B & H & \boxed{D} & H & -
\end{bmatrix}$$

After the fifth iteration we obtain the following matrices:

$$\mathbf{V}^5$$

$$\begin{bmatrix}
0 & 20 & 35 & 15 & 29 & 7 & 28 & \boxed{37} \\
20 & 0 & 15 & 6 & 9 & 11 & 19 & \boxed{17} \\
35 & 15 & 0 & \boxed{20} & 6 & \boxed{25} & \boxed{20} & 12 \\
15 & 6 & \boxed{20} & 0 & 14 & 5 & 13 & \boxed{22} \\
29 & 9 & 6 & 14 & 0 & 19 & 14 & 8 \\
7 & 11 & \boxed{25} & 5 & 19 & 0 & 18 & \boxed{27} \\
28 & 19 & \boxed{20} & 13 & 14 & 18 & 0 & 20 \\
\boxed{37} & \boxed{17} & 12 & \boxed{22} & 8 & \boxed{27} & 20 & 0
\end{bmatrix}$$

$$\mathbf{P}^5$$

$$\begin{bmatrix}
- & A & B & A & B & A & D & \boxed{E} \\
B & - & B & B & B & D & D & \boxed{E} \\
B & C & - & \boxed{E} & C & D & \boxed{E} & C \\
D & D & \boxed{E} & - & D & D & D & \boxed{E} \\
B & E & E & E & - & D & E & E \\
F & D & \boxed{E} & F & D & - & D & \boxed{E} \\
D & D & \boxed{E} & G & G & D & - & G \\
B & \boxed{E} & H & \boxed{E} & H & D & H & -
\end{bmatrix}$$

After the sixth iteration we obtain the following matrices:

$$\mathbf{V}^6$$

$$\begin{bmatrix}
0 & \boxed{18} & \boxed{32} & \boxed{12} & \boxed{26} & 7 & \boxed{25} & \boxed{34} \\
\boxed{18} & 0 & 15 & 6 & 9 & 11 & 19 & 17 \\
\boxed{32} & 15 & 0 & 20 & 6 & 25 & 20 & 12 \\
\boxed{12} & 6 & 20 & 0 & 14 & 5 & 13 & 22 \\
\boxed{26} & 9 & 6 & 14 & 0 & 19 & 14 & 8 \\
7 & 11 & 25 & 5 & 19 & 0 & 18 & 27 \\
\boxed{25} & 19 & 20 & 13 & 14 & 18 & 0 & 20 \\
\boxed{34} & 17 & 12 & 22 & 8 & 27 & 20 & 0
\end{bmatrix}$$

$$\mathbf{P}^6$$

$$\begin{bmatrix}
- & \boxed{D} & \boxed{E} & \boxed{F} & \boxed{D} & A & D & E \\
\boxed{F} & - & B & B & B & D & D & E \\
\boxed{F} & C & - & E & C & D & E & C \\
\boxed{F} & D & E & - & D & D & D & E \\
\boxed{F} & E & E & E & - & D & E & E \\
F & D & E & F & D & - & D & E \\
\boxed{F} & D & E & G & G & D & - & G \\
\boxed{F} & E & H & E & H & D & H & -
\end{bmatrix}$$

After the seventh iteration we obtain the following matrices:

$$\mathbf{V}^7$$

$$\begin{bmatrix}
0 & 18 & 32 & 12 & 26 & 7 & 25 & 34 \\
18 & 0 & 15 & 6 & 9 & 11 & 19 & 17 \\
32 & 15 & 0 & 20 & 6 & 25 & 20 & 12 \\
12 & 6 & 20 & 0 & 14 & 5 & 13 & 22 \\
26 & 9 & 6 & 14 & 0 & 19 & 14 & 8 \\
7 & 11 & 25 & 5 & 19 & 0 & 18 & 27 \\
25 & 19 & 20 & 13 & 14 & 18 & 0 & 20 \\
34 & 17 & 12 & 22 & 8 & 27 & 20 & 0
\end{bmatrix}$$

The Out-of-Kilter Flow Algorithm

$$\mathbf{P}^7$$

$$\begin{bmatrix}
- & D & E & F & D & A & D & E \\
F & - & B & B & B & D & D & E \\
F & C & - & E & C & D & E & C \\
F & D & E & - & D & D & D & E \\
F & E & E & E & - & D & E & E \\
F & D & E & F & D & - & D & E \\
F & D & E & G & G & D & - & G \\
F & E & H & E & H & D & H & -
\end{bmatrix}$$

Note that there were no changes during this iteration. After the final iteration we have the following matrices:

$$\mathbf{V}^8$$

$$\begin{bmatrix}
0 & 18 & 32 & 12 & 26 & 7 & 25 & 34 \\
18 & 0 & 15 & 6 & 9 & 11 & 19 & 17 \\
32 & 15 & 0 & 20 & 6 & 25 & 20 & 12 \\
12 & 6 & 20 & 0 & 14 & 5 & 13 & 22 \\
26 & 9 & 6 & 14 & 0 & 19 & 14 & 8 \\
7 & 11 & 25 & 5 & 19 & 0 & 18 & 27 \\
25 & 19 & 20 & 13 & 14 & 18 & 0 & 20 \\
34 & 17 & 12 & 22 & 8 & 27 & 20 & 0
\end{bmatrix}$$

$$\mathbf{P}^8$$

$$\begin{bmatrix}
- & D & E & F & D & A & D & E \\
F & - & B & B & B & D & D & E \\
F & C & - & E & C & D & E & C \\
F & D & E & - & D & D & D & E \\
F & E & E & E & - & D & E & E \\
F & D & E & F & D & - & D & E \\
F & D & E & G & G & D & - & G \\
F & E & H & E & H & D & H & -
\end{bmatrix}$$

Again, there were no changes. Note that the interpretation of row 1 (the row corresponding to node A) of the matrices allows us to retrieve the shortest path tree computed above using Dijkstra's algorithm. Other rows would allow us to derive the shortest path trees rooted at other nodes. The reader is encouraged to develop one or more of these shortest path trees based on these matrices.

2.7 THE OUT-OF-KILTER FLOW ALGORITHM

The final class of special linear programming problems that we discuss are more general minimum cost network flow problems. We discuss these problems in the context of the *out-of-kilter flow algorithm*. Before discussing this algorithm, we need to define a *circulation flow*. A circulation flow in a network is one in which the flow into every node exactly equals the flow out of every node. The out-of-kilter flow algorithm finds a minimum cost circulation flow through a network that satisfies lower and upper bounds that may be imposed on each of the link flows (Fulkerson, 1961).

Many network flow problems including the transportation problem and the shortest path problem may be structured as minimum cost network flow problems and may be solved using the out-of-kilter flow algorithm.[5] The out-of-kilter flow algorithm is but one of many network flow algorithms. The reader interested in such algorithms should consult any one of a number of texts including: Ahuja, Magnanti, and Orlin (1993), Bertsekas (1991), Jensen and Barnes (1980), Kennington and Helgason (1980), Minieka (1978), and Phillips and Garcia-Diaz (1981). We have elected to discuss the out-of-kilter flow algorithm here because of the intimate relationship between the algorithm and duality theory in linear programming.[6]

The out-of-kilter flow problem may be formulated as a linear programming problem using the following notation:

Inputs

c_{ij} = unit cost of traversing link (i, j)

l_{ij} = minimum required flow on link (i, j) [i.e., the *lower bound* on the flow on link (i, j)]

u_{ij} = maximum allowed flow on link (i, j) [i.e., the *upper bound* on the flow on link (i, j)]

Decision Variables

X_{ij} = flow on link (i, j)

With these definitions, the primal may be formulated as follows:

Out-of-Kilter Flow Problem — Primal

MINIMIZE $\quad \sum_i \sum_j c_{ij} X_{ij} \quad$ (2.21a)

SUBJECT TO: $\quad \sum_j X_{ji} - \sum_k X_{ik} = 0 \quad \forall\, i \quad$ (2.21b)

$\qquad\qquad\qquad X_{ij} \leq u_{ij} \quad\qquad \forall\, i, j \quad$ (2.21c)

$\qquad\qquad\qquad X_{ij} \geq l_{ij} \quad\qquad \forall\, i, j \quad$ (2.21d)

---

[5]Glover, Klingman, and Phillips (1992) offer an outstanding review of problems that can be structured as linear network flow problems. The reader interested in such topics is strongly encouraged to consult this text which focuses on modeling such problems and not on algorithms.

[6]Also, an implementation of the out-of-kilter flow algorithm—MENU-OKF—is included in the software that accompanies this text. The reader is referred to Appendix E for a description of this program.

The Out-of-Kilter Flow Algorithm

The objective function (2.21a) minimizes the total cost of the flow on the links. Constraint (2.21b) stipulates that the flow constitute a circulation flow. Constraint (2.21c) is the upper-bound constraint on the flow on each link, while constraint (2.21d) is the lower-bound constraint on the flow on each link.

The dual of this linear programming problem is

Out-of-Kilter Flow Problem—Dual

MAXIMIZE $$\sum_i \sum_j l_{ij} \alpha_{ij} - \sum_i \sum_j u_{ij} \beta_{ij} \qquad (2.22a)$$

SUBJECT TO:
$$\pi_j - \pi_i + \alpha_{ij} - \beta_{ij} \leq c_{ij} \qquad \forall\, i,j \qquad (2.22b)$$

$$\pi_i \text{ unrestricted} \qquad \forall\, i \qquad (2.22c)$$

$$\alpha_{ij} \geq 0 \qquad \forall\, i,j \qquad (2.22d)$$

$$\beta_{ij} \geq 0 \qquad \forall\, i,j \qquad (2.22e)$$

where

π_i = the dual variable for constraint (2.21b)
β_{ij} = the dual variable for constraint (2.21c)
α_{ij} = the dual variable for constraint (2.21d)

Again, our interest in this problem stems from the fact that many network problems can easily be stated in terms of out-of-kilter flow problems. For example, to formulate a transportation problem as an out-of-kilter flow problem, we add a super source node and a super sink node. We connect the super source node to each supply node with a lower bound of 0; a cost of 0; and an upper bound equal to the suply of the supply node. Similarly, we connect each demand node to the super sink node. The lower bound on a link between a demand node and the super sink node equals the demand at that node; the upper bound is infinite; and the cost is 0. We also connect every supply node to every demand node. These links have lower bounds of 0; upper bounds of ∞; and costs equal to the unit transport cost between the supply node and demand node being connected. Finally, we add a link from the super sink to the super source with a lower bound of 0; an upper bound of ∞; and a cost of 0. Figure 2.22 illustrates such a network for the transportation problem solved in Section 2.5 above. (The lower and upper bounds and costs are shown only for selected links.) Many other network problems can also be formulated as out-of-kilter flow problems. Therefore, it is useful to know how to transform a problem into this format.

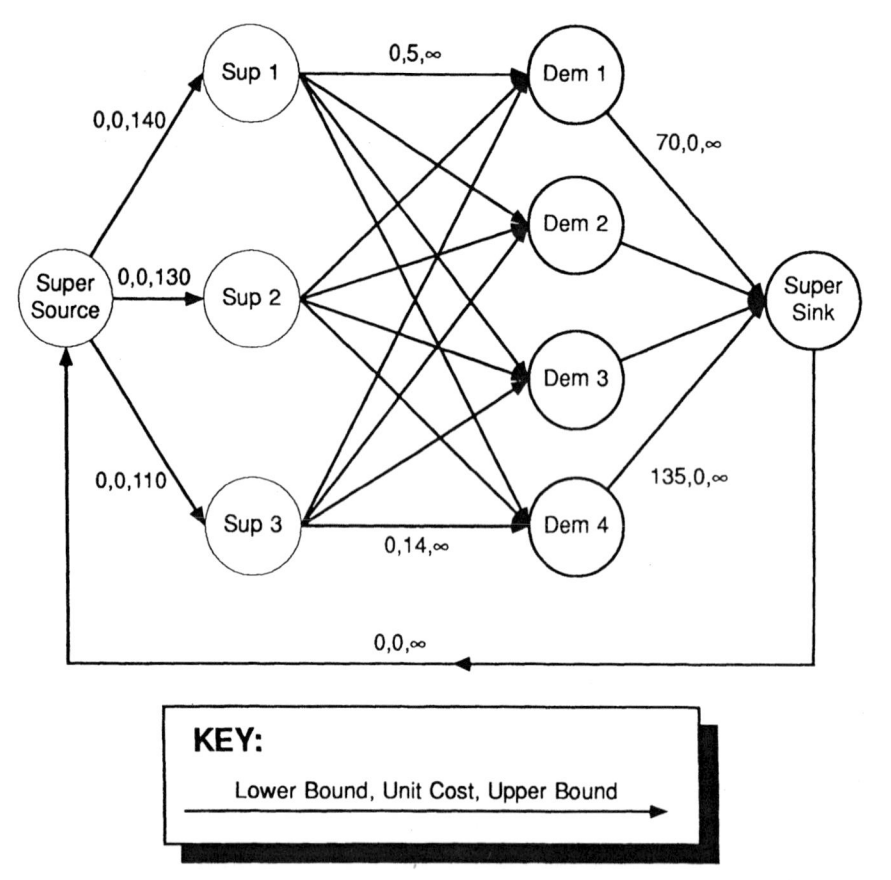

Figure 2.22. Sample out-of-kilter flow network for the transportation problem solved in section 2.5.

We now turn to the complementary slackness conditions for this problem. They are

$$X_{ij}(c_{ij} - \pi_j + \pi_i - \alpha_{ij} + \beta_{ij}) = 0 \quad \forall\, i, j \quad (2.23a)$$

$$(X_{ij} - l_{ij})\alpha_{ij} = 0 \quad \forall\, i, j \quad (2.23b)$$

$$(u_{ij} - X_{ij})\beta_{ij} = 0 \quad \forall\, i, j \quad (2.23c)$$

Again, any solution that satisfies primal feasibility [(2.21b)–(2.21d)], dual feasibility [(2.22b)–(2.22e)], and complementary slackness [(2.23a)–(2.23c)] is optimal.

Given optimal flows, X_{ij}, and node values, π_i, we can infer optimal values for the dual variables α_{ij} and β_{ij}. We begin by noting that in any optimal solution either $\alpha_{ij} = 0$ or $\beta_{ij} = 0$ (or both equal 0) for every link (i, j). Clearly, if $l_{ij} < u_{ij}$, conditions (2.23b) and (2.23c) ensure that at least one of

The Out-of-Kilter Flow Algorithm

the two values must be 0. If $l_{ij} = u_{ij}$ and we have a solution in which both $\alpha_{ij} > 0$ and $\beta_{ij} > 0$, we can reduce both values until one is equal to 0. This will not affect the dual objective function when $l_{ij} = u_{ij}$. It will also not affect the feasibility of the dual constraint (2.22b) or the complementary slackness condition (2.23a). Thus, we can always find a solution in which at least one of the two variables α_{ij} and β_{ij} equals 0.

Next, we show that we can infer the values of α_{ij} and β_{ij} from knowledge of the node values, π_i. To see how this is done, let us replace $c_{ij} - \pi_j + \pi_i$ by $\bar{c}_{ij}$. Conditions (2.23) then imply:

If $\bar{c}_{ij} > 0$, then $\beta_{ij} = 0$. [To see that this must be true, note that if we had $\beta_{ij} > 0$, we would need to have $X_{ij} = u_{ij}$ to satisfy (2.23c). However, if $\bar{c}_{ij} > 0$ and $\beta_{ij} > 0$ and $X_{ij} = u_{ij} > 0$, we would need $\alpha_{ij} = \bar{c}_{ij} + \beta_{ij} > 0$ to satisfy (2.23a). However, we just argued that α_{ij} and β_{ij} cannot simultaneously be greater than 0.] We also will have $X_{ij} = l_{ij}$ and $\bar{c}_{ij} = c_{ij} - \pi_j + \pi_i = \alpha_{ij}$. (2.24a)

Similarly, if $\bar{c}_{ij} < 0$, then $\alpha_{ij} = 0$, $X_{ij} = u_{ij}$, and $-\bar{c}_{ij} = \pi_j - \pi_i - c_{ij} = \beta_{ij}$. (2.24b)

Finally, if $\bar{c}_{ij} = 0$, then $\alpha_{ij} = 0$, $\beta_{ij} = 0$, and $l_{ij} \le X_{ij} \le u_{ij}$. (2.24c)

The implications of these three conditions are that if we know $\bar{c}_{ij}$ [which implies knowing the values of the dual variables (π_i) associated with the nodes], then we can infer the values of the dual variables associated with the links (α_{ij} and β_{ij}). This greatly reduces the amount of information we need to keep track of.

Conditions (2.24a)–(2.24c) may be summarized in a *kilter diagram* in which we plot the flow, X_{ij}, on the X axis and the value of $\pi_j - \pi_i$ on the Y axis. Figure 2.23 is a typical kilter diagram.

Any link whose flow and $\pi_j - \pi_i$ values plot on the kilter line is *in kilter* and satisfies all of the optimality conditions (provided we ensure, as we will, that we always have a circulation flow). Any link that does not plot on the kilter line is *out of kilter* and violates one or more of the optimality conditions. To facilitate the discussion, we often associate kilter numbers and kilter states with different regions of the kilter diagram. These are shown in Figure 2.24 and Table 2.1.

Note that all of the kilter numbers are positive. These numbers measure (in some loose sense) the extent to which the solution violates the optimality conditions. Also, note that if we could change the flow as indicated in Table 2.1 on any link which is out of kilter, the link would then satisfy the optimality conditions.

At the optimal solution, we note that $\bar{c}_{ij}$ is the *reduced cost* associated with link (i, j). It gives the rate of change of the objective function with respect to changes in l_{ij} or u_{ij}. In particular, if $\bar{c}_{ij} > 0$, then $\bar{c}_{ij}$ gives the amount by which the objective function will increase if we increase l_{ij} by 1.

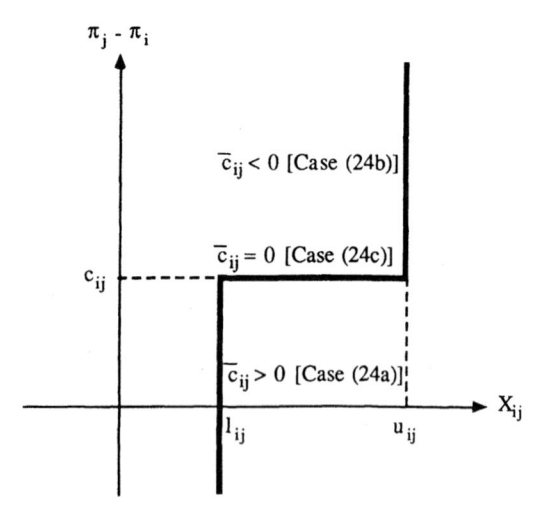

Figure 2.23. Typical kilter diagram.

Similarly, if $\bar{c}_{ij} < 0$, then $\bar{c}_{ij}$ gives the amount by which the objective function will decrease if we increase u_{ij} by 1. Finally, if $\bar{c}_{ij} = 0$, then $\bar{c}_{ij}$ indicates that the objective function will not change if we make small changes in l_{ij} or u_{ij}. As always, we need to exercise care in interpreting these values and in applying them to anything more than the smallest changes. Again, the reader is referred to Rubin and Wagner (1990) for a discussion of the interpretation of such shadow prices.

Finally, we note that if l_{ij} and u_{ij} are integers for all links (i, j), then the optimal solution will be all integer.

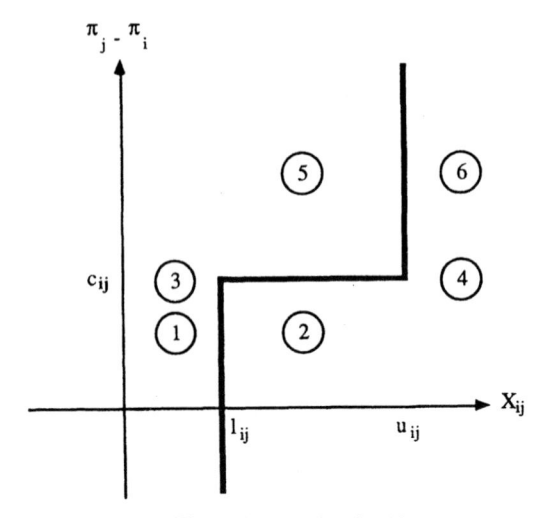

Figure 2.24. Kilter diagram showing kilter states.

Table 2.1 Summary of Kilter States and Kilter Numbers

State	Flow, X_{ij}	Reduced Cost, $\bar{c}_{ij}$	Kilter Number	Want to (Increase/Decrease) Flow
1	$< l_{ij}$	> 0	$l_{ij} - X_{ij}$	Increase to l_{ij}
2	$> l_{ij}$	> 0	$\bar{c}_{ij}(X_{ij} - l_{ij})$	Decrease to l_{ij}
3	$< l_{ij}$	$= 0$	$l_{ij} - X_{ij}$	Increase to l_{ij}
4	$> u_{ij}$	$= 0$	$X_{ij} - u_{ij}$	Decrease to u_{ij}
5	$< u_{ij}$	< 0	$\bar{c}_{ij}(X_{ij} - u_{ij})$	Increase to u_{ij}
6	$> u_{ij}$	< 0	$X_{ij} - u_{ij}$	Decrease to u_{ij}

The solution strategy that we adopt is as follows:

Step 0: Begin with any circulation flow (feasible or not) and any node values π_i. The initial flow $X_{ij} = 0$ for all links (i, j) and node values $\pi_i = 0$ for all nodes i are fine, for example. Often, we may be able to do much better than this, however.

Step 1: Compute kilter numbers for all links and kilter states for all out-of-kilter links. If all links are in kilter, *stop*, the solution is optimal. If not, for links (p, q) in kilter states 1, 3, or 5, we want to *increase* the flow. This means we need to find a feasible flow augmentation path from q to p. (See below for a discussion of flow augmentation paths.) For links (p, q) in kilter states 2, 4, or 6 of the kilter diagram, we want to *decrease* the flow. Here we need a flow augmentation path from p to q.

Step 2: Select an out-of-kilter link and *try* to find a flow augmentation path from q to p if the selected link is in states 1, 3, or 5 or from p to q if the selected link is in states 2, 4, or 6. If such a path can be found, augment the flow by the appropriate amount [the minimum of (i) the amount needed to bring link (p, q) into kilter and (ii) the maximum allowable flow increase on the flow augmentation path]. Go to step 1. If no such path exists, go to step 3 to revise the node numbers to try to bring a link into kilter.

Step 3: Revise the node numbers of all unlabeled nodes (nodes that are not already on the potential or emerging flow augmentation path)[7] so that an additional node can be reached from this labeled set of nodes. If the node numbers cannot be revised in this way, *no feasible solution exists*. If the node numbers can be revised, recompute the kilter numbers for all links from the labeled to the unlabeled nodes and vice versa. If link (p, q) was brought into kilter by this change in node numbers, go to step 1. If not, go to step 2 and continue trying to find a flow augmentation path.

---

[7]The details of the labeling process for building the emerging flow augmentation path are beyond the scope of this text. The interested reader is referred to Minieka (1978) or Jensen and Barnes (1980).

Outbound Links

Flow may be added to outbound links in shaded zone or on horizontal line, excluding circled point

Inbound Links

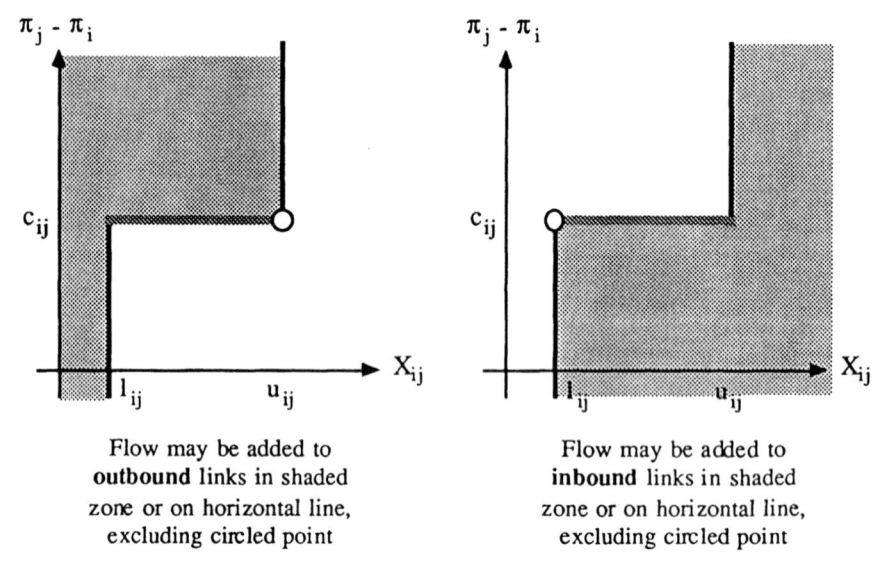

Flow may be added to inbound links in shaded zone or on horizontal line, excluding circled point

Figure 2.25. Kilter diagrams showing where a link can plot and be added to a flow augmenting path.

To understand this algorithm completely, we need to explain the concepts of *flow augmentation paths* and *revising node labels*. A *flow augmentation path* is a connected set of links from an origin (call it node s) to a destination (call it node t). Links that are directed from s to t are called *outbound* links; links that are oriented from t to s are called *inbound* links. In step 2, an outbound link may be added to a flow augmentation path if either of the following conditions hold:

i. $X_{ij} < u_{ij}$ and $\bar{c}_{ij} \leq 0$ or
ii. $X_{ij} < l_{ij}$ and $\bar{c}_{ij} > 0$.

These two conditions say that link (i, j) is to the left of the kilter line or on the horizontal part. An inbound link may be added to the flow augmentation path if either of the following conditions hold:

i. $X_{ij} > l_{ij}$ and $\bar{c}_{ij} \geq 0$ or
ii. $X_{ij} > u_{ij}$ and $\bar{c}_{ij} < 0$.

These conditions imply that link (i, j) is to the right of the kilter line or on the horizontal part. Figure 2.25 summarizes the regions in which a link may plot in the kilter diagram and be added to the flow augmentation path.

Note that by adding (or subtracting) flow only on these links, (i) in-kilter links will remain in kilter as long as the flow increment is chosen appropri-

The Out-of-Kilter Flow Algorithm

ately and (ii) out-of-kilter links will move toward the kilter line. Thus, all kilter numbers will either stay the same or will decrease. Further details on flow augmenting paths and the out-of-kilter flow algorithm may be found in Phillips and Garcia-Diaz (1981) as well as Jensen and Barnes (1980).

If we cannot add any additional links to an emerging flow augmentation path (i.e., none of the links with one node labeled and one unlabeled fall into the shaded regions identified on the kilter diagrams above), node labels need to be revised. We may think of adding (or subtracting) flow on a flow augmentation path as moving links *horizontally* to bring them into kilter. When we revise node numbers, we try to shift a link *vertically* to move it onto the kilter line. We are only concerned with links from a labeled node (one on the emerging flow augmentation path) to an unlabeled node or vice versa. In particular,

i. for links from a labeled node to an unlabeled node (outbound links) such that $X_{ij} \leq u_{ij}$ and $\bar{c}_{ij} > 0$, we compute $\bar{c}_{out} = \text{MIN}_{\text{all such links}}\{\bar{c}_{ij}\}$. If no such links exist, we set $\bar{c}_{out} = \infty$.

ii. for links from an unlabeled node to a labeled node (inbound links) such that $l_{ij} \leq X_{ij}$ and $\bar{c}_{ij} < 0$, we compute $\bar{c}_{in} = \text{MIN}_{\text{all such links}}\{|\bar{c}_{ij}|\}$. If no such links exist, we set $\bar{c}_{in} = \infty$.

Cases (i) and (ii) are shown in Figure 2.26.

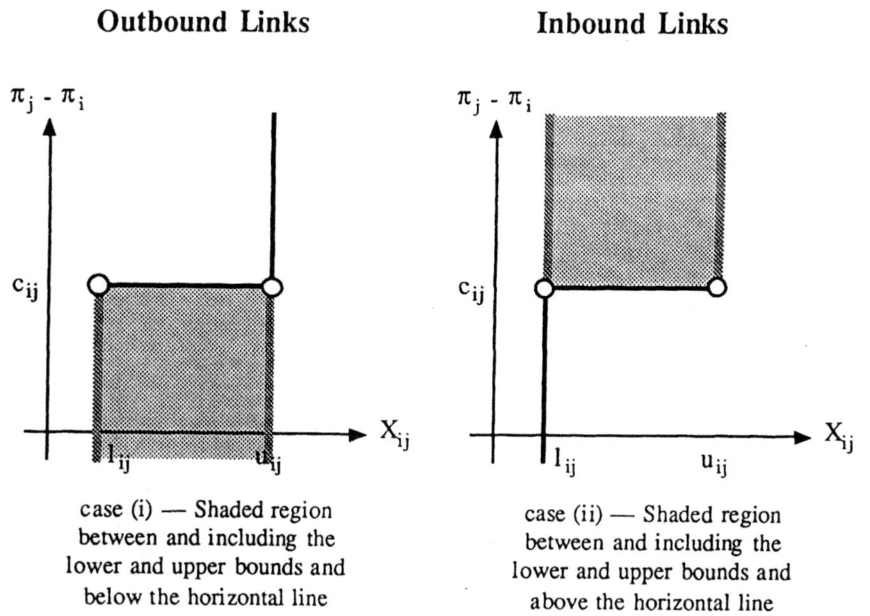

Outbound Links   **Inbound Links**

case (i) — Shaded region between and including the lower and upper bounds and below the horizontal line

case (ii) — Shaded region between and including the lower and upper bounds and above the horizontal line

Figure 2.26. Regions in which a link between a labeled and an unlabeled node may plot to allow the node label on the labeled node to be revised.

Finally, we compute $\bar{c}^* = \text{MIN}\{\bar{c}_{in}, \bar{c}_{out}\}$. If both $\bar{c}_{in}$ and $\bar{c}_{out}$ equal ∞, then *no feasible solution exists*; that is, no flow augmentation path from s to t could be found when one was needed. If $\bar{c}^* < \infty$, then we add $\bar{c}^*$ to all π_i for unlabeled nodes i. This will

  i. bring link (t, s) into kilter [where (t, s) is the link which was out of kilter and which caused us to initiate the search for the flow augmentation path], in which case we return to step 1 as indicated above; or
  ii. allow at least one more node to be labeled, in which case we go to step 2; or
  iii. bring some link into kilter but in such a way that the associated unlabeled node can still not be labeled, in which case we go to step 2 and immediately return to step 3. The reader should think about the conditions under which this can occur.

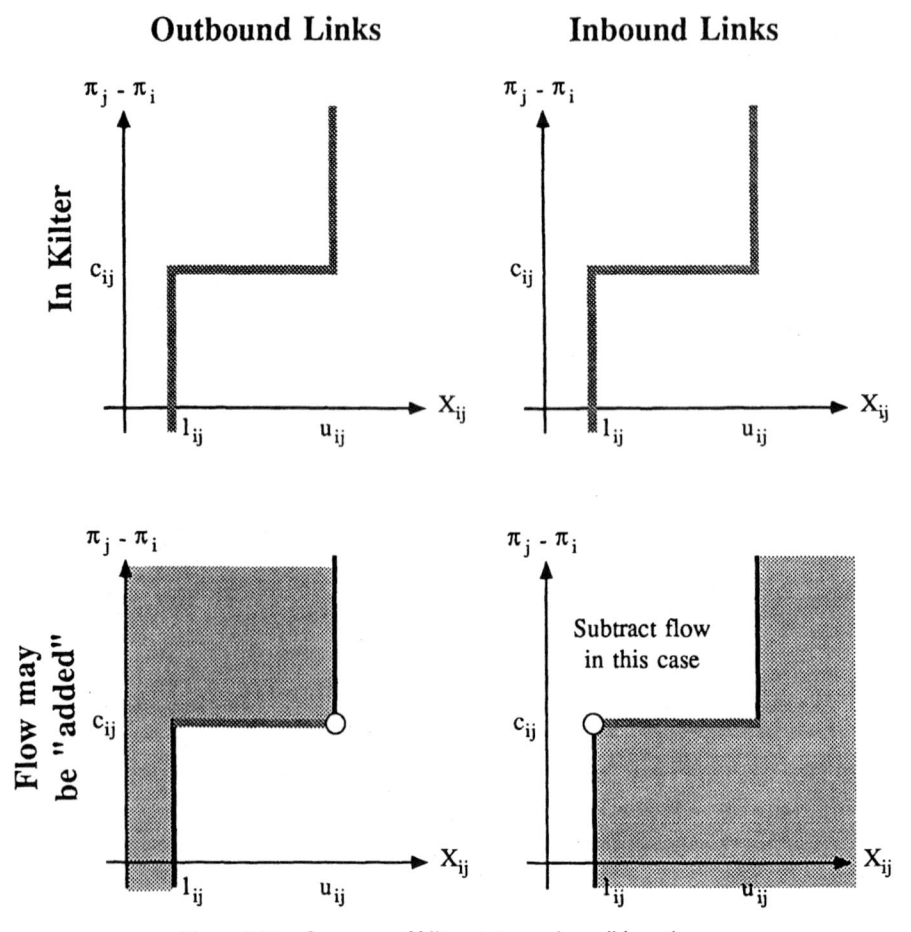

Figure 2.27a. Summary of kilter states and possible actions.

The Out-of-Kilter Flow Algorithm

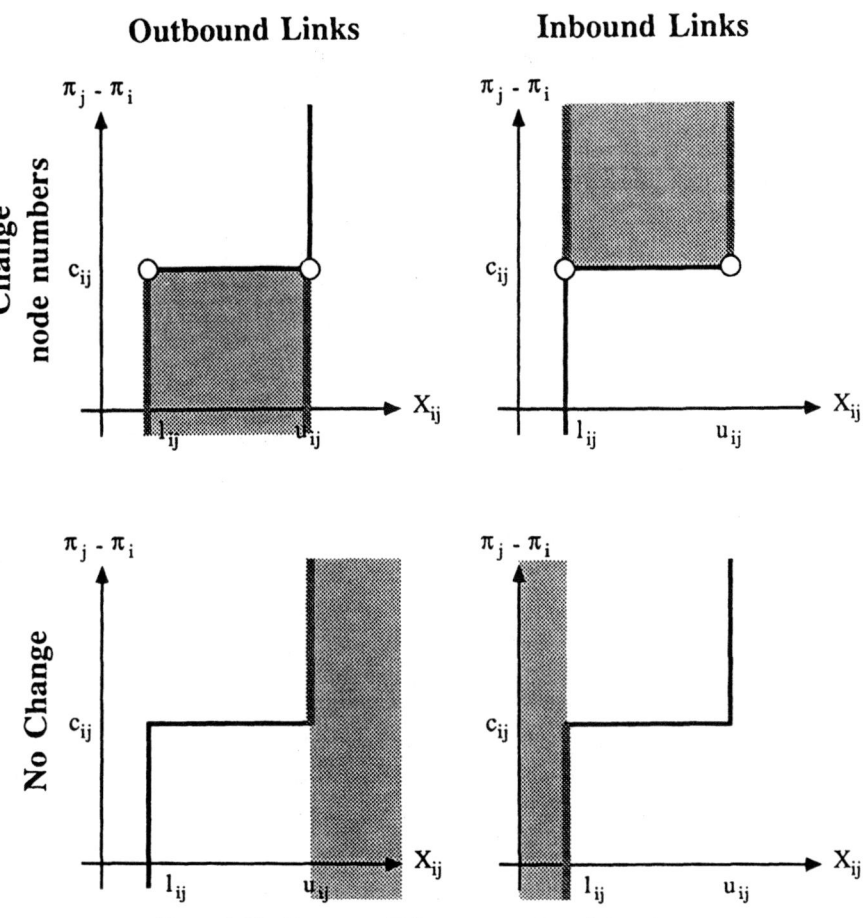

Figure 2.27b. Summary of kilter states and possible actions.

Figure 2.27 summarizes the various conditions in which a link may be found and the resulting actions we can take.

While some understanding of the solution algorithm for the out-of-kilter flow problem is valuable, it is probably more important for the reader to consider the range of network problems that can be formulated and solved using this approach. These include, but are not limited to,

a. the transportation problem
b. the assignment problem
c. the shortest path problem
d. the maximum flow problem
e. the minimum-cost flow problem
f. the transshipment problem

In many of these cases (e.g., the transportation problem and the shortest path problem), there are more efficient means of solving the problem than using the more general-purpose out-of-kilter flow algorithm. However, the out-of-kilter flow algorithm is just that—very general purpose and useful for a large number of problems. As such, it is useful to understand and to have available. Appendix E is a user's guide to MENU-OKF, a menu-based implementation of the out-of-kilter flow algorithm.

2.8 SUMMARY

This chapter has reviewed linear programming. In particular, we have stressed the relationship between the primal and dual formulations and the complementary slackness conditions. Most linear programming problems that arise in location contexts have special structures. We have outlined solution algorithms that take advantage of these special structures for three such problems: the shortest path problem, the transportation problem, and the out-of-kilter flow problem which can be used to represent many network flow problems.

EXERCISES

2.1 Consider the following linear programming problem:

MINIMIZE $\quad 100U + 70W - 10X$

SUBJECT TO: $\quad U - 3V + 0.5W + 0.25X \geq 1$
$\quad U + V + W - X \geq 2$
$\quad U \geq 0, V \geq 0, W \geq 0, X \geq 0$

(a) Formulate the dual of this problem.

(b) Solve the linear programming problem *graphically*. In presenting the solution, be sure to give the values of all primal and dual decision variables, all slack and surplus variables, and the value of the objective function.

2.2 Consider the following linear programming problem:

MAXIMIZE $\quad 20X_1 + 18X_2 - 13X_3 - 5X_4$

SUBJECT TO: $\quad X_1 + X_2 + X_3 + X_4 \geq 200$
$\quad 5X_1 + 10X_2 = 450$
$\quad X_1 - 2X_2 + 7X_3 + 2X_4 \leq 650$
$\quad X_1, X_2, X_3, X_4 \geq 0$

Exercises

(a) Transform this formulation into canonical form.

(b) Write out the dual to the problem as reformulated in part (a). Clearly indicate which dual variables are associated with each of the constraints in the primal.

(c) Solve the original problem by inspection. (*Hint:* Ignore the third constraint initially, find a solution that optimizes the objective function subject to the first two constraints and nonnegativity, and then check that the solution satisfies the third constraint as well.)

(d) What are the values of the dual variables for the original problem shown above? Justify your answer.

2.3 Show that the dual variables corresponding to equality constraints in a primal problem are unrestricted in sign. (*Hint:* Convert the primal equality constraints into two inequality constraints in canonical form. Associate dual variables with each of the corresponding constraints. Show that the new dual variables appear as pairs that can be converted into one unconstrained variable.)

2.4 Complete the solution of the transportation problem discussed in the text, beginning with the tableau shown in Figure 2.8. Be sure you obtain the optimal solution shown in Figure 2.9.

2.5 (a) Compute the dual variables and reduce costs for the following *optimal* transportation problem solution:

	15	18	26	Supply
	150			150
	31	24	16	
			225	225
	22	19	35	
	50	130		180
	25	12	18	
		245	15	260
Demand	200	375	240	

Total Cost = 12,630

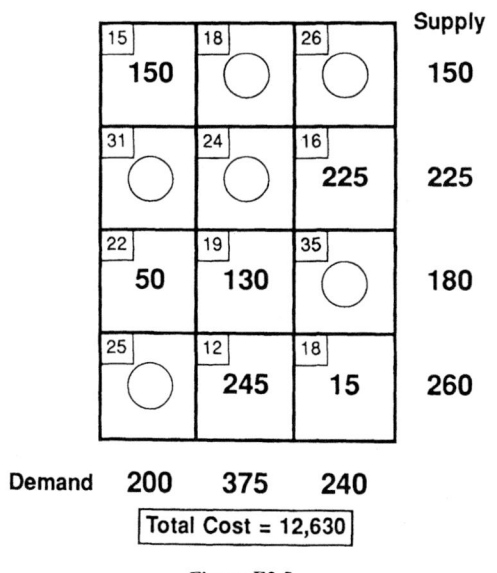

Figure E2.5

(b) Suppose we were required to ship 10 units from supply point 2 to demand node 1 (the cell with a unit cost of 31). All total demands and supplies remain unchanged. What will the new total cost be?

(c) For the case described in part (b), how will the flows change?

(d) Going back to the solution shown in part (a), supply the supply at node 2 increased to 230 and the supply at node 3 decreased to 175 simultaneously. What will the new total cost be?

(e) For the case described in part (d), how will the flows change?

2.6 (a) Below, you are given the information for a transportation problem. This problem is partially solved. *Solve* this problem to optimality. Compute the total cost of the solution and show it next to the final tableau.

(b) Based on the optimal solution you obtained in part (a), what would the impact be on (1) the total cost and (2) the optimal flow pattern of a *5-unit increase* in the demand at demand node 2 coupled with a *5-unit decrease* in the demand at demand node 3?

(c) Can you use the information in the optimal solution to predict the impact of a *40-unit* decrease in the supply at supply node 3 coupled with a *40-unit* increase in the supply at supply node 1? If so, give the new cost and the new flows. If not, briefly explain why not.

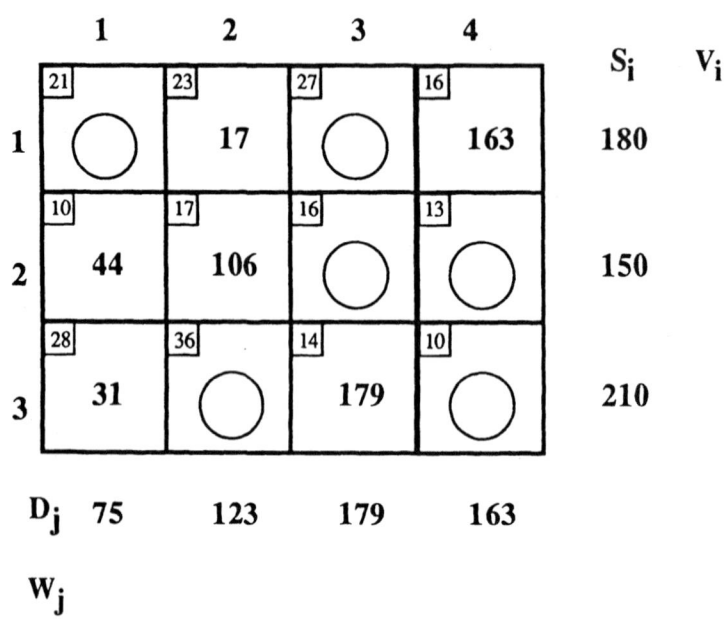

	1	2	3	4	S_i	V_i
1	[21] ○	[23] 17	[27] ○	[16] 163	180	
2	[10] 44	[17] 106	[16] ○	[13] ○	150	
3	[28] 31	[36] ○	[14] 179	[10] ○	210	
D_j	75	123	179	163		
W_j						

Figure E2.6

2.7 **(a)** Consider the following transportation problem. Solve for the optimal solution. In each successive tableau, show the values of (1) all flow variables, (2) all dual variables, (3) all reduced costs, and (4) the objective function.

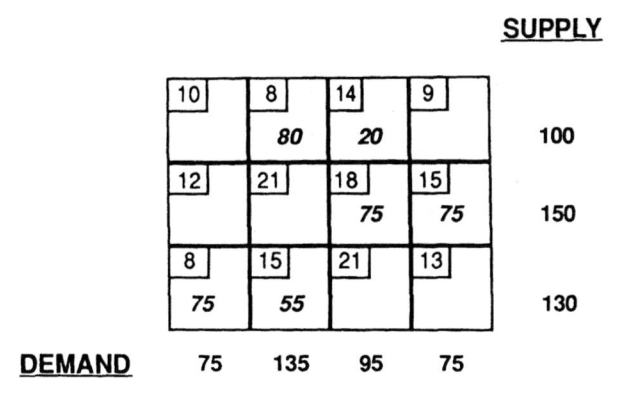

				SUPPLY
10	8 80	14 20	9	100
12	21	18 75	15 75	150
8 75	15 55	21	13	130
DEMAND 75	135	95	75	

OBJECTIVE FUNCTION = 4,820

Figure E2.7

(b) If we were required to ship 10 units from supply point 1 to demand point 1 (the cell with the unit cost of 10), by how much would the objective function change from its optimal value? Would it increase or decrease?

(c) How would the flows change if we had to ship 10 units from supply point 1 to demand point 1?

2.8 Consider the following transportation problem:

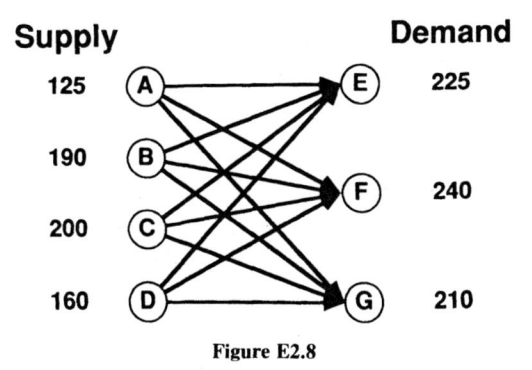

Supply: 125 (A), 190 (B), 200 (C), 160 (D)
Demand: 225 (E), 240 (F), 210 (G)

Figure E2.8

The unit costs are given by the following matrix:

From	To		
	E	F	G
A	15	10	8
B	18	14	6
C	10	13	17
D	9	11	16

(a) Solve the problem for the optimal flows (using any manual or computerized means that you have at your disposal. That is, you can solve it by hand, use MENU-OKF, or any other linear programming package available to you.) No matter how you solve the problem, present your solution in the form of a table as shown in Figure 2.9. Be sure to include not only the optimal flows, but also the dual variables and the dual slack values.

(b) Suppose the supply at node B is increased to 191 and the supply at node D is decreased to 159. What is the change in the objective function? In general, if the supply at node i is increased by a small amount and the supply at node j is decreased by an equal amount, what is the change in the objective function in terms of the supplies, the costs, the dual variables, and the magnitude of the change in supplies? (Assume that the change is small enough that the basic variables in the primal solution—those that are greater than zero—remain basic after the change.) Clearly define any notation that you use.

2.9 For the following network, find the shortest path trees rooted at nodes A, E, and H.

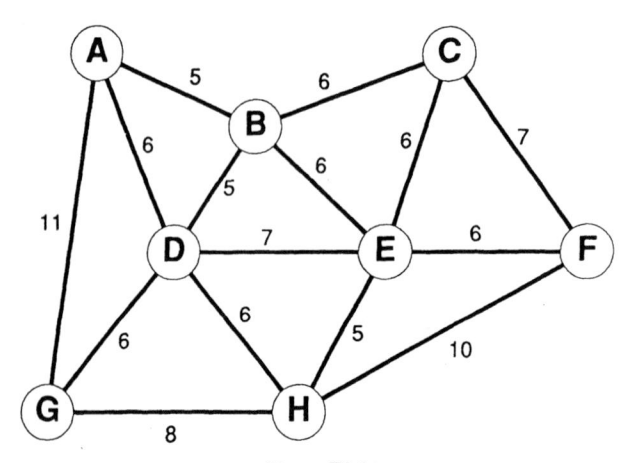

Figure E2.9

Exercises

2.10 For the following network, find the shortest path trees rooted at nodes $B, C, D,$ and E.

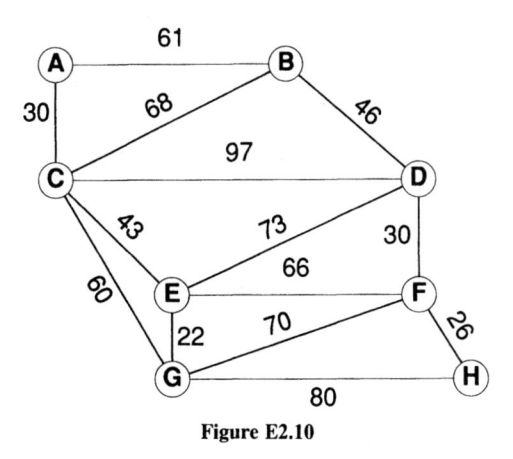

Figure E2.10

2.11 Consider the following network:

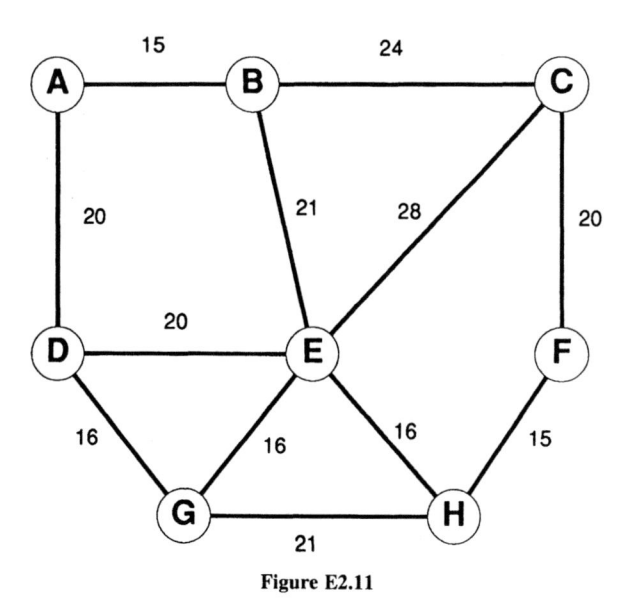

Figure E2.11

Find the shortest path trees rooted at each of the nodes. For each tree, clearly indicate the links on the tree and the distances to each of the other nodes.

2.12 For the following network, find the shortest path trees rooted at each node and show those trees, along with the distances from the root node to each other node.

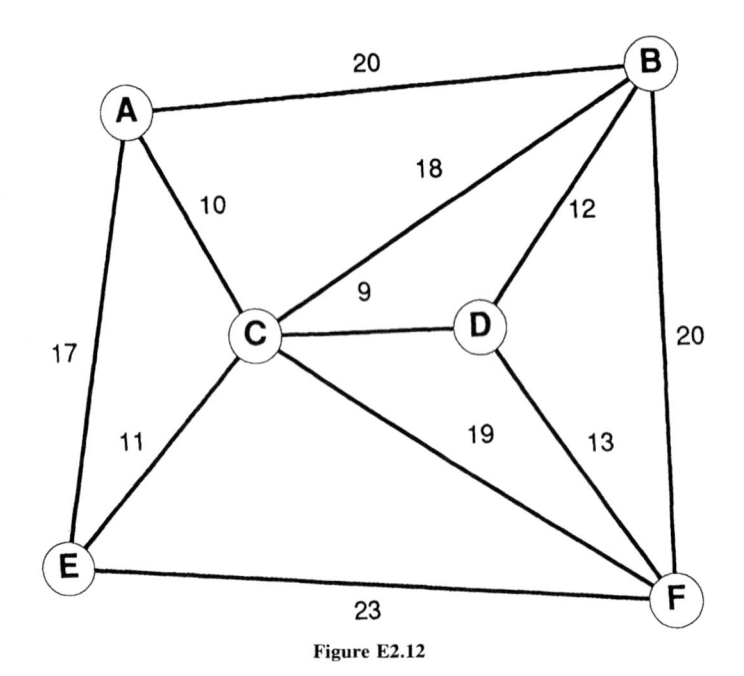

Figure E2.12

2.13 (a) Consider the problem of finding a shortest path tree rooted at a particular node. Show how this problem can be formulated as an out-of-kilter flow problem. *Note that we would almost certainly never solve this problem in this way, but it is useful to know how to*

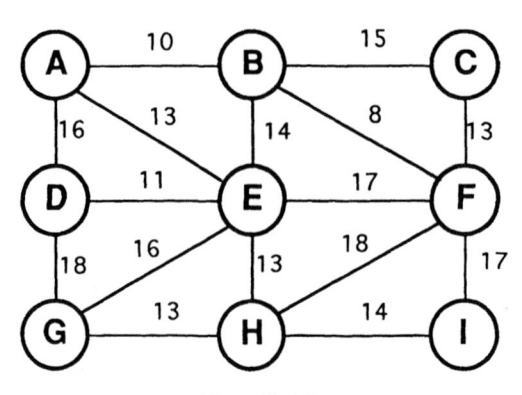

Figure E2.13

Exercises

transform a problem into this format. Illustrate your approach on the network in Figure E2.13 for the shortest path tree rooted at node A.

(b) Formulate the problem of finding the shortest path from node s to *all other nodes in a network* as an optimization problem. Clearly define all decision variables and all inputs. State the objective function and all constraints in both words and using the notation you have defined.

2.14 Use the V^8 and P^8 matrices computed in Section 2.6 to show that the shortest path tree rooted at node D (for the problem at hand) is given by the highlighted links in the following figure:

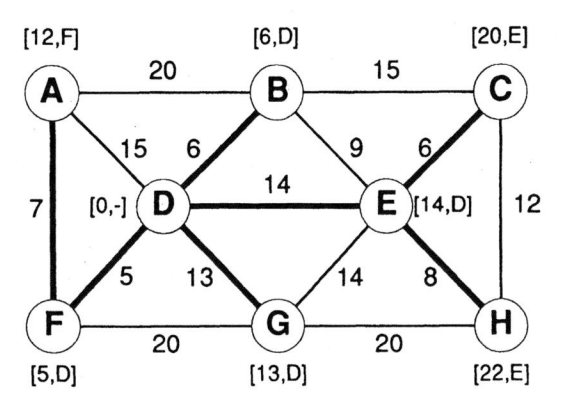

Figure E2.14

2.15 In the matrix approach to finding the shortest paths in a network, what is the interpretation of the V^k matrix (i.e., the matrix of distances you obtain after the kth iteration of the algorithm)?

2.16 A *spanning tree* is defined as a tree that "spans" or connects together all nodes in a graph. Alternatively (and equivalently), a spanning tree is a connected acyclic graph. A spanning tree for a network with N nodes will have $N - 1$ links and no cycles. A *minimum spanning tree* is a spanning tree having minimum total arc length. A number of very efficient algorithms exist for finding minimum spanning trees given a set of arc lengths for links in a network. The minimum spanning tree is useful in a number of contexts including network planning in developing countries and as an input to other algorithms including heuristics for the traveling salesman problem.

Consider the following network. The minimum spanning tree is shown by the bold links. Note that the total length of the links in the minimum spanning tree is 29 units. Find the six shortest path trees (one rooted at each node).

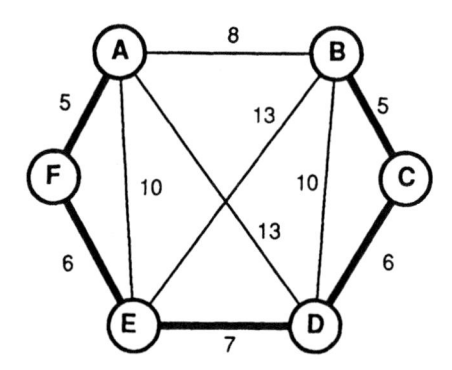

Figure E2.16

What if anything is the relationship between the minimum spanning tree and the shortest path trees on a network?

2.17 Consider the problem of finding the maximum flow through a network from a set of supply points to a set of demand points. If there are alternate optima (different ways of maximizing the flow), you want to find the way that minimizes the total shipment cost. Associated with each link in the network is: a lower bound, l_{ij}; and upper bound, u_{ij}, and a unit cost, c_{ij}. An example network follows.

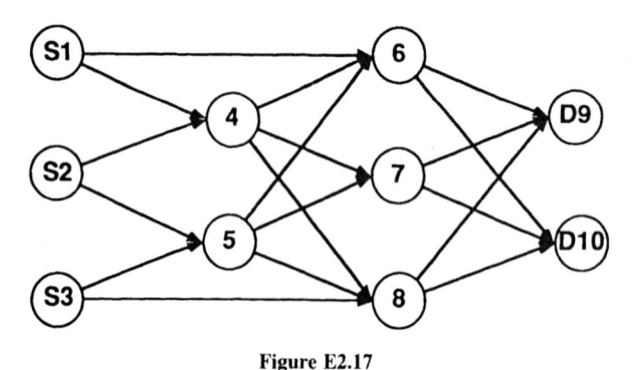

Figure E2.17

The problem is characterized by the inputs shown on the following page. Nodes $S1$, $S2$, and $S3$ are supply points; nodes $D9$ and $D10$ are demand points. Nodes 4 through 8 are transshipment points. Note that no flow is permitted between nodes $S1$ and 5 or between nodes $S3$ and 4. However, flow is permitted between nodes $S1$ and 6 and between nodes $S3$ and 8.

(a) Show how this problem—that of maximizing the flow from the source nodes to the destination nodes at minimum total cost—can be structured as an out-of-kilter flow problem. That is, *draw* the network that would be used in an out-of-kilter flow problem, clearly identifying the lower bounds, upper bounds, and unit costs

Exercises

on all links in the original network (such as the one shown above). If you elect to add any nodes or links, clearly label all link costs associated with these links as well.

(b) Suppose you want to find the minimum cost way of shipping as much as possible through the network subject to the condition that no shipment costs more than C_{max} to ship. (That is, all shipments must cost C_{max} or less to ship.) Discuss how this can be formulated as an out of-kilter-flow problem. *Be careful here!* You may need to change some of the costs in the original network to be sure you get the right answer!

(c) For the network shown above, find the minimum cost way of shipping the maximum flow from the supply nodes to the demand nodes using the MENU-OKF algorithm.

Hint: Before typing the inputs into MENU-OKF, read part (d). Careful specification of the inputs will allow you to solve both problems with relatively few changes to the network.

(d) For the network shown above, find the minimum cost way of shipping the maximum possible flow subject to the additional condition that no shipment costs more than 25 units.

Data for Sample Network

From	To	Lower Bound	Unit Cost	Upper Bound
S1	4	0	10	100
	5	0	0	0
	6	0	18	50
S2	4	0	5	70
	5	0	7	120
S3	4	0	0	0
	5	0	8	140
	8	0	17	55
4	6	0	9	100
	7	0	8	50
	8	0	12	45
5	6	0	4	30
	7	0	5	50
	8	0	9	20
6	D9	0	7	95
	D10	0	10	30
7	D9	0	8	125
	D10	0	3	20
8	D9	0	9	155
	D10	0	7	15

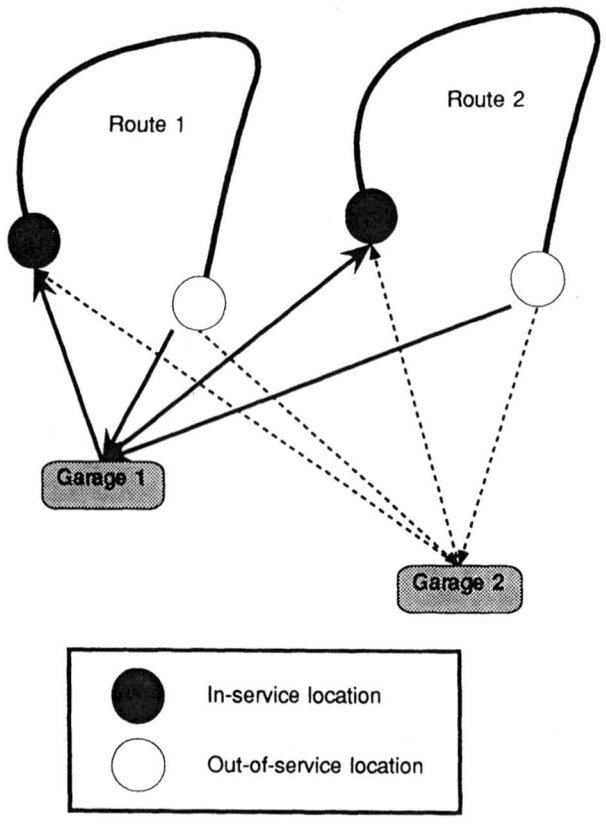

Figure E2.18

2.18 One problem that arises in public transit authorities is that of assigning routes to garages. In its simplest form, the problem may be thought of as follows. The locations of the bus garages are given. For each route, there is an "in-service" location and an "out-of-service" location. These are the locations at which the vehicles that traverse the route begin serving the route and the locations at which they go out of service. Vehicles must deadhead, or move empty, from the garage to in-service locations at which they begin their routes (typically at the beginning of the day) and from out-of-service locations at which the routes end to the garage (at the end of the day). The figure above illustrates this situation. Two routes are shown in this figure along with two garages. Candidate assignments of routes to garage 1 are shown with solid lines, while candidate assignments to garage 2 are shown using dotted lines.

The objective in assigning vehicles from a garage to a particular route is to minimize the total deadheading distance of all vehicles

Exercises

subject to the following constraints:
- the number of buses assigned out of any garage cannot exceed the capacity of the garage;
- the number of buses assigned to each route must be at least the number that are required to serve the route; and
- the vehicle must return to its home garage (the one from which it departed for the in-service location at the beginning of the day) from the out-of-service location (at the end of the day).

(a) Formulate this problem as a linear programming problem. Clearly define all inputs and decision variables. Clearly sate the objective function and all constraints in words and in notation.

(b) Solve this problem for the following 3 garage, 12 route problem:

Route/Garage Distances and Buses per Route

	Garage 1		Garage 2		Garage 3		
Route	In	Out	In	Out	In	Out	Buses
1	3	4	6	7	9	3	10
2	10	8	5	4	3	1	11
3	4	2	7	8	5	9	8
4	3	4	7	9	11	10	6
5	2	4	8	9	12	5	7
6	1	5	7	3	4	9	12
7	8	8	5	5	9	9	9
8	3	3	7	7	8	8	4
9	3	6	5	3	6	7	5
10	5	5	8	8	3	3	8
11	8	8	6	7	4	4	9
12	4	4	8	8	11	11	7

Garage Capacities

	Garage 1	Garage 2	Garage 3
Capacity	30	30	40

(c) Often, transit authorities want all buses assigned to a route to originate from the same garage. This facilitates vehicle dispatching. Would the solution to the problem that you formulated in part (a) ensure that all buses assigned to a route were garaged at the same location? If so, why? If not, how can you reformulate the problem to ensure that this condition is met.

(d) Many transit authorities operate different types of vehicles. Associated with a bus route will be the in-service location, the out-of-service location, and the number and type of vehicle to be used on the route (e.g., a standard transit vehicle, a minibus, an articulated bus). Buses of different types will require different amounts of parking space at each garage. Since the garage capacity is generally measured in terms of the number of standard bus parking spaces available, assigning different bus types to a garage will utilize different amounts of the capacity at the garage. For example, a large articulated bus might count as two buses. In this case, the authority must simultaneously determine how many of each vehicle type should be assigned to each garage (subject to capacity constraints) and the assignment of vehicles to routes to minimize the total deadheading distance. Ignoring the issue outlined in part (c)—that of requiring that all vehicles assigned to a route originate from the same garage—formulate this problem as an optimization problem. Again, clearly define all notation separating inputs from decision variables. Also, clearly state the objective function and the constraints in words as well as notation.

(e) Briefly discuss why the model formulated in part (d) may or may not be solved using the same algorithm(s) that can be used for the problem of part (a).

Note: A number of additional concerns must be addressed in assigning bus routes to garages. Maze et al. (1981, 1982), Daskin and Jones (1993), and Vasudevan, Malini, and Victor (1993) all discuss models for this problem.

2.19 Captain Motors Corporation produces two different models of a particular vehicle at each of two plants. The models are similar except for some minor styling differences (slightly different sheet metal designs, color combinations, and the availability of leather seats on the more expensive model). At each plant, CM can produce 1000 vehicles per day (split in any way between the two vehicle models).

Finished vehicles are shipped to each of five different regions of the country. Shipping costs to each region are the same for the two vehicle models, but differ by plant since the plants are located in different parts of the country. These shipping costs are as follows:

To Region	From Plant 1	From Plant 2
Northeast	200	300
Southeast	250	350
Central	150	250
Northwest	310	130
Southwest	370	90

Exercises

Production costs for the two models by plant are as follows:

Model	Cost to Produce At Plant 1	Cost to Produce At Plant 2
Basic Model	14,600	15,300
Deluxe Model	16,800	17,200

These costs differ slightly because of differing suppliers. Also, many parts are supplied by the same supplier and most of the firm's suppliers are closer to plant 1.

During one week in January with five production days for a total production capacity of 10,000 vehicles, the firm is planning to build the following types of cars for each of the five markets.

To Region	Basic Model	Deluxe Model
Northeast	1200	1400
Southeast	1100	1200
Central	850	700
Northwest	650	1000
Southwest	750	800

(a) Show that this problem can be structured as an out-of-kilter flow problem. Draw the OKF network. In a table give the lower bounds, upper bounds, and unit costs for all links. Describe in words what each class of links does.

(b) Use the MENU-OKF program to solve this problem. (*Note you will need to divide all costs by 10 since unit costs cannot exceed 10,000 in the program.*)

(c) By how much would the total cost go down if we increased capacity at plant 1 by one unit?

(d) By how much would the total cost go down if we increased capacity at plant 2 by one unit?

(e) For production planning and scheduling reasons and issues related to work flow balancing on the assembly line, the firm does not want the mix of vehicle models at either of the plants to exceed a 55:45 ratio in either direction. In other words, the percentage of production of either model cannot exceed 55 percent of the total production at either plant. (Note that this is actual production and not production capacity.) Again, use the MENU-OKF algorithm to find a new optimal solution in the face of this added constraint.

(f) What is the percentage increase in total cost as a result of adding the constraint outlined in part (e)?

2.20 Off-The-Wall Drugs has three distribution centers in a region from which it supplies four stores. Each distribution center stocks all of the supplies that are sold at each of the stores. When orders come in, the individual items being requested by the stores are packaged at the distribution centers into standard boxes and then shipped from the distribution center to the store requesting the material. Thus, under these assumptions, we can treat demand as being in terms of boxes requested by the stores.

While each distribution center stocks all of the items in inventory at each of the stores, each distribution center has a limited packaging capability (measured in terms of boxes per week). These capacities are as follows:

Distribution Center	Capacity (Boxes/Week)
1	200
2	150
3	225

The weekly demands of the stores are as follows:

Store	Demand (Boxes/Week)
A	140
B	130
C	165
D	100

The unit costs of shipping a box between each of the distribution centers and the stores are as follows:

Distribution Center	Store A	B	C	D
1	15	10	8	12
2	9	11	7	5
3	13	14	3	4

(a) Find the optimal distribution strategy for Off-The-Wall Drugs (i.e., determine which distribution centers should ship how much each week to which stores). Solve the problem as a transportation

Exercises

problem, showing your solution (including dual variables and reduced costs) in the tableau format used in this chapter.

(b) What would the change in the objective function be if we increased the demand at store A by one unit? How would the flows change in response to this increase in demand?

(c) What would the change in the objective function be if we increased the supply at distribution center 1 by one unit? How would the flows change in response to this increase in supply?

(d) What would the change in the objective function be if we increased the supply at distribution center 2 by one unit? How would the flows change in response to this increase in supply?

(e) What would the change in the objective function be if we increased the supply at distribution center 3 by one unit? How would the flows change in response to this increase in supply?

Note: For parts (b) through (e) of this exercise, you should be able to answer the questions without having to resolve the problem.

2.21 Extend formulation (2.16a)–(2.16d) to include facility location decisions and fixed facility location costs. Clearly define all new notation you use separating inputs from decision variables. State the objective function and constraints in both words and with notation.

3

An Overview of Complexity Analysis

3.1 INTRODUCTION

In discussing algorithms, we are often interested in how long the algorithm takes to solve a problem. There are a variety of ways to evaluate this. The most obvious is to code the algorithm in some computer language, to execute the code on a data set, and to record the execution time. This approach tells us how long a *particular* implementation of the algorithm took to solve one *particular* instance of the problem on a *particular* machine. Generalizing this information to other implementations, to other data sets and problem instances, and to other computer environments would be difficult using this approach. Nevertheless, this approach is often adopted, though most of the better papers along these lines report results using a variety of test problems. Also, reputable journals and authors try to emphasize that the results are illustrative and may be difficult to generalize to other problem instances, computers, and implementations.

An alternative approach is to develop a theory regarding problems and algorithms and the time and storage requirements required to solve the problems using alternative algorithms. This is what *complexity theory* does. In doing so, the execution time is given as a function of the size of the problem. The time is not measured in seconds or minutes or hours, since such times depend critically on the computer environment in which the code is developed and executed. Instead, complexity theory tells us how fast the execution time of an algorithm may increase as the size of the problem increases. Specifically, complexity theory (1) defines clearly what solving a problem "efficiently" means, (2) categorizes problems into those that can be solved efficiently and those that cannot, and (3) estimates the amount of time (or storage) needed to solve these problems.

The focus of much of complexity theory has been on the *worst case* performance of an algorithm. In this chapter we adopt this perspective. However, it is important to understand the downside of adopting the worst

Basic Concepts and Notation

case perspective. In some sense, focusing excessively on the worst case performance of an algorithm may be just as misleading as focusing on the worst case performance of aspirin. In the worst case, aspirin will kill you since there are some people who are allergic to aspirin. However, this is not the performance that we expect from aspirin. We expect that it will alleviate our headache (if we have a headache) or reduce our pain in a more general sense. Similarly, there are some algorithms for which the worst case performance is not very good, but, in practice, the algorithms perform quite well on average. For example, the worst case performance of the simplex algorithm for solving linear programming problems is quite bad. (In fact, the simplex algorithm is not a polynomial time algorithm for linear programming. See Section 3.2 for a definition of polynomial time algorithms.) However, in practice, the algorithm does very well. It is only recently that better algorithms for linear programming have been developed, and most researchers suggest that these algorithms only have a strong advantage over linear programming for very large problems at this point in time.

3.2 BASIC CONCEPTS AND NOTATION

We begin by characterizing a bit more rigorously what we mean by a problem. A *problem* is a general description of a class of numerical problems to be solved. Thus, for example, one statement of the *traveling salesman problem* (TSP) is the following:

Traveling Salesman Optimization Problem

> *Given*: A graph $G(N, A)$ with node set N and link set A. Associated with each link (i, j) in A is a nonnegative link length d_{ij}.
> *Find*: A circuit that visits all nodes and is of minimum total length.

An *instance of a problem* is specified by providing actual numbers for the required inputs. Thus, by specifying the actual graph on which the traveling salesman problem is to be solved, along with the distances, we are specifying an instance of the problem.

When we speak of the *size of an instance of a problem*, we are referring to a way of characterizing how big the problem is. Thus, for example, a TSP with 100 nodes is intuitively a larger problem than one with only 10 nodes. For most purposes in this text, the number of *nodes* and the number of *links* in a problem will, together, constitute an adequate description of the size of a problem. In some cases, we may also have to say something about bounds on the costs and bounds on the link flows. For these purposes, let C be the largest link cost and let U be the largest upper bound on a link flow.

Table 3.1 Growth in Solution Times as a Function of Problem Size for Different Complexities

Complexity	$n = 10$	$n = 20$	$n = 40$
$O(n)$	10^{-5} sec	2×10^{-5} sec	4×10^{-5} sec
$O(n^2)$	10^{-4} sec	4×10^{-4} sec	0.0016 sec
$O(n^3)$	10^{-3} sec	8×10^{-3} sec	0.064 sec
$O(2^n)$	10^{-3} sec	1.05 sec	12.7 days
$O(e^n)$	0.022 sec	8.08 min	74.6 centuries!!

From: *Computers and Intractability: A Guide to the Theory of NP-Completeness* by Garey and Johnson. Copyright © 1979 by W. H. Freeman and Company. Used with Permission.

As noted above, we are most interested in the worst case behavior of an algorithm. To denote the worst case behavior, we say that an algorithm is order $f(n)$ [which we denoted by $O(f(n))$] if there exists some constant a such that the number of basic operations (comparisons, assignments, and basic arithmetic operations) required to run the algorithm for a problem of size n is less than or equal to $af(n)$ for all values of n.[1] Note that by counting steps and by sidestepping the issue of what the constant a is, we obtain a description of an algorithm that does not depend on the particular computer on which the algorithm is implemented.[2]

An algorithm is said to be a *polynomial* time algorithm if $f(n)$ is a polynomial function of n. For example, n, n^2, and n^3 are all polynomial functions of n. Similarly, $n \log(n)$, $n^2 \log(n)$, and so on are also considered polynomial functions of n. This may be contrasted with *exponential* time algorithms for which $f(n)$ grows exponentially with n. Exponential functions include 2^n, e^n, 3^n, and $n!$.[3]

Polynomial time algorithms are viewed as efficient, while exponential algorithms are not efficient. The difference between these two types of algorithms is illustrated in the following tables. Table 3.1 shows the computation time for a set of algorithms assuming the computer being used can execute 10^6 operations per second. The tables further assume that the constant multiplying the function of the problem size, n, is 1. Thus, the tables assume that a linear time $[O(n)]$ algorithm will require exactly n operations and not a number that simply grows linearly with n, as is generally assumed.

---

[1] Note that n is the size of the problem in this discussion and may actually be a vector of inputs including the number of nodes and the number of links.

[2] This, too, needs to be qualified, since the complexity analysis for parallel machines is complicated. Throughout this text, we are referring only to sequential—nonparallel—processors.

[3] In describing algorithms, we will want to distinguish between cases in which the appropriate descriptor of the algorithm involves using $\log C$ or $\log U$ on the one hand and C or U on the other hand (where C and U are upper bounds on the link costs and link flows, respectively, or more generally, on the magnitude of the problem inputs). If $\log C$ or $\log U$ may be used, then the algorithm will still be considered *polynomial* in the size of the problem instance; however, if C or U must be used directly, the algorithm will be considered *pseudo-polynomial*. Most of our interest will focus on whether an algorithm is polynomial or exponential.

Basic Concepts and Notation

Table 3.2 Growth in Size of Problems That Can Be Solved in Same Time as a Function of Speed of Computer and Complexity

Complexity	Speed = 1	Speed = 10	Speed = 100	Speed = 1000
$O(n)$	N_1	$10N_1$	$100N_1$	$1000N_1$
$O(n^2)$	N_2	$3.16N_2$	$10N_2$	$31.6N_2$
$O(n^3)$	N_3	$2.15N_3$	$4.64N_3$	$10N_3$
$O(2^n)$	N_4	$N_4 + 3.32$	$N_4 + 6.64$	$N_4 + 9.97$
$O(e^n)$	N_5	$N_5 + 2.30$	$N_5 + 4.61$	$N_5 + 6.91$

Even very large increases in computational speed will now allow us to solve significantly larger problems to optimality (in all cases) if the worst case time behavior goes up exponentially for these problems.

From: *Computers and Intractability: A Guide to the Theory on NP-Completeness* by Garey and Johnson. Copyright © 1979 by W. H. Freeman and Company. Used with Permission.

Table 3.1 shows that, as the problem size grows, the required computation time grows for all algorithms. However, the growth in computation time is explosive for the exponential algorithms. For a given problem size, the table suggests that a linear time [$O(n)$] algorithm is faster than a quadratic [$O(n^2)$] algorithm which, in turn, is faster than a cubic [$O(n^3)$] algorithm. Again, this is so only because we have assumed that the constant multiplying the function of the problem size, n, is 1. In the more general case in which the constant might differ from 1, this relationship might not hold for a given problem size. For example, an algorithm whose number of operations is given by $20,000n$ is linear, while an algorithm which requires $2n^2$ operations is quadratic. However, for values of n less than 10,000, the quadratic algorithm will be faster.

At first glance, it may seem as though the issue of the speed of an algorithm need not really concern us since computers get faster every day. Table 3.2 shows that this is not the case. As indicated in the table, if the algorithm is polynomial, the size of the problem that can be solved in a given amount of time grows "multiplicatively" with the speed of the computer. Thus, for example, for an $O(n^2)$ algorithm, if we had a computer that was 1000 times as fast as the one we were now using, we could solve a problem that was 31.6 times as big as the one we are now solving in the same amount of time. By contrast, for the exponential algorithms, the size of the problem grows only "additively" with the speed of the computer. Thus, if we could solve a problem with N_4 nodes using an $O(2^n)$ algorithm on our present computer, using a computer that was 1000 times faster we could only solve a problem with $N_4 + 10$ nodes in the same amount of time. Thus, even dramatic improvements in the speed of the computers we are using are not likely to allow us to solve dramatically larger problem if the algorithms involved are exponential in nature.

Finally, by way of introduction, we need to outline the notion of a *decision problem*. The statement of the TSP outlined above is that of an *optimization problem*, in that we are trying to find the *best* or shortest distance circuit through all of the nodes. In a decision problem version of the TSP, we would simply ask a yes/no question: Does there exist a tour with length less than or equal to B, where B is an input (a characterization of an instance of the TSP decision problem)? For most of our purposes, it is worth noting that the decision problem associated with an optimization problem is generally no easier nor any harder than the original optimization problem. Thus, for example, if we could solve the TSP decision problem in polynomial time, we could, using a binary search algorithm over the input constant B, find an optimal solution to the TSP optimization problem in polynomial time. Unfortunately, it appears that this cannot be done for either problem, as outlined below.

3.3 EXAMPLE COMPUTATION OF AN ALGORITHM'S COMPLEXITY

Before going much further, let us illustrate the process of estimating the worst case time of an algorithm. We do so using Dijkstra's algorithm for finding the shortest path tree rooted at a particular node when all link distances are nonnegative. (See Section 2.6 for a discussion of this algorithm.) The problem and the algorithm are outlined as follows:

Given: A graph with nonnegative link distances or costs c_{ij} associated with links (i, j); n is the number of nodes in the graph. A root node S.

Find: The shortest path tree rooted at node S.

Note: Nodes will be labeled with a two-part label. Thus, the label for node j would be $[V_j, P_j]$, where V_j gives the minimum cost way of getting to node j, while P_j specifies the predecessor node to node j on the path from S to j.

Step 1: Initialization
    a. Label node S $[0, -]$ $O(1)$
    b. Label all other nodes $[\infty, -]$ $O(n)$
    c. Set node S as scanned $O(1)$

Step 2: Label Updates
    a. Call the last scanned node, node m $O(1)$
    b. For all links (m, j), compute $O(n)$
        b.1. $T_j = V_m + c_{mj}$ $O(1)$
        b.2. If $T_j < V_j$, relabel node j with $[T_j, m]$ $O(1)$

Step 3: Scan a Node
    a. Find the unscanned node with smallest label V_j $O(n)$
    b. Scan this node $O(1)$

Step 4: Termination Check
    a. All are nodes scanned? $O(1)$
        a.1. YES—stop $O(1)$
        a.2. NO—go to step 2 $O(1)$

Beside each step and substep, we indicate the complexity (number of basic operations) involved in that step. Thus, initializing a data element takes $O(1)$ time (constant time). In step 1.b, we do this for $O(n)$ data elements, so the entire step takes $O(n)$ time. Step 2.b calls for us to do something for each link that originates at node m. There are potentially n such links, and so we will need to do something $O(n)$ times. The something we need to do is specified in steps 2.b.1 and 2.b.2 which both take $O(1)$ time, thus all of step 2.b takes $O(n)$ time *each time step 2 is executed*. Steps 2, 3, and 4 will be executed a total of $n - 1$ times (until all nodes are scanned). This means that they will be executed $O(n)$ times and the entire algorithm therefore has complexity $O(n^2)$. Note that in computing the complexity of the algorithm, we are only worried about the dominant terms. Thus, technically, the algorithm will repeat steps 2, 3, and 4 $n - 1$ times and the longest running time of any of these three steps is $O(n)$. We might conclude that the complexity of the algorithm should be stated as $O(n(n - 1))$ or $O(n^2 - n)$. In fact, the n^2 term will dominate the term for n for sufficiently large values of n and we do not need to worry about it.

Section 3.4 is quite technical in nature. It is included here for the sake of completeness. For much of what we will do in the remainder of this text, the key concept that the reader needs can be gleaned from the chapter summary, Section 3.5.

3.4 THE CLASSES P AND NP (AND NP-HARD AND NP-COMPLETE)[4]

The class P of problems is the class of decision problems that can be *solved* in *polynomial* time. The class of problems NP (which stands for nondeterministic polynomial) is the set of decision problems such that a candidate solution (which we may guess) may be *verified* to be a "yes" solution or instance in polynomial time.[5] Clearly, P $\subseteq$ NP since if a problem can be solved in polynomial time, we must be able to verify in polynomial time that a solution or an instance is a "yes" instance.

At first glance, it might seem that all decision problems should be in the class NP. In fact this is not so. Part of the reason for this is that the class NP is defined in terms of being able to verify that a particular instance to a problem is a "yes" instance. The co-TSP problem is similar to the TSP problem, but the question asked is: Are there no tours of length less than or equal to B? The only way to answer this with a "yes" answer involves enumerating all possible tours and showing that each such tour has a length

---

[4]This material is quite technical and can readily be skipped by many readers.
[5]Such problems are said to have a *certificate* which is verifiable in polynomial time. Clearly, the *length* of such a certificate must be a polynomial function of the size of the problem (for otherwise it would take an exponential amount of time just to read the certificate!). The term *nondeterministic* is used because we are not saying where the candidate solution comes from. In fact, we can simply guess the solution.

greater than B. However, the number of tours is $O(n!)$, an exponential function of n, the number of nodes in the problem.

A key open question is whether the two classes are equivalent; that is, does P = NP? While we do not yet have a definitive answer to this, it is widely believed that the answer is no.

At this point, we introduce a classic problem that has played a key role in complexity theory. This is known as the *satisfiability problem*.

Satisfiability Problem

Given: A Boolean expression—a function of true/false variables.

Question: Is there an assignment of truth values (TRUE or FALSE) to the variables such that the expression is TRUE?

To illustrate this problem, let $\bar{x}_i =$ not x_i. Now consider the following Boolean expression: x_1 or $\bar{x}_1$. This is clearly TRUE for any assignment of TRUE or FALSE to x_1. Therefore, the answer to the satisfiability problem in this instance is YES. However, now let us consider the Boolean expression, $(x_1$ or $x_2)$ and $(\bar{x}_1)$ and $(\bar{x}_2)$. Clearly, there is no way of assigning TRUE and/or FALSE to the variables x_1 and x_2 in a way that will allow this entire expression to be TRUE. Therefore, the answer to the satisfiability problem in this instance is NO.

Cook's theorem states that *satisfiability is in the class* P *if and only if* P = NP. This says that the satisfiability problem can be solved using a polynomial time algorithm (is in the class P) if and only if the class of problems that can be *solved* using a polynomial time algorithm (the class P) and the class of problems such that a candidate solution can be *verified* as being optimal or a "yes" solution in polynomial time (the class NP) are equivalent classes of problems (P = NP). This is a key result and will, as we shall see, allow us to categorize problems.

We now introduce a means of showing that two problems are equally difficult in some technical sense. Qualitatively, the basic idea is that we will take a problem that is known to be difficult to solve in this technical sense (e.g., the satisfiability problem). We will then show that we can transform any instance of that problem into an instance of the problem whose difficulty we want to demonstrate. If the transformation may be done in a polynomial number of steps, then if we could solve the new problem in polynomial time we would have found a means of solving the original (previously thought-to-be difficult) problem in polynomial time. We will define a class of problems known as the class of *NP-complete problems* all of which may be transformed in polynomial time into instances of other NP-complete problems. Thus, if we have a new problem whose difficulty we want to demonstrate, to show that the new problem is also NP-complete, we select a problem in the NP-complete class and show that any instance of the NP-complete problem may be transformed in polynomial time into an instance of the new problem.

The Classes P and NP (and NP-Hard and NP-Complete)

To formalize the concept of NP-complete problems, we introduce the notion of *polynomial reducibility*. Problem L_1 polynomially reduces to problem L_2 if and only if there is a way to solve L_1 by a polynomial time algorithm A_1 "using" another algorithm A_2 that solves problem L_2 (counting each "use" of A_2 as a unit operation). In determining whether or not algorithm A_1 is polynomial, we would normally have to determine the complexity of algorithm A_2 and then assess how many times algorithm A_1 uses A_2. However, in thinking about polynomial reducibility, we count each "use" of algorithm A_2 by A_1 as a *unit* operation. Now, if it happens that algorithm A_2 is a polynomial time algorithm, then clearly we can solve problem L_1 in polynomial time. If A_2 is not polynomial, then the use of A_1 (with its embedded use of A_2) does not result in a polynomial time algorithm for solving L_1.

Independent of whether A_2 is or is not polynomial, algorithm A_1 is called a polynomial reduction of problem L_1 to L_2. Clearly, polynomial reducibility is transitive (i.e., if L_1 is polynomially reducible to L_2 and L_2 is polynomially reducible to L_3, then L_1 is polynomially reducible to L_3). A special case of polynomial reducibility occurs when algorithm A_2 is used only once at the end of algorithm A_1. In that case, we say A_1 is a *polynomial transformation* of L_1 to L_2. In essence, a polynomial transformation is a way of converting one problem (L_1) into an instance of another problem (L_2) which we will then solve using an algorithm A_2 for the new instance of problem L_2.

To illustrate the notion of polynomial reducibility, consider again the problem of finding the shortest paths from *all* nodes to *all* other nodes in a network. Let this be problem L_1 whose complexity we wish to analyze. Let problem L_2 be the problem of finding the shortest path tree rooted at some particular node. An algorithm (A_1) for solving the all-pairs shortest path problem (L_1) is to call Dijkstra's algorithm (A_2), for finding the shortest path tree rooted at a particular node, once for each node of the network. A_2 is polynomial as shown in Section 3.3, A_1 uses A_2 $O(n)$ times. Thus, the all-pairs shortest path problem polynomially reduces to the shortest path tree problem and can be solved in polynomial time.

We can now formally define the class of NP-complete problems. A problem L is NP-complete if and only if (a) L is in the class NP and (b) some other problem in the NP-complete class may be polynomially transformed into L. Cook's theorem states the satisfiability problem is NP-complete (thus giving us the first such problem).[6] Since polynomial transformations are transitive, any NP-complete problem may be polynomially transformed into any other NP-complete problem.[7]

A key characteristic of NP-complete problems is that if a polynomial time algorithm can be found for any such problem, then it will also solve all NP-complete problems in polynomial time. If we could find such an algorithm we would have shown that P = NP.

---

[6]See Garey and Johnson (1979) or Papadimitrious and Steiglitz (1982) for a proof of this.

We now illustrate the process of showing that a problem is NP-complete by showing that the TSP-decision problem is NP-complete. We begin by defining the Hamiltonian cycle problem.

Hamiltonian Cycle Problem (HCP)

Given: A graph $G(N, A)$ where N is the set of nodes or vertices and A is the set of links.

Question: Does the graph contain a cycle that visits every vertex (i.e., a path that visits each node exactly once except the first node which is also visited at the last node on the path)?

This problem was shown to be NP-complete by Karp (1972).

To show that the TSP-decision problem is NP-complete, we need to show two things: (i) that the TSP-decision problem is in the class NP and (ii) that a known NP-hard problem (in this case the Hamiltonian cycle problem) reduces to the TSP-decision problem. To show (i), we note that, given any cycle, we can compute the cost of the cycle in polynomial time and therefore determine in polynomial time if the cycle has length less than or equal to B (in which case it would be a "yes" instance to the TSP-decision problem). Thus the TSP-decision problem is in the class NP. To show (ii), we construct a complete graph with the same vertex set as that found in the HCP. For each link in this new graph, if the corresponding link exists in the instance of the HCP, let the link length be 1; otherwise, let the link length be 2. Clearly, the HCP has a solution if and only if the TSP on this complete graph has a solution with value less than or equal to n (where n is the number of nodes in the vertex set).

Finally, we sometimes speak of the following subclasses of algorithms[8]:

Strongly Polynomial: The time is proportional to a polynomial function of the input size only and does not depend on the magnitude of the input values.

Polynomial: The time is proportional to a polynomial function of the input size and also depends on the maximum number of bits needed to store the largest data inputs. In other words, the complexity is a polynomial function involving terms like $\log C$ or $\log U$.

---

[7]We also sometimes speak of problems that are *NP-hard*. Such problems are ones such that an NP-complete problem polynomially reduces to the problem in question, but the problem under study is not provably in the class NP. Formally, the term NP-hard is also used to describe the optimization versions of the decision problems that are NP-complete. In this text, however, we use the looser terminology and refer to optimization problems as being NP-complete when the decision version of the problem is NP-complete.

[8]See Ahuja, Magnanti, and Orlin (1993), pp. 60–62.

Summary

Pseudo-Polynomial: The time is proportional to a polynomial function of the input size and the absolute magnitude of the data inputs. In other words, the complexity may be a function of such terms as C or U, directly, as well as the number of nodes or links in the problem instance.

Linear programming is known to be polynomially solvable. Strongly polynomial algorithms exist for network LP problems.

We also speak of the following subclass of problems:

NP-Complete in the Strong Sense: A problem is NP-complete in the strong sense if it is NP-complete and if it is not solvable in pseudo-polynomial time unless P = NP.

There are some problems that are NP-complete, but not NP-complete in the strong sense. For example, the *partition problem*[9] and the 0/1 *knapsack problem*[10] are both NP-complete, but neither is NP-complete in the strong sense. Both problems can be solved in pseudo-polynomial time using dynamic programming (Garey and Johnson, 1979, pp. 223, 247). Problems that are NP-complete in the strong sense include the *TSP-decision problem* and the 3-*partition problem*.[11] For the remainder of this text, we will generally not need to distinguish between problems that are the NP-complete in the strong sense and those that are NP-complete but that can be solved in pseudo-polynomial time. Most of the problems we will encounter are, in fact, NP-complete in the strong sense.

3.5 SUMMARY

This chapter has provided an introduction to complexity theory—a theory that categorizes problems into those that can be solved efficiently and those that cannot (yet) be solved efficiently. Problems that can be solved efficiently

---

[9] The *partition problem* is the following. Given a set of integers $b_1, \ldots, b_n$, is there a subset A of the indices $1, \ldots, n$ such that the sum of the integers whose indices are in the subset equals the sum of the integers whose indices are not in the subset? In other words, can the integers $b_1, \ldots, b_n$ be divided into two mutually exclusive and collectively exhaustive groups such that the sum of the elements in each group is equal?

[10] The 0/1 *knapsack problem* is the following. Given a set of items with benefits $b_1, \ldots, b_n$ and weights $w_1, \ldots, w_n$ and a maximum allowable weight W that can be placed in the knapsack, find the set of items that maximize the total benefit associated with the selected items. Each item may either be selected once or not selected at all.

[11] The 3-*partition problem* is the following. Given a set of $3m$ integers, $b_1, \ldots, b_{3m}$, and a bound B such that $\Sigma_j b_j = mB$ and $B/4 < b_i < B/2$, is there a partition of the $3m$ integers into m mutually exclusive and collectively exhaustive sets, $A_1, \ldots, A_m$, such that $\Sigma_{j \in A_i} b_j = B$ for all sets i? The reader should note that if there is such a partition of the $3m$ integers, each set A_i will have exactly three elements.

are those for which a polynomial time algorithm exists. A polynomial time algorithm is an algorithm whose worst case execution time increases as a polynomial function of the size of the problem instance. Such problems are said to be in the class P. However, many location problems fall into a class of problems for which no polynomial time algorithm exists (as yet) and it is widely believed that no such algorithm will ever exist (though this has not been proven either). We do know, however, that if a polynomial time algorithm can be found for any such problem (in the class of problems that are called NP-complete), then a polynomial time algorithm must exist for all such problems.

For more information on complexity theory and NP-complete problems, the reader is referred to the seminal text by Garey and Johnson (1979) as well as Ahuja, Magnanti, and Orlin (1993), Karp (1972), Papadimitriou and Steiglitz (1982), and Sahni and Horowitz (1978). Berkeley, Homer, and Kanamori (1993) provide a nice qualitative introduction to complexity theory.

EXERCISES

3.1 (a) What is the complexity of Floyd's matrix approach (described in Section 2.6) for finding the minimum paths from all nodes to all other nodes?

(b) How does this compare to the complexity of using Dijkstra's algorithm for finding the shortest path tree rooted at a single node once for each root node of the network?

3.2 (a) The best algorithms for sorting a set of n numbers have a complexity that is $O(n \log n)$. Add a row to Table 3.1 corresponding to such an algorithm.

(b) Is such an algorithm a polynomial or nonpolynomial time algorithm?

3.3 (a) A total enumeration algorithm for the traveling salesman problem would be an $O(n!)$ algorithm. Add a row to Table 3.1 corresponding to such an algorithm.

(b) Is such an algorithm a polynomial or nonpolynomial time algorithm?

3.4 Suppose a step in an algorithm must examine every possible (nonempty) subset or *combination* of n nodes. Show that the complexity of this step is $O(2^n)$.

3.5 Suppose a step in an algorithm must examine every possible *permutation* of n nodes. Show that the complexity of this step is $O(n!)$.

3.6 In dealing with the complexity of an algorithm, we do not worry about the absolute number of elementary operations (comparisons, additions, subtractions, and assignments) that need to be performed at any step as long as the number does not depend on the size of the problem.

(a) Suppose the actual running times (in milliseconds) of two algorithms are given by the following equations:

$$\text{Algorithm 1 time:} \quad 1000(n^2)$$
$$\text{Algorithm 2 time:} \quad 10^{-8}(2^n)$$

What is the smallest value of n, the number of nodes in the network, for which the execution time of algorithm 2 exceeds the execution time of algorithm 1.

(b) If the number of nodes in the network is only two larger than the number found in part (a), what is the ratio of the execution times of the two algorithms?

(c) If the number of nodes in the network is 10 more than the number found in part (a), what is the ratio of the execution times of the two algorithms? How long does each algorithm take? Do you really want to wait for algorithm 2 to finish?

4

Covering Problems

4.1 INTRODUCTION AND THE NOTION OF COVERAGE

In many location contexts, service to customers (from the facilities that are being located) depends on the distance between the customer and the facility to which the customer is assigned. Customers are generally, though not always, assigned to the nearest facility as discussed in Chapter 1. Often, service is deemed adequate if the customer is within a given distance of the facility and is deemed inadequate if the distance exceeds some critical value.

This leads to the notion of *coverage*. Associated with each demand node i is a subset $\mathbf{N}_i$ of the candidate facility nodes j that can serve or *cover* the demand node. This set may also be specified in terms of binary coefficients a_{ij} which take a value of 1 if a facility at candidate site j can cover demands at demand node i; 0 otherwise. Often, demand nodes are said to be covered if the *shortest path* distance between the demand node and the facility is less than or equal to a coverage distance. A single coverage distance may be used for all demand nodes; alternatively, the coverage distance may depend on either (or both) the demand node being covered and the candidate facility site in question.

This chapter is devoted to the formulation and solution of facility location covering models. In particular, we will discuss the set covering location model, the maximum covering location model, and extensions of these basic models. In the course of this discussion, a number of model properties will be highlighted and a variety of solution algorithms will be outlined. Schilling, Jayaraman, and Barkhi (1993) provide a recent review of covering models as applied to facility location problems.

4.2 THE SET COVERING MODEL

We will begin the discussion of covering models with the set covering model. As noted in Chapter 1, this is perhaps the simplest of facility location models. The set covering problem is to find a minimum cost set of facilities from among a finite set of candidate facilities so that every demand node is covered by at least one facility. This may be formulated mathematically using

The Set Covering Model

the following notation:

Inputs

$$a_{ij} = \begin{cases} 1 & \text{if candidate site } j \text{ can cover demands at node } i \\ 0 & \text{if not} \end{cases}$$

f_j = cost of locating a facility at candidate site j

Decision Variables

$$X_j = \begin{cases} 1 & \text{if we locate at candidate site } j \\ 0 & \text{if not} \end{cases}$$

With this notation, we can formulate the set covering problem as follows:

$$\text{MINIMIZE} \quad \sum_j f_j X_j \quad (4.1a)$$

$$\text{SUBJECT TO:} \quad \sum_j a_{ij} X_j \geq 1 \quad \forall\, i \quad (4.1b)$$

$$X_j = 0, 1 \quad \forall\, j \quad (4.1c)$$

The objective function (4.1a) minimizes the total cost of the facilities that are selected. Constraints (4.1b) stipulate the each demand node i must be covered by at least one facility. Note that the left-hand side of (4.1b) gives the number of located facilities that can cover demand node i. These constraints may be rewritten in terms of the set $\mathbf{N}_i$ as follows:

$$\sum_{j \in \mathbf{N}_i} X_j \geq 1 \quad \forall\, i \quad (4.1b')$$

where $\mathbf{N}_i$ is the set of candidate locations j that can cover demand node i. The two forms of the constraint are equivalent. Constraints (4.1c) are the integrality constraints.

As noted in Chapter 1, if all of the facility costs are identical (e.g., $f_j = 1$ for all candidates sites j), or if we simply want to minimize the number of selected facilities, the objective function may be simplified to become

$$\text{MINIMIZE} \quad \sum_j X_j \quad (4.1d)$$

In the remainder of this section, we will focus on this simpler version of the problem.

Figure 4.1. Example network illustrating suboptimality of nodal locations.

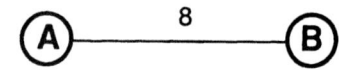

In location problems, the sets $\mathbf{N}_i$ (or, equivalently, the coefficients a_{ij}) are often defined in terms of the distance between a demand node i and the candidate facilities. If we let D_c be the coverage distance, then we will have $a_{ij} = 1$ if $d_{ij} \leq D_c$, or, equivalently, $\mathbf{N}_i = \{j | d_{ij} \leq D_c\}$.

Before proceeding further, we should note that the set covering problem on a general graph is NP-complete.[1] This is true for either objective function (4.1a) or objective function (4.1d).

We note that the set of demand nodes need not be the same as the set of candidate facility sites. Often, however, they will be the same. Also, we note that restricting the candidate sites to a (poorly selected) finite set of nodes (e.g., the demand nodes) may result in our needing more facilities than we would need if we could locate anywhere on the network. This is illustrated in Figure 4.1.

If the coverage distance is 5 (i.e., $D_c = 5$) and we restrict the set of candidate locations to the set of demand nodes (A and B), then we need two facilities, one at each site. However, if we allow facilities to be located anywhere on the network, then we can locate a *single* facility midway between nodes A and B and the facility will cover both nodes A and B. Church and Meadows (1979) have shown, however, that if the set of demand nodes is augmented by a finite set of *network intersection points*,[2] then the solution to the set covering problem in which the candidate facilities are restricted to the augmented set (demand nodes plus network intersection points) will contain the same number of facilities as would the solution to the problem in which facilities were permitted to be located anywhere on the network. They also show how to compute the locations of the network intersection points. Augmenting the set of candidate locations in this way enlarges the problem since it increases the number of columns. The column (and row) reduction techniques outlined below can subsequently be used to reduce the size of the problem.

To illustrate the formulation of the set covering problem, we consider the network shown in Figure 4.2.

---

[1] See Garey and Johnson (1979), p. 222. The proof of this is beyond the scope of this text as it requires the definition of a number of other NP-complete problems which have little direct relation to location modeling other than the fact that they allow us to prove that the set covering problem is NP-complete.

[2] The network intersection points are points chosen on the links of the network such that the distance between any network intersection point and at least one node is exactly equal to the coverage distance.

The Set Covering Model

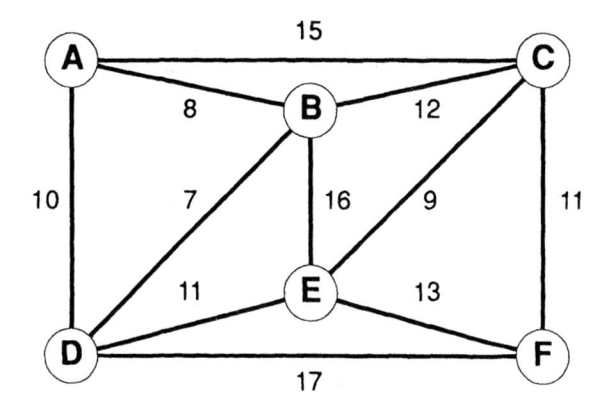

Link distances not drawn to scale.

Figure 4.2. Example set covering network.

If we use a coverage distance of 11 units and restrict the candidate facilities to be on the demand nodes, we obtain the following set covering model:

MINIMIZE
(No. Selected) $\quad X_A + X_B + X_C + X_D + X_E + X_F$

SUBJECT TO:
(Node A Covered) $\quad X_A + X_B + X_D \geq 1$
(Node B Covered) $\quad X_A + X_B + X_D \geq 1$
(Node C Covered) $\quad X_C + X_E + X_F \geq 1$
(Node D Covered) $\quad X_A + X_B + X_D + X_E \geq 1$
(Node E Covered) $\quad X_C + X_D + X_E \geq 1$
(Node F Covered) $\quad X_C + X_F \geq 1$
(Integrality) $\quad X_A, \; X_B, \; X_C, \; X_D, \; X_E, \; X_F = 0, 1$

One solution to this problem is $X_C = X_D = 1$, $X_A = X_B = X_E = X_F = 0$. The objective function equals 2.

When all of the fixed costs are equal (i.e. $f_j = 1$), we can often reduce the size of the problem using a variety of reduction rules. We begin with a *column reduction* rule. Consider two columns j and k. If $a_{ij} \leq a_{ik}$ for all demand nodes i and $a_{ij} < a_{ik}$ for at least one demand node i, then location k covers all demands covered by location j. We say that location k *dominates* location j. In this case, column j may be eliminated, since, if we were to locate at node j, we could always do at least as well by locating at node k. In addition, we can then set $X_j = 0$. For example, in the problem above, candidate site D dominates nodes A and B (since a facility located at D will cover nodes A, B, D, and E, while a facility located at A will only cover nodes A, B, and D, and a facility at B will only cover nodes A, B, and D). Thus, we can set $X_A = X_B = 0$. Similarly, candidate site C dominates site F (since a facility located at C covers demand nodes C, E, and F, while a facility at F covers only nodes C and F). We therefore set $X_F = 0$. (Note that we can also eliminate all but one of a set of equivalent sites, where sites j and k are said to be equivalent if $a_{ij} = a_{ik}$ for all demand nodes i.) After the column reductions described above, the problem becomes

MINIMIZE
(No. Selected) $X_C + X_D + X_E$

SUBJECT TO:
(Node A Covered) $X_D \geq 1$

(Node B Covered) $X_D \geq 1$

(Node C Covered) $X_C + X_E \geq 1$

(Node D Covered) $X_D + X_E \geq 1$

(Node E Covered) $X_C + X_D + X_E \geq 1$

(Node F Covered) $X_C \geq 1$

(Integrality) $X_C, X_D, X_E = 0, 1$

We now consider *row reduction* techniques which allow us to eliminate rows from the problem. Consider row i. If $\Sigma_j a_{ij} = 1$, then there is only one facility site that can cover node i. In that case, we find the location j^* such that $a_{ij^*} = 1$ and set $X_{j^*} = 1$. We can then eliminate any row in which X_{j^*} appears, since, with $X_{j^*} = 1$, those constraints will be satisfied (i.e., those nodes will be covered by the facility at location j^*). For example, in the problem above, there is only one nonzero coefficient in the first constraint. Therefore, we know that X_D must equal 1. We set $X_D = 1$ and then eliminate the constraints corresponding to rows A, B, D, and E, since these demand nodes will all be covered by the facility at location D. Similarly, the constraint for node F has only one nonzero coefficient. Thus, we can set $X_C = 1$ and remove the rows corresponding to nodes C and F. At this point, there are no remaining rows, and the problem becomes the trivial problem of minimizing X_E subject to the integrality constraint that X_E equals either 0 or 1. Clearly, we set $X_E = 0$. At this point, we know the optimal value of all of the decision variables and the problem is solved.

Note that despite the fact that this is an NP-complete problem whose optimal solution is technically difficult to obtain, we have been able to solve

The Set Covering Model

the problem without resorting to any formal optimization technique (such as linear programming).

The row and column reduction rules outlined above may be used iteratively until neither rule allows us to eliminate a column or a row. Often, application of these rules will allow us to solve the problem completely. This is not always the case, however, as the following example shows. This example also allows us to introduce an additional row reduction rule. In this example, we consider the network shown in Figure 4.2, but now use a coverage distance of 18 (i.e., $D_c = 18$). The optimization problem becomes

MINIMIZE
(No. Selected) $\quad X_A + X_B + X_C + X_D + X_E + X_F$

SUBJECT TO:
(Node A Covered) $\quad X_A + X_B + X_C + X_D \geq 1$
(Node B Covered) $\quad X_A + X_B + X_C + X_D + X_E \geq 1$
(Node C Covered) $\quad X_A + X_B + X_C + X_E + X_F \geq 1$
(Node D Covered) $\quad X_A + X_B + X_D + X_E + X_F \geq 1$
(Node E Covered) $\quad X_B + X_C + X_D + X_E + X_F \geq 1$
(Node F Covered) $\quad X_C + X_D + X_E + X_F \geq 1$
(Integrality) $\quad X_A, X_B, X_C, X_D, X_E, X_F = 0, 1$

After using the column reduction rule outlined above, we can eliminate the columns corresponding to candidate sites A and F (and set $X_A = X_F = 0$). The resulting problem is

MINIMIZE
(No. Selected) $\quad X_B + X_C + X_D + X_E$

SUBJECT TO:
(Node A Covered) $\quad X_B + X_C + X_D \geq 1$
(Node B Covered) $\quad X_B + X_C + X_D + X_E \geq 1$
(Node C Covered) $\quad X_B + X_C + X_E \geq 1$
(Node D Covered) $\quad X_B + X_D + X_E \geq 1$
(Node E Covered) $\quad X_B + X_C + X_D + X_E \geq 1$
(Node F Covered) $\quad X_C + X_D + X_E \geq 1$
(Integrality) $\quad X_B, X_C, X_D, X_E = 0, 1$

Since no row has only a single element, we cannot use the row reduction rule outlined above to eliminate any rows. However, we can use a *second row reduction* rule. Consider two rows m and n. If $a_{mj} \leq a_{nj}$ for all candidate sites j and $a_{mj} < a_{nj}$ for at least one candidate site j, then we can eliminate row n. This is so because the requirement that demand node m be covered will guarantee that node n is also covered. (Any facility site that covers demand node m also covers demand node n). This rule allows us to eliminate the rows corresponding to nodes B and E, since any facility that covers node A (i.e., a facility located at node B, C, or D) will also cover nodes B and E. (As before, we can eliminate all but one of a set of equivalent demand nodes, where demand nodes m and n are said to be equivalent if $a_{mj} = a_{nj}$ for all candidate sites j.) After the row reductions outlined above, the problem becomes

MINIMIZE
(No. Selected) $\qquad X_B + X_C + X_D + X_E$

SUBJECT TO:
(Node A Covered) $\qquad X_B + X_C + X_D \qquad\quad \geq 1$

(Node C Covered) $\qquad X_B + X_C + \qquad\quad X_E \geq 1$

(Node D Covered) $\qquad X_B \qquad\quad + X_D + X_E \geq 1$

(Node F Covered) $\qquad\qquad X_C + X_D + X_E \geq 1$

(Integrality) $\qquad X_B, \ X_C, \ X_D, \ X_E = 0, 1$

Unfortunately, we cannot reduce the size of this problem any further. Repeated application of the column reduction rule and the two row reduction rules to this problem will not eliminate any more rows or columns. Thus, we must find some other way to solve this problem. One way of doing so is to ignore the integrality constraint and replace it by a nonnegativity constraint, as follows:

MINIMIZE
(No. Selected) $\qquad X_B + X_C + X_D + X_E$

SUBJECT TO:
(Node A Covered) $\qquad X_B + X_C + X_D \qquad\quad \geq 1$

(Node OJC Covered) $\qquad X_B + X_C + \qquad\quad X_E \geq 1$

(Node D Covered) $\qquad X_B \qquad\quad + X_D + X_E \geq 1$

(Node F Covered) $\qquad\qquad X_C + X_D + X_E \geq 1$

(Nonnegativity) $\qquad X_B, \ X_C, \ X_D, \ X_E \geq 0$

If we solve the resulting linear programming problem, we find $X_B = X_C = X_D = X_E = 1/3$. The objective function is $4/3$. Clearly, this is not an

The Set Covering Model

all-integer solution; it does not solve the original set covering problem. However, since the objective function exceeds 1, all the coefficients in the objective function are integers, and all the decision variables in the objective function of the integer programming problem must be integers, we know that the optimal integer solution will have an objective function value that is at least 2. In some cases, we can obtain an all-integer solution by appending to the linear programming problem a constraint that forces the sum of the variables to be greater than or equal to the lower bound on the all-integer solution computed in this manner. The resulting linear programming problem would be

MINIMIZE
(No. Selected) $\quad\quad\quad X_B + X_C + X_D + X_E$

SUBJECT TO:
(Node A Covered) $\quad\quad X_B + X_C + X_E \geq 1$

(Node C Covered) $\quad\quad X_B + X_C + X_D \geq 1$

(Node E Covered) $\quad\quad X_B + X_D + X_E \geq 1$

(Node F Covered) $\quad\quad X_C + X_D + X_E \geq 1$

(Bound on Obj. Fcn.) $\quad X_B + X_C + X_D + X_E \geq 2$

(Nonnegativity) $\quad\quad\quad X_B, \; X_C, \; X_D, \; X_E \geq 0$

Appending this constraint to this problem will cause many linear programming packages to find an all-integer solution (e.g., $X_B = X_C = 1$, $X_D = X_E = 0$). However, other noninteger solutions are also possible such as $X_B = X_C = X_D = X_E = 0.5$. In this case, all of the original four constraints are nonbinding, so the corresponding dual variables (we can call them U_A, U_C, U_E, and U_F) will all be 0 by the complementary slackness conditions. The additional constraint corresponding to the bound on the objective function will be binding. Therefore, the dual variable associated with this constraint may be positive. In particular, we can let the dual variable associated with this constraint (call it U_{obj}) equal 1. This solution will satisfy all of the primal and dual feasibility conditions as well as the complementary slackness conditions. Thus, it is also an optimal solution which some linear programming routines may find.

To ensure that we obtain an all-integer solution, additional techniques will generally be required. One approach is to use *branch and bound*. The easiest way to begin to understand branch and bound is to illustrate its use on the example above. We begin with the fractional solution $X_B = X_C = X_D = X_E = 1/3$. From this solution, we select one of the noninteger variables and branch on it. For example, we know that X_B which equals 1/3 in the linear programming solution must be either 0 or 1 in the all-integer solution. Thus, we branch on X_B and create two new problems; one in which we add the

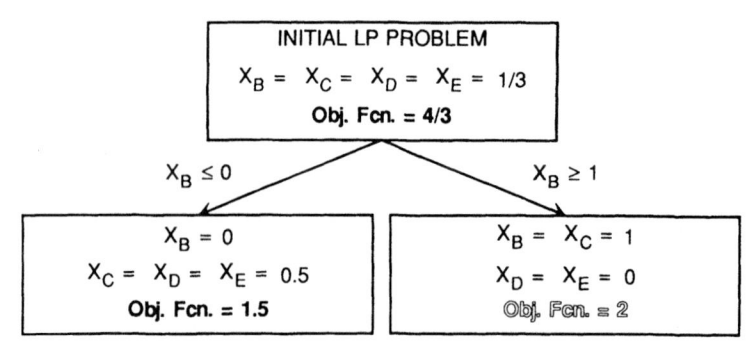

```
┌─────────────────────────────┐
│      INITIAL LP PROBLEM     │
│  X_B = X_C = X_D = X_E = 1/3│
│       Obj. Fcn. = 4/3       │
└─────────────────────────────┘
        /                \
   X_B ≤ 0            X_B ≥ 1
   /                          \
┌──────────────────┐    ┌──────────────────┐
│    X_B = 0       │    │  X_B = X_C = 1   │
│ X_C = X_D = X_E  │    │  X_D = X_E = 0   │
│     = 0.5        │    │  Obj. Fcn. = 2   │
│ Obj. Fcn. = 1.5  │    │                  │
└──────────────────┘    └──────────────────┘
```

Figure 4.3. Results of branching on X_B.

constraint $X_B \geq 1$, and one in which we add the constraint $X_B \leq 0$. We solve each of these two problems. The results are shown in Figure 4.3.

After we branch on X_B, we find that forcing $X_B \geq 1$ results in an all-integer solution ($X_B = X_C = 1$, $X_D = X_E = 0$) with an objective function of 2. Thus, we now know that the optimal all-integer solution cannot have a value that exceeds 2. If we add the constraint $X_B \leq 0$, we obtain another fractional solution in which $X_C = X_D = X_E = 0.5$ and the objective function equals 1.5.[3] Since this is not an all-integer solution, we again select one of the noninteger variables (e.g., X_C) and branch on this variable. We again create two new problems in which we add the constraints $X_C \leq 0$ and $X_C \geq 1$, respectively. Note that the constraint $X_B \leq 0$ also applies to these two new problems; that is, all constraints applied further up the branch-and-bound tree continue to apply. The results are shown in Figure 4.4. Adding either of these constraints results in an all-integer solution whose objective function value is equal to 2. No further branching is required. In this case, we have found three alternate optima. The reader can readily verify that locating at any two of the four nodes B, C, D, and E is optimal. Thus, we have found three of the six alternate optima represented by the possible combinations of two of these four nodes.[4]

One approach to using branch and bound to solve an integer programming problem (in which only the integrality constraints are relaxed) is described below. The algorithm is described in terms of a minimization problem. The

---

[3]As before, since we know that the sum of any number of integers must be an integer, we can infer that the optimal all-integer solution must have a value of at least 2, rounding 1.5 up to the next larger integer. This would allow us to stop the branch-and-bound algorithm since we would now know that the optimal all-integer solution value must be at least 2 and must be no greater than 2, and we have a solution that satisfies the integrality constraints. However, we will ignore this consideration for the moment so that we can illustrate an extra branch in the tree.

[4]In fact, there are 13 alternate optima: $\{A,C\}$, $\{A,D\}$, $\{A,E\}$, $\{A,F\}$, $\{B,C\}$, $\{B,D\}$, $\{B,E\}$, $\{B,F\}$, $\{C,D\}$, $\{C,E\}$, $\{C,F\}$, $\{D,E\}$, and $\{D,F\}$. Of the 15 possible ways of selecting two nodes from the six nodes in Figure 4.2, only $\{A,B\}$ and $\{E,F\}$ fail to solve the set covering problem with a coverage distance of 18.

The Set Covering Model

```
                    INITIAL LP PROBLEM
              X_B = X_C = X_D = X_E = 1/3
                     Obj. Fcn. = 4/3
```

Branching: $X_B \leq 0$ (left), $X_B \geq 1$ (right)

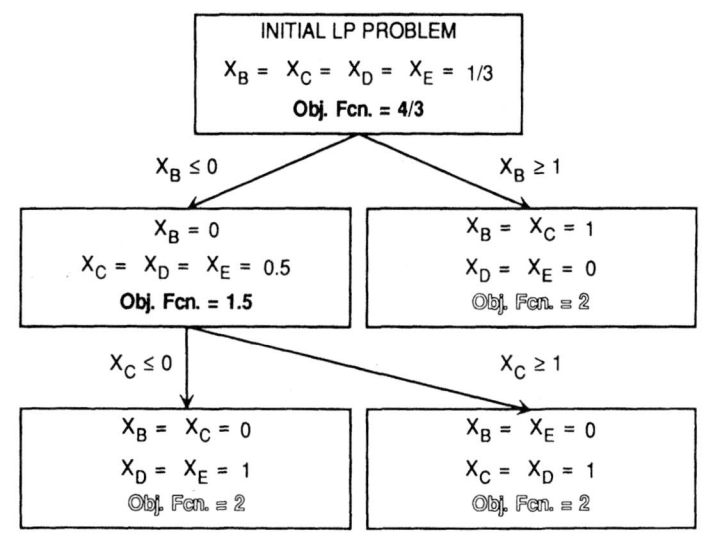

Left child:
$X_B = 0$
$X_C = X_D = X_E = 0.5$
Obj. Fcn. = 1.5

Right child:
$X_B = X_C = 1$
$X_D = X_E = 0$
Obj. Fcn. = 2

From left child, branching: $X_C \leq 0$ (left), $X_C \geq 1$ (right)

Left:
$X_B = X_C = 0$
$X_D = X_E = 1$
Obj. Fcn. = 2

Right:
$X_B = X_E = 0$
$X_C = X_D = 1$
Obj. Fcn. = 2

Figure 4.4. Results of branching on X_C (as well as X_B).

same discussion applies to a maximization problem except that the terms *lower bound* and *upper bound*, *smaller* and *larger*, as well as *minimization* and *maximization* clearly need to be interchanged. The algorithm involves creating a branch-and-bound tree as shown in Figure 4.4. The first or top node is the *root* node. Nodes immediately below another node will be referred to as the *child* nodes of the node above them; similarly, the node immediately above another node (heading toward the root node) will be referred to as the *parent* node of the child node. Associated with each node will be an optimization problem that must be solved. The problem at a child node will be identical to that at the parent node except that one additional constraint will be imposed on the problem at the child node.

The optimization problem at each node represents a relaxation of the original integer programming problem along with some additional constraints that have been imposed on the problem between the root node of the tree and the node in question. Since the optimization problem is a relaxation of the original problem, the value of the objective function at each node represents a *lower bound* on the value of any solution that may be found at the node or below that node in the tree.

Nodes in the branch-and-bound tree are said to be *fathomed* if we have sufficient information at that point in the tree to know that further branching from that node will not be fruitful. As indicated above, this happens in one of three cases:

a. The solution at that node of the tree satisfies all of the relaxed constraints of the original problem. In this case, the addition of extra constraints is clearly not warranted. If the objective function value at

such a node is smaller than that of any other solution which is known to satisfy all of the constraints of the original problem, the node's solution value becomes the new upper bound on the optimal value of the objective function.

  b. The optimization problem (e.g., the linear programming problem) that must be solved at the node is infeasible. In this case, the node may also be fathomed, because adding more constraints to the problem will not make the problem feasible. Thus, further branching from such a node would be fruitless. It is clear that the optimal solution to the original problem cannot be found at a node further down the branch-and-bound tree from a node whose optimization problem is infeasible.

  c. The optimization problem that must be solved at the node is feasible, but has an objective function value that is larger than that of a solution which is known to satisfy all of the constraints of the original problem. Adding some constraints to the node's optimization problem will only further degrade or increase the objective function value. Thus, we can again conclude that the optimal solution to the original problem cannot be found by branching further from such a node in the tree. Such a node may also be fathomed.

With this background, we can now describe one approach to using a branch-and-bound algorithm together with the linear programming relaxation of an integer programming problem to solve the integer programming problem. The step-by-step procedure is given below, followed by a discussion of a number of the strategic issues that must be faced in using this approach.

Step 0: Set the best known solution value to infinity.

Step 1: Relax all of the integer requirements. Add upper bounds on the resulting real decision variables if necessary. (This occurs if the integer variables are bounded—e.g., 0 or 1—and the linear programming relaxation of the integer programming problem is such that one of the relaxed variables could exceed the upper bound on the variable if such an upper bound is not explicitly imposed. This is not necessary in the case of the set covering problem since there will not be any incentive for one of the relaxed integer variables to exceed its integer upper bound of 1.) Solve the resulting linear programming problem. If the solution satisfies all of the relaxed integrality constraints, stop, the solution is optimal; otherwise, proceed to step 2.

Step 2: Select an unfathomed node (parent) in the branch-and-bound tree. (Initially, we will only have the original or root node if the linear programming relaxation of the original problem did not satisfy all of the integrality constraints.) If no unfathomed nodes exists, stop, the best known solution is optimal. If an unfathomed node exists, select a decision variable whose value in the relaxed problem at that node is noninteger but whose value in the original problem must be integer. Let X be this variable and let γ be its noninteger value.

Create two child nodes of the unfathomed node. At the left child node, append the following constraint to the optimization problem solved at the parent node and solve the resulting problem: $X \leq \lfloor \gamma \rfloor$, where $\lfloor \gamma \rfloor$ denotes the largest integer less than γ. If the child node can be fathomed for any of the three reasons outlined above, fathom the node. If the node is fathomed because an improved solution to the original optimization problem has been found, update the upper bound on the best known solution, record the solution at this node as being the best known solution, and fathom any nodes in the tree whose objective function value is not less than the value of the solution just found.

At the right child node, append the following constraint to the optimization problem solved at the parent node and solve the resulting problem: $X \geq \lceil \gamma \rceil$, where $\lceil \gamma \rceil$ denotes the smallest integer greater than γ. If the child node can fathomed for any of the three reasons outlined above, fathom the node. If the node is fathomed because an improved solution to the original optimization problem has been found, update the upper bound on the best known solution, record the solution at this node as being the best known solution, the fathom any nodes in the tree whose objective function value is not less than the value of the solution just found.

Step 3: Return to step 2.

In using this approach to solving an integer programming problem, there are at least two key issues that need to be addressed. First, we must establish a rule for selecting the next unfathomed node from which we want to branch. One of two approaches is generally adopted. In the first approach, we branch from the unfathomed node whose linear programming problem has the smallest objective function value (i.e., the node with the smallest lower bound on the objective function of the original problem). The rationale for this branching rule is that a descendent of the node with the smallest lower bound is likely to be good solution to the original problem and may allow us to fathom additional nodes in the tree. The problem with this approach is that it may necessitate our jumping around from node to node in the tree. In doing so, the tree is likely to grow very large before we find the first feasible solution to the original integer programming problem. The second approach to selecting a node from which to branch is simply to branch from the rightmost unfathomed node at all times. Using this approach, we are more likely to find a feasible solution to the original problem quickly. Also, the size of the tree is likely to remain relatively small throughout the process. However, we may ultimately need to explore more of the tree using this approach. A hybrid approach in which we branch from the node with the smallest lower bound until we begin to approach the memory limits of the

Table 4.1 Solution to the Set Covering Problem for the Network of Figure 4.2 for Various Coverage Distances

Coverage Distance	Location	Number
Less than 7	A, B, C, D, E, F	6
7	A, B, C, E, F	5
8	B, C, E, F	4
9, 10	B, E, F	3
11–18	C, D	2
19 or more	C	1

Note that facilities are constrained to be on the nodes in the solutions outlined above.

computer on which we are running and then adopt the second (branch from the rightmost node) rule might combine the best of both approaches.

The second issue that must be faced in implementing this approach to solving integer programming problems is that of selecting the noninteger decision variable for branching at each node. Often, we select the variable whose value is closest to being an integer. More sophisticated rules that exploit the structure of the problem can and should be developed for specific problem contexts.

Before leaving this section on the set covering model, we present the solution to the example problem for a range of coverage distances in Table 4.1 and Figure 4.5. Note that the number of required facilities decreases in a step-function manner as the coverage distance increases.

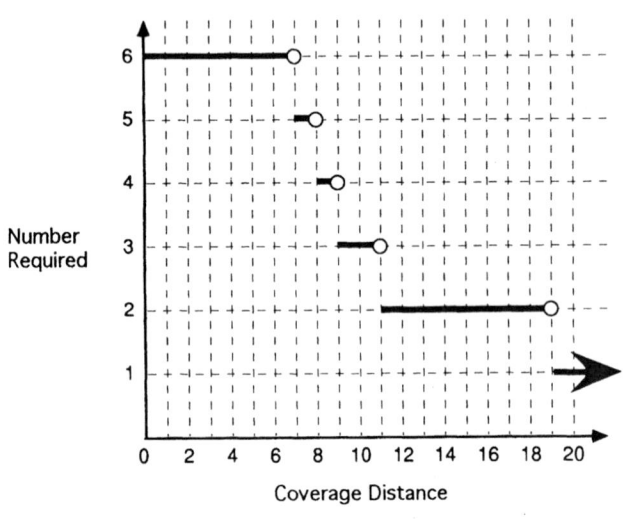

Figure 4.5. Graph of solution to the set covering problem for the network of Figure 4.2 for various distances.

Applications of the Set Covering Model

Figures such as that shown in Figure 4.5 will be very important as we solve center problems (as discussed in Chapter 5).

4.3 APPLICATIONS OF THE SET COVERING MODEL

The set covering problem has been applied in a broad range of contexts. In this section we outline a number of applications of the set covering model to problems outside the scope of location analysis. In Section 4.4, we summarize a number of extensions of the location set covering model that allow the model to incorporate additional (secondary) location concerns.

Applications of the set covering model range from airline crew scheduling (Desrochers et al. 1991) to tool selection in flexible manufacturing systems (Daskin, Jones, and Lowe, 1990). In airline crew scheduling problems, we are given a set of flight legs i. In addition, we can create a very large number of schedules or tours of duty for airline crews. Each schedule consists of a sequence of flight legs. We define the inputs and decision variables as follows:

Inputs

$$a_{ij} = \begin{cases} 1 & \text{if flight leg } i \text{ is part of schedule } j \text{ (tour of duty } j) \\ 0 & \text{if not} \end{cases}$$

f_j = cost of using schedule j (including such costs as crew deadheading costs, hotel and meal costs for crews and so on)

Decision Variables

$$X_j = \begin{cases} 1 & \text{if schedule } j \text{ is selected} \\ 0 & \text{if not} \end{cases}$$

With this notation, the problem of staffing all flight legs at minimum cost can, in principle, be formulated as a set covering problem. The key problem associated with this approach, however, is that the number of possible schedules that meet all of the workrule constraints is astronomically large for practical problems. Large problems of this form are often used for testing new linear programming algorithms and codes. Such problems often have thousands of rows and millions of columns. Even so, the test problems do not include all possible schedules.

The airlines generally try to solve the problem with constraint (4.1b) converted to an equality constraint. The resulting problem is known as the *set partitioning* problem. Airlines prefer set partitioning to set covering solutions because the partitioning solutions do not require deadheading crews between airports. Crew members who must deadhead between airports

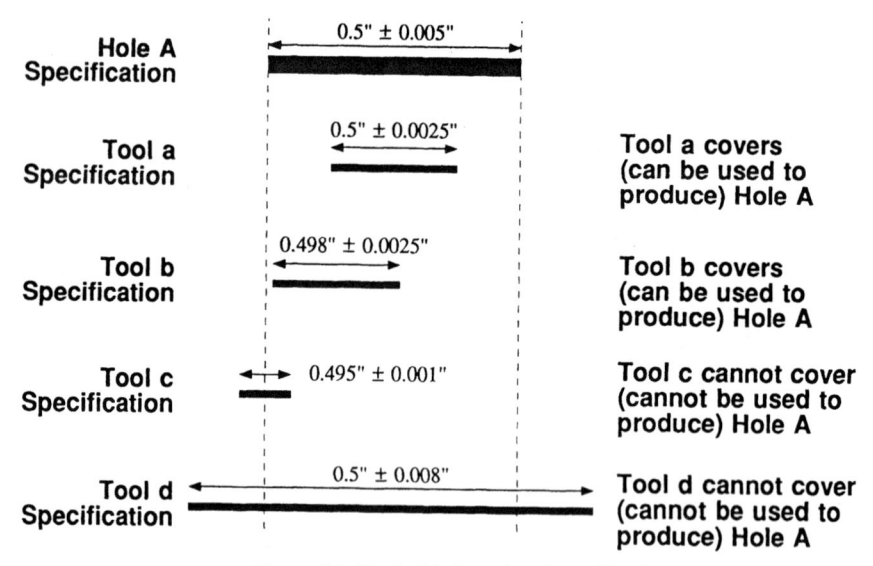

Figure 4.6. Typical hole and tool specifications.

occupy seats that customers might pay for. Consequently, such movements represent potential revenue losses for the airlines.

The set covering model may also be used to identify tools for inclusion in a tool magazine in a flexible manufacturing environment. Consider the problem of punching holes in a piece of flat sheet metal. Associated with each hole i is a set of specifications (e.g., the diameter of the hole and the tolerance on the diameter). Each candidate tool j can produce holes of a given diameter with a particular tolerance. Typical hole and tool specifications are shown graphically in Figure 4.6. Let

d_j = nominal diameter of tool j
t_j = tolerance of tool j
D_i = desired diameter of hole i
T_i = tolerance of hole i

If $(D_i - T_i) \le (d_j - t_j)$ and $(d_j + t_j) \le (D_i + T_i)$, then tool j can produce (can cover) hole i; otherwise it cannot. This allows us to define the coefficients a_{ij} of a set covering model. Daskin, Jones, and Lowe (1990) show that this coefficient matrix has a standard greedy form (see Hoffman, Kolen, and Sakarovitch, 1985; Broin and Lowe, 1986, Nemhauser, and Wolsey, 1988). Using this fact, Daskin, Jones, and Lowe develop a linear time algorithm for solving this variant of the set covering model; that is, an algorithm whose execution time increases only linearly with the number of holes and tools.

4.4 VARIANTS OF THE SET COVERING LOCATION MODEL

The set covering model has been extended to include a number of secondary objectives that are often important in facility location modeling. Many of these extensions are motivated by the observation that there are often alternate optima for a given coverage distance. For example, for the network shown in Figure 4.2, Table 4.2 lists the number of alternate optima for coverage distances between 11 and 18, for which only two sites are needed to cover all demand nodes. As the coverage distance increases within the range over which the number of required facilities is constant, the number of alternate optima increases. Note that of 15 combinations of 2 facilities out of 6 candidate sites, 10 combinations cover all nodes within 15 distance units.

For a coverage distance of 15 units, Table 4.3 lists all the solutions that entail locating two facilities on the nodes of the network. All 10 combinations of facility sites cover all nodes at least once. However, combinations (A, C), (B, C), (C, D), and (C, E) cover three nodes twice. Thus, if all nodes are

Table 4.2 Number of Alternate Optima versus Coverage Distance for the Network of Figure 4.2

Coverage Distance	Number of Alternate Optima
11	3
12	4
13	9
14	9
15	10
16	10
17	13
18	13

Table 4.3 Combinations of Facility Locations That Cover All Nodes at Least Once for the Network of Figure 4.2 with a Coverage Distance of 15 Units

Facility Locations	Nodes Convered Twice
A, C	A, B, C
A, E	C, D
A, F	C
B, C	A, B, C
B, E	C, D
B, F	C
C, D	A, B, E
C, E	C, E, F
D, E	D, E
D, F	E

given equal weight, any one of these four configurations provides backup coverage to more locations that do any of the other six combinations of sites.

One extension of the set covering problem is to select the combination of sites that maximizes the number of demand nodes covered twice from among the alternate optima to the set covering problem. This problem may be formulated using the following notation along with the notation defined above:

Input

I = number of demand nodes

Decision Variable

$$S_i = \begin{cases} 1 & \text{if demand node } i \text{ is covered at least twice} \\ 0 & \text{if not} \end{cases}$$

With this additional notation, the model may be formulated as follows:

MINIMIZE $\quad (I + 1) \sum_j X_j - \sum_i S_i \quad$ (4.2a)

SUBJECT TO: $\quad \sum_j a_{ij} X_j - S_i \geq 1 \quad \forall i \quad$ (2.4b)

$$X_j = 0, 1 \quad \forall j \quad (4.2c)$$

$$S_i = 0, 1 \quad \forall i \quad (4.2d)$$

The objective function (4.2a) minimizes the number of selected facilities, $\sum_j X_j$, weighted by $(I + 1)$ minus the total number of demand nodes that are covered multiple times, $\sum_i S_i$. Recall that minimizing a negative quantity is equivalent to maximizing the quantity. Thus, the model minimizes the number of required facilities and maximizes the number of demand nodes that are covered at least twice. By weighting the first term of the objective function by $(I + 1)$, we can guarantee that the solution will never select more facilities than the minimum number required to cover all demand nodes once. [See Benedict (1983) or Daskin and Stern (1981) for a proof of this property in models similar to this. Daskin, Hogan, and ReVelle (1988) summarize a variety of related models.]

As before, this model may not always terminate in an all-integer solution. For example, if we solve this problem for the network of Figure 4.2 and a coverage distance of 15 units, we find $X_C = X_D = X_E = 0.5$ and $S_C = 0.5$ (all other decision variables are equal to 0) and the objective function equals 10.

Variants of the Set Covering Location Model

If we append the following constraint:

(Need at least 2) $X_A + X_B + X_C + X_D + X_E + X_F \geq 2$

to the problem, the solution to the linear programming problem becomes $X_C = X_E = 1$ and $S_C = S_E + S_F = 1$ (all other decision variables are equal to 0) and the objective function equals 11. As before, if the linear programming relaxation does not result in an all-integer solution even after this constraint is appended, we must use a more powerful technique such as branch and bound to arrive at an all-integer solution.

In some cases, we want to count the actual number of extra times each node is covered and not just whether or not it is covered extra times. In this case, we relax the binary constraint on the S_i values [constraint (4.2d)] and simply allow

$$S_i \geq 0 \quad \forall\, i \qquad (4.2d')$$

As before, if we enforce the integrality constraint on the location variables, X_j, the S_i variables will automatically be integers. In this case, we will need to increase the value of the weight on the first term, $\Sigma_j X_j$, of the objective function to ensure that the minimum number of facilities is actually located. The reader is encouraged to find the smallest value of the weight on this term that will ensure that the minimum number of facilities is sited.

In still other instances, we want to minimize the total number of selected facilities while selecting from among the alternate optima that solution which maximizes the use of existing facilities. For example, in the example shown in Table 4.3, if a facility already existed at node F, we would prefer solution (A,F), (B,F), or (D,F) to any of the other seven solutions. Plane and Hendrick (1977) reformulate the set covering model objective function to account for this concern using the notation outlined above along with the definitions of the following two sets and one constant:

Additional Input Sets and Constant

$\mathbf{J}_e$ = set of existing facility sites
$\mathbf{J}_n$ = set of new candidate facility sites
ε = a very small number as outlined below

The reformulated objective function is

MINIMIZE $\qquad \sum_{j \in \mathbf{J}_e} X_j + (1 + \varepsilon) \sum_{j \in \mathbf{J}_n} X_j \qquad (4.1d')$

We can show that if $\varepsilon < 1/|\mathbf{J}_n|$, then the solution will use the minimum number of facilities [i.e., the number of selected sites will be the same as that found using objective function (4.1d)] and the solution will maximize the number of existing facilities that are selected. The reader is encouraged to try to prove this property.

4.5 THE MAXIMUM COVERING LOCATION MODEL

One of the key problems associated with the set covering model is that the number of facilities that are needed to cover *all* demand nodes is likely to exceed the number that can actually be built (for budgetary and other reasons). Furthermore, the set covering model treats all demand nodes identically. It is equally important in the set covering model to cover a demand node that generates 10 calls for service per year as it is to cover a node that generates 10,000 demands for service per year.

These two concerns lead us to consider fixing the number of facilities that are to be located and maximizing the number of covered demands (as opposed to the number of covered demand nodes). This is exactly what the *maximum covering location model* does. To formulate this model [originally proposed by Church and ReVelle (1974)], we define the following additional notation:

Inputs

h_i = demand at node i
P = number of facilities to locate

Decision Variables

$$Z_i = \begin{cases} 1 & \text{if node } i \text{ is covered} \\ 0 & \text{if not} \end{cases}$$

With this additional notation, the maximum covering location model may be formulated as follows:

MAXIMIZE $\quad \sum_i h_i Z_i \quad$ (4.3a)

SUBJECT TO: $\quad Z_i \leq \sum_j a_{ij} X_j \quad \forall i \quad$ (4.3b)

$$\sum_j X_j \leq P \quad (4.3c)$$

$$X_j = 0, 1 \quad \forall j \quad (4.3d)$$

$$Z_i = 0, 1 \quad \forall i \quad (4.3e)$$

The objective function (4.3a) maximizes the number of covered demands. Constraints (4.3b) state that demand at node i cannot be covered unless at

The Maximum Covering Location Mod

least one of the facility sites that cover node i is selected. Recall that the right-hand side of (4.3b), $\sum_j a_{ij} X_j$, which is identical to the left-hand side of constraint (4.1b), gives the number of selected facilities that can cover node i. Constraint (4.3c) stipulates that we locate no more than P facilities. Note that unless P exceeds the number of facilities needed to cover all demand nodes, constraint (4.3c) will be binding in the optimal solution. Finally, constraints (4.3d) and (4.3e) are the integrality constraints on the decision variables.

In this problem, we are constraining the facility sites to be located at one of a finite number of sites. As before, the optimal solution to the problem in which the set of candidate sites is the same as the set of demand nodes may be inferior to the optimal solution to the maximum covering problem had we allowed the facilities to be located anywhere on the network. Again, the demand nodes may be augmented by a set of network intersection points (Church and Meadows, 1979) such that constraining the facility sites to be on the demand nodes or network intersection points will result in a solution which is as good as a solution in which facilities could be located anywhere on the network.

To illustrate the formulation of the maximum covering location problem, we formulate the problem for the network shown in Figure 4.7, which is identical to Figure 4.2 except that we have added demands at the nodes.

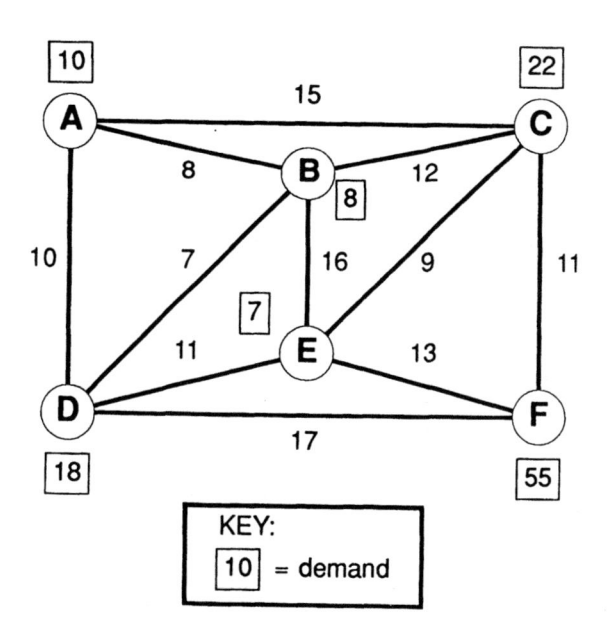

Link distances not drawn to scale.

Figure 4.7. Small example network for the maximum covering problem.

For a coverage distance of 11 (i.e., $D_c = 11$) and $P = 1$, we have the following formulation:

MAXIMIZE
(Covered Demands) $\quad 10Z_A + 8Z_B + 22Z_C + 18Z_D + 7Z_E + 55Z_F$

SUBJECT TO:
(Node A Coverage) $\quad X_A + X_B + X_D \geq Z_A$

(Node B Coverage) $\quad X_A + X_B + X_D \geq Z_B$

(Node C Coverage) $\quad X_C + X_E + X_F \geq Z_C$

(Node D Coverage) $\quad X_A + X_B + X_D + X_E \geq Z_D$

(Node E Coverage) $\quad X_C + X_D + X_E \geq Z_E$

(Node F Coverage) $\quad X_C + X_F \geq Z_F$

(No. to Locate) $\quad X_A + X_B + X_C + X_D + X_E + X_F \leq 1$

(Integrality) $\quad X_A, X_B, X_C, X_D, X_E, X_F = 0, 1$

$\quad\quad\quad\quad\quad\quad Z_A, Z_B, Z_C, Z_D, Z_E, Z_F = 0, 1$

We can use the column reduction technique outlined above to eliminate columns A, B, and F, thereby setting $X_A = X_B = X_F = 0$. The resulting problem is

MAXIMIZE
(Covered Demands) $\quad 10Z_A + 8Z_B + 22Z_C + 18Z_D + 7Z_E + 55Z_F$

SUBJECT TO:
(Node A Coverage) $\quad X_D \geq Z_A$

(Node B Coverage) $\quad X_D \geq Z_B$

(Node C Coverage) $\quad X_C + X_E \geq Z_C$

(Node D Coverage) $\quad X_D + X_E \geq Z_D$

(Node E Coverage) $\quad X_C + X_D + X_E \geq Z_E$

(Node F Coverage) $\quad X_C \geq Z_F$

(No. to Locate) $\quad X_C + X_D + X_E \leq 1$

(Integrality) $\quad X_C, X_D, X_E = 0, 1$

$\quad\quad\quad\quad\quad\quad Z_A, Z_B, Z_C, Z_D, Z_E, Z_F = 0, 1$

Unfortunately, neither of the row reduction procedures outlined above will work on the maximum covering location model. Nevertheless, we can

The Maximum Covering Location Mod

readily solve this small problem (by total enumeration for example). The resulting solution is

$$X_C = 1$$

$$(X_A = X_B =) X_D = X_E (= X_F) = 0$$

(where solutions obtained by the column reduction are shown in parentheses)

$$Z_A = Z_B = Z_D = 0$$

$$Z_C = Z_E = Z_F = 1$$

and the objective function equals 84.

4.5.1 The Greedy Adding Algorithm: A Heuristic Algorithm for Solving the Maximum Covering Location Model

A number of algorithms have been proposed for solving the maximum covering model. In this subsection we outline a heuristic algorithm for solving the model. This algorithm and its variants may be used to solve (at least approximately) a large number of other location problems. The algorithm is known as a *greedy algorithm* since it does what is best at each step of the algorithm without looking ahead to see how the current decisions will impact on later decisions and alternatives.

If we were to locate only one facility (i.e., $P = 1$), we could solve the problem optimally by simply evaluating how many demands each candidate site covers (candidate site j covers $\Sigma_i a_{ij} h_i$ demands) and selecting the site that covers the most demands. Note that this can be done in $O(IJ)$ time, where I is the number of demand nodes and J is the number of candidate facility sites. In theory, this total enumeration approach could be used to solve problems with any number of facilities. However, the time needed to solve the problem with P facilities being selected out of J candidate sites is at least $O\left(I\binom{J}{P}\right)$, where $\binom{J}{P}$ gives the number of combinations of P sites out of J. Actually, if we count only elementary operations (additions, subtractions, and comparisons), we are likely to need $O\left(IP\binom{J}{P}\right)$ operations, since evaluating each combination of sites requires P comparisons for each demand node to see if any of the P sites can cover the demand node in question. For example, if $I = 100$, $J = 50$, and $P = 1$, we would need 5000 operations; for $P = 3$, we would need 5,880,000 operations; finally, for $P = 10$, we would need 1.027×10^{13} operations! If we could evaluate 10^7 operations per second, we would need almost 12 days to solve this problem

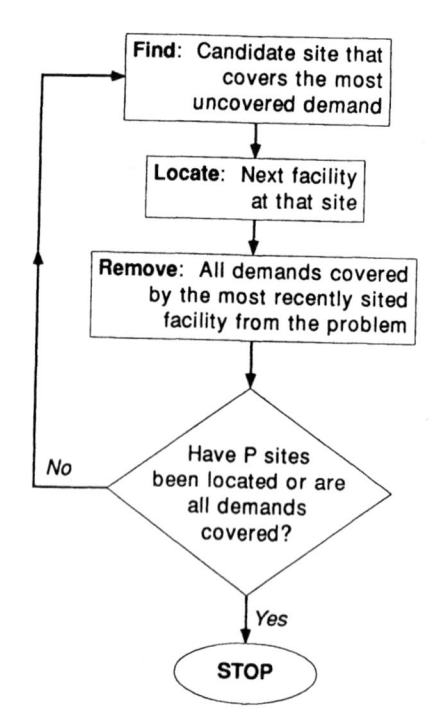

Figure 4.8. Flowchart for the greedy adding algorithm for the maximum covering problem.

[Flowchart:
- Find: Candidate site that covers the most uncovered demand
- Locate: Next facility at that site
- Remove: All demands covered by the most recently sited facility from the problem
- Decision: Have P sites been located or are all demands covered? No → back to Find; Yes → STOP]

with $P = 10$. To solve a problem with 15 sites being located, we would need over 10 years. Clearly, a total enumeration approach is not very attractive.[5]

However, if we want to locate two facilities and fix the first facility at the location determined by the total enumeration algorithm outlined above, we can evaluate the best place to locate the second facility (given the location of the first) in $O(IJ)$ time again. We could repeat this procedure for locating the third site (conditional on the locations of the first two facilities), locating the fourth site (conditional on the location of the first three sites), and so on, locating the Pth site (conditional on the locations of the first $P - 1$ sites). In this way, we can reduce the complexity of the algorithm to $O(IJP)$. The time required to locate 15 facilities on a computer capable of evaluating 10^7 operations per unit time would be about 1 second. What we give up is the guarantee of obtaining an optimal solution. Nevertheless, this approach seems to work quite well, particularly if a large number of facilities are to be located. Figure 4.8 is a flowchart of this algorithm.

If the coverage distance for Figure 4.7 is 9 (i.e., $D_c = 9$), Table 4.4 lists the demand nodes covered by each candidate site. Clearly, the best location is node F, which covers only itself, but which generates 55 demands. Removing

---

[5] In fact, Megiddo, Zemel, and Hakimi (1983) show that the maximum covering problem on a general network is NP-complete. For tree networks, however, they present a polynomial time algorithm for the problem.

The Maximum Covering Location Mod 115

Table 4.4 Coverage by Each Candidate Site with a Coverage Distance of 9 in Figure 4.7

Candidate Site	Nodes Covered	Demand Covered
A	A, B	18
B	A, B, D	36
C	C, E	29
D	B, D	26
E	C, E	29
F	F	55

Table 4.5 Coverage by Each Candidate Site After Locating at Nodes B and F

Candidate Site	Previously Uncovered Nodes Now Covered	Previously Uncovered Demand Now Covered
A	—	0
B	—	0
C	C, E	29
D	—	0
E	C, E	29
F	—	0

all of the demand nodes covered by node F (node F itself) from the problem does not alter the nodes or number of demands covered by any other node in this example. Therefore, the second facility should be located at node B, covering nodes A, B, and D and a total of 36 demands. Removing the nodes covered by node B results in the coverage shown in Table 4.5. Either node C or E may be selected. Following this selection, all demand nodes are covered and we can stop the algorithm.

Consider now the problem shown in Figure 4.9. In this problem, suppose the coverage distance is 9 (i.e., $D_c = 9$). Table 4.6 gives the demand nodes and the number of demands covered by each node. The solution to the maximum covering problem using the greedy adding algorithm would involve locating the first facility at node C (since it covers 21 demands). Demand nodes B, C, and D are now covered (but in general should remain as

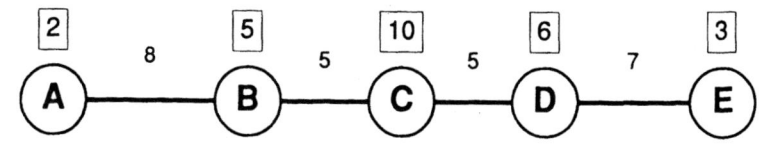

Figure 4.9. Example network to illustrate the need for substitution.

Table 4.6 Coverage by Each Candidate Site with a Coverage Distance of 9 in Figure 4.9

Candidate Site	Nodes Covered	Demand Covered
A	A, B	7
B	A, B, C	17
C	B, C, D	21
D	C, D, E	19
E	D, E	9

candidate facility locations).[6] The second facility can be located at either node D or E (since either one covers three of the remaining five demands). After one of these sites is selected, demand node E will be covered. To cover all demands, we would need to locate a *third* facility at either node A or B to cover the remaining two demands. It should be clear, however, that had we located at nodes B and D, we could have covered all of the demands with only two facilities. Thus, the coverage with two facilities given by the greedy adding algorithm (i.e., 24 demands) is *suboptimal*.

Suppose that after locating the second facility we had considered the possibility of moving one of the facilities to a different node to see if the total covered demand could be increased. Let us assume that we located the first facility at node C, the only choice and that the second facility was located at node E. Moving the facility at node C to node B would increase the total coverage from 23 demands to 25 demands and all nodes would be covered. This leads to the *greedy adding and substitution algorithm* whose flowchart is shown in Figure 4.10. Figure 4.11 shows the substitution part of the algorithm. In essence, the substitution algorithm considers removing every selected candidate site and replacing it with every nonselected candidate site. If any such exchange or swap improves the objective function (the total number of covered demands), the exchange is made. The algorithm continues until there is no exchange of a selected site (to be removed from the tentative solution) and a nonselected site (to be inserted into the tentative solution) which increases the total number of covered demands. The reader should note that there are a number of different ways of implementing such an algorithm. Figure 4.11 suggests that as soon as an exchange that improves the solution is found, the exchange is made. An alternative approach would be to find the best node to substitute for a given node that is in the solution and to perform that swap. This would entail moving the exchange block outside the loop over n, the index of nonselected sites, and would require that we keep the track of the best node n^* (if any) to substitute for node m, the node that is currently in the solution. Yet another approach would be to find the best

---

[6]The need to retain demand nodes that are already covered as candidate facility locations is particularly critical if the distance matrix is not symmetric or if facilities cannot cover demands at their own locations. (See Exercise 4.16 for an example of such a problem.)

The Maximum Covering Location Mod

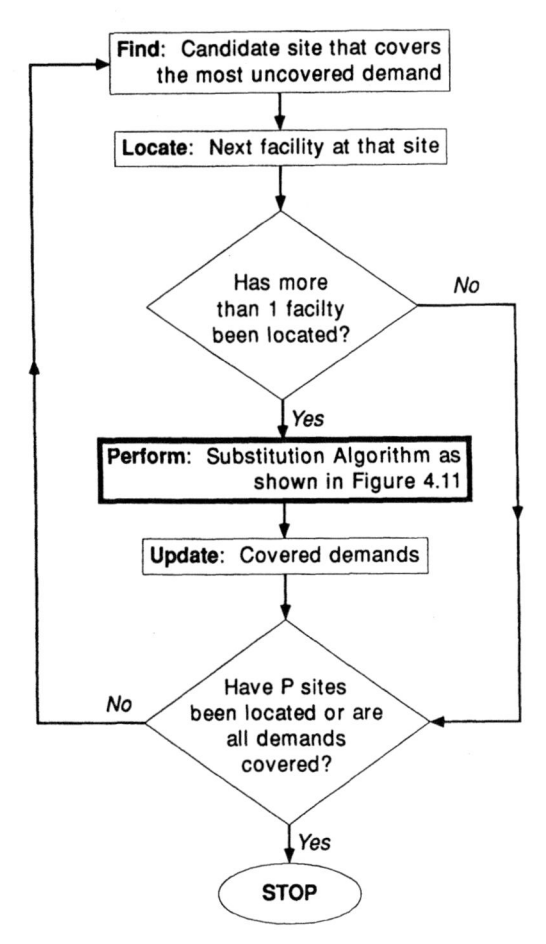

Figure 4.10. Flowchart for the greedy adding and substitution algorithm for the maximum covering problem.

exchange from among all possible exchanges. This would entail moving the exchange block outside the loop over m. As before, we would need to keep track of the indices m^* and n^* of the best node to remove (m^*) and the best node (n^*) to substitute for that node. Other implementations can be created by altering the order in which nodes are processed and the point in the algorithm at which we reset the indices m and n. Each such implementation is likely to result in a different solution.

It is important to note that even with the substitution algorithm (which adds significantly to the execution time of the procedure), there is no guarantee that the resulting solution will be optimal.

To illustrate the importance of using the substitution algorithm, we consider the problem generated by the need to cover the 48 capitals of the

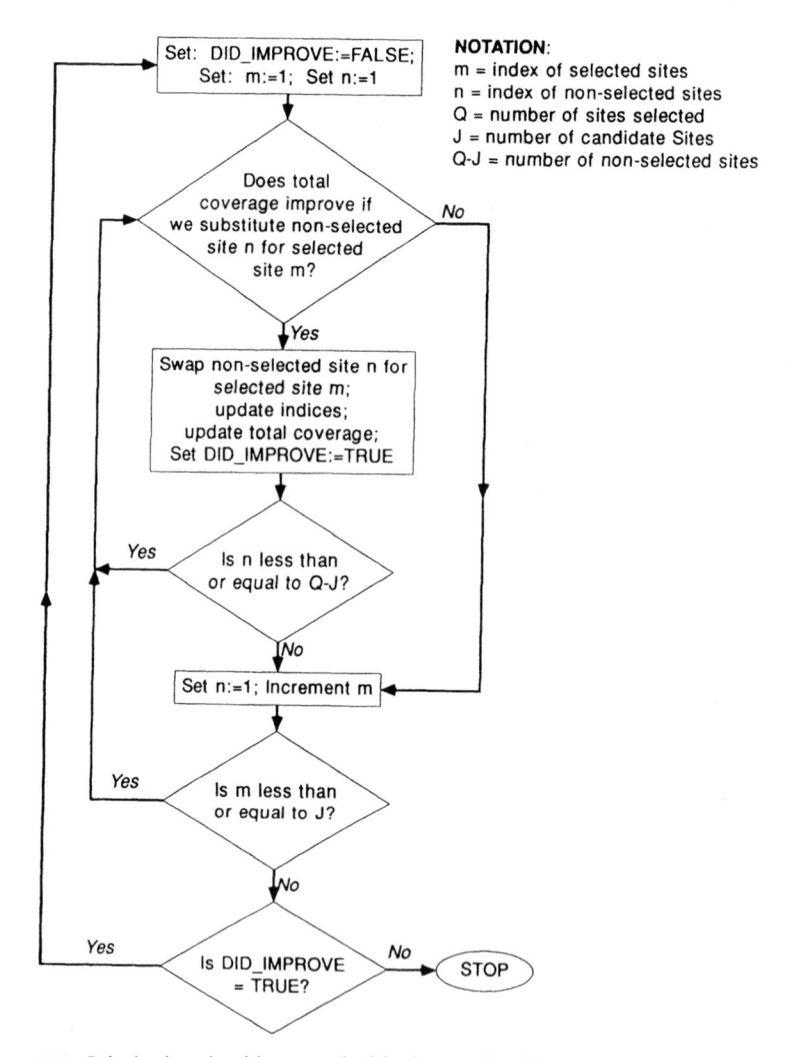

Figure 4.11. Substitution algorithm to embed in the greedy adding and substitution algorithm.

continental United States, Washington, DC, and the 50 most populous cities in the continental United States. After duplicate cities are eliminated, this results in 88 cities. The longitudes, latitudes, and demands associated with these cities are given in Appendix G. Using a coverage distance of 720 miles (and great circle distances), using only the greedy adding algorithm, we obtain the solution shown in Table 4.7. as we increase the number of facilities to be located. Note that the algorithm suggests that five sites are needed to cover all 88 cities. With three facilities selected, Boston, MA;

The Maximum Covering Location Mod

Table 4.7 Solution to Covering Problem on the Continental United States Using Greedy Adding Algorithm

Number of Sites	Cities Selected	Percentage Coverage
1	Indianapolis	60.65
2	Indianapolis, El Paso	88.61
3	Indianapolis, El Paso, Salt Lake City	97.42
4	Indianapolis, El Paso, Salt Lake City, Raleigh	99.95
5	Indianapolis, El Paso, Salt Lake City, Raleigh, Detroit	100.00

Providence, RI; Concord, NH; Augusta, ME; Montpelier, VT (all in New England); and Miami, FL are not covered. A fourth facility in Raleigh, NC is added to cover all demands except Augusta, ME. Finally, a fifth facility in Detroit, MI is added to cover Augusta, ME.

If we use the greedy adding and substitution algorithm, we obtain the solution shown in Table 4.8. Now, only three cities are needed to cover all demands. The table shows that after a second city is added in El Paso, the substitution algorithm substitutes Cincinnati for Indianapolis. The algorithm then adds a facility in Salt Lake City. Next, it substitutes New Orleans for El Paso, thereby covering all locations except Boston, MA; Concord, NH; and Augusta, ME. Next, it substitutes Detroit for Cincinnati. After this substitution, all 88 cities are covered and the algorithm stops.

Finally, by way of example, Figure 4.12 shows the tradeoff curve that results from using the greedy adding and substitution algorithm and a

Table 4.8 Solution to Covering Problem on the Continental United States Using Greedy Adding and Substitution Algorithm

Number of Sites	Cities Selected	Percentage Coverage
1	Indianapolis	60.65
2	Indianapolis, El Paso	88.61
2	*Cincinnati*, El Paso	88.98
3	Cincinnati, El Paso, Salt Lake City	97.79
3	Cincinnati, *New Orleans*, Salt Lake City	98.59
3	*Detroit*, New Orleans, Salt Lake City	100.00

The city shown in *italics* is the one that was substituted for another city in the solution shown on the previous line.

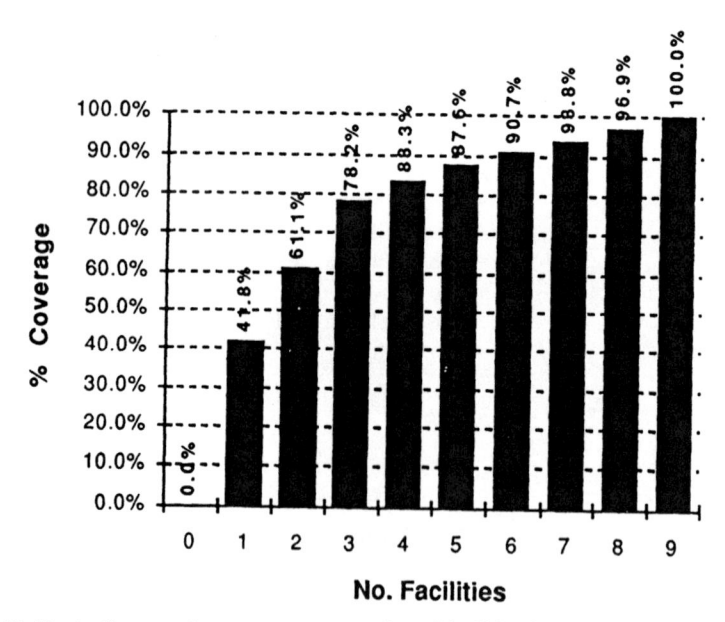

Figure 4.12. Tradeoff curve of coverage versus number of facilities for CITY1990.GRT problem and a coverage distance of 410 miles.

coverage distance of 410 miles. Note that the curve generally exhibits decreasing marginal coverage with each additional facility located. In other words, the additional coverage obtained by adding the mth facility is generally less than the additional coverage that is obtained by adding the $(m-1)$th facility. One exception to this general rule results as we go from seven to nine facilities. As we go from seven to eight facilities, the percentage coverage increases 3.036 percent from 93.842 to 96.878 percent. If we add a ninth facility, the percentage coverage increases to 100.00 percent, a 3.122 percent increase. Note that this is greater than the percentage increase found when we went from seven to eight facilities. We can show that the optimal curve will *usually* exhibit decreasing marginal coverage throughout the range of the number of facilities located. This indicates that the solution with eight sites is likely to be suboptimal. The heuristic solution covers 43,440,613 demands (96.878 percent). In fact, we can find a solution (using the procedure outlined in the next subsection) that covers all 44,840,571 demands that is provably optimal. Finally, the reader should compare the heuristic results shown in Figure 4.12 with the optimal results for the same problem as shown in Figure 1.1. Note that the greedy adding and substitution algorithm is suboptimal when we are locating four or more facilities in this case. Also, the greedy adding and substitution algorithm requires one more facility to cover all demands than does the optimal solution.

The Maximum Covering Location Mod 121

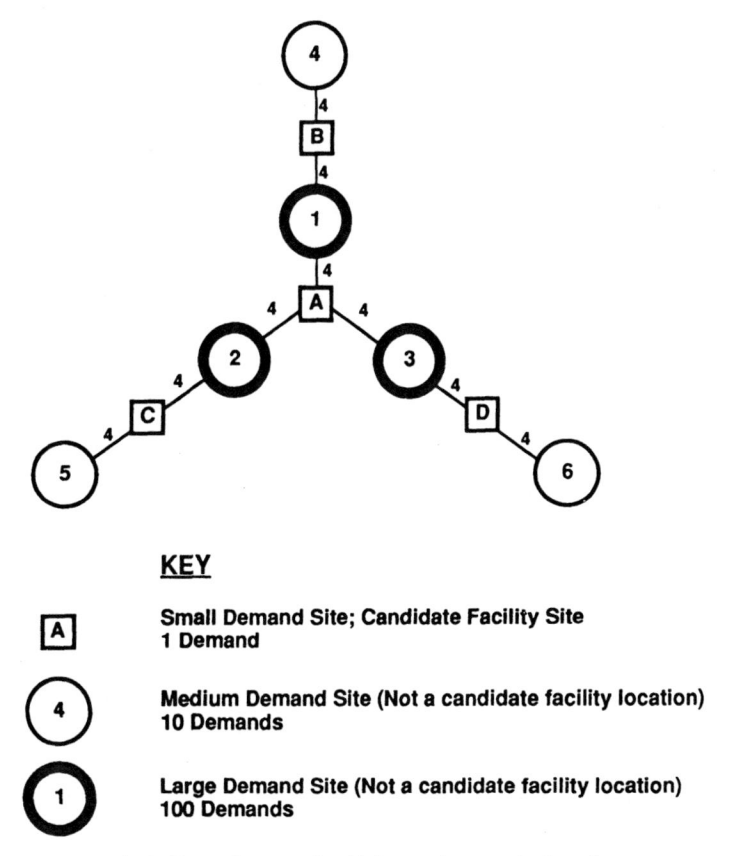

Figure 4.13. Example network with increasing marginal total coverage.

While most networks and graphs will usually exhibit decreasing marginal coverage throughout the range of the number of facilities being located, this is not always the case as the example in Figure 4.13 indicates. [This example was motivated by an example in Berman, Larson, and Fouska (1992).] Lettered nodes A, B, C, and D are candidate sites and generate one demand each. Nodes 1, 2, and 3 are large demand nodes, generating 100 demands each. Nodes 4, 5, and 6 are medium demand nodes, generating 10 demands each. All links are 4 units long. The coverage distance is 10 units.

Node A covers 304 demands; nodes B, C, and D cover 112 demands each. The optimal single facility sites is clearly at node A for a total coverage of 304. The optimal solution for two facilities is to locate one facility at node A and the other facility at either B, C, or D for a total coverage of 314. This represents an increase in coverage of 10 units. The optimal solution for three facilities is to locate at nodes B, C, and D, resulting in a total coverage of 334. This represents an increase of 20 demand units, showing that the marginal coverage need not be decreasing. Despite this example, we note again that it is *usually* true that marginal coverage decreases with the location of additional facilities.

4.5.2. Lagrangian Relaxation: An Optimization-Based Heuristic Algorithm for Solving the Maximum Covering Location Model

As noted above, there is no guarantee that the solution that results from the greedy adding (or greedy adding and substitution) algorithm will be optimal. In fact, not only might it not be optimal, but we have no way of knowing how far it actually is from an (unknown) optimal solution. The approach outlined in this subsection, Lagrangian relaxation, provides us with an upper bound on the value of the objective function. When followed by a substitution algorithm, Lagrangian relaxation performs exceptionally well on the maximum covering problem (Daskin, Haghani, and Malandraki, 1986).

Lagrangian relaxation is an approach to solving difficult problems (such as integer programming problems). The approach outlined below is cast in terms of solving a maximization problem. The approach involves the following general steps:

1. Relax one or more constraints by multiplying the constraint(s) by Lagrange multiplier(s) and bringing the constraint(s) into the objective function. The relaxed problem should be such that it can be solved very easily for fixed values of the Lagrange multipliers. In addition, there are a number of other desirable features of the relaxed problem that are beyond the scope of this text. [For example, one can show that if Lagrangian relaxation is being used to solve integer programming problems and if the relaxed problem has a natural all-integer solution (e.g., it is a network flow problem), then the best upper bound that can be obtained using Lagrangian relaxation will not be better than the upper bound that would have resulted from solving the linear programming relaxation of the integer programming problem. Thus, for such problems, it would be useful to find a relaxed problem whose linear programming solution is not guaranteed to be all integer, but for which an all-integer solution can readily be found. In such a case, the Lagrangian upper bound will be better (tighter) than the linear programming bound.]
2. Solve the resulting relaxed problem to find the optimal values of the original decision variables (in the relaxed problem).
3. (Optional) Use the resulting decision variables from the solution to the relaxed problem found in step 2 to find a feasible solution to the original problem. This can often be done fairly easily. Update the lower bound on the best feasible solution known for the problem.

The Maximum Covering Location Model

4. Use the solution obtained in step 2 to compute an upper bound on the best value of the objective function.
5. Examine the solution obtained in step 2 and determine which of the relaxed constraints are violated. Use some method to modify the Lagrange multipliers in such a way that the violated constraints are less likely to be violated on the subsequent iteration. The method outlined below is that of *subgradient optimization*. After new Lagrange multipliers have been identified, return to step 2.

We now consider the application of Lagrangian relaxation to the maximum covering problem. Recall that the maximum covering problem was formulated as follows:

$$\text{MAXIMIZE} \quad \sum_i h_i Z_i \quad (4.3a)$$

$$\text{SUBJECT TO:} \quad Z_i \leq \sum_j a_{ij} X_j \quad \forall\, i \quad (4.3b)$$

$$\sum_j X_j \leq P \quad (4.3c)$$

$$X_j = 0, 1 \quad \forall\, j \quad (4.3d)$$

$$Z_i = 0, 1 \quad \forall\, i \quad (4.3e)$$

Constraint (4.3b) complicates the analysis as it links the location variables, X_j, and the coverage variables, Z_i. Therefore, we elect to relax this constraint. (Note that in other problems it may not be advisable to relax the constraints that link decision variables. In fact, in many problems, there are multiple relaxations and many need to be tested to determine which performs best.) After relaxing constraint (4.3b), using Lagrange multipliers, λ_i, we obtain the following problem:

$$\underset{\lambda}{\text{MIN}}\ \underset{X,Z}{\text{MAX}} \quad \sum_i h_i Z_i + \sum_i \lambda_i \left(\sum_j a_{ij} X_j - Z_i \right) \quad (4.4a)$$

$$\text{SUBJECT TO:} \quad \sum_j X_j \leq P \quad (4.4b)$$

$$X_j = 0, 1 \quad \forall\, j \quad (4.4c)$$

$$Z_i = 0, 1 \quad \forall\, i \quad (4.4d)$$

We will be trying to *maximize* the objective function with respect to the decision variables Z_i and X_j and *minimize* the objective function with respect to the Lagrangian variables λ_i. Note also that we need one Lagrange multiplier for each of the constraints that is relaxed. Since constraint (4.3b) applies to all values of the demand node index i, we need Lagrange multipliers λ_i that are indexed by i.

Combining the terms in Z_i in the objective function, we obtain the following problem:

$$\text{MIN}_{\lambda} \text{MAX}_{X,Z} \quad \sum_i (h_i - \lambda_i) Z_i + \sum_j \left(\sum_i a_{ij} \lambda_i \right) X_j \quad (4.4a')$$

SUBJECT TO:
$$\sum_j X_j \leq P \quad (4.4b)$$

$$X_j = 0, 1 \quad \forall j \quad (4.4c)$$

$$Z_i = 0, 1 \quad \forall i \quad (4.4d)$$

$$\lambda_i \geq 0 \quad \forall i \quad (4.4e)$$

Since we are relaxing an inequality constraint, the Lagrange multipliers λ_i are constrained to be nonnegative in (4.4e).

Before outlining how to solve this problem, we need to justify that its solution will give us valid bounds on the original maximum covering problem (4.3a)–(4.3e). Let $(\mathbf{X}^{L*}(\boldsymbol{\lambda}), \mathbf{Z}^{L*}(\boldsymbol{\lambda}))$ be the optimal solution to the Lagrangian problem for some given values of the Lagrange multipliers $\boldsymbol{\lambda}$.[7] Since this is an optimal solution to the Lagrangian problem for the given values of the Lagrange multipliers, we know that the Lagrangian objective function evaluated with these values of the decision variables is greater than or equal to the Lagrangian objective function for any other set of decision variables that satisfy constraints (4.4b)–(4.4e). Specifically, it is greater than or equal to the Lagrangian objective function evaluated at the optimal solution to the original maximum covering problem (4.3a)–(4.3e), since any solution that satisfies constraints (4.3b)–(4.3e) also satisfies (4.4b)–(4.4e). In other words, we know that

$$\sum_i h_i Z_i^{L*} + \sum_i \lambda_i \left(\sum_j a_{ij} X_j^{L*} - Z_i^{L*} \right)$$

$$\geq \sum_i h_i Z_i^* + \sum_i \lambda_i \left(\sum_j a_{ij} X_j^* - Z_i^* \right)$$

$$\geq \sum_i h_i Z_i^*$$

where X_j^* and Z_i^* are the optimal solutions to the original maximum covering problem, (4.3a)–(4.3e). The first inequality follows from the optimality of the Lagrangian solution and the second inequality follows from the

---

[7] We are using boldfaced characters to denote vectors. Thus, $\mathbf{X}^{L*}(\boldsymbol{\lambda})$ denotes a vector of location variables, X_j^{L*}, that depend on the vector of Lagrange multipliers, λ_i.

The Maximum Covering Location Model

feasibility of the optimal solution and the nonnegativity of the Lagrange multipliers (4.4e). Specifically, we know that $\sum_j a_{ij} X_j^* - Z_i^* \geq 0$ by constraint (4.3b). The last term, $\sum_i h_i Z_i^*$, is simply the optimal value of the original maximum covering problem. Thus, for any feasible, nonnegative values of the Lagrange multipliers, the Lagrangian objective function maximized over the location variables and the coverage variables will provide an upper bound on the objective function of the maximum covering problem.

4.5.2.1 Solving the Relaxed Problem

For fixed values of the Lagrange multipliers, λ_i, the problem decomposes into problems in Z_i and X_j, each of which may be readily solved. The solution for Z_i is

$$Z_i = \begin{cases} 1 & \text{if } h_i - \lambda_i > 0 \\ 0 & \text{if not} \end{cases}$$

To solve for X_j, we need to find the P largest coefficients of the X_j terms. But the coefficient of the X_j term in (4.4a′) is simply the λ-demand that is covered by a facility located at candidate site j. Thus, we find the P sites that cover the most demand when demand at node i is given by λ_i. For those P sites, we set $X_j = 1$, and for the remaining sites, we set $X_j = 0$. In other words, we replace the actual demand, h_i, at each demand node i by the λ value associated with the node, λ_i. We then compute the total λ-demand covered by each candidate site. We do *not* need to worry about demands that are double counted. We then set the X_j variables for the P candidate sites that cover the most λ-demand to 1 and all other X_j values to 0.

Thus, we can now readily solve the relaxed problem. Note that the relaxed problem has a natural all-integer solution and so the upper bound that we compute will not be any better than that which could be obtained by using linear programming to solve the LP relaxation of (4.3a)–(4.3e). However, for large problems, solving the LP relaxation alone may be difficult. We now have a means of solving the relaxed problem that uses only sorting (of the $\sum_i a_{ij} \lambda_i$ terms).

4.5.2.2 Finding a Feasible Solution and a Lower Bound

The values of Z_i and X_j that result from solving the subproblems are not likely to be feasible for the original problem. In particular, they are likely to violate constraint (4.3b), the constraint that was relaxed. However, we can readily find a feasible solution from these values by simply finding the total demand that is covered by the P sites whose X_j values are equal to 1. This value will be a *lower bound* on the value of the original objective function. Let LB^n be the value of this lower bound where n is the index of the iteration number (through steps 2 through 5 above). Also, let LB be the best lower bound (the one with the largest value) that we have found so far.

4.5.2.3 Finding an Upper Bound For any given values of λ_i, the solution to (4.4a'), (4.4b)–(4.4d) gives an *upper bound* on the solution to the original problem (4.3a)–(4.3e). Denote the bound that we compute on the nth iteration by $\mathscr{L}^n$. Note that the values of $\mathscr{L}^n$ need not decrease from iteration to iteration. However, if we look at the best (smallest) of these over all iterations, then this value will be nonincreasing.

4.5.2.4 Updating the Lagrange Multipliers The method outlined below for updating the Lagrange multipliers is known as subgradient optimization. The basic idea is that, for fixed values of the decision variables, Z_i and X_j, we want to find values of λ_i that minimize the Lagrangian function (4.4a'). In the equations that follow, we have superscripted decision variables by the iteration number, n. The values used are those that solve the relaxed problem (4.4a'), (4.4b)–(4.4d). We begin by computing a stepsize t^n as follows:

$$t^n = \frac{\alpha^n(\mathscr{L}^n - \text{LB})}{\sum_i \left\{ \left(\sum_j a_{ij} X_j^n \right) - Z_i^n \right\}^2}$$

Note that the denominator is just the square of the difference between the number of times node i is covered as indicated by the X_j variables ($\sum_j a_{ij} X_j^n$) and the number of times node i is covered as indicated by Z_i. Also, note that in the numerator we use the upper bound computed using the variables on the nth iteration and the *best* lower bound that is available. The term α^n is simply a constant that will be changed from iteration to iteration as outlined below. The value of α^1 is usually 2.

With this stepsize, the values of λ_i are updated using the following relationship:

$$\lambda_i^{n+1} = \max\left[0, \lambda_i^n - t^n \left(\sum_j a_{ij} X_j^n - Z_i^n \right) \right]$$

Note that if more sites cover node i than indicated by the Z_i^n variable (i.e., $\sum_j a_{ij} X_j^n > Z_i^n$), then we reduce λ_i, thereby making it more likely to have $Z_i^{n+1} = 1$ and less likely to select nodes j that cover node i. Similar (but reverse) statements are true if $Z_i^n = 1$, but none of the nodes j that can cover node i have been selected. In that case, we increase λ_i, making it less likely to have $Z_i^{n+1} = 1$ and also increasing the likelihood that one of the nodes j that cover node i will be selected on the $n + 1^{st}$ iteration.

4.5.2.5 Modifying the Constant α^n As indicated above, we usually begin with $\alpha^1 = 2$. The value of α^n is generally halved if the upper bound, $\mathscr{L}^n$, has not decreased in a given number of consecutive iterations. This is often done

The Maximum Covering Location Model

if $\mathscr{L}^n$ has not decreased in four consecutive iterations, though other values are also used. (In the example below, we will halve α^n whenever the upper bound fails to go down from one iteration to the next, just to illustrate the process of reducing α^n.)

4.5.2.6 Termination The algorithm terminates (or is terminated) when one of the following conditions is true:

1. We have done a prespecified number of iterations.
2. The lower bound equals the upper bound (LB = $\mathscr{L}^n$) or is close enough to the upper bound.
3. α^n becomes small. When α^n is very small, the changes in λ_i will also be very small. Such small changes are not likely to help solve the problem.

4.5.2.7 Further Improvements in the Solution The procedure outlined above is *not* guaranteed to provide an optimal solution. Thus, it may be possible to improve on the solution found using this procedure using a number of heuristic approaches including trying single node substitutions. When this was done using a very large number of problems, the average difference between the upper bound found using Lagrangian relaxation and the best feasible solution was less than 0.25 percent of the value of the solution (Daskin, Haghani, and Malandraki, 1986).

4.5.2.8 Example Problem The procedure outlined above is illustrated in the following tables using the network shown in Figure 4.7 and a coverage distance of 10. The initial values of λ_i were selected using the following equation:

$$\lambda_i = \bar{h} + 0.5(h_i - \bar{h})$$

Table 4.9 First Iteration of Lagrangian Relaxation Calculations

Node	h_i	λ_i	Z_i	$(h_i - \lambda_i)Z_i$	$\Sigma_i a_{ij}\lambda_i$	X_j	$(\Sigma_i a_{ij}\lambda_i)X_j$	$\Sigma_j a_{ij}X_j$	$\Sigma_j a_{ij}X_j - Z_i$
A	10	15	0	0	48	1	48	2	2
B	8	14	0	0	48	1	48	2	2
C	22	21	1	1	34.5	0	0	0	−1
D	18	19	0	0	48	0	0	2	2
E	7	13.5	0	0	34.5	0	0	0	0
F	55	37.5	1	17.5	37.5	0	0	0	−1

Best upper bound = 114.5
$\bar{h} = 20$

Upper bound = 114.5
Best lower bound = 36.0
$\alpha = 2$
$t^n = 11.2143$

Table 4.10 Second Iteration of Lagrangian Relaxation Calculations

Node	h_i	λ_i	Z_i	$(h_i - \lambda_i)Z_i$	$\Sigma_i a_{ij}\lambda_i$	X_j	$(\Sigma_i a_{ij}\lambda_i)X_j$	$\Sigma_j a_{ij}X_j$	$\Sigma_j a_{ij}X_j - Z_i$
A	10	0	1	10	0	0	0	0	−1
A	8	0	1	8	0	0	0	0	−1
C	22	32.21	0	0	45.71	1	45.71	1	1
D	18	0	1	18	0	0	0	0	−1
E	7	13.5	0	0	45.71	0	0	1	1
F	55	48.71	1	6.29	48.71	1	48.71	1	0

Best upper bound = 114.5

Upper bound = 136.71
Best lower bound = 84.0
$\alpha = 1$
$t^n = 10.5429$

where $\bar{h}$ is the average demand (averaged over all nodes). Alternatively, we could have initialized $\lambda_i = 0$ for all nodes i. The choice above means that on the initial iteration, nodes with demands greater than the average demand will have $Z_i = 1$, while other nodes will have $Z_i = 0$. Also, by setting λ_i proportional to the demand at the node, we are more likely to select candidate sites that cover a lot of the real demand (as opposed to the demand as measured by λ_i). Tables 4.9 through 4.11 illustrate the first three iterations of the Lagrangian relaxation calculations, while Table 4.12 summarizes the entire procedure. Note that we can terminate the procedure when the difference between the lower and upper bounds is less than 1, since all of the demands are integer valued.

Appendix A is a user's guide to SITATION, a personal computer program that solves maximum covering location problems using both the greedy adding and substitution algorithm and Lagrangian relaxation.

Table 4.11 Third Iteration of Lagrangian Relaxation Calculations

Node	h_i	λ_i	Z_i	$(h_i - \lambda_i)Z_i$	$\Sigma_i a_{ij}\lambda_i$	X_j	$(\Sigma_i a_{ij}\lambda_i)X_j$	$\Sigma_j a_{ij}X_j$	$\Sigma_j a_{ij}X_j - Z_i$
A	10	10.54	0	0	31.63	1	31.63	1	1
B	8	10.54	0	0	31.63	0	0	1	1
C	22	21.67	1	0.33	24.63	0	0	0	−1
D	18	10.54	1	7.46	31.63	0	0	1	0
E	7	2.96	1	4.04	24.63	0	0	0	−1
F	55	48.71	1	6.29	48.71	1	48.71	1	0

Best upper bound = 98.4571
$\bar{h} = 20$

Upper bound = 98.4571
Best lower bound = 91.0
$\alpha = 1$
$t^n = 1.86429$

The Maximum Covering Location Model

Table 4.12 Summary of Lagrangian Relaxation Calculations

Iteration	Z_1-Z_6	X_1-X_6	LB	$\mathscr{L}^n$	Best UB	α^n	t^n
1	001001	110000	36	114.5	114.5	2	11.214
2	110101	001001	84	136.71	114.5	1	10.543
3	001111	100001	91	98.46	98.46	1	1.864
4	100111	001001	91	94.31	94.31	1	1.105
5	100111	100001	91	93.86	93.86	1	1.429
6	110111	100001	91	91.75	91.75	1	0.75

4.5.3 Other Solution Approaches and Example Results

The greedy adding and substitution algorithm and Lagrangian relaxation are heuristic approaches; they are not guaranteed to result in an optimal solution. The advantage of Lagrangian relaxation is that it provides an upper bound on the optimal solution from which we can often assess whether or not it is worth spending additional time and effort to find an optimal solution.

When an optimal solution is absolutely needed, branch and bound may again be used. Branching should only be performed on the location variables, X_j, since the coverage variables, Z_i, will be integer when the location variables are integer. However, we now have two ways of solving the problem

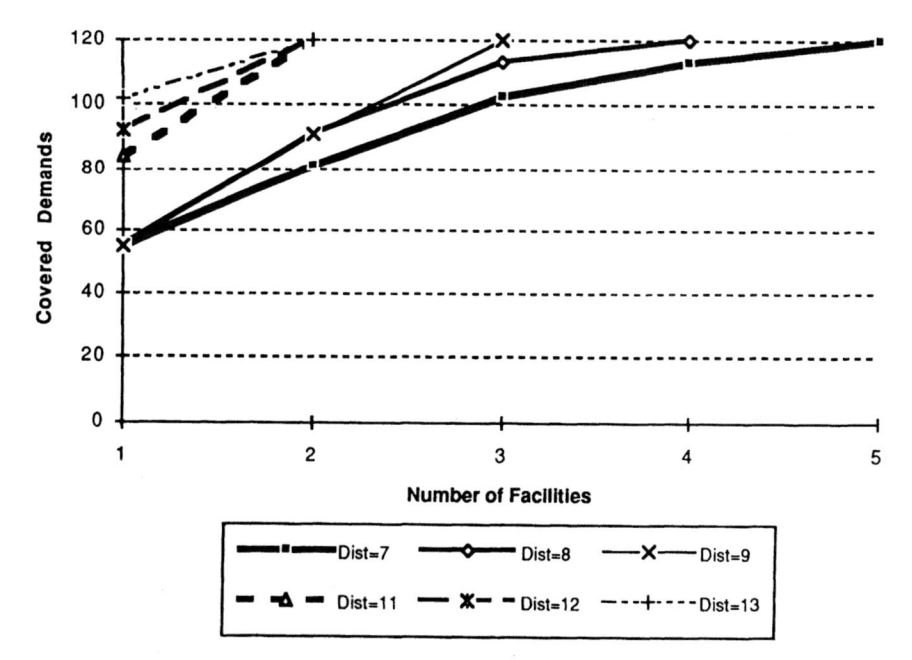

Figure 4.14. Coverage versus number of facilities located: Tradeoffs for various coverage distances.

at each of the nodes in the branch-and-bound tree. As before, we can relax the integrality constraints, impose upper-bound constraints of 1 on all of the coverage variables, Z_i, and solve the resulting linear programming problem. We can also relax constraints (4.3b) and solve the resulting Lagrangian relaxation problem, using the upper bound to bound the problem at each node. Adding constraints for $X_j = 0$ or $X_j = 1$ does not complicate the Lagrangian solution procedure. We simply preset these variables in the subproblem for the **X** variables and reduce the number of free (or not preset) **X** variables by the number that were forced to 1. If we are at a node in the branch-and-bound tree at which the number of **X** variables being forced to 1 equals P, the number of facilities to select, then we need only evaluate the original objective function at that node. This value is both a lower and an upper bound on the value of the problem at that node of the tree.

Figure 4.14 plots the optimal solution values to the maximum covering problem of Figure 4.7 for various coverage distances and number of facilities. Note that each curve exhibits decreasing marginal benefits of adding extra facilities. Also, as expected for a given number of facilities, the number of covered demands decreases (or does not increase) as the coverage distance decreases.

4.6 THE MAXIMUM EXPECTED COVERING LOCATION MODEL

In many cases, the facilities we are locating are subject to congestion or to being busy. In such instances, we are interested not only in locating facilities so that each demand is covered by one of the facilities being located, but is covered by a facility that is *available* when a demand for service arises. Thus, for example, in locating ambulances, we would like to know that one of the nearby ambulances is available when we call for service. This leads to the development of the *maximum expected covering location model* in which we locate a fixed number of facilities to maximize the expected number of demands that are covered by an available facility (Daskin, 1982, 1983).

Suppose that we can estimate a system-wide average probability that a facility will be busy when a demand for service arises. Let this probability be q. We further assume that the probability that a facility at node j_1 is busy is independent of the probability that a facility at some other node j_2 is busy. Now let us consider the probability that demands at node i are covered by an available facility *given* that we have located n_i facilities that could cover node i if they (the facilities) were available. Under these assumptions, this probability is given by

$$1 - q^{n_i}$$

since q^{n_i} gives the probability that all of the n_i facilities that can cover node i are busy. If we now locate an additional facility capable of covering node i,

The Maximum Expected Covering Location Model

the probability that the node is covered increases to

$$1 - q^{(n_i+1)}$$

The increase in coverage (or incremental coverage that results from adding one more facility that can cover node i) is given by the difference between these two terms or

$$\{1 - q^{(n_i+1)}\} - \{1 - q^{n_i}\} = q^{n_i}(1 - q)$$

Thus,

$$h_i\{q^{n_i}(1 - q)\}$$

gives the expected number of additional demands at node i that are covered by a nonbusy facility as a result of increasing the number of facilities that cover node i from n_i to $n_i + 1$.

With this background, we can now formulate the maximum expected covering location model using the following additional inputs and decision variables:

Inputs

q = system-wide average probability that a facility is busy

Decision Variables

X_j = number of facilities to locate at node j

$Z_{ik} = \begin{cases} 1 & \text{if demands at node } i \text{ are covered at least } k \text{ times} \\ 0 & \text{if not} \end{cases}$

Note that the location variable X_j has been redefined from a binary (0, 1) variable to a more general integer variable reflecting the fact that it may be optimal to locate several facilities at the same site. With these additional definitions, we can formulate the maximum expected covering location model as follows:

MAXIMIZE $\quad (1 - q) \sum_i h_i \left\{ \sum_{k=1}^{P} q^{k-1} Z_{ik} \right\} \quad$ (4.5a)

SUBJECT TO: $\quad \sum_k Z_{ik} \leq \sum_j a_{ij} X_j \qquad \forall\, i \quad$ (4.5b)

$\qquad\qquad\quad \sum_j X_j \leq P \qquad\qquad\qquad$ (4.5c)

$\qquad\qquad\quad X_j \geq 0 \text{ and Integer} \qquad \forall\, j \quad$ (4.5d)

$\qquad\qquad\quad Z_{ik} = 0, 1 \qquad\qquad\quad \forall\, k, i \quad$ (4.5e)

The objective function (4.5a) maximizes the expected number of covered demands. The term in the inner summation, $h_i\{\sum_{k=1}^{P} q^{k-1} Z_{ik}\}$, when multiplied by $(1-q)$, represents the expected number of covered demands at demand node i. When summed over all demand nodes i, we obtain the expected number of covered demands. Constraint (4.5b) states that node i can be counted as being covered at least k times (i.e., $Z_{ik} = 1$) only if at least k facilities are located at nodes that cover node i. Recall that the right-hand side of (4.5b) represents the number of facilities that are located that can cover node i. Constraint (4.5c) stipulates that at most P facilities are located. Constraint (4.5d) is the integrality constraint on the location variables. Again, note that the location variables must be nonnegative and integer, though they are not required to be binary. Constraint (4.5e) states that each demand node can be counted as being covered k times at most once.

Note that the coefficients of the objective function are decreasing functions of k for any node i since q is less than 1. Therefore, if $Z_{ik} = 1$, then $Z_{i,k-1} = 1$. Similarly, if $Z_{ik} = 0$, then $Z_{i,k+1} = 0$ as well. In other words, we are guaranteed that the Z_{ik} variables will be introduced into the solution in the appropriate order. We will never have a solution in which a node is counted as being covered k times, but the same node is not counted as being covered $k-1$ times. The reader should also note that, in this model, it is entirely reasonable to use a value of P, the number of facilities to locate, that exceeds the number required to cover all demand nodes exactly once. In other words, we may well want to locate more facilities for any given coverage distance than are indicated as being optimal in the set covering location model.

Most of the solution techniques outlined above [e.g., the greedy adding and substitution algorithm and branch and bound (branching only on the location variables with the coverage variables, Z_{ik}, constrained by upper-bound constraints of 1)] can be readily adapted to solve the maximum expected covering location problem.

To illustrate the solution to this problem, Table 4.13 presents the solution to the maximum expected covering location model for the network shown in Figure 4.7 with $D_c = 10$, $P = 3$, and $q = 0.6$. The solution is $X_D = 1$, $X_F = 2$, $Z_{A1} = 1$, $Z_{B1} = 1$, $Z_{D1} = 1$, $Z_{F1} = 1$, $Z_{F2} = 1$, and all other X_j and Z_{ik} variables are equal to 0. Note that two facilities are located at node F, and node F is covered twice, while nodes C and E are not covered at all. The objective function value is 49.6. This indicates that of the 120 demands, only 49.6 demands will be covered by an available facility within a coverage distance of 10 when three facilities are located and the probability of a facility being busy is 0.6. Note that if we ignore the probability of a facility being busy, we should locate three facilities with one at either node A, B, or D, a second facility at node C or E, and a third facility at node F. In that case, all six demand nodes will be covered exactly once.

We should note that the maximum expected covering location model requires that a number of rather strong assumptions be made. First, the

Table 4.13 Example Solution of the Maximum Expected Covering Location Model for the Network of Figure 4.7

Facility Node	Number of Facilities
A	0
B	0
C	0
D	1
E	0
F	2

Demand Node	Demand	Number of Times Covered	Contribution to Objective Function
A	10	1	4.0
B	8	1	3.2
C	22	0	0.0
D	18	1	7.2
E	7	0	0.0
F	55	2	35.2
Total	120		49.6

Solution for $P = 3$; $q = 0.6$

model assumes that the probability of a facility being busy is the same for all candidate facility sites. This system-wide average probability of a facility being busy can be computed from historical data using the following equation:

$$q = \frac{\text{Total number of hours all vehicles are busy during historical period}}{\left(\begin{array}{c}\text{Number of hours}\\\text{during historical period}\end{array}\right)\left(\begin{array}{c}\text{Number of vehicles on duty}\\\text{during historical period}\end{array}\right)}$$

While it may be relatively easy to estimate a value of q from historical data, in fact facilities are likely to differ dramatically in the fraction of time that they are busy responding to demands. The model also assumes that the probability of finding a facility busy is independent of the state of the other facilities (i.e., is independent of whether or not the other facilities are busy). This assumption is also not likely to be valid. Larson (1974, 1975) develops a hypercube queueing model which, while making a number of other assumptions, allows for interactions and dependencies in the probabilities of facilities being busy.

A number of authors (e.g., Goldberg et al., 1990a, 1990b; Goldberg and Paz, 1991; Goldberg and Szidarovszky, 1991a, 1991b) have proposed extensions of the maximum expected covering location model that attempt to capture the simplicity of the maximum expected covering location model and the more accurate modeling assumptions inherent in the hypercube queuing model of Larson. These modeling attempts, however, demand considerably greater computational effort in evaluating the objective function and result in nonlinear formulations. Other authors (e.g., ReVelle and Hogan, 1989; ReVelle and Marianov, 1991) have incorporated busy probabilities that depend on the demand nodes instead of the candidate facility locations.

In summary, a number of attempts have been made to relax some of the rather stringent assumptions inherent in the maximum expected covering location model. None has met with unqualified success. Extending deterministic covering models to account for facility availability is an ongoing and active research area.

4.7 SUMMARY

Many facility location problems are formulated in terms of demand coverage. Typically, demands at node i can be covered by a facility at node j if the distance (or travel time or travel cost) between nodes i and j is less than some exogenously specified value, D_c, which is known as the coverage distance. This chapter has outlined a number of variants of this problem.

In the simplest variant, we wanted to find the number and location of the minimum number of facilities so that each demand node was covered at least once. This problem is known as the location set covering model. We showed that the optimal solution to this problem may, in general, involve locating on the links of the network and not just on the nodes of the network. However, we can identify a finite number of additional points, known as network intersection points (Church and Meadows, 1979), such that at least one optimal solution to the set covering problem in which facilities can be located anywhere on the network consists of locating facilities on the nodes or network intersection points. This allowed us to talk about the problem in terms of a finite discrete set of candidate locations. (We often impose the additional restriction that facilities be located only on the demand nodes.) Row and column reduction techniques were outlined that allow us to force candidate sites either into or out of the solution. Often, these rules allow us to solve the problem completely. If they do not, we can solve a linear programming relaxation of the resulting integer programming problem. If the solution is all integer, we have solved the problem. If it is not, we outlined a branch-and-bound technique for solving the problem.

The solution to the set covering problem is often not unique. We formulated a number of extensions to the problem that attempt to capture other objectives. These objectives included (i) the need or desire to provide backup coverage to demand nodes to account for the possibility of facilities being busy or (ii) the desire to locate as many facilities as possible at the locations of existing facilities.

Exercises

Two important limitations of the location set covering model were identified. First, the model tends to require more facilities than are permitted by a budget. Second, the model fails to differentiate between high demand nodes and low demand nodes; all demand nodes are treated identically in the location set covering model. To resolve these limitations, we formulated the maximum covering location problem whose objective is to locate a fixed number of facilities to maximize the number of demands (as opposed to demand nodes) that are covered at least once.

Three solution approaches were outlined for the maximum covering location problem: a greedy adding and substitution heuristic, a heuristic based on Lagrangian relaxation, and branch and bound. The greedy adding and substitution heuristic is perhaps the easiest approach to understand and to implement. However, we do not get any information from this approach about how close (or far) the solution is to (or from) an optimal solution. Lagrangian relaxation provides a lower bound on the solution. Computational experiments with the Lagrangian relaxation approach suggest that the solutions obtained from this algorithm, when it is followed by a substitution procedure, are very close to provably optimal solutions. Also, the solutions to this approach tend to be slightly better than those obtained from the greedy adding and substitution algorithm.

To account for the possibility of facilities being busy when called upon to serve demands, we formulated the maximum expected covering location model. Most of the solution approaches outlined for the maximum covering location model can readily be modified for use in solving the maximum expected covering location model. The chapter concluded with a number of references to other extensions of the maximum covering and maximum expected covering location models.

EXERCISES

4.1 For the network of Figure 4.2, show that only one facility is needed to cover all demands if the coverage distance is 19 or more. At what two nodes can the facility be located?

4.2 Consider the network shown in Figure E4.2.

(a) Solve the set covering location problem with all fixed costs equal to 1 ($f_j = 1$ for all j) and a critical distance of 7 ($D_c = 7$). Assume that facilities may only be located on the nodes of the network. Give the set of (node) locations that cover all other nodes. Also, clearly show which candidate facility sites are dominated by which other sites.

(b) Resolve the set covering location problem with all fixed costs equal to 1 ($f_j = 1$ for all j) and a critical distance of 10 ($D_c = 10$).

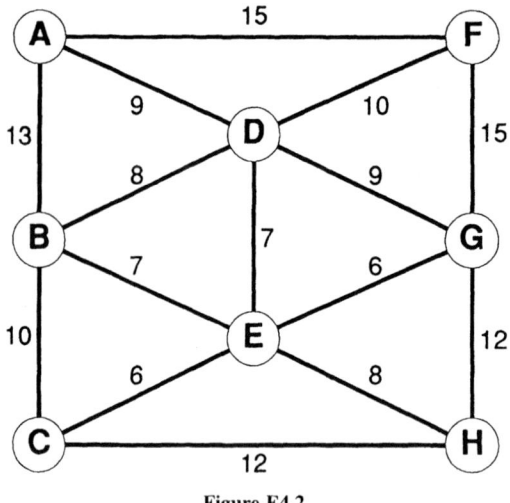

Figure E4.2

Again, assume that facilities may only be located on the nodes of the network. Give the set of (node) locations that cover all other nodes. Also, clearly show which candidate facility sites are dominated by which other sites.

4.3 Consider the following network:

(a) Write out the set covering problem formulation for this network, if the coverage distance is 25. Recall that facilities can cover demands at nodes that are not directly linked to the node at which

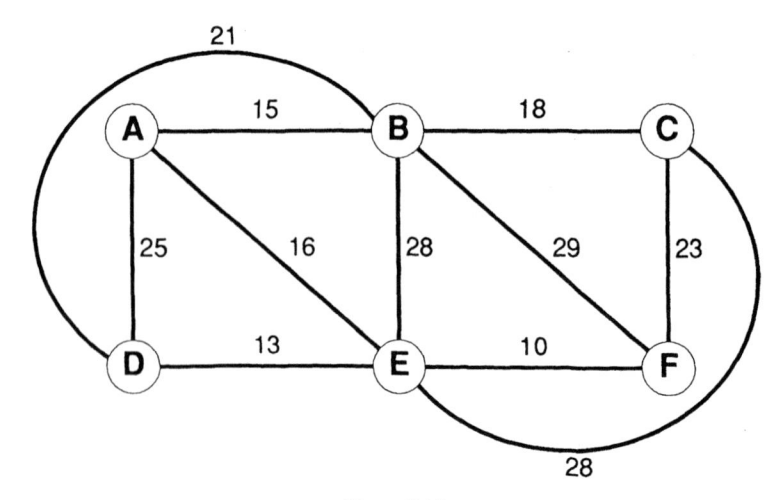

Figure E4.3

the facility is located as long as the shortest path distance between the facility and the demand node is less than or equal to the coverage distance.

(b) With a coverage distance of 25, *clearly* indicate which, if any, locations can be eliminated because they are dominated by other locations (and specify the identity of the dominating node for each dominated node). Also, *clearly* indicate which, if any, rows of the constraint matrix may be eliminated and indicate why they may be eliminated.

(c) Based on the reduced problem that you obtain in part (b), solve the set covering problem (with a distance of 25). *Justify* your answer either in words or by stating how you would solve the optimization problem that results from part (b).

(d) Identify all alternate optima for the problem.
   Hint: Go back to the original formulation in part (a); this should help you answer this question.

4.4 Consider the following network:
   (a) Solve for the set covering solution with $D_c = 20$. Be sure
      (i) to show clearly the initial constraint matrix,
      (ii) to explain why any rows and/or columns can be removed,
      (iii) to indicate which nodes are forced into the solution,
      (iv) to state how you solve the problem once all possible row and column reductions have been done, and
      (v) to show where you actually locate facilities.
   (b) Repeat part (a) using $D_c = 18$.

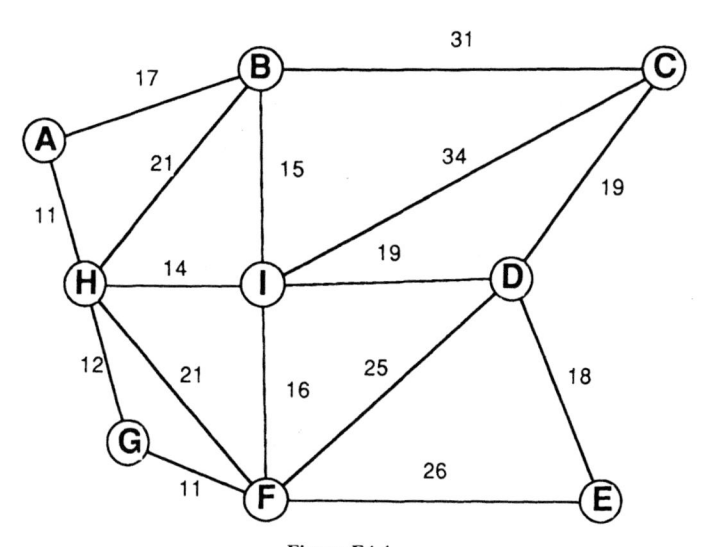

Figure E4.4

4.5 (a) For the following network, write out the set covering problem formulation if the coverage distance is 19.

(b) Solve the set covering problem for this network with a coverage distance of 19. Clearly indicate which, if any, locations can be eliminated because they are dominated by other locations (and specify the identity of the dominating node for each dominated node). Also, clearly indicate which, if any, nodes must be in the solution and why they must be included.

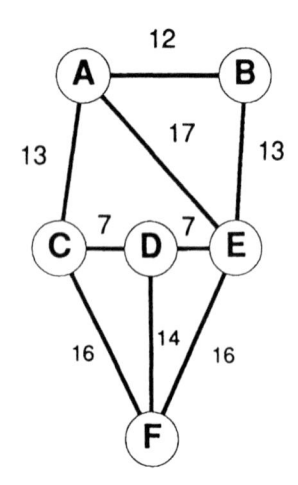

Figure E4.5

4.6 (a) For the following network, write down the objective function and constraints for a set covering problem (with facilities located only on the nodes) using a coverage distance of 10.

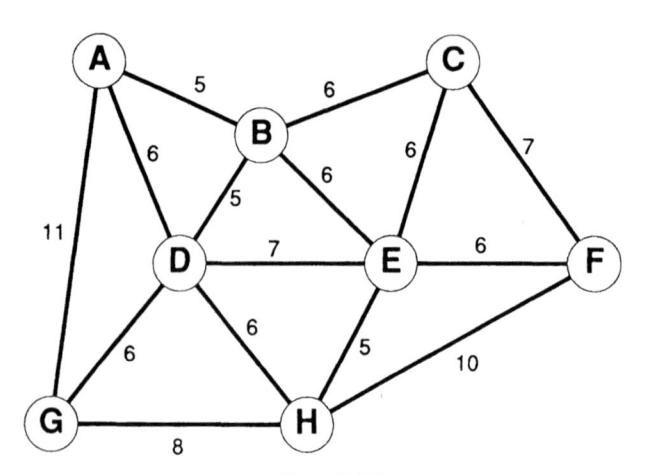

Figure E4.6

Exercises 139

(b) Solve the set covering problem that you formulated. Clearly indicate which locations are dominated by which other locations. Also, clearly indicate how many facilities are needed and where they should be located.

4.7 Consider the following network:

(a) Solve the set covering location problem with all fixed costs equal to 1 ($f_j = 1$ for all j) and a critical distance of 14 ($D_c = 14$). Assume that facilities may only be located on the nodes of the network. Give the set of (node) locations that cover all other nodes. Also, clearly show which candidate facility sites are dominated by which other sites.

(b) Now consider the set covering location problem with all fixed costs equal to 1 ($f_j = 1$ for all j) and a critical distance of 30 ($D_c = 30$). Again, assume that facilities may only be located on the nodes of the network.
  (i) Clearly show which candidate facility sites are dominated by which other sites.
  (ii) Clearly indicate which rows of the constraint matrix can be eliminated.
  (iii) Solve the resulting problem as a *linear programming* problem. Note that the problem should be very small and you should be able to solve it by *hand* if you do not have any other linear programming solver available.

  Note: Solving this problem may be more difficult than solving part (a) was. First, you must consider cases in which nodes are covered by facilities that are not directly linked to the facility node. For example, demands at node *J* can be covered by a facility at node *A* with a coverage distance of 30.

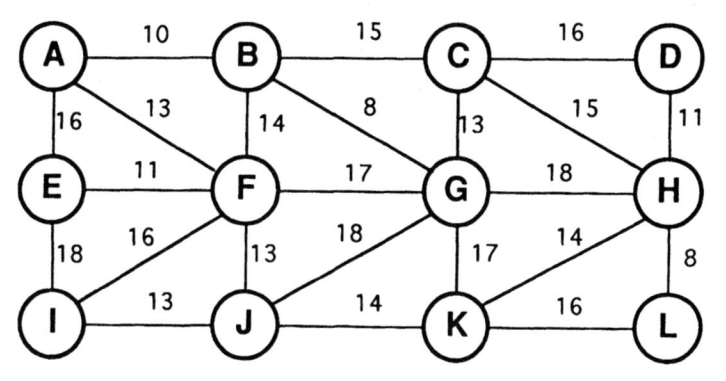

Figure E4.7

Second, if you do this right, the resulting linear programming problem will not give you an all-integer solution.

(iv) Find an all-integer solution to the problem. Explain any additional constraints that you add to the linear programming relaxation to force the solution to give you an all-integer solution.

4.8 Consider the following network. Numbers beside each node enclosed in a box (e.g., $\boxed{10}$) are the demands associated with the node.

(a) Write out the objective function and constraints for the set covering model for this network when the coverage distance is 18. Assume that facilities can only be located on the nodes of the network.

(b) Which candidate sites can be excluded from the formulation? Which rows (corresponding to the need to cover specific demands) can be excluded? In both cases, justify your answer.

(c) After you have reduced the problem as suggested in part (b), are there any sites at which you must locate facilities? If so, where? Why? If any sites are now forced into the solution, which demand nodes are now covered?

(d) Write out the remaining covering problem. That is, write out the constraint matrix for the remaining candidate sites and uncovered demand nodes.

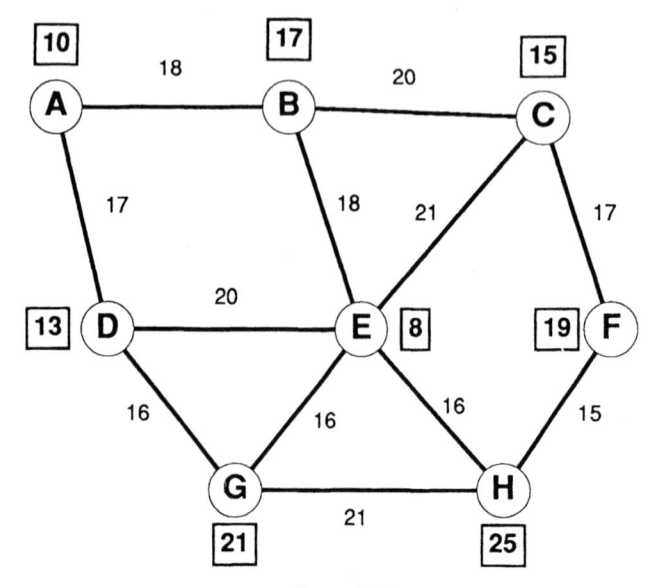

Figure E4.8

Exercises

(e) What is the linear programming relaxation solution to the original problem? Note that you should be able to solve for this using the matrix you obtain in part (d).

(f) What is the solution to the (integer programming) set covering problem? Specifically, where do you locate facilities and what is the objective function value?

4.9 As indicated in Section 4.4, and as should be evident from Exercise 4.8, there are often multiple alternate optima for the set covering location problem. This suggests that we can append secondary objectives to the problem to select from among the alternate optima to the primary objective problem (that of minimizing the number of facilities needed to cover all demands) a solution that best attains some secondary objective.

(a) For the network shown in Exercise 4.8, write out as many alternate optima as possible to the set covering problem with a coverage distance of 18.

(b) Formulate the following problem:

Primary Objective: MINIMIZE the number of facilities needed to cover all demand *nodes* at least once with a coverage distance D_c^1

Secondary Objective: MAXIMIZE the number of *demands* (as opposed to nodes) that are covered at least twice with a coverage distance D_c^2 (which may be different from D_c^1)

Be sure to define all inputs and decision variables clearly. Separate inputs from decision variables. State the objectives and constraints in both words and using notation.

(c) Suggest a means of solving this problem.

(d) Solve the problem for the network shown in Exercise 4.7 using coverage distances of $D_c^1 = 18$ and $D_c^2 = 18$.

4.10 (a) Write out the formulation of the maximum covering problem for the network shown in Exercise 4.8 using a coverage distance of 18 if we are to locate P facilities.

(b) For $P = 1$, what is the optimal location? Which demand nodes are covered? What is the total covered demand

(c) For $P = 2$, what is the solution that you obtain using the greedy adding algorithm (*without* substitution)? Again, which demand nodes are covered? What is the total covered demand?

(d) What is the optimal solution to the maximum covering problem for $P = 2$?

Note: For this exercise as well as Exercises 4.8 and 4.9, you may want to input the problem into the SITATION software. This can be done using NET-SPEC.

4.11 Consider the following network. Link distances are shown beside the links and the demands are shown in boxes beside each node.

(a) Use the greedy adding algorithm to solve the maximum covering problem for this network with a coverage distance of 34 and a single facility. Assume you can only locate on the nodes of the network. In particular, where do you locate? How many demands are covered? Which nodes are covered and which are not?

(b) For a coverage distance of 34 use the greedy adding algorithm (*without* substitution) for the maximum covering problem to complete the following table:

Number of Facilities	Facility Locations	Total Covered Demand	Covered Demand Nodes
1			
2			
3			
4			
5			

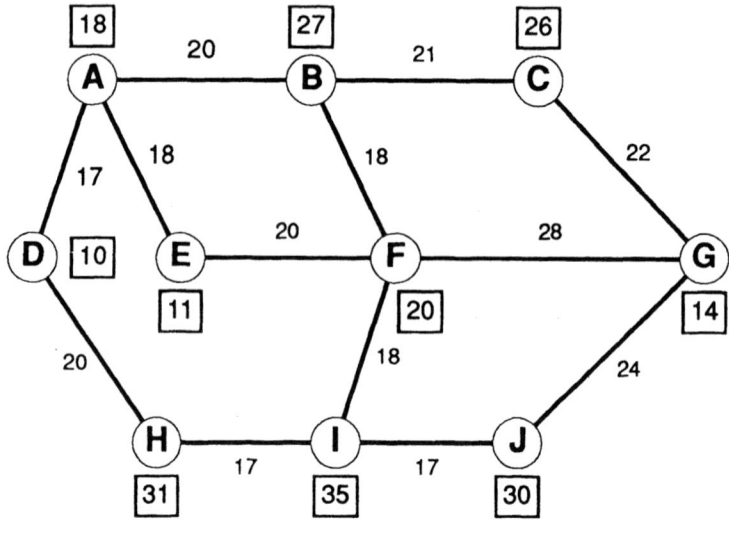

Figure E4.11

Exercises

Notes: 1. You may want to set up a table showing how much demand each candidate location covers.
2. Remember that even once a demand node is covered it may still be considered as a candidate facility site.
3. You may not need all rows of the table.

4.12 (a) For the network shown above in Exercise 4.11, write out the set covering problem formulation for a coverage distance of 34.

(b) For the problem formulated in part (a), can you eliminate any columns (e.g., force out any nodes)? Can any nodes be forced in? Can any rows be eliminated? After forcing in and out any nodes that can be forced in this way and eliminating any rows, find an optimal solution to the problem.

(c) What if anything does your solution tell you about the results you obtained in part (b) of Exercise 4.11?

The vertex P-center problem is the problem of finding P locations on the nodes of the network to minimize the maximum distance between any demand node and the nearest facility to it. The objective function of the P-center problem is to minimize this maximum (implied coverage) distance. This model is discussed in detail in Chapter 5. However, you should be able to answer the following two questions:

(d) What, if anything, does your solution tell you about the objective function for the 3-center problem?

(e) What, if anything, does your solution tell you about the objective function for the 2-center problem?

4.13 (a) Use the SITATION program with the 49-node data set (SORT-CAP.GRT) and the first set of demands representing state populations to find the tradeoff between the number of facilities needed to cover all demands and the coverage distance. In particular, plot a bar chart showing the number of facilities needed to cover all demands for coverage distances of $200, 250, \ldots, 600$ miles.

(b) Find the smallest (integer) distance such that all demands can be covered by five facilities. Where are these facilities located? (Give both the node numbers and the city names which can be obtained from Appendix H.)

4.14 In this exercise, you will use the SITATION program and the Lagrangian relaxation algorithm for solving the maximum covering location problem that is built into that package to do a simple branch-and-bound analysis. The data set for the problem is the 88-node problem for the United States given in Appendix G. This data set is called CITY1990.GRT.

One of the purposes of this exercise is to explore the implications of different solution procedures on the quality of the solution. In general, the Lagrangian procedure will give better results than those obtained by the greedy adding algorithm (or the greedy adding and substitution algorithm). This, however, is not always the case.

(a) For a coverage distance of 300, use the greedy adding algorithm with substitution allowed after each iteration (excluding dominated sites) to find the (approximate) maximum number of demands that can be covered by locating three facilities. Use the first demand data set. How many demands are covered in total? Where does the algorithm suggest locating facilities? *Note that you can get the city names from the list of names in Appendix G. The city numbers on the data set and those in the table are the same.*

(b) Now do the same thing, but use the Lagrangian relaxation algorithm. Again, exclude dominated sites and allow substitution at the end of the procedure. Let the model locate three facilities. (Do not change any of the Lagrangian options on the OPTION SETTING MENU.) Again, how many demands are covered in total? Where does the algorithm suggest locating facilities? How does the total covered demand in this case compare with:

   (i) the upper bound on the coverage provided by the Lagrangian procedure and

   (ii) the coverage found by the greedy adding and substitution algorithm in part (a)?

(c) Verify that the three sites found in part (b) must be optimal. Do so by using the upper bounds available from the Lagrangian relaxation algorithm when you exclude each of these three sites *one by one* from the solution. Note that you will have to use the Force Specific Sites In/Out of Soln. option that is available on the main menu to force each site *in turn* out of the solution and then run the Lagrangian relaxation procedure. In all you will need to run the Lagrangian procedure a number of different times. *Clearly show* the upper bound obtained from each of these runs. *Briefly* explain why these runs show that the optimal solution is to locate at the three nodes you found in part (b).

(d) Return to the MAIN OPTION MENU and be sure that all candidate sites are allowed in the solution. (To do so, use the **P** option from the FORCE NODES OPTION MENU.) Now generate the covering tradeoff curve using the **T** option from the MAIN OPTION MENU. Allow substitution after each iteration and exclude dominated nodes.

   (i) How many facilities does the heuristic think are needed to cover all 88 nodes?

   (ii) Looking carefully at the tradeoff curve developed by the algorithm and at the incremental coverages, which of the solutions are likely to be suboptimal and why?

Exercises

(e) Based on what you have done so far and learned so far about the differences between the greedy adding and substitution algorithm and Lagrangian relaxation, find the *true minimum* number of facilities needed to cover all demands. Note that to do so, you will need to change some of the options on the Lagrangian OPTION SETTING MENU that you are shown when you run the Lagrangian relaxation algorithm. You may also need to force some nodes into or out of the solution. *In short, this will require considerable experimentation and work.* Clearly document how you obtained the optimal solution and how you can show that fewer facilities cannot cover all of the demands.

4.15 In general, we would expect that the tradeoff curve of the percentage of the total demand covered versus the number of facilities located will be concave. In other words, we generally expect that we will have decreasing marginal or incremental coverage. However, this is not always the case.

In this exercise, you will solve a series of maximum covering problems using SITATION. You will use the 88-node data set, CITY1990.GRT, described in Appendix G. In addition, you will limit the candidate sites to the capitals of the 48 states in the continental United States plus Washington, DC. You will use a coverage distance of 550 miles. The easiest way to do this is to:

(i) *Load CITY1990.GRT* (by pressing **L** at the MAIN OPTION MENU, **G** at the DISTANCE OPTIONS MENU, typing **CITY1990.GRT** as the file name to read, **F** at the DEMAND OPTIONS MENU, and **0.001** when prompted for the cost per mile);

(ii) *Limit the candidate sites to those in the file CAPITALS.ONL* (by pressing **F** at the MAIN OPTION MENU, **R** at the FORCE NODES OPTION MENU, **CAPITALS.ONL** when prompted for the file name to read, and **Q** at the FORCE NODES OPTION MENU to return to the MAIN OPTION MENU);

(iii) *Set the coverage distance to 550 miles* (by typing **C** at the MAIN OPTION MENU and then **550** when prompted for the coverage distance).

(a) Find the optimal solution to the maximum covering problem locating three facilities. Do so using the Lagrangian algorithm. Set the Lagrangian options as follows to force the algorithm to obtain very tight bounds:

Critical percentage difference	0.0000001
Maximum number of iterations	2000
Minimum α value allowed	0.0000001
Number of failures before changing α	12

In addition, allow substitution at the end of the Lagrangian algorithm and exclude dominated nodes. What are the values of the lower and upper bounds on the coverage? Where do you locate facilities? Is the solution provably optimal?

(b) Find the optimal solution to the maximum covering problem locating five facilities. Use the same parameter values for the Lagrangian problem that you used in part (a). What are the values of the lower and upper bounds on the coverage? Where do you locate facilities? Is the solution provably optimal?

(c) If you did parts (a) and (b) correctly, you should have obtained provably optimal solutions (since all of the demands are integer valued). Assuming the tradeoff curve does exhibit decreasing marginal coverage, what is the minimum coverage that we must attain with four facilities?

(d) Use the Lagrangian procedure to solve the problem locating four facilities. Again, use the same Lagrangian parameters you used in part (a). What are the lower and upper bounds on the coverage? Where do you locate the four facilities? Is the solution provably optimal? How many demands are covered by each of the four facilities?

(e) Exclude the node that covers the most demand in the four-facility solution found in part (d) from the solution (by pressing **F** at the MAIN OPTION MENU, **E** at the FORCE NODES OPTION MENU, the **node number** when prompted for the node to exclude, and **Q** at the FORCE NODES OPTION MENU).

Run the Lagrangian algorithm again locating four facilities. What are the lower and upper bounds on the coverage now? Comparing the bounds to the value you computed in part (c), what can you conclude about whether the node you excluded must be in or out of the solution if the tradeoff curve is to exhibit decreasing marginal coverage? Why?

(f) Force the node that covers the most demand in the four-facility solution found in part (d) into the solution (by pressing **F** at the MAIN OPTION MENU, **F** at the FORCE NODES OPTION MENU, the **node number** when prompted for the node to include, and **Q** at the FORCE NODES OPTION MENU).

Run the Lagrangian algorithm again locating four facilities. Now what are the lower and upper bounds on the coverage? Comparing the bounds to the value you computed in part (c) and using the knowledge you obtained in part (e), what can you conclude about whether the tradeoff curve exhibits decreasing marginal coverage? Why?

(g) In parts (d), (e), and (f) you essentially walked through part of a branch-and-bound tree. Draw the tree showing the lower and upper bounds on the problem in each node of the tree and the facility location that was forced into or out of the solution on each link of the tree.

4.16 The SITATION program plots a tradeoff curve showing the percentage of the total demand that is covered as a function of the number of facilities that are located. In many practical decision contexts, we would also like to know the tradeoff between the minimum number of facilities that are needed to attain a given level of service (e.g., 90 percent coverage) as a function of the coverage distance. In other words, we would like a tradeoff curve that plots the number of facilities needed on the Y axis and the coverage distance on the X axis for a specified level of service.

(a) For a level of service of 100 percent (i.e., total coverage), find the tradeoff curve between the minimum number of facilities needed and the coverage distance. Use distances of 100, 200, 300, 400, 500, 600, and 700 miles with the 88-node data set, CITY1990.GRT (and the first set of demands representing the 1990 city populations). Note that for a level of service of 100 percent you are implicitly using the maximum covering algorithm to solve a set covering problem.

(b) Repeat part (a) using a level of service of 90 percent.

(c) Repeat part (a) using a level of service of 80 percent.

Hints for parts (a), (b), and (c): For each coverage distance in question, first use the ability of the *SITATION* software to plot a tradeoff curve between the percentage of the total demand that is covered and the number of facilities located. These results should give you an *upper bound* on the number of facilities needed to attain any given level of service for the specified coverage level. For example, with a coverage distance of 300, this approach suggests you will need at *most* 16 facilities for total (100 percent) coverage, 8 facilities for 90 percent coverage, and 6 facilities for 80 percent coverage. Since the total demand in the data set is 44,840,571, to attain 90 percent coverage we need to cover at least 40,356,514 demands and to attain 80 percent coverage we need to cover at least 35,872,457 demands.

Next, use the Lagrangian relaxation approach to find the true minimum number of facilities that are needed for any given level of service and coverage distance. Note that the upper bound on the coverage provided by the Lagrangian procedure should be used here. If this upper bound (for a given coverage distance and a specified number of facilities) is less than the number of demands that must be covered (given in the paragraph above), then you need more than that number of facilities to cover the necessary number of demands.

Additional important suggestion: You may need to change the Lagrangian default parameters for the critical percentage difference between the lower and upper bounds, the minimum α value, the number of failures before changing α, and the maximum number of iterations to the values indicated in part (a) of Exercise 4.15. This is very important for some of the Lagrangian runs that you will have to make.

(d) Plot the results you obtained in parts (a), (b), and (c) using a good spreadsheet program.

(e) Briefly explain how the results obtained in parts (a) through (d) of this exercise (or similar results) might be useful to a firm in the clothing business in locating warehouses that supply retail stores by truck deliveries from the warehouses. In particular, how might these results be used in determining the number of warehouses, the level of service to provide to the stores, and the fraction of the stores that should be promised this level of service? To solve this problem for such a firm, what additional information would you want to have?

4.17 On Monday, January 17, 1994, an earthquake of magnitude 6.6 on the Richter scale hit Los Angeles killing over 30 people and causing major damage to buildings, roads, and other parts of the urban infrastructure. In partial response to this, President Clinton ordered a 60-vehicle mobile emergency response unit to be dispatched from Denver to Los Angeles. Included in the response unit were telecommunications and electric generation facilities.

(a) Of critical concern in such a situation is the distance between the responding units and the emergency. Also, we are concerned that the units themselves not be damaged in the emergency. Use the SITATION software and the 88-node data set (CITY1900.GRT described in Appendix G) to find the minimum number of facility bases needed to ensure that all demand nodes are covered within a distance of 500 miles. Assume that a facility base at a particular city *cannot* cover emergencies in that city since there is some reasonable likelihood that the vehicles in the affected city will be disabled. Thus, cities in which a facility base is located must be covered by a facility in some *other* city.

How many facilities are needed? Where are facilities located? Which cities cover the cities in which facilities are located?

Hint: To solve this problem, you will need to modify the MDST88.GRT distance file to reflect the fact that a facility in city i *cannot* cover city i. Use MOD-DIST to make the necessary changes. Be sure to replace the revised MDST88.GRT file with the original file which MOD-DIST will probably rename as MDST88.G00 after you are finished doing this exercise and Exercise 4.18.

(b) Print a map showing the locations of the facilities and the assignments of cities to facilities.

(c) How many more facilities are needed for the solution to the problem as posed in part (a) than would be needed if we allowed facilities based in city i to cover city i itself?

4.18 The director of FEMA (the Federal Emergency Management Agency) feels that the number of facilities you have found in Exercise 4.17 is excessive. Find the smallest integer coverage distance such that FEMA can get by with only four facility bases.

Note: You may need to change some of the default conditions on the Lagrangian procedure here and you may also need to walk through the branch-and-bound process by forcing sites in and out of the solution.

Clearly show how you obtained the solution you did. In other words, discuss what values of the coverage distance you tried and how you had to solve each case. Under the restriction that you locate only four facilities, where should they be located if the coverage distance is to be minimized and all demands are to be covered?

4.19 Throughout most of the discussion in this chapter related to the maximum covering model, we have implicitly assumed that all facilities have identical costs. In particular, it is that assumption that allowed us to limit the number of facilities that are located. In many contexts, however, construction costs differ depending on the location. In addition, there might be an explicit cost associated with not covering demands.

Let f_j be the cost of building a facility at candidate location j. Also, let B be the maximum amount that can be spent on building facilities. Finally, let φ_i be a unit cost of not covering demands at demand node i.

(a) Formulate the problem of minimizing the combined cost of locating facilities and not covering demands. Clearly define any additional notation that you use. Also, clearly state the constraints and the objective function both mathematically and in words.

Note: You should not have a constraint like constraint (4.3c) which limits the number of facilities to P. The optimal number of facilities will be determined endogenously by this model.

(b) Relax an appropriate constraint and formulate a Lagrangian problem related to the one you formulated in part (a). (You really should have only one candidate constraint to relax.)

(c) For fixed values of the Lagrange multipliers in the problem you formulated in part (b), clearly state how you would solve the Lagrangian optimization problem. In other words, how would you find the optimal values of the location variables, X_j, and the coverage variables, Z_i?

(d) For fixed values of the Lagrange multipliers, how can you extract or construct a feasible solution to the problem of part (a) from the solution you obtained in part (c)? Note that there may well be more than one way to do this. Determining the "best" way is likely to require extensive computational experiments.

4.20 In the maximum covering problem, we assumed that all facilities cost the same amount to build. We also did not worry about demands that could not be covered. Often, these are not good assumptions. First, in many cases, it makes sense to consider a model in which we minimize the sum of the construction costs (e.g., the construction costs for fire stations) and a penalty cost for not covering demands in the desired time, T^0. Assume that the costs and penalties (f_j and p_i) are defined in commensurable terms. Second, we may require that *all* demands be covered within a time T^1, where $T^1 > T^0$.

Formulate such a model using the following:

Inputs

T^0 = desired coverage time (demands not covered within this time will be penalized in the objective function)

T^1 = required coverage time ($T^1 > T^0$ and all demands must be covered by at least one facility within time T^1)

t_{ij} = travel time from demand node i to candidate site j

$$a_{ij}^0 = \begin{cases} 1 & t_{ij} \leq T^0 \\ 0 & \text{if not} \end{cases}$$

$$a_{ij}^1 = \begin{cases} 1 & t_{ij} \leq T^1 \\ 0 & \text{if not} \end{cases}$$

h_i = demand at node i

f_j = fixed cost of locating a facility at node j

p_i = penalty cost per unit demand not covered at demand node i

Decision Variables

$$X_j = \begin{cases} 1 & \text{if a facility is located at candidate site } j \\ 0 & \text{if not} \end{cases}$$

$$Y_i = \begin{cases} 1 & \text{if demand node } i \text{ is covered} \\ & \text{by at least one facility within time } T^0 \\ 0 & \text{if not} \end{cases}$$

Formulate the following problem:

MINIMIZE

Total facility location costs + Penalty costs

SUBJECT TO:

Relationship between coverage within T^0 and location decisions

All demands covered within T^1

Integrality

4.21 You are faced with locating ambulances in a city. You would like to find the locations of a fixed number of ambulance bases. Let P be the number you want to locate. You would like to maximize the number of demands (demands occur on nodes of the network) that can be covered within 5 minutes of the nearest ambulance. However, you require that the selected sites be such that *all* demand nodes can be covered by an ambulance within 10 minutes.

(a) Using the notation we have defined for the set covering and maximum covering problems as well as any other notation you require, formulate this problem as an integer linear programming problem. Clearly define any new notation that you use. Clearly state in words what each of the constraints and the objective function is doing.

(b) What can you say about the minimum value of P (the number of facilities to be located) which is an *input*, such that the problem will have a feasible solution?

4.22 Consider the following problem. You are faced with locating fire stations in a city. You would like to find the locations of a fixed number of stations. Let P be the number you want to locate. You would like to maximize the number of demands that can be covered within 4 minutes of the nearest station. Demands occur on the nodes of the network. You require that the selected sites be such that 90 percent the demand *nodes* can be covered by an ambulance within 8 minutes.

(a) Using the notation defined in the text for the set covering and maximum covering problems as well as any other notation you require, formulate this problem as an integer linear programming problem. Clearly define any new notation that you use. Clearly state in words what each of the constraints and the objective function is doing.

(b) Qualitatively describe how the objective function (maximizing the number of demands that can be covered within 4 minutes) will change if we change the requirement that 90 percent of the demand nodes be covered within 8 minutes to a requirement that 90 percent be covered within 10 minutes. Justify your answer briefly.

4.23 In many facility location contexts, it is important to have more than one facility able to cover demands at a node. Thus, in an inventory system that is subject to frequent stockouts, we might like to have two warehouses capable of replenishing any store in a timely manner. In emergency services such as fire departments, the need for multiple coverage should be clear.

Suppose we want to find the locations of P facilities to maximize the number of demands that are covered *by at least two facilities*. Use the following notation to show how this problem can be formulated. Also indicate the indices over which summations apply and the indices over which each constraint set applies (e.g., for all i).

Inputs

h_i = demand at node i

$a_{ij} = \begin{cases} 1 & \text{if a facility at candidate site } j \text{ can cover demands at node } i \\ 0 & \text{if not} \end{cases}$

P = number of facilities to locate

Decision Variables

X_j = number of located at candidate site j. Note that this can be any integer number greater than or equal to 0

$Y_{i1} = \begin{cases} 1 & \text{if demand node } i \text{ is covered at least } once \\ 0 & \text{if not} \end{cases}$

$Y_{i2} = \begin{cases} 1 & \text{if demand node } i \text{ is covered at least } twice \\ 0 & \text{if not} \end{cases}$

Hint: In words the problem may be formulated as follows:

MAXIMIZE

Total number of demands covered at least twice

SUBJECT TO:

Locate at most P facilities

Integrality

And at least one other constraint set which you should define in both words and notation.

4.24 Consider the following variant of the maximum expected covering model. In this case, we want to maximize the number of demands that can be covered within the coverage distance D_c with a probability of at

Exercises

least Q^0, when the probability of finding an individual vehicle busy is q. In other words, the probability of at least one *available* vehicle being able to cover the demand at a node, given that K vehicles are located within a distance D_c of the node, is, as before, $1 - q^K$. For the node to be covered with a probability of Q^0, we need this quantity to be at least Q^0.

(a) For $q = 0.25$ and $Q^0 = 0.98$, find the minimum value of K (i.e., the minimum number of facilities needed to be located within D_c of a node for the node to be covered with probability 0.98).

(b) Using the notation defined in Section 4.6, formulate the problem of locating P vehicles (on the demand nodes of the network) so that we maximize the number of demands that are covered with a probability of at least 0.98 (when $q = 0.25$). State the constraints and objective function in words as well as mathematical notation.

5

Center Problems

5.1 INTRODUCTION

In the covering problems discussed in Chapter 4, the coverage distance between a demand and the nearest facility was specified exogenously. In the set covering model, we attempted to find the locations of the minimum number of facilities needed to cover all demand nodes. We found that in many practical contexts the number of facilities needed to cover all demand nodes within the exogenously specified distance was prohibitively large. In addition, the set covering model failed to account for the fact that the demands at the nodes differ. To alleviate these problems, we formulated and discussed the maximum covering location problem. In that model, we associated a demand level with each demand node and found the locations of a fixed number of facilities to maximize the number of covered demands. In essence, in the maximum covering location model, we relaxed the requirement that all demand nodes be covered.

In this chapter we adopt a different strategy for addressing the shortcomings of the set covering model. In particular, as in the set covering model, we still require that all demands be covered. Now, however, instead of using an exogenously specified coverage distance and asking the model to minimize the number of facilities needed to cover all demand nodes, we will ask the model to *minimize the coverage distance such that each demand node is covered within the endogenously determined distance by one of the facilities.* The model is known as the *P-center problem* or a *minimax* problem, since we are minimizing the maximum distance between a demand and the nearest facility to the demand. Figure 5.1 summarizes the relationship between the set covering, maximum covering, and center problems.

We need to distinguish between problems in which the facilities can be located anywhere on the network (i.e., on the nodes and on the links of the network) and problems in which facilities can be located only on the nodes of the network. The former category of problems, those in which facilities can be located anywhere on the network, are known as *absolute center problems*. Problems in which the facilities can only be located on the nodes of the network are known as *vertex center problems*. As in the case of the set covering model, we can readily show that the solution to the absolute center

Introduction

```
┌─────────────────────────────────────────────┐
│              SET COVERING                   │
│               PROBLEM                        │
│                                              │
│   GIVEN:  Demand Nodes                      │
│           Candidate Sites                    │
│           Demand Node to Candidate Site     │
│             Distances                        │
│           Coverage Distance                  │
│                                              │
│   FIND:   Minimum Number (and Location of)  │
│           Sites to Cover ALL Demand Nodes   │
└─────────────────────────────────────────────┘
```

OBSERVATIONS: (a) Many Sites Often Needed
(b) Need To Relax Problem Specifications

Relax Total Coverage Requirement → **MAXIMUM COVERING PROBLEM**

GIVEN: Set Covering Inputs PLUS
Number to Locate, P
Demand Levels at Nodes

FIND: Locations of P Facilities to MAXIMIZE Number of Covered Demands

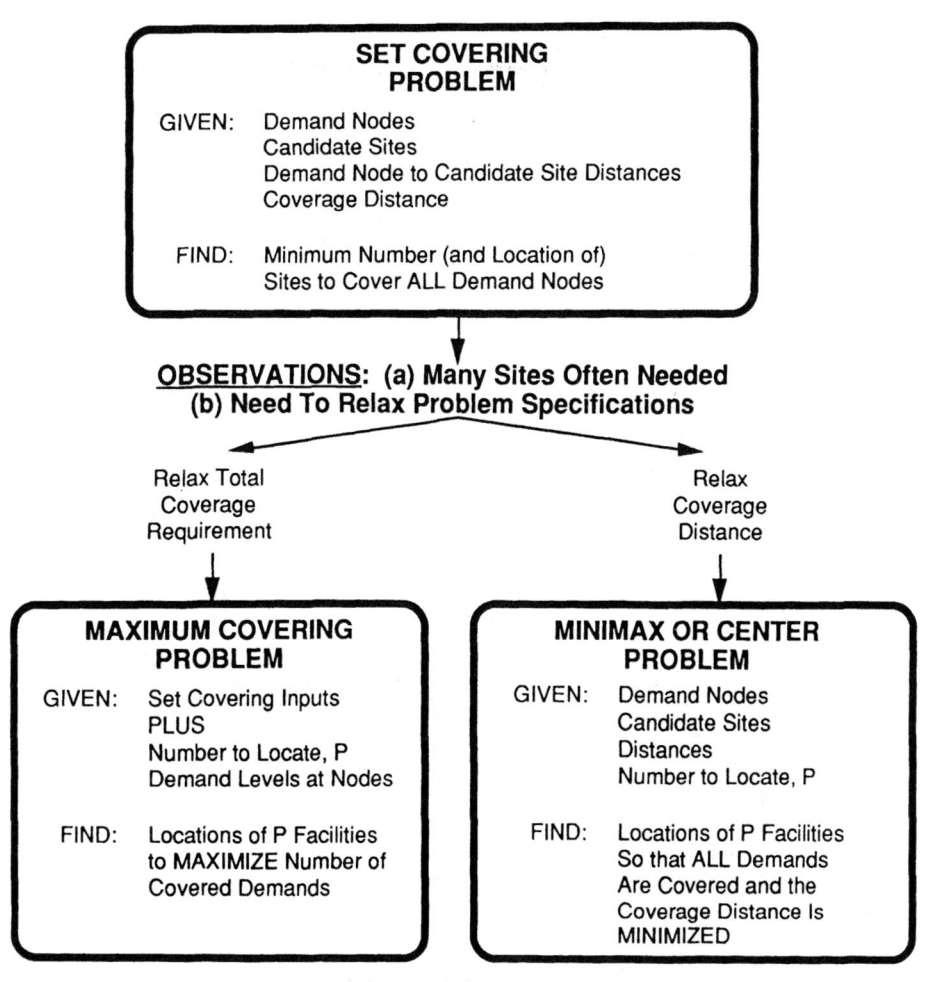

Relax Coverage Distance → **MINIMAX OR CENTER PROBLEM**

GIVEN: Demand Nodes
Candidate Sites
Distances
Number to Locate, P

FIND: Locations of P Facilities So that ALL Demands Are Covered and the Coverage Distance Is MINIMIZED

Figure 5.1. Relationships among the set covering, maximum covering, and center problems.

problem may be better than the solution to the vertex center problem. Consider Figure 5.2 (which is identical to Figure 4.1). If facilities can be located only on the nodes (as in the vertex center problem) and we can locate only one facility ($P = 1$), either node is optimal and the maximum distance from the other demand to the facility is 8. However, if we can locate anywhere on the network, locating the single facility midway between nodes

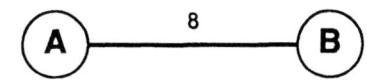

(A) ——8—— (B)   **Figure 5.2.** Example network illustrating suboptimality of nodal locations.

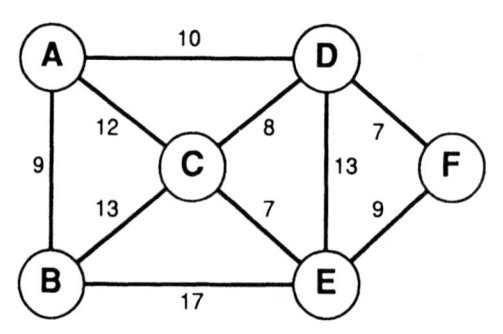

Figure 5.3. Example network to show the relationship between the set covering and P-center problems.

A and B is optimal and the maximum distance from either demand node to the facility is only 4.

To illustrate further the relationship between the set covering model and the P-center problem, consider the network shown in Figure 5.3. Table 5.1 shows solutions to the set covering model (in which we constrain facilities to be only on the nodes) for all relevant coverage distances from under 7 (when six facilities are needed) to a distance of 15 (when only one facility is needed). These results are also shown in Figure 5.4. The solutions to the vertex P-center problem for various values of P, the number of facilities to locate, may be found at the *left* end of the five horizontal lines in Figure 5.4. For example, for $P = 6$, one facility is located on each of the nodes and the value of the vertex P-center problem objective function is 0. For $P = 4$, four facilities are located (at nodes A, B, D, and E for example). The maximum distance between a facility and a demand node is 7, the distance between the facility at node E and demand node C (or between the facility at node D and demand node F).

Note that Figure 5.4 does not include a line representing five facilities. What would be the solution to the P-center problem with $P = 5$? The answer is that the solution would be the same as that obtained for $P = 4$. In

Table 5.1 Solutions to Set Covering Problem of Figure 5.3 for Various Coverage Distances

Coverage Distance	Number of Facilities	Sample Locations
< 7	6	A, B, C, D, E, F
7, 8	4	A, B, C, F
9	3	A, C, F
10–14	2	A, E
15	1	C

Introduction

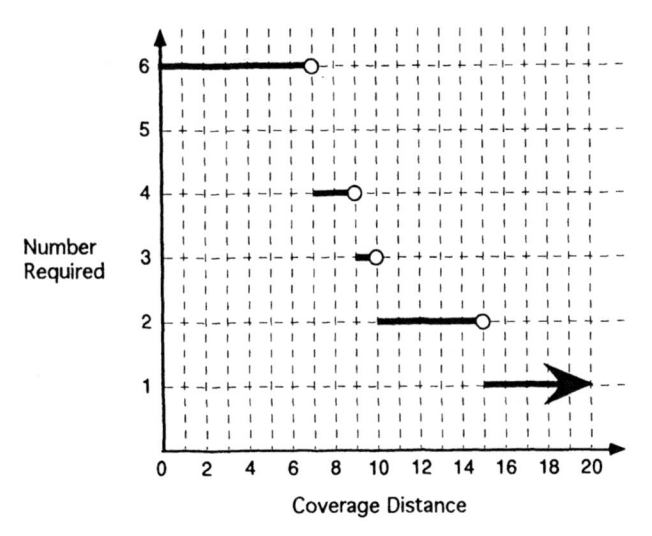

Figure 5.4. Number of facilities needed versus coverage distance for Figure 5.3.

Table 5.2 Solutions to *P*-Center Problem of Figure 5.3 for Various Numbers of Facilities

Number of Facilities, P	Coverage Distance	Sample Locations
1	15	C
2	10	A, E
3	9	A, D, E
4	7	A, B, D, E
5	7	A, B, D, E, ?
6	0	A, B, C, D, E

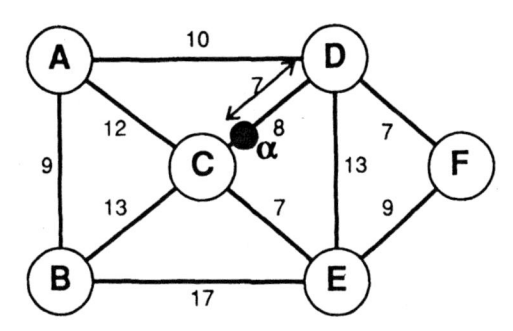

Figure 5.5. Optimal solution to the absolute 1-center problem for Figure 5.3.

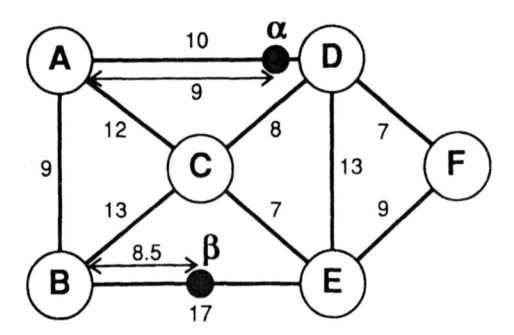

Figure 5.6. Optimal solution to the absolute 2-center problem for Figure 5.3.

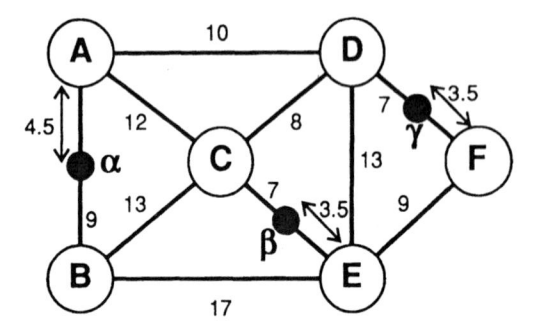

Figure 5.7. Optimal solution to the absolute 3-center problem for Figure 5.3.

this case, adding a fifth facility that must be located on the nodes does not allow us to reduce the maximum distance between a demand node which does not have a facility located on it and the nearest facility.

Table 5.2 summarizes the P-center results for this problem. Note that the coverage distance in this problem is the value of the objective function and is determined *endogenously*.

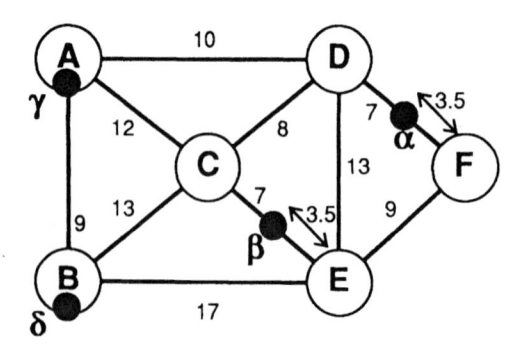

Figure 5.8. Optimal solution to the absolute 4-center problem for Figure 5.3.

Introduction

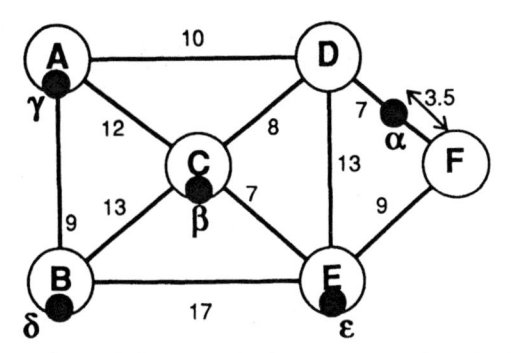

Figure 5.9. Optimal solution to the absolute 5-center problem for Figure 5.3.

Finally, by way of introduction, we again note that the solution can often be improved significantly if facilities can be located on the links as well as the nodes. Figures 5.5 through 5.9 show the optimal solutions for one through five facilities when facilities can be located anywhere on the network shown in Figure 5.3. Facility locations are shown by black dots and are denoted by Greek letters (α, β, γ, δ, and ε). In all cases, the facility at node α is the one whose distance to a demand node defines the absolute P-center objective

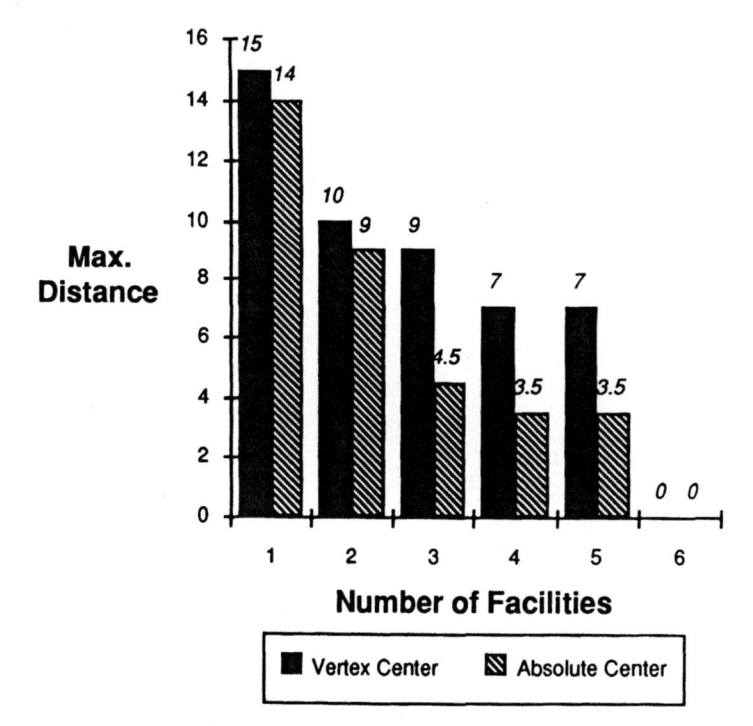

Figure 5.10. Comparison of absolute and vertex P-center solutions for Figure 5.3.

function value. Figure 5.10 compares the objective function values for the vertex and absolute center problems for the network of Figure 5.3.

5.2 VERTEX P-CENTER FORMULATION

In this section we formulate the vertex P-center problem. We define the following notation:

Inputs

d_{ij} = distance from demand node i to candidate facility site j
h_i = demand at node i
P = number of facilities to locate

Decision Variables

$X_j = \begin{cases} 1 & \text{if we locate at candidate site } j \\ 0 & \text{if not} \end{cases}$

Y_{ij} = fraction of demand at node i that is served by a facility at node j
W = maximum distance between a demand node and the nearest facility

With this notation, we can formulate the vertex P-center problem as follows:

MINIMIZE W (5.1a)

SUBJECT TO: $\sum_j Y_{ij} = 1$ $\forall i$ (5.1b)

$\sum_j X_j = P$ (5.1c)

$Y_{ij} \leq X_j$ $\forall i, j$ (5.1d)

$W \geq \sum_j d_{ij} Y_{ij}$ $\forall i$ (5.1e)

$X_j = 0, 1$ $\forall j$ (5.1f)

$Y_{ij} \geq 0$ $\forall i, j$ (5.1g)

The objective function (5.1a) minimizes the maximum distance between a demand node and the closest facility to the node. Constraints (5.1b) state that all of the demand at node i must be assigned to a facility at some node j for all nodes i. Constraint (5.1c) stipulates that P facilities be located. Constraints (5.1d) state that demands at node i cannot be assigned to a facility at node j unless a facility is located at node j. Constraints (5.1e) state

Vertex P-Center Formulation

that the maximum distance between a demand node and the nearest facility to the node (W) must be greater than the distance between *any* demand node i and the facility j to which it is assigned. Constraints (5.1f) and (5.1g) are the integrality and nonnegativity constraints, respectively. Note that while we do not explicitly require the allocation variables Y_{ij} to be integers, in the optimal solution they will naturally be integers (or we can find an alternate optimum in which they are all integers). This is so because the facilities are not capacitated and there is no reason to allocate demands at node i to a facility other than the closest one.

In some cases, we want to consider the demand-weighted distance. In that case, constraint (5.1e) can be replaced by

$$W \geq h_i \sum_j d_{ij} Y_{ij} \quad \forall\, i \qquad (5.1e')$$

While this generalization of the problem is of mathematical interest, in most applied contexts, if we are interested in the P-center problem we want to minimize the worst case level of service of the system, independent of the number of demands that experience that level of service.[1]

For fixed values of the number of facilities to locate, P, both the absolute and vertex P-center problems may be solved in polynomial time. To see this for the vertex P-center problem, all we need to realize is that for a network with n nodes, we need only evaluate each of the $O\binom{N}{P} = O(N^P)$ possible combinations of P facility sites. Evaluating each of these can be done in polynomial time, so finding the optimal solution to the vertex P-center problem, for *fixed* values of P, can be done in polynomial time. As indicated below in Section 5.5, we can identify a finite polynomial number of points to add to the network such that the solution to the absolute P-center problem will consist of locating on a subset of the original nodes and/or the additional points. Thus, the absolute P-center problem can also be solved in a polynomial amount of time for fixed values of P.

For a general graph and for variable values of P, the P-center problem is NP-complete (Kariv and Hakimi, 1979a; Garey and Johnson, 1979, pp. 219–220). This is true for both the vertex and absolute P-center problems. To see why this is so, note that the time required to solve all vertex P-center problems by enumeration for $P = 1$ to $P = N$ for any given value of N is

$$\sum_{j=1}^{N} \binom{N}{j} = 2^N - 1 = O(2^N)$$

---

[1] Since the solution techniques for the P-center problem depend in part on whether or not we want to solve a demand-weighted or unweighted version of the problem, we will retain this distinction throughout this chapter.

which is exponential in N. The fact that this problem is NP-complete means that there is no known polynomial time optimal algorithm for solving these problems on a general network. However, when the network is a tree, we can often find optimal solutions in polynomial time to problems which are NP-complete on more general graphs. This is the case here. Section 5.3 outlines a solution approach to the absolute 1-center and absolute 2-center problems on a tree. Section 5.4 summarizes a solution approach to the vertex P-center problem on a general graph. Section 5.5 presents characteristics of the solution to the absolute P-center problem on a general graph and outlines a solution procedure for absolute P-center problem on a general graph.

5.3 THE ABSOLUTE 1- AND 2-CENTER PROBLEMS ON A TREE

While the absolute P-center problem is NP-complete on a general graph, if we are only locating a single facility on a tree, there are very efficient algorithms for solving the problem optimally. In this section we outline a number of those approaches. We begin by considering the absolute 1-center problem on a tree in which all of the demands are equal. We call such a tree an *unweighted* tree. Since all of the demands are equal, we can normalize them so that they are all equal to 1. We then show how that algorithm can be extended to the case in which we want to find two centers on an unweighted tree. This section concludes by considering the case of a single absolute center on a weighted tree.

5.3.1 Absolute 1-Center on an Unweighted Tree

To find the absolute 1-center on an unweighted tree, we can use the following algorithm:

Step 1: Pick any point on the tree and find the vertex that is farthest away from the point that was picked. Call this vertex e_1.

Step 2: Find the vertex that is farthest from e_1 and call this vertex e_2.

Step 3: The absolute 1-center (X^*) of the unweighted tree is at the midpoint of the (unique) path from e_1 to e_2. The vertex 1-center of the unweighted tree is at the vertex of the tree that is closest to the absolute 1-center. (Note that there may be two vertices that satisfy this condition if the absolute 1-center is at the midpoint of a link.)

We can illustrate this algorithm using Figure 5.11.

We begin by picking any point on the tree. Suppose we pick node C. We then find the distances between the point (node C) and all vertices in the tree. Figure 5.12 shows the result of this calculation (with the distances shown in small squares beside the node letters). Node G is the farthest node

The Absolute 1- and 2-Center Problems on a Tree

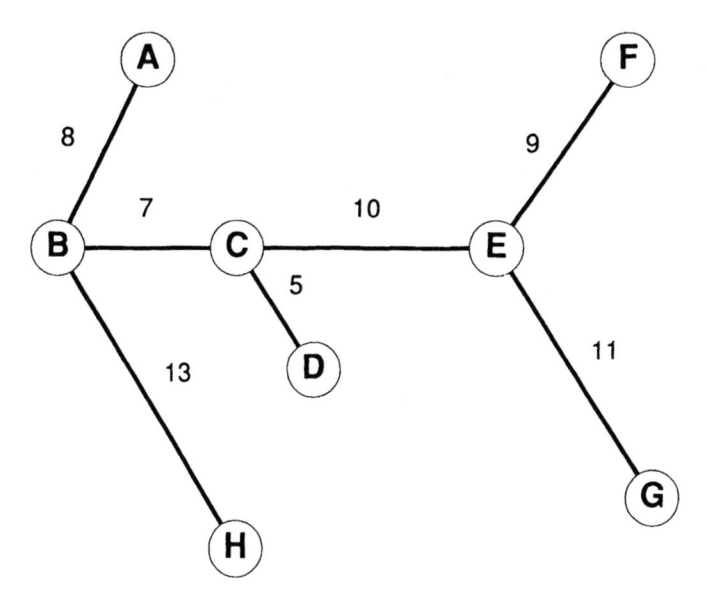

Figure 5.11. Example tree.

from node C, and so we let node G be e_1. Figure 5.13 shows the distance from node G to all other nodes in the tree. Node H is the farthest node from node G, and so we call node H e_2. The absolute 1-center is at the point midway between nodes G and H on the unique path from node G to node H, or 9.5 units from node E on the link connecting nodes C and E, as shown in Figure 5.14. The objective function equals 20.5, the maximum distance

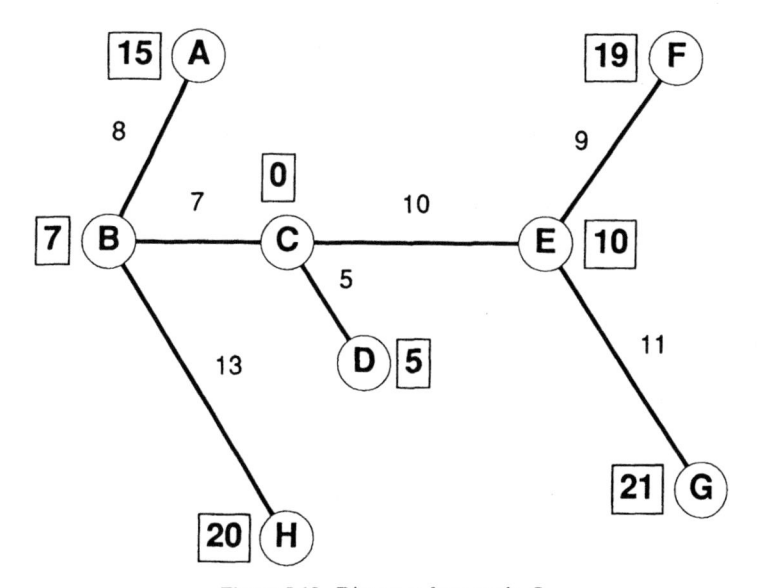

Figure 5.12. Distances from node C.

164 Center Problems

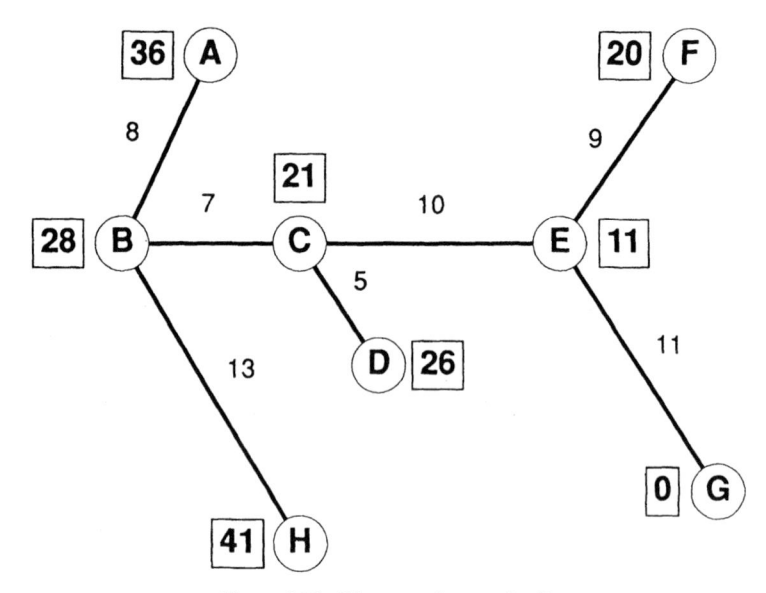

Figure 5.13. Distances from node G.

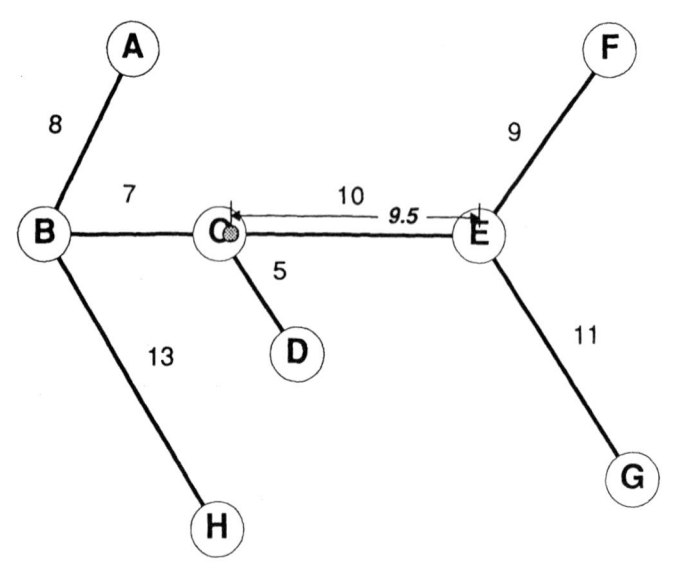

Center is located 9.5 units from node E on the link from node C to node E

Figure 5.14. Absolute 1-center of the tree.

from the absolute 1-center to either node G or node H. The vertex 1-center of the tree would be at node C (found by moving the absolute center to the nearest vertex if it is not already located on a vertex). The objective function for the vertex 1-center problem on the tree is 21.

While finding shortest path distances from a given node to all other nodes in a general network requires $O(n^2)$ work, on a tree the shortest path distances from any node to all other nodes may be found in $O(n)$ time, where n is the number of nodes in the graph or tree. In step 1 of the algorithm, we need to find the shortest path distances from a node selected at random to every other node. This requires $O(n)$ time. Similarly, step 2 requires $O(n)$ time since there, too, we are finding the shortest path from a node (node e_1) to all other nodes. Finally, step 3 involves enumerating the path from e_1 to e_2 which can also be done in $O(n)$ time. Thus, the entire algorithm requires $O(n)$ time. In other words, the worst case execution time of the algorithm increases only *linearly* with the number of nodes in the tree.

Having demonstrated the operation of the algorithm, we now prove that the point X^* found by the algorithm is optimal. To do so, we need only prove that there is no point e_3 that is farther from X^* than either e_1 or e_2. We will prove that no such point exists by assuming that such a point *does* exist and showing that this assumption leads to a contradiction of known information.

If a point e_3 exists which is farther from X^* than either e_1 or e_2, then there are two cases to consider. In the first case, node e_3 is on the same side of X^* on the unique path from e_1 to e_2 as is e_2, the point identified in step 2 of the algorithm above. If this is the case, $\alpha < \beta$ as shown in Figure 5.15. where node e_4 is the node at which the path from X^* to e_2 diverts to node e_3, α is the distance between nodes e_2 and e_4, and β is the distance between nodes e_3 and e_4. However, if node e_3 is farther from e_4 than is node e_2, then it is also farther from e_1 than is e_2. Therefore, node e_2 cannot be the farthest node from e_1 as required by step 2 of the algorithm.

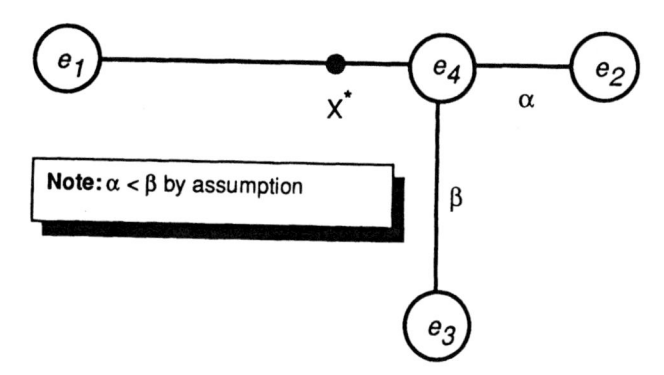

Figure 5.15. Case 1: Node e_3 is farther from X^* than is node e_2.

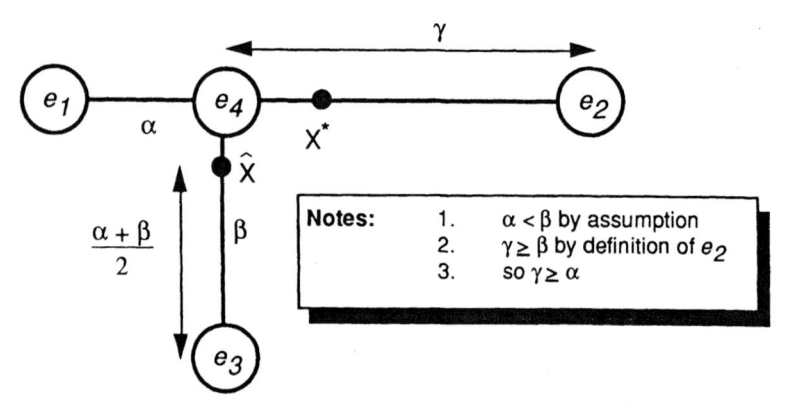

Figure 5.16. Case 2: Node e_3 is farther from X^* than is node e_1.

The second case we need to consider is that there exists a point e_3 which is farther from X^* than either e_1 or e_2 and that point e_3 is on the same side of X^* as is node e_1, the node found in step 1 of the algorithm. This situation is illustrated in Figure 5.16. For any initial point anywhere on the tree except on the path from e_3 to $\hat{X}$, node e_3 is farther than node e_1, which contradicts the definition of node e_1 (i.e., that is the farthest node from the initial node). Similarly, for any initial point on the path from e_3 to $\hat{X}$, node e_2 is farther than e_1, which means that for any initial point, node e_1 would not have been the farthest node from the initial point.

These two cases both lead to contradictions. Since there are no other possible cases, we have proven that there are no points that are farther from X^* than either e_1 or e_2 and by construction these two points are equidistant from X^*.

In a similar manner we can prove that the location X^* is unique. That is, we can assume that it is not unique and show that this assumption leads to a series of contradictions. This is left as an exercise to the reader.

5.3.2 Absolute 2-Centers on an Unweighted Tree

To solve the absolute 2-center problem on an unweighted tree, we can modify the algorithm outlined in Section 5.3.1 as follows:

Step 1: Using the algorithm for the absolute 1-center of Section 5.3.1, find the absolute 1-center.

Step 2: Delete from the tree the arc containing the absolute 1-center. (If the absolute 1-center is on a node, delete one of the arcs incident on the center which is on the path from e_1 to e_2.) This divides the tree into two disconnected subtrees.

Step 3: Use the absolute 1-center algorithm to find the absolute 1-center of each of the subtrees. These locations constitute a solution to the absolute 2-center problem.

The Absolute 1- and 2-Center Problems on a Tree

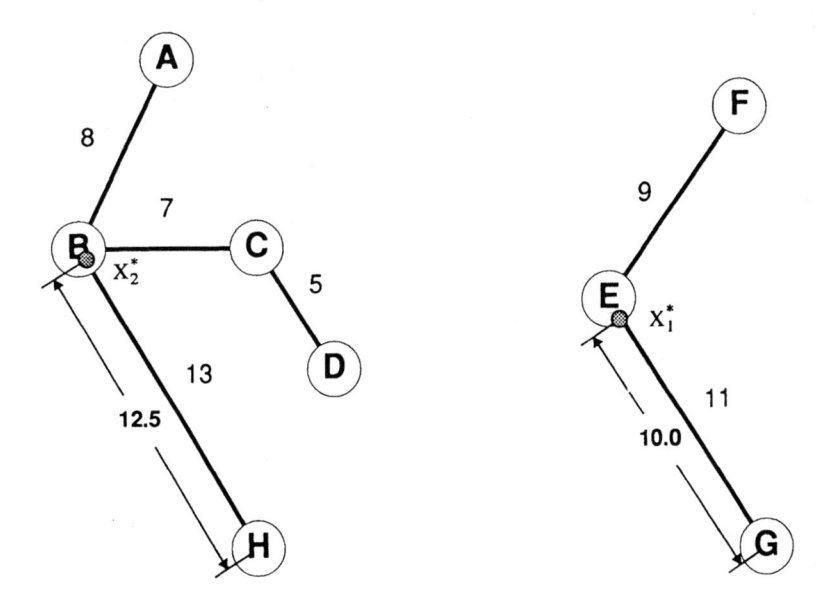

Center 1 is located 10 units from node G on the link from node E to node G.
Center 2 is located 12.5 units from node H on the link from node B to node H.

Figure 5.17. Absolute 2-centers of the tree.

To illustrate this algorithm, consider again the tree shown in Figure 5.11. As shown in Figure 5.14, the absolute 1-center lies on link *CE*. Removing this link results in the two trees shown in Figure 5.17. After applying the absolute 1-center algorithm to each of these trees, we obtain the locations X_1^* and X_2^* as shown. The value of the objective function is 12.5 (the larger of the objective functions for the two individual subtrees). This corresponds to the distance between center X_2^* and nodes *D* and *H*.

5.3.3 Absolute 1-Center on a Weighted Tree

We now turn to the problem of locating a single absolute center on a *weighted* tree or a tree in which the weights associated with each of the nodes are not equal.[2] To motivate the approach to solving this problem, consider the simple network shown in Figure 5.18. In the weighted case, we want to minimize the maximum demand-weighted distance between the facility being located and the nodes. In the simple case of Figure 5.18, we will locate at a point *X* units from node *A* such that

$$3X = 2(10 - X)$$

---

[2] Recall that while this problem is of academic interest, in many practical contexts, if we are concerned with minimizing the worst level of service experienced by any customers in the network, we will not be interested in weighting the nodes differentially.

$h_A = 3$ (A) ——10—— (B) $h_B = 2$

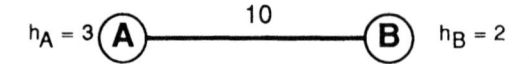

Figure 5.18. Simple network for the weighted problem on a tree.

$h_A = 3$ (A) ——10—— (B) $h_B = 2$
       ← 4 →
Maximum demand-weighted distance = 12

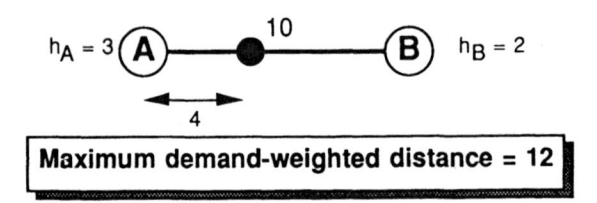

Figure 5.19. Solution to the simple network for the weighted problem on a tree.

or $X = 4$. That is, we locate 4 distance units from node A. The demand-weighted distance between the facility and either node is then 12. This solution is shown in Figure 5.19. Note that the solution involves locating closer to the larger demand node.

For a tree with only two nodes, computing the absolute 1-center is easy. For a more complicated tree, the problem becomes slightly more difficult. Consider the tree shown in Figure 5.20. In this case, it is not immediately clear whether the center should be on the path between nodes A and B, the path between nodes A and C, or the path between nodes B and C. One way to approach this problem is to try each of the three pairs of nodes and then evaluate the impact of the trial location on the third node. For example, if we locate at a distance X from node A on the path between nodes A and B, then we must solve the following problem:

$$10X = 6(20 - X)$$

The solution is $X = 7.5$. Nodes A and B are both 75.0 weighted distance

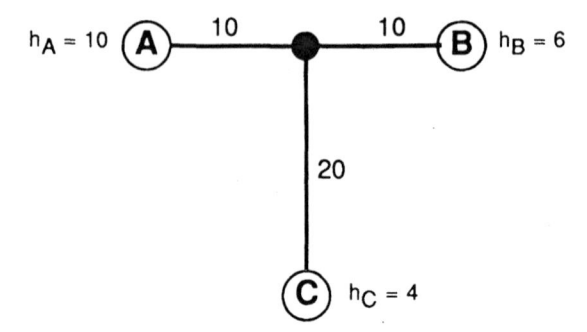

$h_A = 10$ (A) ——10—— • ——10—— (B) $h_B = 6$
                        |
                        20
                        |
                        (C) $h_C = 4$

Figure 5.20. A more complicated weighted tree.

The Absolute 1- and 2-Center Problems on a Tree

units from the center. Node C, however, is 22.5 distance units away and has a weight of 4. Thus, it contributes 90 to the objective function. This indicates that the optimal location cannot be on the path between nodes A and B. If we now locate on the path between nodes A and C, we must solve the following equation:

$$10X = 4(30 - X)$$

The solution is $X = 8.571428$. Both nodes are a weighted distance of 85.71428 units from the center. Node B is 11.42857 units away and its demand-weighted distance is 68.571428. Since this is less than either node A or node C contribute, this location is a candidate for the optimal location. Finally, if we consider the path from B to C, we must solve

$$6X = 4(30 - X)$$

The solution is $X = 12$. Both nodes B and C are 72 weighted distance units from the candidate center. However, node A is 120 demand-weighted distance units away. Thus, this candidate location is not the optimal location.

In summary, the optimal location is 8.571428 units from node A on the path from A to C. The value of the objective function is 85.71428.

Clearly, this approach would be overly tedious for a more general network. However, we can generalize the approach as follows. Consider two nodes i and j. If the center is located on the path from i to j, then we must solve the following *pair* of equations for the location X on the path between the two nodes:

$$h_i d(i, X) = h_j d(j, X)$$

and

$$d(i, X) + d(X, j) = d(i, j)$$

Substituting the second equation into the first, we obtain

$$h_i d(i, X) = h_j [d(i, j) - d(i, X)]$$

Solving for $d(i, X)$, we obtain

$$d(i, X) = \frac{h_j d(i, j)}{(h_i + h_j)}$$

Finally, if we were to locate at this location, nodes i and j would both be

$$\frac{h_i h_j d(i, j)}{(h_i + h_j)}$$

demand-weighted distance units from the center located $d(i, X)$ units from node i on the path from i to j. If we were to compute this value for *every pair* of nodes in the tree and select the pair that results in the *largest* value, we would find the optimal solution. This approach is summarized as follows:

Compute:

$$\beta_{ij} = \frac{h_i h_j d(i,j)}{(h_i + h_j)}$$

for every pair of nodes i and j.

Find: $\beta_{ST} = \max_{ij}(\beta_{ij})$ and let S and T be the nodes which correspond to β_{ST}.

Locate: At a point $[h_T/(h_S + h_T)] d(S, T)$ from node S on the unique path from S to T or, equivalently, at a point $[h_S/(h_S + h_T)] d(S, T)$ from node T on the unique path from S to T.

This approach involves computing $O(n^2)$ terms β_{ij}, where n is the number of nodes in the tree. We can simplify the approach by computing only a portion

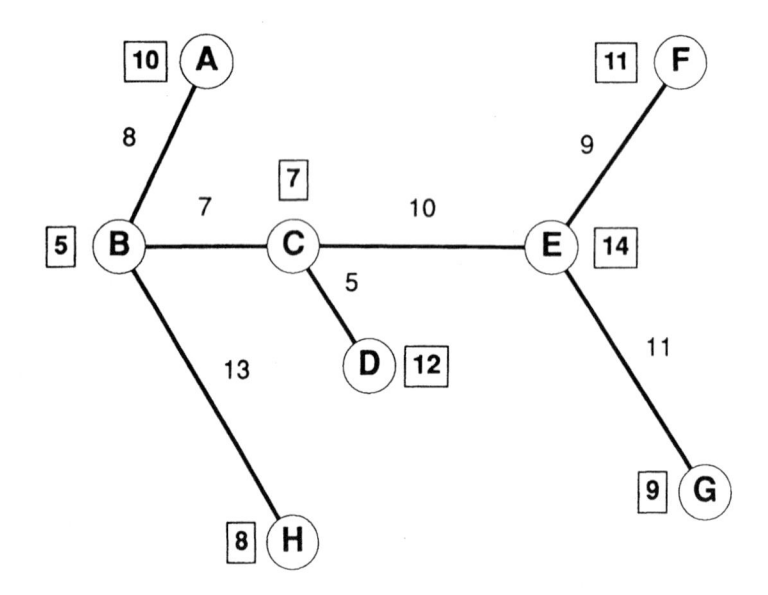

Figure 5.21. A weighted tree.

The Absolute 1- and 2-Center Problems on a Tree

	To							
From	**A**	**B**	**C**	**D**	**E**	**F**	**G**	**H**
A		26.67	61.76	109.09	145.83	**178.09**	170.53	93.33

Figure 5.22. A partially computed matrix of β_{ij} terms.

of the matrix of β_{ij} terms as follows:

Step 1: Compute one row of the β_{ij} elements.
Step 2: Find the maximum element in the row that was just computed. (If the maximum element is in a column that was already computed, stop.)
Step 3: Compute the elements β_{ij} in the column in which the maximum β_{ij} element occurred in step 2.
Step 4: Find the maximum element in the column that was just computed. (If the maximum element is in a row that was already computed, stop.)
Step 5: Compute the elements β_{ij} in the row in which the maximum β_{ij} element occurred in step 4. Go to step 2.

We illustrate this approach using the network of Figure 5.21. We begin by computing the elements in the row corresponding to node A. We obtain the partial matrix shown in Figure 5.22. The element in the column corresponding to node F is the largest element so, in step 4, we compute the elements in the column corresponding to node F. This is shown in Figure 5.23. The largest element corresponds to the row associated with node H. We therefore compute the row corresponding to node H as shown in Figure 5.24. The largest element is the element in the column corresponding to node F.

	To							
From	**A**	**B**	**C**	**D**	**E**	**F**	**G**	**H**
A	0.00	26.67	61.76	109.09	145.83	178.09	170.53	93.33
B						89.38		
C						81.28		
D						137.74		
E						55.44		
F						0.00		
G						99.00		
H						**180.63**		

Figure 5.23. A partially computed matrix of β_{ij} terms.

172   Center Problems

				To				
From	A	B	C	D	E	F	G	H
A	0.00	26.67	61.76	109.09	145.83	178.09	170.53	93.33
B						89.38		
C						81.28		
D						137.74		
E						55.44		
F						0.00		
G						99.00		
H	93.33	40.00	74.67	120.00	152.73	**180.63**	173.64	0.00

Figure 5.24. A partially computed matrix of β_{ij} terms.

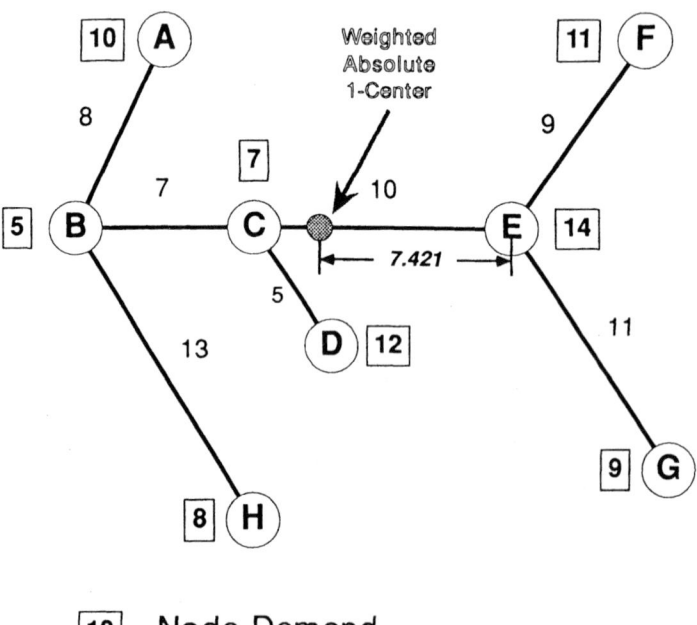

[Figure 5.25 diagram: weighted tree with nodes A(10), B(5), C, D(12), E(14), F(11), G(9), H(8) and edge weights A–B=8, B–C=7, B–H=13, C–D=5, C–E=10, E–F=9, E–G=11; node C has demand 7; weighted absolute 1-center located on link C–E, 7.421 units from E]

|10| Node Demand

The weighted absolute 1-center is located 7.421 units from node E on the link from node C to node E

Figure 5.25. Location of the weighted absolute 1-center for the weighted tree of Figure 5.21.

However, we have already computed the elements in the column corresponding to node F, so we stop. The optimal location is at the point

$$\frac{h_H}{(h_F + h_H)} d(F, H) = \frac{8}{19}(39) = 16.4211 \text{ units}$$

away from node F on the path from node F to node H, or 7.4211 units from node E to node C on the link from node E to node C as shown in Figure 5.25. The objective function is 180.63. Note that we only had to compute 19 of the β_{ij} elements as opposed to 28 elements. [The actual maximum number of elements that must be computed is $n(n - 1)/2$ or 28 in this case.]

5.4 THE UNWEIGHTED VERTEX *P*-CENTER PROBLEM ON A GENERAL GRAPH

In this section we outline a solution approach to solving the unweighted vertex *P*-center problem on a general graph. Throughout this section, we assume that all link distances are integer valued. (Note that this is not a terribly restrictive assumption since all rational distances can be converted to integer values by multiplying them by a sufficiently large number. Distances that are irrational numbers, such as $\sqrt{2}$, can be approximated to any desired level of accuracy by a rational number and then converted to an integer value in a similar manner.) The approach is based on searching over the range of coverage distances for the smallest coverage distance that allows all nodes to be covered. The search procedure used is called a *binary search*. In the next section we indicate how the approach may be generalized, through the addition of a large though finite set of additional points, to allow it to be used in solving the unweighted absolute *P*-center problem of a graph.

The procedure works as follows. Select initial lower and upper bounds on the value of the (unweighted vertex) *P*-center objective function. Solve the set covering problem using the average of the lower and upper bounds on the objective function as the coverage distance (rounding the average down to the largest integer less than or equal to the average). If the number of facilities needed to cover all nodes at that distance (the objective of the set covering model) is less than or equal to *P*, reset the upper bound on the value of the *P*-center objective function to the coverage distance that was just used; if the number of facilities needed is greater than *P*, reset the lower bound to the coverage distance that was just used plus 1. If the lower and upper bounds are equal, stop; if not, solve the set covering problem with a coverage distance equal to the average of the lower and upper bounds (rounded as before) and continue the process.

The algorithm may be thought of as guessing at the value of the *P*-center objective function. Using that value, we enter a figure like Figure 5.4 and find the number of facilities needed to cover all nodes at that coverage distance. If the number is less than or equal to P, then we know that the objective function for the P-center problem can be no larger than this coverage distance and we reset the upper bound on our guess at the objective function value to this number. Similarly, if after entering the figure like Figure 5.4 we find that the number of facilities needed to cover all nodes at the guessed coverage distance is greater than P, then we know that the value of the objective function for the P-center problem must be strictly larger than the guessed value (i.e., at least equal to the guessed value plus 1). Therefore, we reset the lower bound on our guess at the objective function to the previously guessed coverage distance plus 1.

Formally, let us define $P^*(x)$ as the optimal value of the set covering problem when the coverage distance is x. Also, define D_c^L and D_c^H as lower and upper bounds on the value of the P-center objective function. With these definitions, the algorithm may be stated formally as follows:

Algorithm for the Vertex P-Center Problem on a General Graph (with Integer Distances)

Step 1: Set D_c^H to a suitably large number. For example, set $D_c^H = (n - 1) \max_{i,j} \{d_{ij}\}$, where n is the number of nodes in the graph and d_{ij} is the length of link (i, j). Also, set $D_c^L = 0$. (Note that by setting D_c^H in this manner we are sure that the high estimate of the coverage distance is sufficiently large. This is so because a path between any two nodes will have at most $n - 1$ links and $\max_{i,j} \{d_{ij}\}$ is the length of the longest link. Therefore, $(n - 1) \max_{i,j} \{d_{ij}\}$ is an upper bound on the distance between any pair of nodes in the network.)

Step 2: Set $D_c = \lfloor (D_c^L + D_c^H)/2 \rfloor$, where $\lfloor x \rfloor$ denotes the largest integer less than or equal to x.

Step 3: Solve a set covering problem with a coverage distance of D_c. Let the solution be $P^*(D_c)$.

Step 4: If $P^*(D_c) \le P$, reset D_c^H to D_c; otherwise reset D_c^L to $D_c + 1$.

Step 5: If $D_c^L \ne D_c^H$, go to step 2; otherwise, stop, D_c^L is the optimal value of the objective function and the locations corresponding to the set covering solution for this coverage distance are the optimal locations for the P-center problem.

To illustrate this algorithm, we again return to the example network, shown in Figure 5.3. Let $P = 2$. The maximum link distance is 17, and there are six nodes. Therefore, we initially set $D_c^H = 85$ and $D_c^L = 0$. We then set $D_c = 42$ and solve the set covering problem. The solution to the set covering problem with $D_c = 42$ is $P^*(42) = 1$. Since this is less than the value of P, we reset the upper bound to 42 (i.e., set $D_c^H = 42$). We then set $D_c = 21$ and

The Unweighted Vertex P-Center Problem on a General Graph

Table 5.3 Summary of Iterations of the Vertex 2-Center Algorithm for the Graph of Figure 5.3

Iteration	D_c^L	D_c^H	D_c	$P^*(D_c)$
1	0	85	42	1
2	0	42	21	1
3	0	21	10	2
4	0	10	5	6
5	6	10	8	4
6	9	10	9	3
7	10	10	Stop!	

again solve a set covering problem. The solution is $P^*(21) = 1$. Again, we reset the upper bound, this time to 21 (i.e., set $D_c^H = 21$). We then solve a set covering problem with $D_c = 10$ and find $P^*(10) = 2$. Again, we reset the upper bound, this time to 10 (i.e., set $D_c^H = 10$). We then solve a set covering problem with $D_c = 5$ and find $P^*(5) = 6$. Since this is greater than the value of P, we reset the lower bound to $D_c^L = 6$. Again, we solve a set covering problem, this time with $D_c = 8$ and find $P^*(8) = 4$. Again, we reset the lower bound, this time to $D_c^L = 9$. We now solve one more set covering problem, this time with $D_c = 9$ and find $P^*(9) = 3$. Again, we reset the lower bound, this time to $D_c^L = 10$. Since the lower and upper bounds are

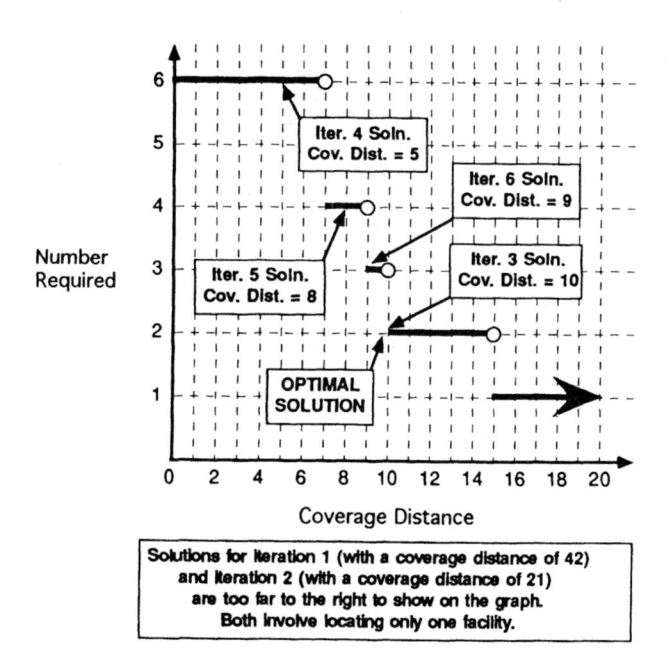

Figure 5.26. Sequence of solutions for the vertex 2-center problem of Figure 5.3.

identical, we stop. The optimal solution to the vertex 2-center problem has an objective function value of 10. Facilities should be located at nodes A (covering nodes A, B, and D) and E (covering nodes C, E, and F).

Table 5.3 and Figure 5.26 summarizes the iterations of this algorithm for this problem.

This algorithm can readily be extended to allow it to solve the *weighted* vertex P-center problem. A number of minor changes need to be made. First, the initial upper bound, D_c^H, needs to be set to account for the demands in the network. One way of doing so is to set

$$D_c^H = (n - 1)\left[\max_{i,j}\{d_{ij}\}\right]\left[\max_i\{h_i\}\right]$$

In other words, we need to multiply the upper bound used in the unweighted case by the largest demand. Second, in solving the set covering problem, candidate site j will be able to cover demand node i if $d_{ij}h_i \leq D_c$.

5.5 THE UNWEIGHTED ABSOLUTE P-CENTER PROBLEM ON A GENERAL GRAPH

This section presents a solution technique for the absolute P-center problem on a general graph. We begin by discussing characteristics of the solution to the problem. We then outline an extension of the algorithm presented in Section 5.4 for the vertex P-center problem that allows it to solve the absolute P-center problem.

5.5.1 Characteristics of the Solution to the Absolute P-Center Problem

We begin by examining the characteristics of a solution to the absolute P-center problem assuming the facility is located on a particular link, link AB. The top portion of Figure 5.27 illustrates such a link along with another node X. The squiggly lines depict paths (which may be composed of any number of links) between node X and link AB. The bottom portion of the figure plots a typical function of the distance between a point on link AB and node X. In drawing this figure, we assume that the distance from nodes A to node X, d_{AX}, plus the length of link AB, l_{AB}, is strictly greater than the distance from node B to node X, d_{BX}, and vice versa. This means that for some facility locations along link AB, the minimum path from the facility to node X will go through node A, while for others, the minimum path will go through node B. The slope of the distance function is either 1 or -1, as shown. Figure 5.28 shows two other possible forms for the distance function in which we relax the assumption outlined above. In the top figure, the shortest path from X to node A goes through node B; in the bottom figure,

The Unweighted Absolute P-Center Problem on a General Graph 177

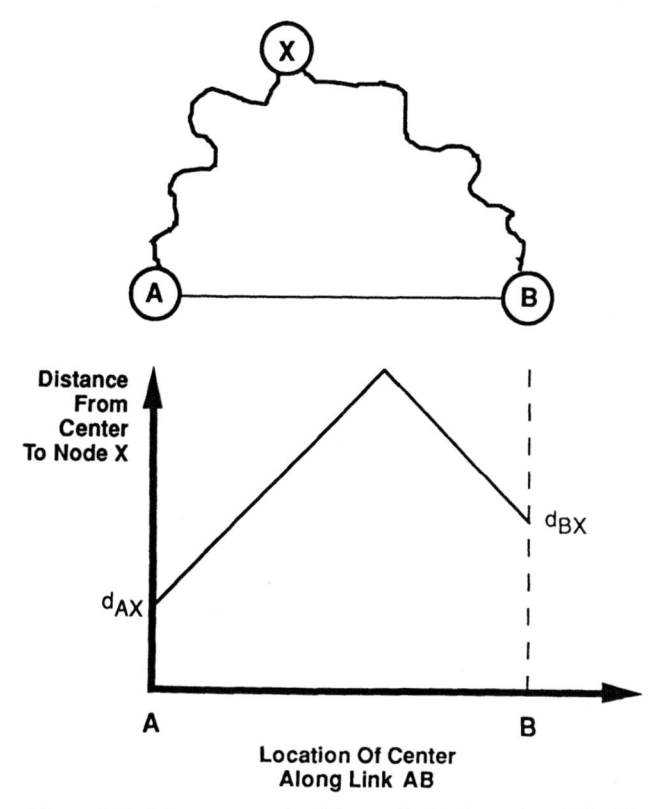

Figure 5.27. Distance to node X from a facility located on link AB.

the shortest path from X to node B goes through node A. Again, the absolute values of the slopes of these functions are 1.

Figure 5.29 plots typical functions of this form for a number of hypothetical nodes (nodes $\alpha, \beta, \gamma, \partial$ and ε). Drawing a vertical line through the curves gives the distance between a candidate site located at that position on link AB and the five nodes. If we were to locate on link AB, we would like to select the location on link AB that minimizes the maximum distance to any node. Thus, we are interested in the upper envelope of the curves shown in Figure 5.29. This is shown in Figure 5.30 along with the optimal location of a center on link AB. To find the optimal 1-center on a general graph, in principle, all we need to do is to repeat the steps that led to Figure 5.30 for each link in the network and then select the best (over all links) of the centers that are identified with that procedure.

We can exclude certain links from further consideration if we know the values of the vertex center objective function associated with locating at each node (Larson and Odoni, 1981). The upper envelope of the distance functions along link AB must have a value $V(A)$ at node A, where $V(A)$ is the

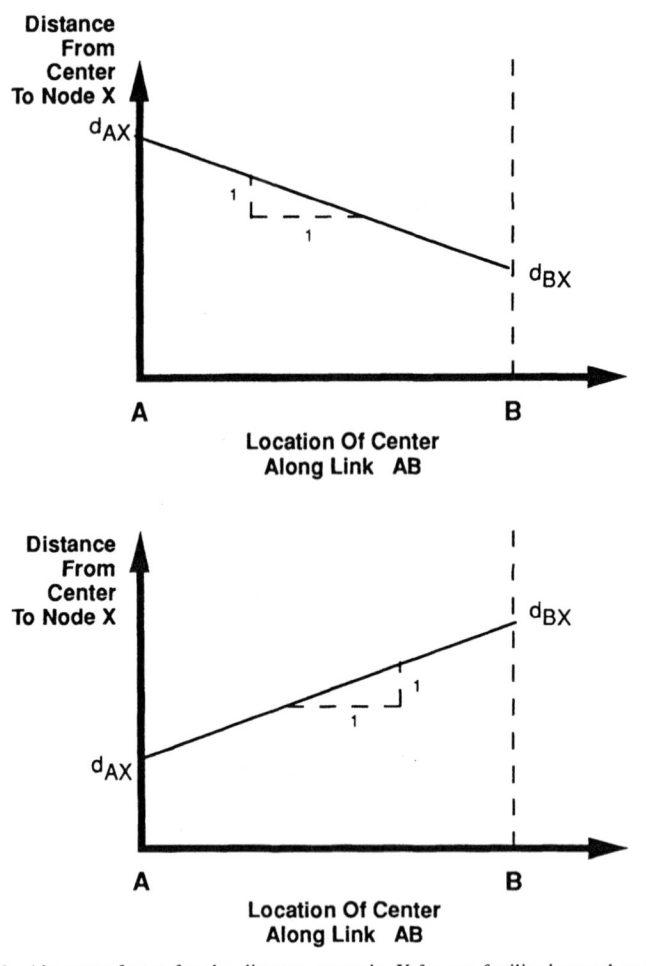

Figure 5.28. Alternate forms for the distance to node X from a facility located on link AB.

value of the vertex 1-center if we locate at node A. Similarly, the upper envelope of the distance functions along link AB must have a value $V(B)$ at node B. Since the slope of the distance function is either 1 or -1, a lower bound on the value of the best location on link AB can be found by solving the following equation for x, the distance from node A:

$$V(A) - x = V(B) - l_{AB} + x \qquad (5.2a)$$

yielding

$$x = \frac{V(A) + l_{AB} - V(B)}{2} \qquad (5.2b)$$

The Unweighted Absolute P-Center Problem on a General Graph

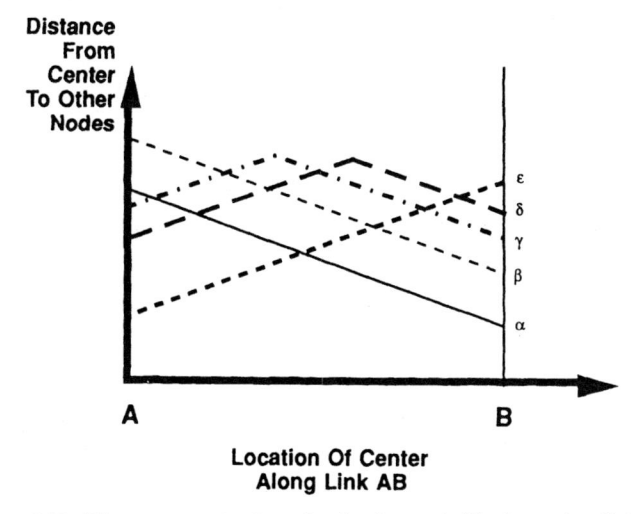

Figure 5.29. Distances to a number of nodes from a facility located on link AB.

Equation (5.2*a*) assumes (optimistically) that the upper envelope of the distance functions decreases as we move from node A and node B along link AB. A position x away from node A on link AB is therefore the location on the link at which the upper envelope would, under these optimistic assumptions, be a minimum. At this position, under these assumptions, the value of the absolute center objective function on the link would be $[V(A) + V(B) -$

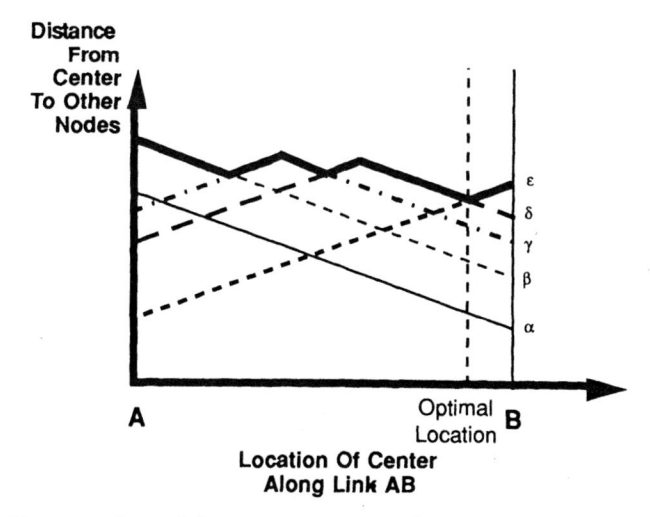

Figure 5.30. Upper envelope of distances to a number of nodes from a facility located on link AB and optimal center location on link AB.

Table 5.4 Vertex 1-Center Objective Function Values for the Network of Figure 5.3

Node	Vertex Center Value
A	19
B	26
C	15
D	19
E	19
F	26

$l_{AB}]/2$. Clearly, if a solution with an objective function value less than or equal to this bound is known, link *AB* can be excluded from further consideration.

To illustrate the use of this bounding procedure, Table 5.4 gives the value of the vertex center objective function for each of the six nodes of Figure 5.3. Using this information, Table 5.5 gives the bound computed for each of the 10 links of Figure 5.3. At least 3 of the 10 links can be excluded from further analysis since the vertex 1-center solution is 15.

Kariv and Hakimi (1979a) developed a polynomial time algorithm that can be used to find the local center on a given link. By searching over all links in a network, the algorithm may be used to find the absolute 1-center on a general graph. Kariv and Hakimi (1979a) also provided a polynomial time algorithm for solving the weighted absolute 1-center problem on a general network. The details of these algorithms are beyond the scope of this text. The reader interested in these algorithms is referred to their outstanding (though difficult) paper.

Table 5.5 Lower Bounds on the Value of a Center for Each Link of Figure 5.3

Link (AB)	Link Length	V(A)	V(B)	Bound	Exclude?
AB	9	19	26	18.00	Exclude
AC	12	19	15	11.00	Need to analyze
AD	10	19	19	14.00	Need to analyze
BC	13	26	15	14.00	Need to analyze
BE	17	26	19	14.00	Need to analyze
CD	8	15	19	13.00	Need to analyze
CE	7	15	19	13.50	Need to analyze
DE	13	19	19	12.50	Need to analyze
DF	7	19	26	19.00	Exclude
EF	9	19	26	18.00	Exclude

5.5.2 An Algorithm for the Unweighted Absolute *P*-Center on a General Graph

The algorithm outlined in Section 5.4 can be extended to solve for the absolute *P*-center on a graph. In doing so, observe that there are a finite number of candidate locations (in addition to the nodes) at which an absolute center might be located. The key to making this observation is that for each center there will be at least two nodes that are equidistant from the center. Furthermore, the distance to either of these nodes will be greater than or equal to the distance from the center to any other node assigned to the center for service. Finally, it will be impossible to move slightly from the center and get closer to both of the two nodes that are at this maximum distance.

Figure 5.31 illustrates such a candidate location, or local center, on a generic link *AB*. Link *AB* is l_{AB} units long; the distances from *X* to nodes *A* and *B* are $d(X, A)$ and $d(X, B)$ units, respectively; and, similarly, the distances from *Y* to nodes *A* and *B* are $d(Y, A)$ and $d(Y, B)$, respectively. *Assuming a local center with respect to nodes X and Y exists on link AB*, the point that is equidistant from *X* and *Y* is at a distance θ from node *A* on link *AB* where θ satisfies the following relationship:

$$\min\{d(A, X), d(A, Y)\} + \theta = \min\{d(B, X), d(B, Y)\} + (l_{AB} - \theta)$$

or

$$\theta = \frac{\min\{d(B, X), d(B, Y)\} + l_{AB} - \min\{d(A, X), d(A, Y)\}}{2}$$

It may be that nodes *X* and *Y* do not define a point that is interior to link *AB* (strictly between nodes *A* and *B*). Figure 5.32 illustrates such a situation. In this case, node *B* is a local center on link *AB* with respect to nodes *X*

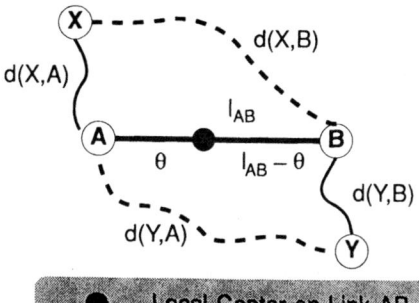

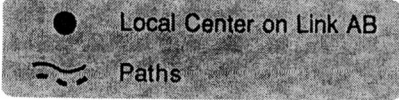

Figure 5.31. Candidate location on link *AB* defined by nodes *X* and *Y*.

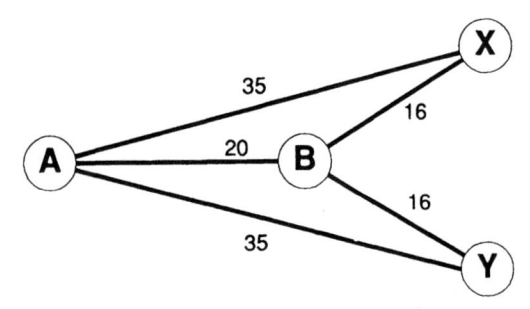

Figure 5.32. Example of when a local center is not interior on a link.

and Y. If we locate anywhere else on the link, we will be able to move in one direction or the other and get closer to both nodes simultaneously. While node B is on link AB, it is not interior to the link. Note that the equation for θ above gives $\theta = 0.5$. However, if we locate at this point, we can move in either direction and get closer to both nodes X and Y.

Figure 5.33 illustrates a case in which nodes X and Y do not define a local center on link AB. In this case, no matter where we locate on link AB, we could move closer to both nodes by moving toward the closer endpoint of link AB.

We can show that at least one solution to the absolute P-center problem consists of locating P facilities at one of the nodes in a set we will call $\mathcal{N}$, where $\mathcal{N}$ is the union of the set of nodes of the network and the set of all possible local nodes identified by the procedure outlined above (and illustrated in Figures 5.31, 5.32, and 5.33). Unfortunately, simple enumeration of the set $\mathcal{N}$ is a formidable task and greatly increases the number of locations that must be considered. In general, there will be $O(n^2)$ pairs of nodes. For each pair of nodes, we need to consider each of the $O(n^2)$ links AB of Figure 5.31. Thus, the cardinality of the set $\mathcal{N}$ is $O(n^4)$. For a network with 50 nodes, this means that there are potentially 6,250,050 candidate locations in the set $\mathcal{N}$. Fortunately, as indicated below (and as shown in Figures 5.32 and 5.33), many of these points do not need to be explicitly enumerated.

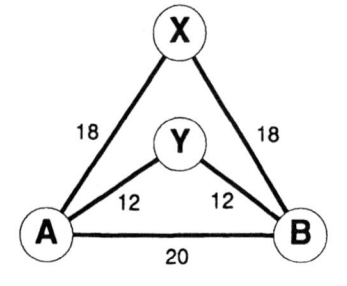

Figure 5.33. Example of the nonexistence of a local center on a link.

The Unweighted Absolute P-Center Problem on a General Graph

In practical contexts, a large number of the candidate locations in the set $\mathcal{N}$ can be eliminated and need not be considered explicitly. As shown in Figure 5.33, nodes X and Y may not define a local center on link AB. Handler and Mirchandani (1979) and Handler (1990) show that when $d(A, X) \neq d(A, Y)$ and $d(B, X) \neq d(B, Y)$, nodes X and Y cannot define a local center on link AB unless the following condition holds:

$$[S^A(X,Y) - d(A,X)][S^B(X,Y) - d(B,X)]$$
$$+ [S^A(X,Y) - d(A,Y)][S^B(X,Y) - d(B,Y)] = 0$$

where

$$S^A(X,Y) = \min\{d(A,X), d(A,Y)\}$$

In other words, $S^A(X,Y)$ is the distance from A to the closer of X and Y. Note that all of the terms in the brackets are less than or equal to 0. Thus, the condition may actually be rewritten as two separate conditions:

$$[S^A(X,Y) - d(A,X)][S^B(X,Y) - d(B,X)] = 0$$

and

$$[S^A(X,Y) - d(A,Y)][S^B(X,Y) - d(B,Y)] = 0$$

The first condition states that either node X is closer than node Y to node A or node X is closer than node Y to node B. Alternatively, we can state this as meaning either node A is closer to node X than it is to node Y or node B is closer to node X than it is to node Y. Node Y cannot be closer to node A than is node X and closer to node B than is node X. The second condition is the same except that the roles of X and Y are reversed. Note that these conditions should not be applied to the network of Figure 5.32 since $d(B, X) = d(B, Y) = 16$. However, in Figure 5.33, we obtain

$$(12 - 20)(12 - 20) \neq 0 \quad \text{for node } X$$

and

$$(12 - 12)(12 - 12) = 0 \quad \text{for node } Y$$

Thus, the condition states that a local center with respect to nodes X and Y does not exist on link AB.

We observe that it is possible for node A to be closer to both nodes X and Y than node B (or vice versa) and for a local center with respect to nodes X and Y to exist on link AB. Figure 5.34 illustrates such a case. Node B is closer to both nodes X and Y than is node A. However, the conditions

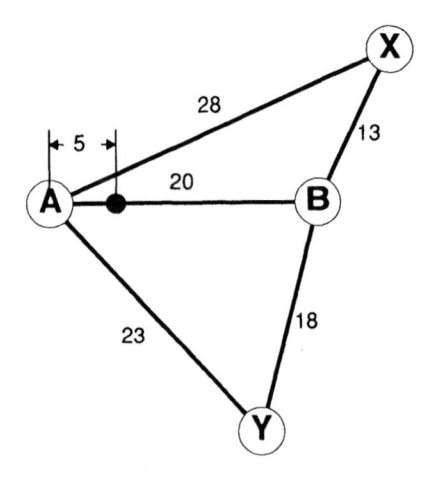

A local center with respect to nodes X and Y exists at a point 5 units from A on link AB.

Figure 5.34. Existence of a local center even when node B (or A) is closer to both nodes X and Y.

described above hold. A local center with respect to nodes X and Y exists at a point 5 units from node A on link AB. Moving slightly from this position will take us further from either node X or node Y.

We also note that while a point 2.5 units from node B on link BY would be better in terms of serving nodes X and Y only, the point shown in Figure 5.34 must be considered as a candidate center. Recall that the link AB and nodes X and Y that we have been drawing are only part of a much larger network. Suppose the complete network of which Figure 5.34 is a part is shown in Figure 5.35. If we failed to consider the local center shown in Figure 5.34 and 5.35, we might fail to find a solution that is better than a known solution. In particular, we would fail to find the optimal solution to the absolute 1-center problem for the network shown in Figure 5.35.

There are other ways in which we can avoid enumerating all of the local centers in the set $\mathcal{N}$. Suppose we know an upper bound on the value of the objective function for the absolute P-center problem. (Note that the value of the solution for the vertex P-center problem is clearly one such upper bound that we can readily obtain using the algorithm outlined in Section 5.4 above.) Let $\hat{R}$ be this upper bound. If the points X and Y define a candidate location on link AB, but $\min\{d(A, X), d(A, Y)\} + \theta > \hat{R}$, then the local center need not be added to the set $\mathcal{N}$, since the distance between either of these two nodes (X and Y) and the local center on link AB exceeds the upper bound on the absolute P-center objective function, $\hat{R}$.

Finally, suppose we have included only a portion of the nodes of the network as rows in a set covering problem. Let **V** be the set of *all* nodes in

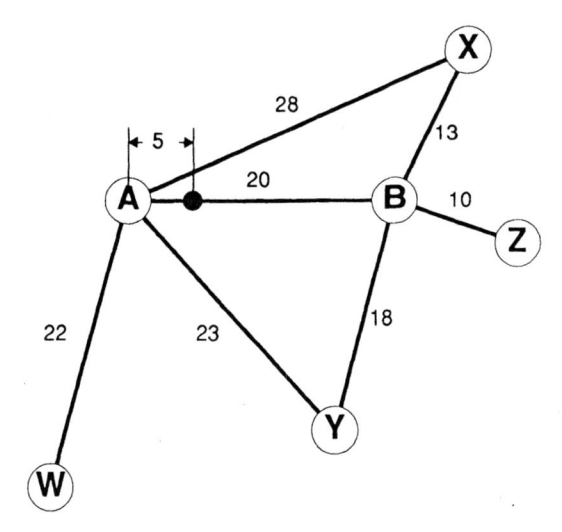

The Unweighted Absolute *P*-Center Problem on a General Graph

Figure 5.35. Complete network of which Figure 5.34 is a part.

the network and let $\hat{V}$ be the subset of nodes actually included in the set covering problem. Suppose also that we have some distance R which we are considering as the value of the objective function for the absolute *P*-center problem. For this subset of nodes that are included as rows in the set covering problem, we enumerate all of the local centers on all links of the original network and include all such local centers whose corresponding distances are less than or equal to R. Imagine that we then solve the resulting set covering problem. As before, let $P^*(R)$ be the optimal number of locations found by the set covering problem. Also, let $\underline{X}$ denote the optimal locations in the set covering problem and let $r(\underline{X})$ be the maximum distance from the nodes in the set $\underline{X}$ to any node in the network. That is,

$$r(\underline{X}) = \max_{i \in V} \left\{ \min_{j \in \underline{X}} [d(i,j)] \right\}$$

Finally, we let $\tilde{R}$ be the maximum distance associated with covering any of the nodes in the set $\hat{V}$ from a facility in the set $\underline{X}$.

Note that the set covering problem we are solving is only a portion of the complete set covering problem we would like to solve. However, we have included all locations that need to be considered to cover all of the nodes in the subset of nodes being considered within a distance R. As such, it is a relaxation of the real problem we want to solve.

A number of outcomes can result from the solution to such a set covering problem. Clearly, if the relaxed set covering problem requires more than P facilities [i.e., $P^*(R) > P$], then there is no *P*-facility solution to the complete problem with a distance R or less, since we cannot even cover all of the nodes in the subset $\hat{V}$ (let alone those not being considered) with only P facilities. Also, if the number of required nodes is less than or equal to P [i.e., $P^*(R) \leq P$] and the maximum distance from any node to the nearest facility is less than $\hat{R}$ [i.e., $r(\underline{X}) < \tilde{R}$], then $\underline{X}$ is an improved solution to the absolute P-center problem and $r(\underline{X})$ is an improved value for $\hat{R}$. If $P^*(R) \leq P$, but $r(\underline{X}) > R$, then there is some node not in the set $\hat{V}$ that is farther from $\underline{X}$ than R. In this case, we need to add such a node (e.g., the farthest node from $\underline{X}$) to the set $\hat{V}$, update the element in the row corresponding to this new node for each existing column in the set covering problem, add new local centers as candidate locations, and resolve the set covering problem.

Handler and Mirchandani (1979) and Handler (1990) describe this algorithm more rigorously. Rather than introduce the additional notation needed to state the algorithm formally, we refer the reader to these references. In the example that we outline below, we assume that we have integer link distances.

To illustrate the use of this approach, we consider the network shown in Figure 5.35. We begin by finding the vertex 1-center of the network. This is at node A with an objective function value of 30. Thus, we **set** $\hat{R} = 30$. We also set $R = 30$. Node Z is the farthest from node A, so we begin by solving a set covering problem with $\hat{V} = \{Z\}$. Clearly, there is only one column in this problem as well. We find that we need one facility located at Z to cover Z. $\tilde{R} = 0$. Not much of a surprise here. For this location, we have $r(\underline{X}) = 52$. Since this exceeds R, we add node W (at a distance of 52 from node Z) to the set $\hat{V}$, to obtain $\hat{V} = \{W, Z\}$. We now solve a set covering problem characterized by the following coefficient matrix:

Node 1	W	W	Z
Node 2	Z	W	Z
Link	AB		
W	1	1	
Z	1		1
Dist.	26.0	0.0	0.0

In each column of the matrix, we show the two nodes and the link that define a local center, the distance from either defining node to the center, and the other nodes in the set $\hat{V}$ that are covered by the local center. The solution to this problem is clearly to set the variable corresponding to the first column equal to 1. This involves locating a facility 4 units from node A on link AB. With this solution, we find that $r(\underline{X}) = 29$. Since this is less than $\hat{R}$, we update $\hat{R}$ and store this location as the best known 1-center solution. We eliminate all columns with associated distances of 29 or more. We set $R = 28.5$.

The Unweighted Absolute P-Center Problem on a General Graph

Since $r(\underline{X}) > \tilde{R} = 26$, we need to add the most remote node from the set $\underline{X}$ to the set of nodes being analyzed. Specifically, node X is the farthest from the facility. We update the set $\hat{V}$ to include node X, thereby obtaining $\hat{V} = \{X, W, Z\}$. Finally, we solve the set covering problem characterized by the following constraint matrix:

Node 1	W	W	W	X	W	X	Z
Node 2	X	Z	X	Z	W	X	Z
Link	AB	AB	AX	BX			
W	1	1	1		1		
X	1		1	1		1	
Z	1	1		1			1
Dist.	27.5	26.0	25.0	11.5	0.0	0.0	0.0

The solution is to select the location corresponding to the first column. This corresponds to locating at a point 5.5 units from node A on link AB. We now find $r(\underline{X}) = 28.5$. Since this is better than our previous best solution value, we set $\hat{R} = 28.5$ and store this solution as the best known 1-center solution. We set $R = 28$. Since $\hat{R} = 28.5 > \tilde{R} = 27.5$, we identify node Y as the farthest from the center. We update set $\hat{V} = \{X, Y, W, Z\}$ and solve the new set covering problem shown below:

Node 1	X	W	Y	W	X	W	W	W	X	Y	X	W	X	Y	Z
Node 2	Y	X	Z	Z	Y	X	Y	Y	Y	Z	Z	W	X	Y	Z
Link	AB	AB	AB	AB	AX	AX	AB	AY	BY	BY	BX				
W	1	1	1	1	1	1	1	1			1				
X	1	1			1	1			1		1		1		
Y	1		1		1		1	1	1	1				1	
Z	1	1	1	1					1	1	1				1
Dist.	28.0	27.5	26.5	26.0	25.5	25.0	25.0	22.5	15.5	14.0	11.5	0.0	0.0	0.0	0.0

The solution to this problem involves locating at the location given by the first column. This location is 5 units from A on link AB (the position shown in Figure 5.35). For this solution, we have $r(\underline{X}) = 28$. Since this is better than the previous best known solution, we update $\hat{R} = 28$ and store this location as the best known 1-center solution. All nodes of the network are covered within this distance, so no nodes need to be added to the set $\hat{V}$. We now eliminate all columns with distances greater than or equal to 28, set $R = 27.5$, and solve the problem again. Clearly, there is no solution with $P = 1$, so 28 is the optimal solution value for the absolute 1-center problem.

The solution to the problem above (with the first column eliminated since its distance is 28 or more) is to locate at node W and the location defined by nodes X and Y on link BY. With this solution, we have $r(\underline{X}) = 22$. Since this is better than our previous value of $\hat{R}$, we update our best guess at the optimal solution value *which now applies to the 2-center problem*. Therefore, we set $\hat{R} = 22$ and record this solution as the best known 2-center solution. Since node A is not covered by the solution within $\tilde{R} = 15.5$ (the maximum distance associated with any of the selected columns in the set covering problem), we add node A to the set $\hat{V}$ to obtain $\hat{V} = \{A, X, Y, W, Z\}$. We eliminate all columns with distances of 22 or more, add new columns resulting from the addition of node A to the set $\hat{V}$, and solve the set covering problem characterized by the following coefficient matrix:

Node 1	A	A	X	A	Y	A	X	A	A	A	W	X	Y	Z
Node 2	Y	X	Y	Z	Z	X	Z	Y	W	A	W	X	Y	Z
Link	AB	AB	BY	AB	BY	AX	BX	AY	AW					
A	1	1		1		1		1	1	1				
W									1		1			
X		1	1			1	1					1		
Y	1		1		1			1					1	
Z	1	1	1	1	1		1							1
Dist.	19.0	16.5	15.5	15.0	14.0	14.0	11.5	11.5	11.0	0.0	0.0	0.0	0.0	0.0

The solution is now to locate at the local centers defined by: (1) nodes X and Y on link BY and (2) nodes A and W on link AW. For this solution, we have $\tilde{R} = 15.5$ and $r(\underline{X}) = 15.5$. Thus, we update $\hat{R} = 15.5$ and record this as the best known solution. Since $r(\underline{X}) = \hat{R}$, no new nodes need to be added to the set $\hat{V}$. We eliminate all columns with distances greater than or equal to 15.5 and solve the set covering problem with the following constraint matrix:

Node 1	Y	A	X	A	A	A	W	X	Y	Z
Node 2	Z	X	Z	Y	W	A	W	X	Y	Z
Link	BY	AX	BX	AY	AW					
A		1		1	1	1				
W					1		1			
X		1	1					1		
Y	1			1					1	
Z	1		1							1
Dist.	14.0	14.0	11.5	11.5	11.0	0.0	0.0	0.0	0.0	0.0

There is clearly no solution with $P = 2$. Thus, the solution identified above is the optimal 2-center solution with an objective function value of 15.5. One $P = 3$ solution to this problem consists of locating at the centers defined by: (1) nodes X and Z on link BX; (2) nodes A and W on link AW; and (3) node Y. For this solution, we have $\tilde{R} = 11.5$ and $r(\underline{X}) = 11.5$. Thus,

The Unweighted Absolute P-Center Problem on a General Graph

we can set $\hat{R} = 11.5$ for a 3-center problem. We now eliminate all columns with distances greater than or equal to 11.5 and solve the set covering problem with the following coefficient matrix:

Node 1	A	A	W	X	Y	Z
Node 2	W	A	W	X	Y	Z
Link	AW					
A	1	1				
W	1		1			
X				1		
Y					1	
Z						1
Dist.	11.0	0.0	0.0	0.0	0.0	0.0

Since no solution exists with $P = 3$, the previous solution is the optimal absolute 3-center solution with an objective function value of 11.5. The solution to this set covering problem is clearly to locate at the local center on link AW, as well as nodes X, Y, and Z. For this $P = 4$ solution, we have $\tilde{R} = 11$ and $r(\underline{X}) = 12$. Note that the old value of $\hat{R} = 11.5$ is still a valid upper bound on the $P = 4$ solution. Node B is now the farthest node that is not covered within a distance $\tilde{R}$ and so we add node B to the set $\hat{V}$ to obtain $\hat{V} = \{A, B, X, Y, W, Z\}$. We now augment the columns in the constraint matrix and solve the set covering problem whose coefficient matrix is shown below:

Node 1	A	A	B	B	B	A	B	W	X	Y	Z
Node 2	W	B	Y	X	Z	A	B	W	X	Y	Z
Link	AW	AB	BY	BX	BZ						
A	1	1				1					
B		1	1	1	1		1				
W	1							1			
X				1					1		
Y			1							1	
Z					1						1
Dist.	11.0	10.0	9.0	6.5	5.0	0.0	0.0	0.0	0.0	0.0	0.0

One solution to this problem is to locate at the local centers defined by: (1) nodes A and W on link AW; (2) nodes B and Z on link BZ; (3) node X; and (4) node Y. This solution has $\tilde{R} = 11$ and $r(\underline{X}) = 11$. Thus, we set $\hat{R} = 11$ and eliminate all columns with distances greater than or equal to 11.

The new set covering problem has the following coefficient matrix:

	Node 1	A	B	B	B	A	B	W	X	Y	Z
	Node 2	B	Y	X	Z	A	B	W	X	Y	Z
	Link	AB	BY	BX	BZ						
A		1				1					
B		1	1	1	1		1				
W								1			
X				1					1		
Y			1							1	
Z					1						1
Dist.		10.0	9.0	6.5	5.0	0.0	0.0	0.0	0.0	0.0	0.0

No solution with $P = 4$ exists for this problem, so the solution identified above is optimal for $P = 4$ with an objective function value of 11. For $P = 5$, one solution is to locate at nodes A, W, X, and Y, as well as the local center on link BZ defined by nodes B and Z. For this solution, we have $\tilde{R} = 5$ and $r(\underline{X}) = 5$, so we set $\hat{R} = 5$ for the $P = 5$ solution. Eliminating all rows with distances greater than or equal to 5 leaves the following identity matrix:

	Node 1	A	B	W	X	Y	Z
	Node 2	A	B	W	X	Y	Z
	Link						
A		1					
B			1				
W				1			
X					1		
Y						1	
Z							1
Dist.		0.0	0.0	0.0	0.0	0.0	0.0

Clearly, no solution exists with $P = 5$, so the previous solution is the optimal 5-center solution with an objective function value of 5 units. Since there are six nodes in the network, the optimal solution for the 6-center problem is simply to locate a facility at each node.

Note that in solving this problem, we only had to enumerate 23 local centers out of a maximum of 105 total possibilities (15 pairs of nodes times 7 links). Also, note that the largest set covering problem we had to solve had only 15 columns (as opposed to a maximum of 111 resulting from 105 local centers and 6 nodes). All set covering problems were easily solvable by inspection.

Again, the reader interested in a more formal statement of this algorithm is referred to the descriptions in Handler and Mirchandani (1979) and

Handler (1990), as well as related papers by Minieka (1970) and Garfinkel, Neebe, and Rao (1977).

5.6 SUMMARY

In this chapter we have discussed center problems, or problems in which we want to locate a fixed number of facilities on either the nodes (for a vertex center) or the nodes and links (for an absolute center) to minimize the maximum (demand-weighted) distance between a node and the nearest facility to the node. We outlined polynomial time algorithms for solving the unweighted and weighted absolute 1-center, the absolute 2-center, and the vertex 1-center problems on trees. For more general graphs, we summarized a procedure that involves "guessing" a value of the solution and then solving a set covering model for that problem. The result of the set covering solution is used to update the "guess" at the optimal value of the objective function. This "guessing" algorithm can be applied to both the vertex and absolute center problems.

A number of authors have considered extensions to the center problems discussed in this chapter. For example, Minieka (1977) extends the notion of coverage to the need to cover all points on all links of the graph. Thus, the problem becomes one of locating a fixed number of facilities to minimize the maximum distance between all nodes as well as the infinite number of intermediate points on each of the arcs and the nearest facility. On the methodological side, Martinich (1988) proposes a vertex closing heuristic for solving the vertex center problem. A vertex closing algorithm may be thought of as a "greedy subtraction" algorithm in the sense that such heuristics begin with facilities located at all candidate locations and proceed to close facilities in an intelligent manner.

In Chapter 6 we examine median problems in which the objective is to minimize the *sum* of the distances between each demand node and the nearest facility to the node. Such problems are of particular value when the average performance of the system is of more interest than is the worst case, extreme performance.

EXERCISES

5.1 Consider the tree shown in Figure E5.1:

(a) Find the absolute 1-center of the tree.

(b) Find the vertex 1-center of the tree.

(c) Find the absolute 2-centers of the tree.

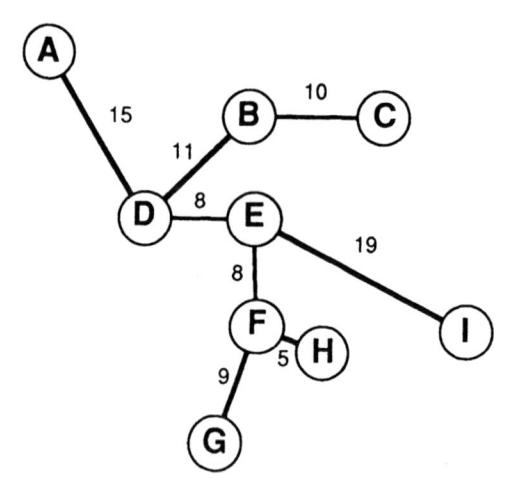

Figure E5.1

5.2 Consider the following tree:
   (a) Find the absolute 1-center of the tree.
   (b) Find the vertex 1-center of the tree.
   (c) Find the absolute 2-centers of the tree.

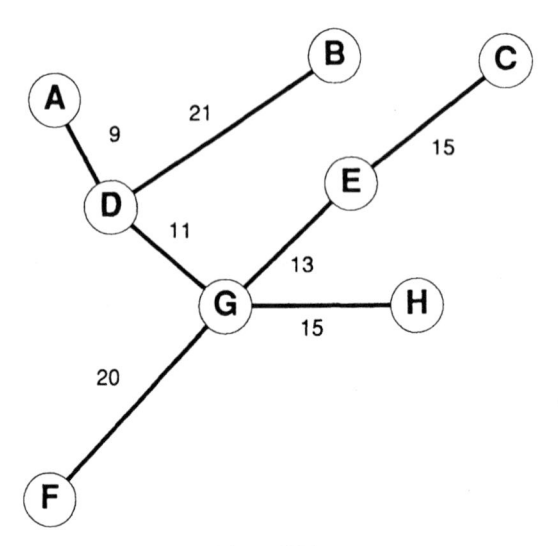

Figure E5.2

5.3 Show that the solution to the absolute 1-center problem on an unweighted tree is unique.

Exercises

5.4 For the following tree, with weights as shown on the tree, solve for the demand-weighted 1-center. Recall that you need not compute all of the $\beta_{ij} = (h_i h_j d_{ij}/(h_i + h_j))$ terms; specifically, you can compute them for one row, find the maximum element in that row, compute them for that column, and so on, as discussed in Section 5.3.3. This should save you considerable calculation time. Clearly show where the 1-center is and compute the objective function for this location.

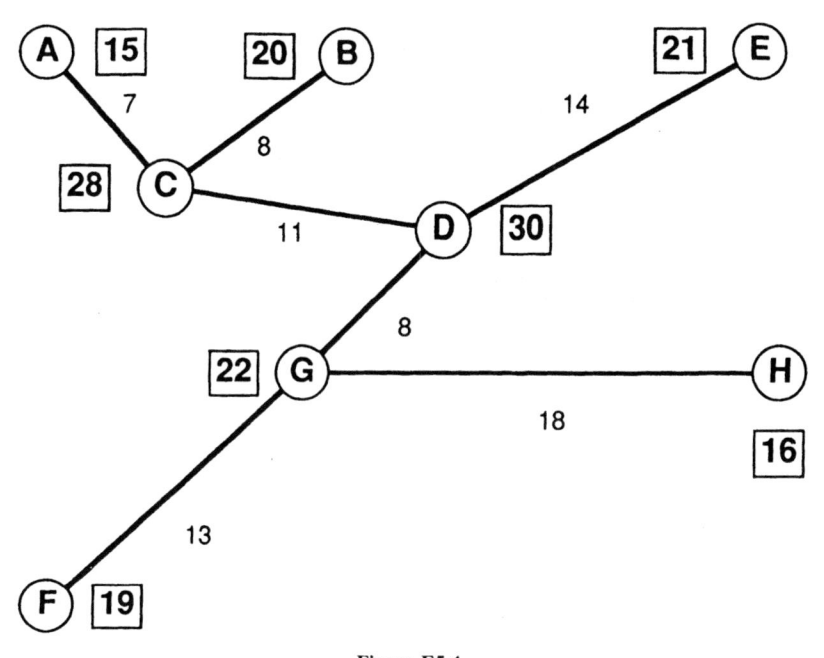

Figure E5.4

5.5 For the tree shown in Figure E5.5, find the demand-weighted 1-center.

5.6 (a) For the tree shown in Figure E5.6, find the demand-weighted 1-center.

(b) How large must the demand at node E become for the location of the weighted absolute 1-center to change? (Give the smallest integer demand such that the location of the 1-center changes.)

(c) For the demand that you identified in part (b), where is the new weighted absolute 1-center location?

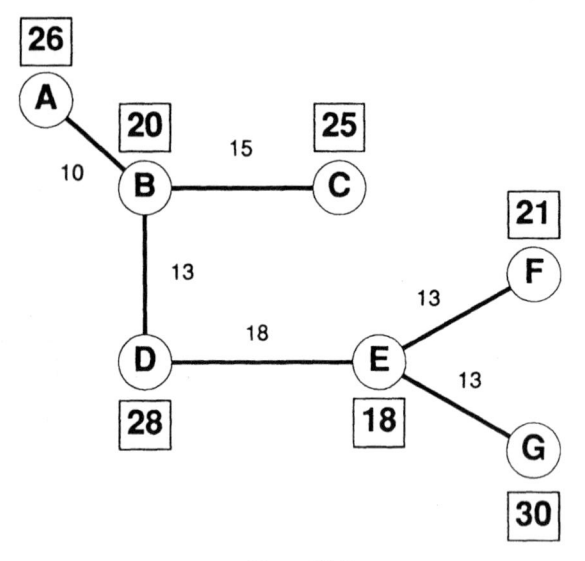

Figure E5.5

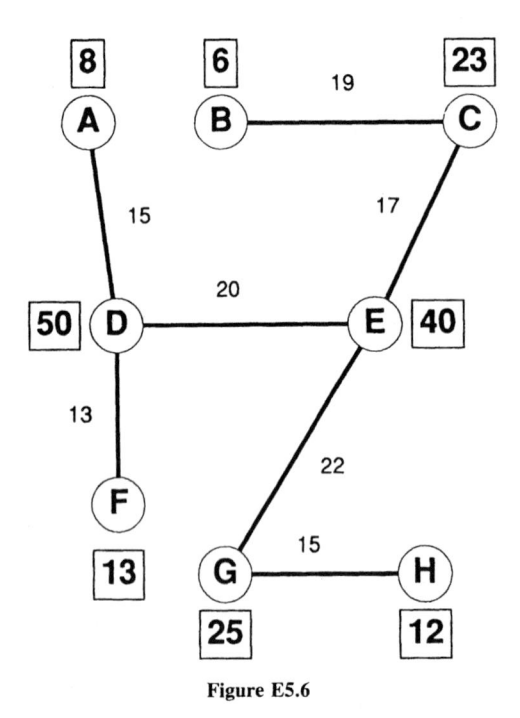

Figure E5.6

Exercises

5.7 For the following network, solve the vertex 2-center problem. Clearly show the results of each iteration of the algorithm including the lower and upper bounds on the implied coverage distance, the coverage distance used in solving each of the set covering problems, and the number of facilities needed to cover all nodes within that distance.

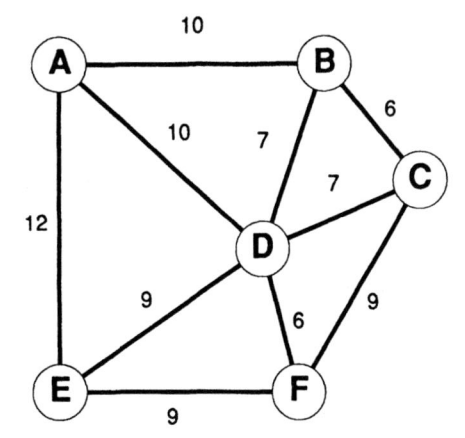

Figure E5.7

5.8 (a) For the following network, solve the vertex 3-center problem. Clearly indicate where you should locate the three facilities.

(b) What is the objective function value for the vertex 3-center problem?

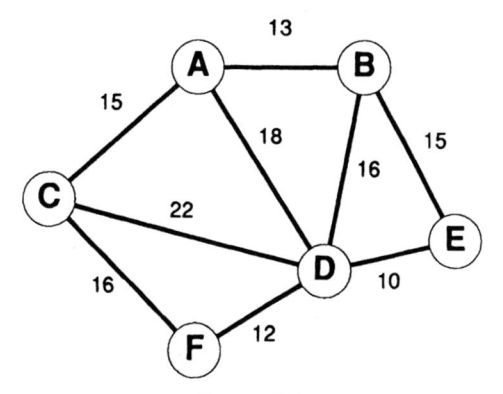

Figure E5.8

5.9 (a) For the following network, solve the vertex 3-center problem. Clearly show the lower and upper bounds that you obtain on each iteration of the algorithm.

(b) What is the objective function for the vertex 3-center problem?

(c) What are the locations of the three nodes that solve the vertex 3-center problem on this network?

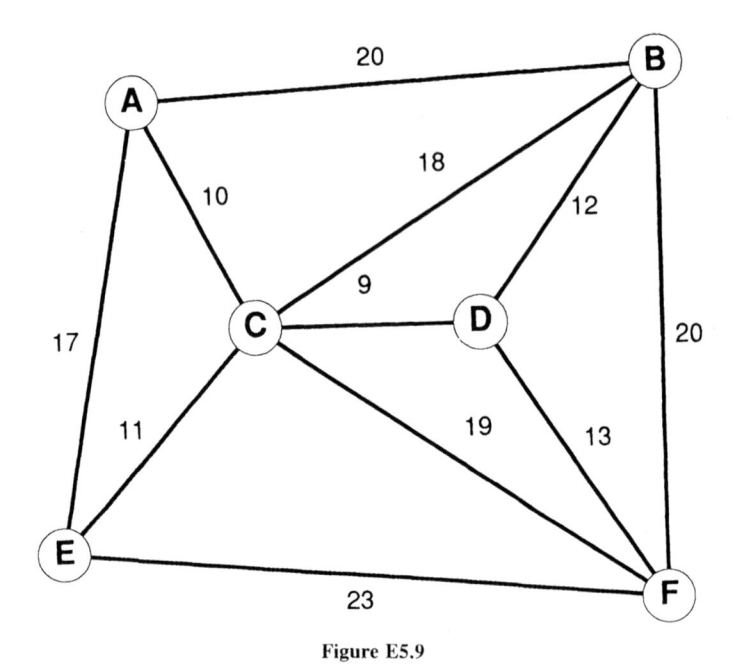

Figure E5.9

5.10 (a) Use the SITATION program to find the solution of the vertex 10-center problem on the 88-node problem (CITY1990.GRT, described in Appendix G). Find the minimum coverage distance such that all demands can be covered by 10 facilities accurate to 1 mile. Use a binary search over an appropriate interval such as 340 to 400 miles. Clearly show your results.

Important note: To solve this exercise, use the Lagrangian relaxation upper bounds. Set the Lagrangian parameters as follows:

Critical percentage difference	0.0000001
Maximum number of iterations	4000
Minimum α value allowed	0.0000001
Number of failures before changing α	12 or 24

Exercises

(b) Repeat part (a), but now find the 8-center solution.

(c) Repeat part (a), but now find the 9-center solution.

(d) What do the solutions to parts (b) and (c) imply about the marginal value of the additional facility in terms of the *P*-center objective as you go from eight to nine facilities?

(e) What do the solutions to parts (a) and (c) imply about the marginal value of the additional facility in terms of the *P*-center objective as you go from 9 to 10 facilities?

5.11 In trying to solve the vertex 4-center problem for the 88-node data set (CITY1990.GRT)—which has a total demand of 44,840,571—I obtained the following results with a coverage distance of 650.

1. With no nodes forced in or out of the solution, the Lagrangian upper bound is 44,840,571 and the lower bound is 44,385,925. Facilities are located at nodes 23, 29, 71, and 78.

2. With node 23 (the node which covers the most demand in the solution above) forced out of the solution, the Lagrangian upper bound is 44,760,066 and the lower bound is 44,757,084. Facilities are located at nodes 48, 59, 74, and 78.

3. With node 23 forced into the solution, the Lagrangian upper bound is 44,630,491 and the lower bound is 44,482,023. Facilities are located at nodes 18, 23, 26, and 78.

With this information what can you say about the optimal solution to the 4-center problem? Be as specific as possible. You might want to confirm your answer using the SITATION program.

6

Median Problems

6.1 INTRODUCTION

Chapters 4 and 5 dealt with covering and center problems, respectively. All of the variants of covering and center problems that we discussed assume that a demand node receives complete benefits from a facility if it is within the coverage distance and no benefits if the distance between the demand node and the nearest facility exceeds the coverage distance. Figure 6.1 illustrates such a benefit curve. (The only exception to this was the maximum expected covering problem in which the incremental benefit or coverage obtained by a demand node was proportional to the probability of a facility being available. Even in this model, however, the full incremental benefit accrued to a demand node if the facility in question was *within the coverage distance* and no benefit accrued to the node if the facility was *beyond the coverage distance*.)

In many cases, however, the benefit (cost) associated with a demand node/facility pair decreases (increases) gradually with the distance between the demand and nearest facility. For example, the cost of serving a retail establishment from a warehouse may depend on the time a driver must spend traveling from the warehouse to the retail store. In this case, the cost depends approximately linearly on the distance between the store and the warehouse. Figure 6.2 illustrates such a *linear* relationship. In other cases, nonlinear relationships may be more appropriate. Figure 6.2 also illustrates both a *convex* relationship between the quantities, in which the cost increases at an increasing rate with distance between the demand node and the facility assigned to serve the demand node, as well as a *concave* relationship, in which the cost increases at a decreasing rate.

In median problems, the topic of this chapter, the relationship between the distance between facilities and demand nodes and the cost associated with the facility/demand pair is usually linear as shown in Figure 6.2. We also allude briefly to the other functional forms shown in Figure 6.2.

Before proceeding with a discussion of median problems, we note that we have transformed the discussion from one of maximizing benefits or the total number of covered demands to one of minimizing costs. For completeness, we note that the maximum covering model may be formulated in terms of

Introduction

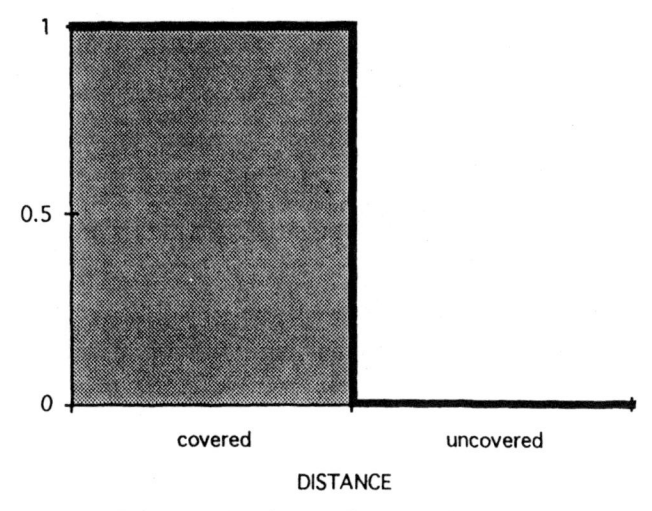

Figure 6.1. Relative benefits in covering models.

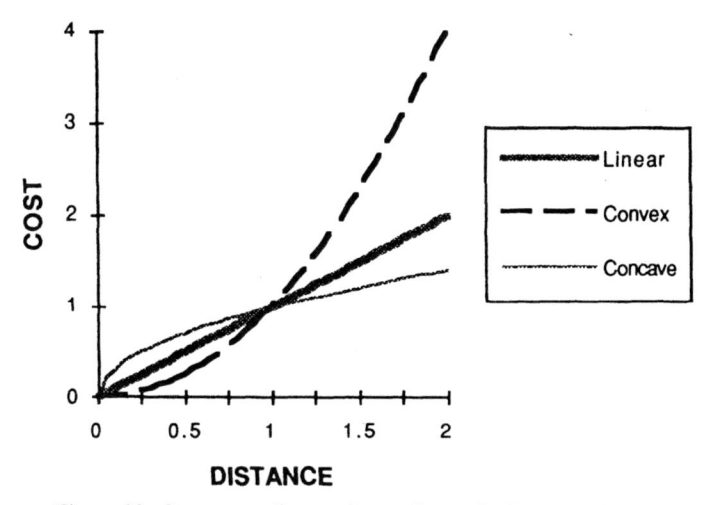

Figure 6.2. Cost versus distance for median and related problems.

minimizing the number of uncovered demands using the following notation (in addition to that defined in Chapter 4):

Decision Variables

$$W_i = \begin{cases} 1 & \text{if node } i \text{ is } not \text{ covered} \\ 0 & \text{if node } i \text{ is covered} \end{cases}$$

With this notation, we can reformulate the maximum covering problem as follows:

MINIMIZE $$\sum_i h_i W_i \qquad (6.1a)$$

SUBJECT TO: $$\sum_j a_{ij} X_j + W_i \geq 1 \qquad \forall\, i \qquad (6.1b)$$

$$\sum_j X_j \leq P \qquad (6.1c)$$

$$X_j = 0, 1 \qquad \forall\, j \qquad (6.1d)$$

$$W_i \geq 0 \qquad \forall\, i \qquad (6.1e)$$

This formulation minimizes the number of uncovered demands (6.1a). Constraint (6.1b) stipulates that either a facility is located that can cover node i ($\sum_j a_{ij} X_j \geq 1$) or node i is not covered ($W_i = 1$). Constraint (6.1c) says that at most P facilities can be located. Constraints (6.1d) and (6.1e) are standard integrality and nonnegativity constraints. Note that the coverage variables (W_i) need only be nonnegative.

Again, the primary motivation for presenting this alternate formulation is simply to indicate that there is really no difference between benefit maximization and cost minimization *in the context of the models we have discussed so far*. However, if the level of demand is sensitive to the quality of the service (e.g., Perl and Ho, 1990, Mukundan and Daskin, 1991), the two approaches are not equivalent. Finally, we note that the elimination of the integrality restriction on the coverage variables may result in some computational advantages for the formulation given above.

6.2 FORMULATION AND PROPERTIES

The P-median problem is to find the location of P facilities on a network so that the total cost is minimized. The cost of serving demands at node i is given by the product of the demand at node i and the distance between demand node i and the nearest facility to node i. This problem may be formulated using the following notation:

Inputs

h_i = demand at node i
d_{ij} = distance between demand node i and candidate site j
P = number of facilities to locate

Formulation and Properties

Decision Variables

$$X_j = \begin{cases} 1 & \text{if we locate at candidate site } j \\ 0 & \text{if not} \end{cases}$$

$$Y_{ij} = \begin{cases} 1 & \text{if demands at node } i \text{ are served by a facility at node } j \\ 0 & \text{if not} \end{cases}$$

With this notation, the P-median problem may be formulated as follows:

$$\text{MINIMIZE} \quad \sum_i \sum_j h_i d_{ij} Y_{ij} \quad (6.2a)$$

$$\text{SUBJECT TO:} \quad \sum_j Y_{ij} = 1 \quad \forall\, i \quad (6.2b)$$

$$\sum_j X_j = P \quad (6.2c)$$

$$Y_{ij} - X_j \leq 0 \quad \forall\, i, j \quad (6.2d)$$

$$X_j = 0, 1 \quad \forall\, j \quad (6.2e)$$

$$Y_{ij} = 0, 1 \quad \forall\, i, j \quad (6.2f)$$

The objective function A(6.2a) minimizes the total demand-weighted distance between each demand node and the nearest facility. Constraint (6.2b) requires each demand node i to be assigned to exactly one facility j. Constraint (6.2c) states that exactly P facilities are to be located. Constraints (6.2d) link the location variables (X_j) and the allocation variables (Y_{ij}). They state that demands at node i can only be assigned to a facility at location j ($Y_{ij} = 1$) if a facility is located at node j ($X_j = 1$). Constraints (6.2e) and (6.2f) are the standard integrality conditions.

Constraint (6.2d) is a strong version of the constraint linking the location and allocation variables. A weaker version is

$$\sum_i Y_{ij} - IX_j \leq 0 \quad \forall\, j \quad (6.2d')$$

where I is the number of demand nodes. If no facility is located at node j ($X_j = 0$), then all of the allocation variables using this facility must be 0 ($\sum_i Y_{ij} = 0$). While this form of the constraint reduces the total number of constraints, the linear programming relaxation of this problem [obtained by replacing the integrality constraints (6.2e) and (6.2f) by nonnegativity restrictions] results in weaker (i.e., smaller, since we are dealing with a minimization problem) estimates of the optimal solution value. In other words, using the stronger constraint form (6.2d) has been shown to be advantageous computationally. Intuitively, one way to see why this is so is to realize that if, for example, two demand nodes are assigned to a facility at node j ($\Sigma_i Y_{ij} = 2$), then, in the strong linear programming relaxation, we will have $Y_{ij} - X_j \leq 0$ for each of the two demand nodes assigned to the facility at node j, and we will have $X_j = 1$. This will allow us to locate only $P - 1$ other facilities. In the weak relaxation, we will have $X_j = 2/I \leq 1$, which will allow us to locate $P - 2/I$ additional facilities. Since this is greater than or equal to $P - 1$ (the remaining number we could locate in the strong relaxation), we will be able to locate more (fractional) facilities in the weak linear programming relaxation and will therefore end up with a smaller demand-weighted distance. In the remainder of this chapter, we will use the original, stronger form of this constraint.

The P-median formulation given above assumes that facilities are located on nodes of the network. As discussed in Chapters 4 and 5, this assumption can lead to suboptimal solutions for the set covering, maximum covering, and P-center problems. However, Hakimi (1965) has shown that *for the P-median problem at least one optimal solution consists of locating P facilities on the network's nodes*. To prove that this is true, we consider a solution in which at least one facility is located on link (i, j) a distance α from node i ($0 < \alpha < d_{ij}$). Let H_i denote the total demand that is served by this facility and that enters link (i, j) through node i. Define H_j similarly with respect to node j. Assume (without loss of generality) that $H_i \geq H_j$. By moving the facility from its position α units away from node i to node i itself (without altering any of the demand allocations), we change the objective function by $(H_j - H_i)\alpha$. Since $H_i \geq H_j$, this quantity is nonpositive. Thus, making this change in the location of the facility will not degrade the solution. This is sufficient to prove that at least one optimal solution consists of locating only on the nodes of the network. [In what follows, we argue that such a change is likely to result in an *improvement* in the objective function. If $H_i > H_j$, the quantity $(H_j - H_i)\alpha$ will be negative and this change will improve the objective function. Furthermore, this quantity does not account for improvements in the objective function that may be obtained by reallocating demands *after* the facility is moved. In particular, some demands may now be closer to node i than they were to whatever node to which they had previously been allocated. Also, some of the demands that were originally allocated to the facility located between nodes i and j and that entered the facility via node j may now be closer to some other facility. Reallocating these demands can further reduce the objective function.]

The property that at least one optimal solution consists of locating only on the nodes has been extended to a number of variants of the P-median problem. The key condition that is necessary for these extensions is that the objective function be concave (see Figure 6.2). Handler and Mirchandani (1979), Mirchandani (1990), and Mirchandani and Odoni (1979), among others, discuss extensions of this property.

1-Median Problem on a Tree

This property limits the number of alternative solutions that must be examined to find a solution to

$$\binom{N}{P} = \frac{N!}{P!(N-P)!}$$

where N is the number of nodes and P is the number of facilities to be located. While this number is $O(N^P)$ (for $P \ll N$) and is therefore polynomial in N for a given value of P, the number of combinations can be very large, indeed. For example, if $N = 20$ and $P = 5$, $\binom{20}{5} = 15{,}504$. For $N = 50$ and $P = 10$, a problem that is not at all large by most standards, $\binom{50}{10} > 10^{10}$. If you could evaluate one million combinations every second, it would take almost 3 hours to enumerate every possible solution. We should note that this is a truly heroic task since evaluating each combination involves finding the closest facility to each of the 40 demand nodes at which a facility is not located. Thus, to evaluate each solution will require at least 400 comparisons, 40 multiplications, and 40 additions. Finally, if $N = 100$ and $P = 15$, it would take over 8 *millennia* to enumerate every possible solution even if we could do so at the rate of one million every second. Most of us do not want to wait that long for the solution!

As indicated above, for fixed P, the P-median problem may technically be solved in an amount of time that is polynomial in the number of nodes. For even moderate values of N and P, however, the number of possible solutions that must be enumerated becomes exceptionally large. Furthermore, for variable P, the problem is still NP-complete (Garey and Johnson, 1979). To see why this is so, note that the time required to solve all P-median problems by enumeration for $P = 1$ to $P = N$ for any given value of N is

$$\sum_{j=1}^{N} \binom{N}{j} = 2^N - 1 = O(2^N)$$

which is exponential in N.

In short, we will again need to find effective heuristic algorithms if we want to solve problems of realistic size in a reasonable amount of time.

6.3 1-MEDIAN PROBLEM ON A TREE

As was true of the P-center problem, if we focus on locating a single median on a tree, we can find a very efficient algorithm for solving the problem optimally. This section outlines that algorithm.

We begin by noting that if half or more of the total demand ($\sum_i h_i = H$) is at any node, then at least one optimal solution consists of locating the facility at that node. Note that if a node has more than half of the total demand, then locating at this node is *the* optimal location. To see that this must be true, consider any solution in which the facility is located at a point on the network (either a node or anywhere on a link, for that matter) other than the node at which at least half of the total demand occurs. Without loss of generality, let the node with at least half of the total demand be node A, with demand h_A. Now consider moving the facility from its current (supposedly optimal) position a distance δ toward node A. The change in the objective function will be given by δ times the difference between the number of demands which are closer to the facility at its new location and the number of demands that are farther from the new location. But since we are moving closer to node A, we will be getting closer to at least half of the total demand, H. The objective function will *decrease by at least* $\delta(2h_A - H) \geq 0$. Thus, moving the facility a distance δ toward node A will not degrade the objective function. If $h_A > H/2$, such a move will strictly decrease the objective function. Thus, we have shown that *at least one optimal solution consists of locating the facility at the node at which half or more of the demand occurs*. It is worth noting that this property is independent of whether or not the underlying network is a tree or not. In other words, if we are locating a single facility on a network and if one node has half or more of the total demand, it is optimal to locate at that node.

Now suppose that half or more of the demand does not exist at any single node, as shown in the example of Figure 6.3. Consider locating at some point X midway between nodes B and C. (Recall that at least one optimal solution to the P-median problem on *any* network must be on a node, so even if such a location were to be optimal, locating on some node would have to be as

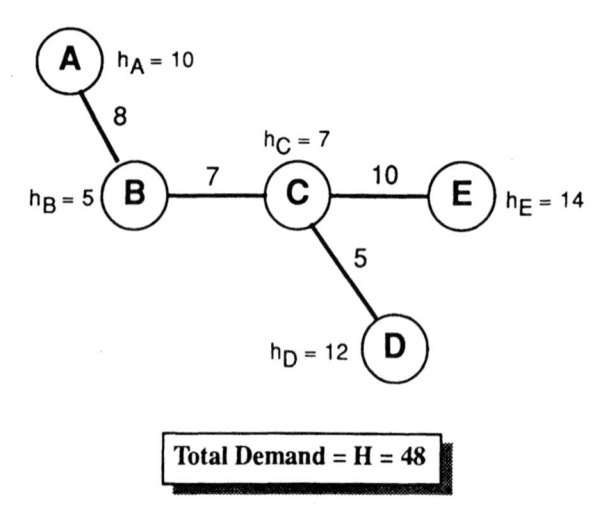

Figure 6.3. Example tree with less than half of the total demand at any node.

1-Median Problem on a Tree

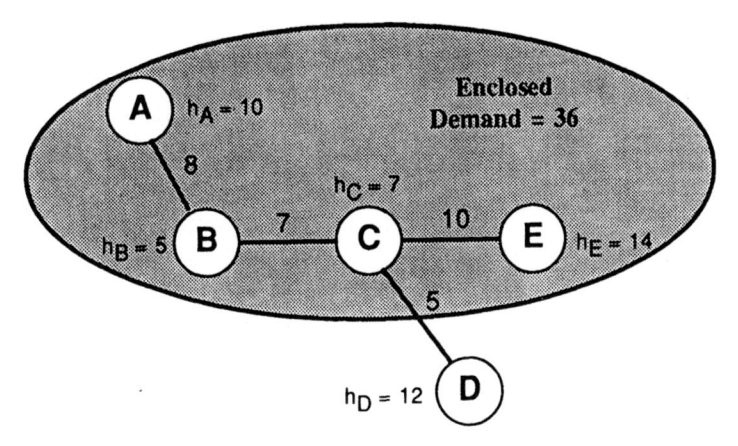

Figure 6.4. Effect of moving the facility from C toward D.

good.) At such a location 15 demands ($h_A + h_B$) are to the left of the facility and 33 demands ($h_C + h_D + h_E$) are to the right of the facility. Moving the facility δ distance units toward node C would reduce the objective function (the demand-weighted total distance) by 18δ units $[(33 - 15)\delta = 18\delta]$. We can continue to reduce the objective function by this amount until we reach node C.

Now consider moving the facility from node C toward either node D or node E. First consider moving it toward node D. As shown in Figure 6.4, moving the facility a distance δ from node C toward node D on link CD would move the facility farther from 36 demands and closer to only 12 demands. Thus, the objective function would *increase* by 24δ. Clearly, such a move is not optimal. Similarly, we can show that moving the facility δ from C toward E on link CE would increase the objective function by 20δ.

Using the qualitative argument outlined above, we can show that an optimal solution to the 1-median problem on a tree may be found by "folding" the demand at a tip node onto the node that is incident on the tip node and deleting the tip node. This process is repeated until a node of the new tree contains half or more of the total demand of the tree. A tip node is any node of the tree that is incident on only one other node. Thus, in the tree shown in Figures 6.3 and 6.4, nodes A, D, and E are tip nodes. By "folding" the demand of a tip node onto the node that is incident on the tip node and deleting the tip node, we mean that we add the demand at the tip node to that of the node that is incident on the tip node. We then remove from the tree the tip node and the link connecting the tip node to the node on which it is incident. Performing this operation on node A of Figure 6.3, we obtain the network shown in Figure 6.5. In this figure, demands are shown as revised demands, which are indicated by a caret (ˆ) over the demand. Note that the revised and original demands are the same at nodes C, D, and E, while the

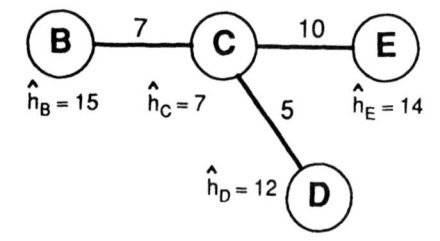

Figure 6.5. Effect of folding tip node *A* onto node *B*.

Figure 6.6. Effect of folding tip node *D* onto node *C*.

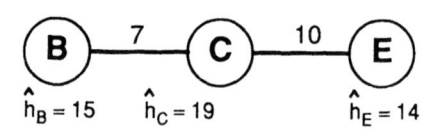

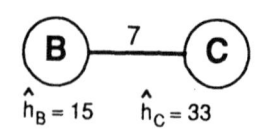

Figure 6.7. Effect of folding tip node *E* onto node *C*.

revised demand at node *B* is the sum of the original demands at nodes *A* and *B* (since we folded node *A* onto node *B*).

In Figure 6.5 none of the nodes has more than half of the total demand (which is still 48 units). Thus, we again select a tip node and fold it onto the node on which it is incident. Figure 6.6 shows the effect of folding tip node *D* onto node *C*. Again, the revised demand at each of the nodes is still less than half the total demand. Therefore, we again select a tip node and fold it onto the node on which it is incident. Figure 6.7 shows the effect of folding node *E* onto node *C*. Now the revised demand at node *C* is over half of the total demand. Thus, the optimal location for the 1-median on the tree is node *C*.

Note that in the discussion above we did not need to know the length of any of the links on the tree; we only needed to know the demand at each of the nodes. To compute the total demand-weighted distance, we *do* need to know these distances. For the network shown in Figure 6.3, the minimum total demand-weighted distance is 385 if we locate at node *C*.

The algorithm outlined above may be formalized as follows:

Algorithm for the 1-Median Problem on a Tree (Goldman, 1971)

Step 1: Set $\hat{h}_i = h_i$ for all nodes *i*.

Step 2: Select any tip node *i*. If $\hat{h}_i \geq \Sigma_j h_j / 2 = H/2$, then locate at node *i* and go to step 3.
If not, add $\hat{h}_i$ to the value of $\hat{h}_k$, where *k* is the unique node that is incident on node *i*; delete node *i* (and the points on the link strictly between *i* and *k*) and repeat step 2.

Step 3: Compute the objective function. (Recall that we now need to employ the link distances.)

1-Median Problem on a Tree

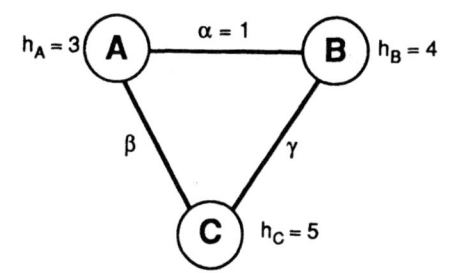

Figure 6.8. Simple cyclic network.

Since each node is examined at most once, this is an $O(n)$—or linear time — algorithm. Kariv and Hakimi (1979b) provide an $O(n^2 P^2)$ algorithm for finding P medians on a tree with n nodes. The details of this algorithm, however, are beyond the scope of this text.

Finally, note that we argued above that if half or more of the total demand was located at a single node, then it was optimal to locate a 1-median at that node. This applied to any network. The process outlined above of folding a tip node onto an incident node applies only to a tree. This is so since, in a

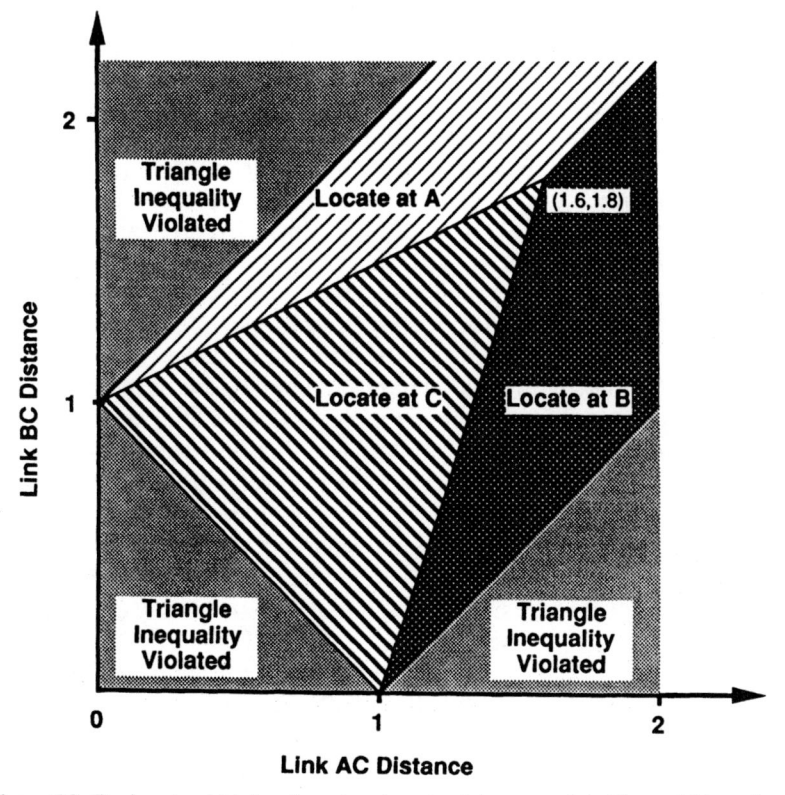

Figure 6.9. Regions in which locating at each node of the network in Figure 6.8 is optimal.

general (cyclic) network, more than one node may be incident on any particular node. Thus, the process of folding demand onto a unique incident node is not possible in a general network. To emphasize this point, consider the simple network shown in Figure 6.8. Since the total demand is 12, no single node has over half of the total demand. Link *AB* is 1 unit long. Links *AC* and *BC* are β and γ units, respectively. Figure 6.9 shows that any one of the three nodes may be optimal depending on the lengths of these links. It also indicates that, while node *C* has more demand than either of the other two nodes, the region in which it is optimal to locate at node *C* is *finite*, while the regions within it is optimal to locate at either node *A* or node *B* (nodes with smaller demand than that found at node *C*) are *infinite*. The reason for this is that as links *AC* and *BC* get longer, locating at node *C* becomes less desirable because doing so necessitates the contribution of two long links (links *AC* and *BC* for the service of nodes *A* and *B*, respectively). Locating at either *A* or *B*, however, necessitates the contribution of only one long link (link *AC* or *BC*, respectively) since the length of link *AB* is fixed at 1 unit in this example.

6.4 HEURISTIC ALGORITHMS FOR THE *P*-MEDIAN PROBLEM

While the *P*-median problem can be solved in polynomial time on a tree network, as indicated in Section 6.2 above, the problem is NP-complete on a general graph (Kariv and Hakimi, 1979b). Thus, a number of heuristic algorithms for the solution of the *P*-median problem have been proposed. In this section we outline three classes of such heuristics: a myopic algorithm, an exchange heuristic, and a neighborhood search algorithm. These heuristics fall into two broad classes of heuristics (Golden et al., 1980): construction algorithms and improvement algorithms. The myopic algorithm is a construction algorithm in which we attempt to build a good solution from scratch. The myopic algorithm is similar in spirit to the greedy adding algorithm for the maximum covering problem (see Section 4.5.2). Both the exchange and neighborhood search algorithms are improvement algorithms which are similar in spirit to the exchange heuristic of the greedy adding and substitution algorithm for the maximum covering problem (see also Section 4.5.2).

In Section 6.5 we summarize a Lagrangian relaxation approach to solving the problem. While technically also a heuristic algorithm, the Largrangian approach that we outline below, when coupled with one or more of the heuristic algorithms of this section, often gives results that are either provably optimal or very close to optimal. Computational results comparing the algorithms are given in Section 6.6.

If we were to locate only a single facility on the network, we could easily find the optimal location by enumerating all possible locations and choosing the best (i.e., by total enumeration). Specifically, since we know that at least one optimal solution to any *P*-median problem consists of locating only on the demand nodes, we could evaluate the 1-median objective function, $Z_j = \sum_i h_i d_{ij}$, that would result if we locate at demand node j, for each demand node. We would then choose the location that results in the smallest value of Z_j. If we only want to locate a single facility, it is clear that this approach would give an optimal solution, since we would have tested each possible location.

Suppose now that we are given the location of $P - 1$ facilities. Let $\mathbf{X}_{P-1}$ denote the set of locations of these $P - 1$ facilities. Also, let $d(i, \mathbf{X}_{P-1})$ be the shortest distance between demand node i and the closest node in the set $\mathbf{X}_{P-1}$. Similarly, we let $d(i, j \cup \mathbf{X}_{P-1})$ be the shortest distance between demand node i and the closest node in the set $\mathbf{X}_{P-1}$ *augmented* by candidate location j. The best place to locate a single new facility, *given* that the first $P - 1$ facilities are located at the sites given in the set $\mathbf{X}_{P-1}$, is at the location j that minimizes $Z_j = \sum_i h_i d(i, j \cup \mathbf{X}_{P-1})$.

This approach leads to the myopic algorithm for constructing a solution to the P-median problem. Formally, we may state the algorithm as follows:

Myopic Algorithm for the P-Median Problem

Step 1: Initialize $k = 0$ (k will count the number of facilities we have located so far) and $\mathbf{X}_k = \emptyset$, the empty set ($\mathbf{X}_k$ will give the location of the k facilities that we have located at each stage of the algorithm).

Step 2: Increment k, the counter on the number of facilities located.

Step 3: Compute $Z_j^k = \sum_i h_i d(i, j \cup \mathbf{X}_{k-1})$ for each node j which is not in the set $\mathbf{X}_{k-1}$. Note that Z_j^k gives the value of the P-median objective function if we locate the kth facility at node j, given that the first $k - 1$ facilities are at the locations given in the set $\mathbf{X}_{k-1}$ (and node j is not part of that set).

Step 4: Find the node $j^*(k)$ that minimizes Z_j^k, that is, $j^*(k) = \text{argmin}_j \{Z_j^k\}$. Note that $j^*(k)$ gives the best location for the kth facility, given the location of the first $k - 1$ facilities. Add node $j^*(k)$ to the set $\mathbf{X}_{k-1}$ to obtain the set $\mathbf{X}_k$; that is, set $\mathbf{X}_k = \mathbf{X}_{k-1} \cup j^*(k)$.

Step 5: If $k = P$ (i.e., we have located P facilities), stop; the set $\mathbf{X}_P$ is the solution to the myopic algorithm. If $k < P$, go to step 2.

Figure 6.10 is a simple flowchart of this heuristic. In this figure we show that one of the improvement algorithms discussed below can be applied to the solution obtained by using the myopic algorithm.

One reason for considering improvement algorithms is that the solution obtained by using the myopic algorithm will not necessarily be optimal, since at each pass through the algorithm, we are holding the locations of the first $k - 1$ facilities fixed. Despite the fact that the solution we obtain in this way may not be optimal, this algorithm is appealing for a number of reasons.

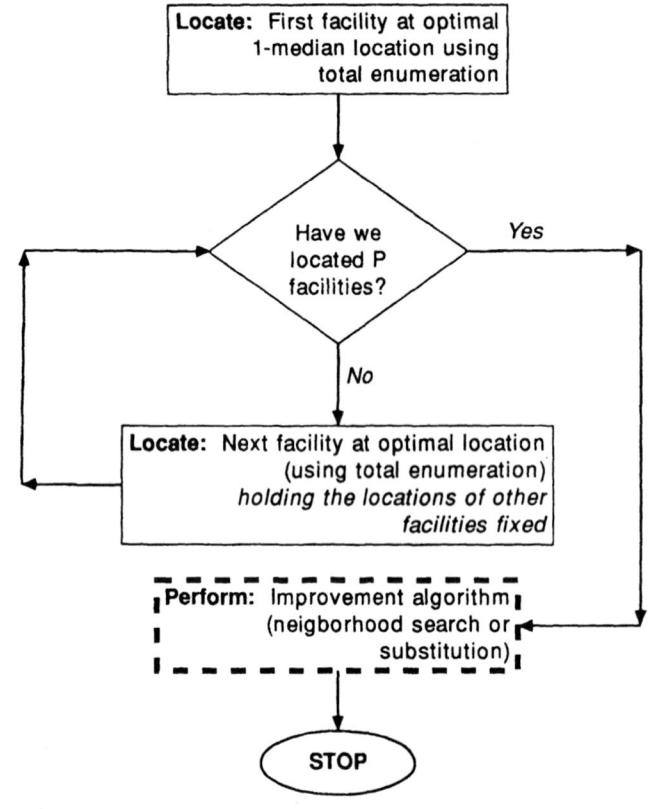

Figure 6.10. Myopic algorithm (with improvement heuristic shown after all facilities are located).

First, it is clearly very simple to understand and to implement. Second, in practice, many decisions are made this way. In practice, we are often given the location of some number of facilities which *cannot* be moved. We are then asked to find the location of a few (often only one or two) new facilities. If we are only required to locate one additional facility and the existing facilities cannot be relocated, this approach will clearly be optimal.

To illustrate the approach, consider the network shown in Figure 6.11. Numbers in boxes next to nodes are the demands, h_i. Table 6.1 gives the distance matrix, while the values of $h_i d_{ij}$ are given in Table 6.2. By summing the entries in each column of Table 6.2, we obtain the values of Z_j^1. The smallest Z_j^1 value corresponds to $j = I$, with a value of 4772. Thus, the optimal total demand-weighted distance if we locate only one median for the network of Figure 6.11 is 4772 resulting in an average distance of 4772/185 or 25.795. (Note that there are 185 demands in the network.)

To locate a second median, we need to compute $h_i \cdot \min\{d(i, I); d(i, j)\}$ for each node/candidate location pair (i, j). Table 6.3 shows the results of this computation. The column totals correspond to Z_j^2. Now it is best to add a facility at node G. The total demand-weighted distance is now 3145

Heuristic Algorithms for the P-Median Problem

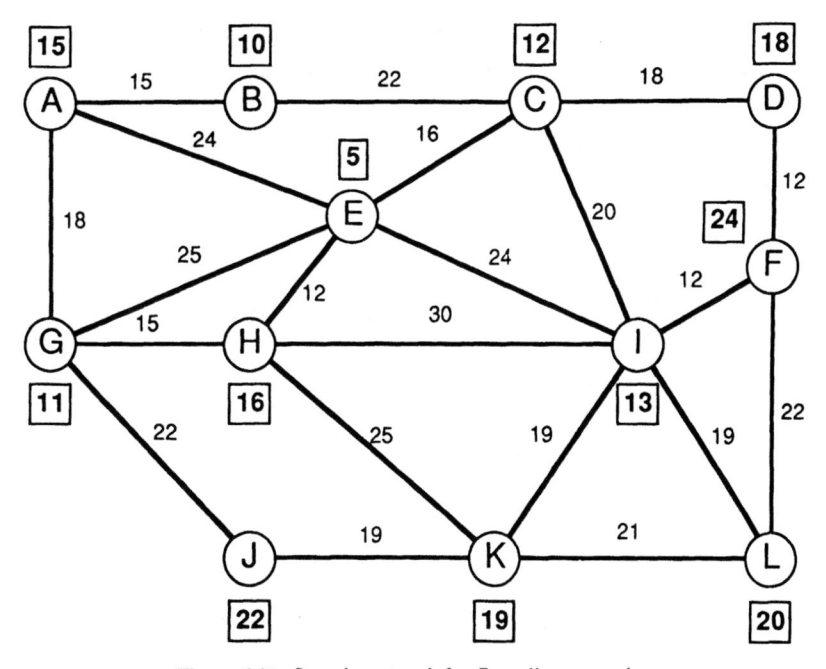

Figure 6.11. Sample network for P-median examples.

resulting in an average distance of 17.000. To add a third facility, we would compute $h_i \cdot \min\{d(i,G); d(i,I); d(i,j)\}$ for each node/candidate location pair (i, j), find the column totals corresponding to Z_j^3, find the column with the smallest total, and locate at the corresponding node. Proceeding in this manner, we obtain the results shown in Table 6.4 for the first five myopic medians.

Table 6.1 Distance Matrix for the Network of Figure 6.11

From	A	B	C	D	E	F	G	H	I	J	K	L
A	0	15	37	55	24	60	18	33	48	40	58	67
B	15	0	22	40	38	52	33	48	42	55	61	61
C	37	22	0	18	16	30	41	28	20	58	39	39
D	55	40	18	0	34	12	59	46	24	62	43	34
E	24	38	16	34	0	36	25	12	24	47	37	43
F	60	52	30	12	36	0	57	43	12	50	31	22
G	18	33	41	59	25	57	0	15	45	22	40	61
H	33	48	28	46	12	42	15	0	30	37	25	46
I	48	42	20	24	24	12	45	30	0	38	19	19
J	40	55	58	62	47	50	22	37	38	0	19	40
K	58	61	39	43	37	31	40	25	19	19	0	21
L	67	61	39	34	43	22	61	46	19	40	21	0

Table 6.2 Demand Times Distance for the Network of Figure 6.11

Node i	A	B	C	D	E	F	G	H	I	J	K	L
A	0	225	555	825	360	900	270	495	720	600	870	1005
B	150	0	220	400	380	520	330	480	420	550	610	610
C	444	264	0	216	192	360	492	336	240	696	468	468
D	990	720	324	0	612	216	1062	828	432	1116	774	612
E	120	190	80	170	0	180	125	60	120	235	185	215
F	1440	1248	720	288	864	0	1368	1008	288	1200	744	528
G	198	363	451	649	275	627	0	165	495	242	440	671
H	528	768	448	736	192	672	240	0	480	592	400	736
I	624	546	260	312	312	156	585	390	0	494	247	247
J	880	1210	1276	1364	1034	1100	484	814	836	0	418	880
K	1102	1159	741	817	703	589	760	475	361	361	0	399
L	1340	1220	780	680	860	440	1220	920	380	800	420	0
Total	7816	7913	5855	6457	5784	5760	6936	5971	4772	6886	5576	6371

Table 6.3 Computations for Second Myopic Median for Figure 6.11

Node i	A	B	C	D	E	F	G	H	I	J	K	L
A	0	225	555	720	360	720	270	495	720	600	720	720
B	150	0	220	400	380	420	330	420	420	420	420	420
C	240	240	0	216	192	240	240	240	240	240	240	240
D	432	432	324	0	432	216	432	432	432	432	432	432
E	120	120	80	120	0	120	120	60	120	120	120	120
F	288	288	288	288	288	0	288	288	288	288	288	288
G	198	363	451	495	275	495	0	165	495	242	440	495
H	480	480	448	480	192	480	240	0	480	480	400	480
I	0	0	0	0	0	0	0	0	0	0	0	0
J	836	836	836	836	836	836	484	814	836	0	418	836
K	361	361	361	361	361	361	361	361	361	361	0	361
L	380	380	380	380	380	380	380	380	380	380	380	0
Total	3485	3725	3943	4296	3696	4268	3145	3655	4772	3563	3858	4392

Table 6.4 Results for First Five Myopic Medians for Figure 6.11

Median Number	Location	Total Demand-Weighted Distance	Average Distance
1	I	4772	25.795
2	G	3145	17.000
3	F	2641	14.276
4	J	2157	11.659
5	A	1707	9.227

Heuristic Algorithms for the P-Median Problem

Given the locations of some facilities (whether the locations are optimal or not), each demand node should be assigned to the nearest facility since the facilities are uncapacitated and we are trying to minimize the demand-weighted total distance. This creates sets of nodes such that all nodes in the same set are assigned to the same facility. We refer to the nodes within a set as being in the *neighborhood* of the facility to which they are assigned. Figure 6.12 displays the neighborhoods associated with the five myopic medians of the network shown in Figure 6.11. Note that in assigning demand nodes to facilities, we can break ties arbitrarily. Thus, demand node E is assigned to the neighborhood of the facility at node A even though it could also have been assigned to the neighborhood of the facility located at node I.

Within each neighborhood, we would expect that the median would be located optimally. In other words, we expect that the facility serving each neighborhood would be located at the optimal 1-median site for the nodes within the neighborhood. Since finding the optimal 1-median may be done simply by enumeration, we can readily ensure that this condition is satisfied. This leads to the *neighborhood search* improvement algorithm. Figure 6.13 is a flowchart of the neighborhood search algorithm. [Maranzana (1964) was the first to propose such an algorithm.]

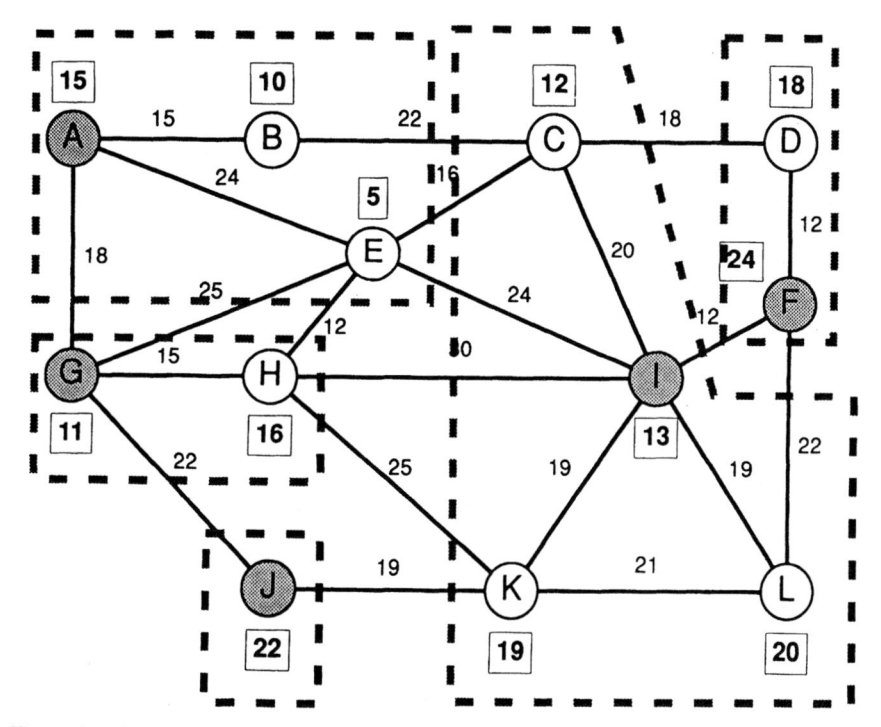

Figure 6.12. Neighborhoods associated with myopic 5-median solution for the network of Figure 6.11.

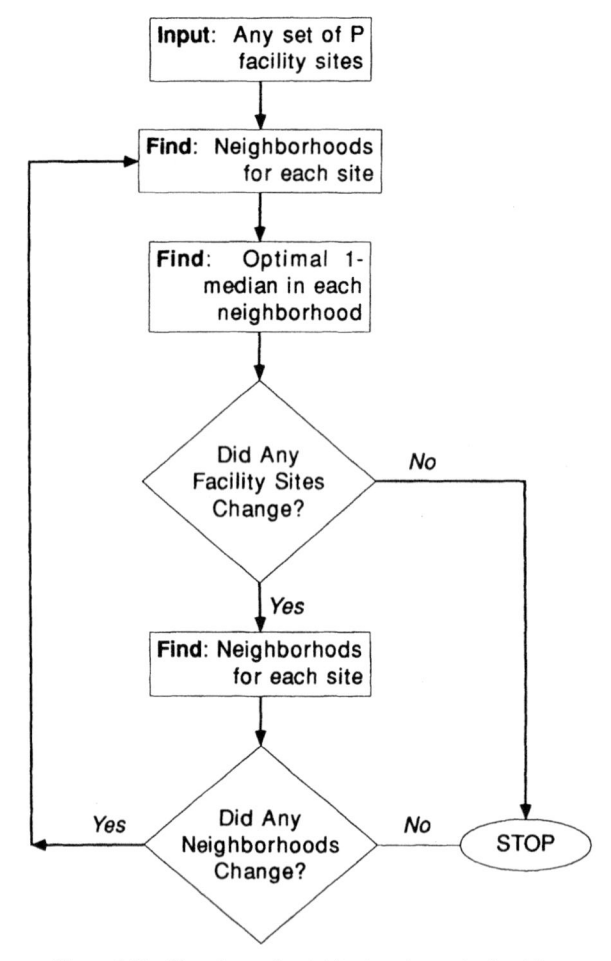

Figure 6.13. Flowchart of neighborhood search algorithm.

The neighborhood search algorithm can begin with any set of P facility sites. For example, we could begin with the P sites identified by the myopic search algorithm. For each facility site, the algorithm identifies the set of demand nodes that constitute the neighborhood around the facility site. Within each neighborhood, the optimal 1-median is found. If any sites have changed, the algorithm reallocates demands to the nearest facility and forms new neighborhoods. If any of the neighborhoods change, the algorithm again finds the 1-median within each neighborhood, and so on.

Returning to the neighborhoods identified for the myopic 5-median solution shown in Figure 6.12, the only neighborhood in which the location of the 1-median is suboptimal is that associated with node G. Specifically, we move the facility from node G to node H. This reduces the total demand-weighted

Heuristic Algorithms for the P-Median Problem

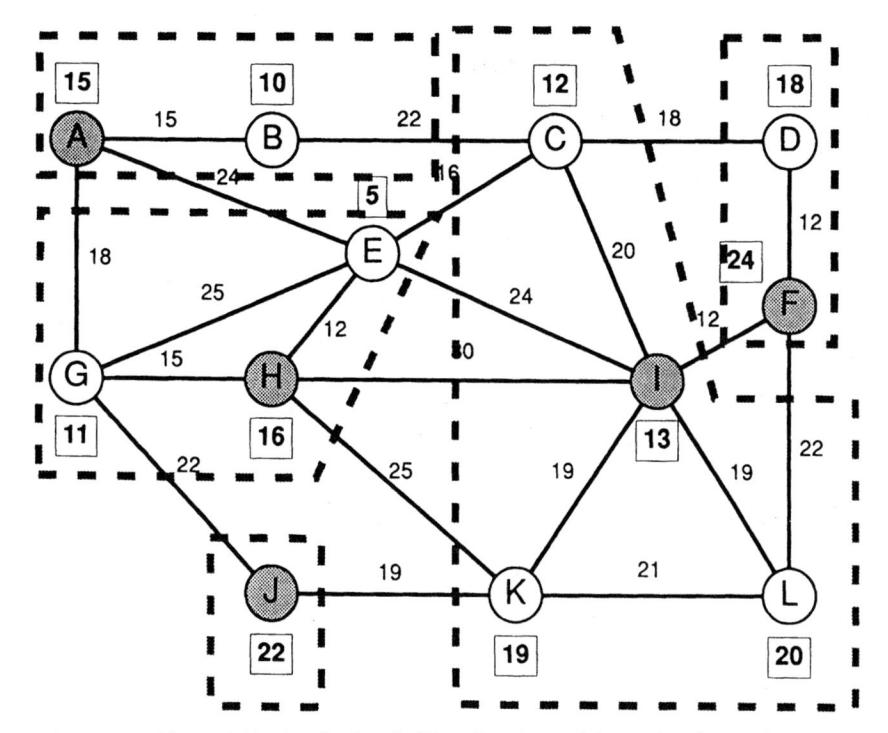

Figure 6.14. New neighborhoods after facility relocation and demand node reassignment.

distance from 1707 to 1632 (reducing the average distance from 9.227 to 8.822 units). We now attempt to reassign demands to the new set of facilities and find that node E should now be assigned to the facility at node H. This further reduces the total demand-weighted distance to 1572 and the average distance to 8.497. The resulting neighborhoods are shown in Figure 6.14. Within each neighborhood, we again find the optimal 1-median. Now the 1-median locations do not change and so the algorithm stops.

One of the limitations associated with the neighborhood search algorithm is that, in evaluating the impact of any relocation decision, only the effects on those nodes in the neighborhood are considered. The potential benefit to nodes outside of the neighborhood is not considered in deciding whether or not a relocation should be made. As an example, the reduction in the distance associated with serving node E (as a result of its being reassigned from the facility at node A to the relocated facility at node H) was not considered in deciding whether or not to move the facility from G to H. As a result of this limitation, some relocations which are beneficial in a global (but not local or neighborhood) sense are not considered.

This leads us to consider an *exchange algorithm* as an alternative improvement procedure. As indicated above, the algorithm is similar in spirit to the substitution algorithm outlined in Chapter 4 for the maximum covering

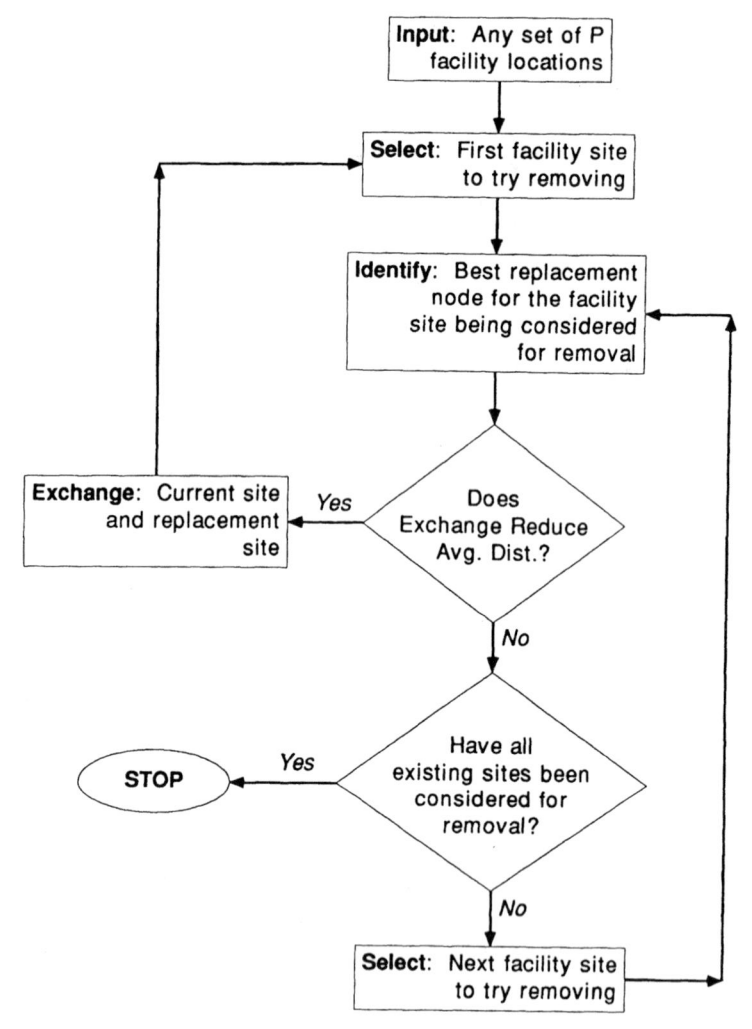

Figure 6.15. Flowchart of exchange or substitution algorithm.

problem. Figure 6.15 is a flowchart of the exchange algorithm. In Chapter 4 we outlined a substitution algorithm in which the first substitution of a node that was not in the solution for a node that is in the solution that improves the total coverage (objective function) was accepted. The algorithm shown in Figure 6.15 is slightly different in that we find the *best* node to enter the solution in place of the one being considered for removal. As indicated in Chapter 4, either approach is valid; either substitution approach could be adopted with either objective function. The solution that would result would depend on the choice of substitution procedure. [Teitz and Bart (1968) were among the first to propose an exchange heuristic for the P-median problem.

Heuristic Algorithms for the P-Median Problem

Table 6.5 Summary of Exchange Opportunities from Neighborhood Search Solution of Figure 6.14 (Facilities at Nodes A, F, H, I, and J with Total Demand-Weighted Distance of 1572)

Remove	Replace With	Total Demand-Weighted Distance	Change*	Average Distance
A	B	1647	75	8.903
F	D	1620	48	8.757
H	E	1689	117	9.130
I	L	1444	−128	7.805
J	K	1629	57	8.805

*Positive values indicate a degradation or worsening of the objective function; negative values indicate an improvement.

Recently, Densham and Rushton (1992) have proposed a number of enhancements to the algorithm as well as specific data structures designed to accelerate the algorithm when applied to large networks.]

To illustrate the exchange algorithm, consider the solution to the 5-median problem for Figure 6.14 obtained using the neighborhood search heuristic. Table 6.5 gives the best substitute node for each of the five nodes in the solution. The solution shown in Figure 6.14 involves locating at nodes A, F, H, I, and J for a total demand-weighted distance of 1572. If we consider removing node A, the best node to insert to replace node A is node B. This *increases* the total demand-weighted distance by 75 units to 1647. Similarly, removing node F and replacing it by its best substitute, node D, increases the objective function by 48 units to 1620. Removing either node H or node J also degrades the objective function. However, if we remove node I and replace it by its best substitute, node L, the objective function decreases by 128 units to 1444 (or an average distance of 7.805). We make this change to obtain a new solution with facilities located at nodes A, F, H, J, and L. This solution is shown in Figure 6.16. Note that following the flowchart shown in Figure 6.15, we would calculate the best possible improvement associated with removing node A, then with removing node F, then with removing node H, and then with removing node I. At that point, we would have found an exchange that improves the solution. We would make that exchange, without evaluating the possible impact of removing node J and replacing it by its best substitute node. Table 6.5, however, shows this impact for completeness.

Also, note that we could not obtain this improved solution using the neighborhood search technique since obtaining the improvement necessitates *simultaneously* moving a facility and reallocating demands to the new sites. The improvement cannot be obtained by either moving a facility (without reallocating demands) or reallocating demands to different facility sites (without moving facilities).

Having made an exchange, we would then return to the second box in the flowchart of Figure 6.15. Table 6.6 shows the result of removing each of the

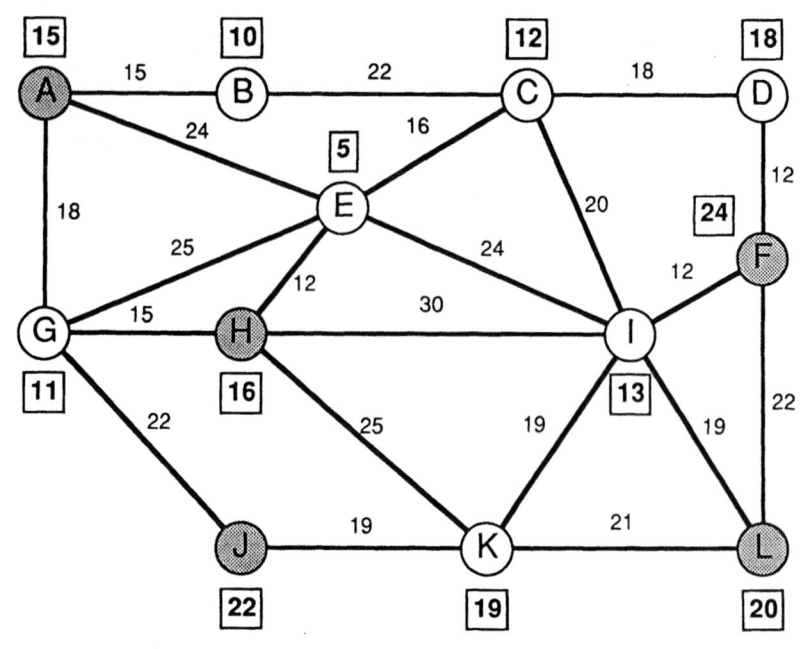

Figure 6.16. Solution after exchange algorithm.

facility sites in the new solution and replacing each with its best substitute node. Each such candidate exchange degrades the solution. Therefore, we retain the solution shown in Figure 6.16 and the algorithm would stop.

The substitutions that we attempted in Table 6.6 failed to identify a solution with a smaller total demand-weighted distance than that found in Table 6.5. Using the Lagrangian procedure discussed in the next section, we will be able to show (in this case) that the solution found in Table 6.5 is optimal. However, the substitutions attempted in Table 6.6 did identify five

Table 6.6 Summary of Exchange Opportunities from Improved Solution of Table 6.5 (Facilities at Nodes A, F, H, J, and L with Total Demand-Weighted Distance of 1444)

Remove	Replace With	Total Demand-Weighted Distance	Change*	Average Distance
A	B	1447	3	7822
F	D	1487	43	8.038
H	E	1465	21	7.919
J	K	1501	57	8.114
L	K	1503	59	8.124

*Positive values indicate a degradation or worsening of the objective function; negative values indicate an improvement.

Heuristic Algorithms for the P-Median Problem

solutions, all of which are within 4.1 percent of the optimal solution. In fact, we found one solution that is only 0.21 percent worse than the optimal solution. Obtaining this solution involves moving the facility from node A to node B and then reallocating the demand at node C from the facility at node H to the new facility at node B and assigning the demand at node A to the facility at node B. Another solution (moving the facility from H to E) degrades the solution by less than 1.5 percent.

In many practical contexts, it is important to be able to identify near optimal solutions. Almost any mathematical formulation will fail to capture all of the complexities of applied problems. Solutions that are slightly suboptimal in a strictly mathematical sense may represent the best set of facility sites when unmodeled factors are considered. In the case of the P-median model, such factors might include, but would not be limited to: fixed facility location costs as discussed in Chapter 7, environmental concerns, and equity measures associated with the distribution of benefits across demands as measured perhaps by the covering and center objectives discussed in Chapters 4 and 5, respectively.

The flowcharts shown in Figure 6.13 (for the neighborhood search algorithm) and 6.15 (for the exchange heuristic) suggest applying these procedures to an initial solution with P facilities. Such a solution may come from any one of a number of sources. One natural source is the myopic algorithm. This is what we did in demonstrating the neighborhood search algorithm. Alternatively, the initial solution may come from another heuristic. This is what we did in illustrating the exchange heuristic. In that case, the initial solution came from the solution obtained after applying the neighborhood search algorithm. Finally, the initial solution may be chosen randomly. In that case, we should try a number of initial random starting solutions and adopt as the final solution the best set of facility sites found from among all of the starting solutions.

Finally, the flowcharts suggest that the improvement algorithms be applied *after* we obtain an initial solution of P facility sites. An alternative implementation would call for applying the improvement algorithm after each new facility is added in the myopic search heuristic. Figure 6.17 is a generic flowchart of such an algorithm. Note that the only difference between this flowchart and that shown in Figure 6.10 is the placement of the box indicating the improvement algorithm. In Figure 6.10, P facilities were located using the myopic algorithm and then an improvement algorithm was employed. In Figure 6.17 the improvement algorithm is used after each new facility is added. In general, the latter approach is more computationally demanding. We would expect the results obtained from this approach to be better than those obtained from the simpler algorithm outlined in Figure 6.10. As discussed in Section 6.6 below, this is not always the case.

We close this section with a brief discussion of the computational complexity associated with each of the algorithms we have outlined in this section. For the myopic algorithm, when we are adding the kth facility, for each

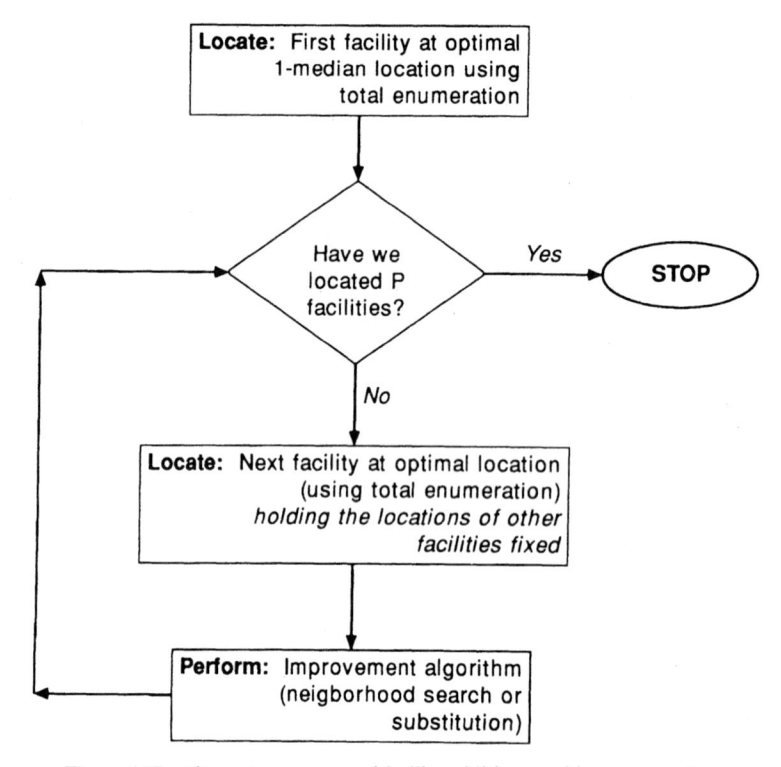

```
┌─────────────────────────────────┐
│ Locate: First facility at optimal│
│         1-median location using  │
│         total enumeration        │
└─────────────────────────────────┘
                │
                ▼
         ╱ Have we ╲      Yes
        ╱  located P ╲───────────▶ STOP
        ╲  facilities?╱
         ╲           ╱
                │ No
                ▼
┌─────────────────────────────────┐
│ Locate: Next facility at optimal │
│         location (using total    │
│         enumeration) holding the │
│         locations of other       │
│         facilities fixed         │
└─────────────────────────────────┘
                │
                ▼
┌─────────────────────────────────┐
│ Perform: Improvement algorithm   │
│          (neighborhood search or │
│          substitution)           │
└─────────────────────────────────┘
```

Figure 6.17. Alternate sequence of facility additions and improvements.

demand node/candidate site pair we need to perform one comparison (to determine whether the candidate site is closer than the closest of the first $k-1$ facilities that have been located) and one multiplication. Thus, for each new facility, we need to do $O(n^2)$ elementary operations. (In addition, we need to add all of the products for each candidate location and to find the minimum of these sums over all candidate locations at which a facility has not yet been located. This does not change the order of magnitude of the number of elementary operations for each new facility). Thus, *to locate P facilities, we need $O(Pn^2)$ operations for the myopic algorithm.*

For the two improvement algorithms, it is nearly impossible to predict how many times the algorithm will need to be executed beginning with any starting solution. However, we can determine the number of computations needed for each major iteration of the algorithm. For the neighborhood search algorithm, we must in essence solve a 1-median problem on each of P smaller networks or neighborhoods. However, since each node is in exactly one neighborhood, *we need $O(n^2)$ operations for each major iteration of the neighborhood search heuristic.*

For the exchange algorithm, we must consider every possible existing-facility/alternate-location pair. There are P existing facilities and $n - P$ alternate locations, or $P(n - P)$ such pairs to be considered. For each such pair, we need to examine each demand node to determine whether the new facility is closer to the demand node than was the facility that was originally assigned to serve the demand node. [If the node being tested for removal was assigned to serve the demand node, we need to determine whether the new facility is closer to the demand node than is the *second* closest of the original facilities. In the absence of sophisticated data structures that keep track of (and update between iterations) the second best, third best, ..., Pth best facilities for each demand node, finding the second closest of the original nodes will require $O(nP)$ comparisons at the beginning of each major iteration of the exchange algorithm. Since this can be done once at the beginning of each iteration and since the time required is less than that required for each major iteration, we can ignore this time in estimating the complexity of the exchange algorithm.] Thus, *each major iteration of the exchange algorithm requires $O(nP(N-P))$ operations.* If P is approximately $n/2$, this becomes $O(n^3)$.

6.5 AN OPTIMIZATION-BASED LAGRANGIAN ALGORITHM FOR THE P-MEDIAN PROBLEM

6.5.1 Methodological Development

In the previous section we outlined a number of heuristic algorithms for solving the P-median problem on a general network. Many of these algorithms have been used successfully in solving applied P-median problems. More importantly, the basic ideas underlying these heuristics—myopic search, neighborhoods, and local improvement procedures—may be readily modified to deal with extensions to the P-median problem. The problem with these and any heuristic approaches is that we do not know how good the solution is. In some cases, the solution may be optimal or very close to optimal; in other cases, the solution may be quite far from optimal. In this section we outline two optimization-based approaches to solving the P-median problem. Since both of these approaches are relaxations of the integer programming formulation of the P-median problem, we begin by restating the formulation of the P-median problem given above:

MINIMIZE $\quad \sum_i \sum_j h_i d_{ij} Y_{ij}$ $\hfill(6.2a)$

SUBJECT TO: $\quad \sum_j Y_{ij} = 1 \quad \forall\, i$ $\hfill(6.2b)$

$\quad\quad\quad\quad\quad\quad \sum_j X_j = P$ $\hfill(6.2c)$

$\quad\quad\quad\quad\quad\quad Y_{ij} - X_j \leq 0 \quad \forall\, i, j$ $\hfill(6.2d)$

$\quad\quad\quad\quad\quad\quad X_j = 0, 1 \quad \forall\, j$ $\hfill(6.2e)$

$\quad\quad\quad\quad\quad\quad Y_{ij} = 0, 1 \quad \forall\, i, j$ $\hfill(6.2f)$

To solve the *P*-median problem using Lagrangian relaxation, we can relax either constraint (6.2b) or constraint (6.2d). Below we outline the solution of the relaxed problem for each of these relaxations. Recall from the discussion of the use of Lagrangian relaxation for the maximum covering problem that, in using Lagrangian relaxation, we need to: (1) be able to solve the relaxed problem easily for fixed values of the Lagrange multipliers, (2) find primal feasible solutions from the relaxed solution, and (3) find improved Lagrange multipliers. For each of the two relaxations that we will consider, we focus first on the problem of solving the relaxed problem for fixed values of the Lagrange multipliers.

If we relax constraint (6.2b), we obtain

$$\text{MAX MIN}_{\lambda \quad X,Y} \quad \sum_i \sum_j h_i d_{ij} Y_{ij} + \sum_i \lambda_i \left[1 - \sum_j Y_{ij}\right]$$

$$= \sum_i \sum_j (h_i d_{ij} - \lambda_i) Y_{ij} + \sum_i \lambda_i \qquad (6.3)$$

SUBJECT TO:

$$\sum_j X_j = P \qquad (6.2c)$$

$$Y_{ij} - X_j \leq 0 \quad \forall\, i, j \qquad (6.2d)$$

$$X_j = 0, 1 \quad \forall\, j \qquad (6.2e)$$

$$Y_{ij} = 0, 1 \quad \forall\, i, j \qquad (6.2f)$$

For fixed values of the Lagrange multipliers, λ_i, we want to minimize the objective function. With the values of λ_i fixed, the second term of the objective function is a constant. To minimize the objective function, we would like to set $Y_{ij} = 1$ if its coefficient $h_i d_{ij} - \lambda_i$ is less than 0, and $Y_{ij} = 0$ otherwise. However, setting $Y_{ij} = 1$ means that we must also set $X_j = 1$ by constraint (6.2d), and constraint (6.2c) states that we can only set P of the X_j values to 1. Thus, to minimize the objective function for fixed values of the Lagrange multipliers, we begin by computing the value of setting each of the X_j values to 1. This value is given by $V_j = \sum_i \min\{0, h_i d_{ij} - \lambda_i\}$ for each candidate location j. We then find the P smallest values of V_j and set the corresponding values of $X_j = 1$ and all other values of $X_j = 0$. We then set

$$\bar{Y}_{ij} = \begin{cases} 1 & \text{if } X_j = 1 \text{ and } h_i d_{ij} - \lambda_i < 0 \\ 0 & \text{if not} \end{cases}$$

An Optimization-Based Lagrangian Algorithm for the P-Median Problem

If we relax constraint (6.2d), we obtain the following problem:

$$\text{MAX}_{\lambda} \text{ MIN}_{X,Y} \quad \sum_i \sum_j h_i d_{ij} Y_{ij} + \sum_i \sum_j \lambda_{ij}(Y_{ij} - X_j)$$

$$= \sum_i \sum_j (h_i d_{ij} + \lambda_{ij}) Y_{ij} - \sum_j \left(\sum_i \lambda_{ij} \right) X_j \quad (6.4a)$$

SUBJECT TO:
$$\sum_j Y_{ij} = 1 \quad \forall \, i \quad (6.2b)$$

$$\sum_j X_j = P \quad (6.2c)$$

$$X_j = 0, 1 \quad \forall \, j \quad (6.2e)$$

$$Y_{ij} \geq 0 \quad \forall \, i, j \quad (6.2f)$$

$$\lambda_{ij} \geq 0 \quad \forall \, i, j \quad (6.4b)$$

Again, for fixed values of the Lagrange multipliers, λ_{ij}, we want to minimize the objective function. In this case, the problem breaks into two separate subproblems: one in the allocation variables, Y_{ij}, and one in the location variables, X_j. These two subproblems are shown below:

Problem in the Allocation Variables Y_{ij} for Fixed Values of λ_{ij}

MINIMIZE
$$\sum_i \sum_j (h_i d_{ij} + \lambda_{ij}) Y_{ij} \quad (6.5)$$

SUBJECT TO:
$$\sum_j Y_{ij} = 1 \quad \forall \, i \quad (6.2b)$$

$$Y_{ij} \geq 0 \quad \forall \, i, j \quad (6.2f)$$

Problem in the Location Variables X_j for Fixed Values of λ_{ij}

MAXIMIZE
$$\sum_j \left(\sum_i \lambda_{ij} \right) X_j \quad (6.6)$$

SUBJECT TO:
$$\sum_j X_j = P \quad (6.2c)$$

$$X_j = 0, 1 \quad \forall \, j \quad (6.2e)$$

Note that the problem in X_j for fixed values of λ_{ij} becomes a maximization problem since in (6.4a) we are minimizing the negative of the sum shown in (6.6).

To solve the problem in Y_{ij}, we note that the problem further decomposes into subproblems for each demand location i. For each demand location, we identify the facility location $j_i^* = \text{argmin}_j\{h_i d_{ij} + \lambda_{ij}\}$. In other words, j_i^* is the facility location that minimizes $h_i d_{ij} + \lambda_{ij}$ for demand node i. We then set $Y_{ik} = 1$ if $k = j_i^*$ and $Y_{ik} = 0$ for all other facility locations k.

To solve for the optimal values of the location variables, X_j (again for fixed values of the Lagrange multipliers, λ_{ij}), we find the P largest values of $\Sigma_i \lambda_{ij}$. We then set the corresponding X_j values to 1 and all other X_j values to 0.

In either relaxation, we can find a primal feasible solution related to the Lagrangian solution by ignoring the allocation variables, Y_{ij}, and siting the facilities at those sites for which $X_j = 1$. We then can let $\mathbf{S} = \{j | X_j = 1\}$; that is, $\mathbf{S}$ is the set of facility locations. For each demand node i, we then find $\hat{j}_i = \text{argmin}_{j \in \mathbf{S}}\{d_{ij}\}$; that is, $\hat{j}_i$ is the *open* facility that is closest to node i. We then set $\hat{Y}_{ik} = 1$ if $k = \hat{j}_i$ and $\hat{Y}_{ik} = 0$ for all other locations k as before. (Note that we are using $\hat{Y}_{ik}$ to denote the feasible allocation of demands that results from the use of the location decisions in the relaxed problem.) We can then evaluate the P-median objective function, $\Sigma_i \Sigma_j h_i d_{ij} \hat{Y}_{ij}$. This value is an upper bound on the solution. Clearly, the best such (smallest) value over all iterations of the Lagrangian relaxation procedure should be used as the upper bound.

In either relaxation, the Lagrange multipliers may be revised using a standard subgradient optimization procedure as was done in the case of the maximum covering location problem. In this case, however, we need to modify the equations given in Chapter 4 to account for the fact that we now have a minimization problem and, when we relax constraint (6.2b), we are relaxing an equality constraint (as opposed to an inequality constraint as was done in the discussion of the maximum covering problem). Specifically, when we relax constraint (6.2b), we need to compute a stepsize, t^n, at the nth iteration of the Lagrangian procedure as follows:

$$t^n = \frac{\alpha^n(\text{UB} - \mathscr{L}^n)}{\sum_i \left\{\sum_j Y_{ij}^n - 1\right\}^2} \tag{6.7}$$

where

t^n = the stepsize at the nth iteration of the Lagrangian procedure

α^n = a constant on the nth iteration, with α^1 generally set to 2

UB = the best (smallest) upper bound on the P-median objective function

$\mathscr{L}^n$ = the objective function of the Lagrangian function (6.3) [or (6.4a) in the case of relaxing constraint (6.2d)] on the nth iteration

Y_{ij}^n = the optimal value of the allocation variable, Y_{ij}, on the nth iteration

An Optimization-Based Lagrangian Algorithm for the P-Median Problem

The Lagrange multipliers are updated using the following equation:

$$\lambda_i^{n+1} = \max\left\{0, \lambda_i^n - t^n\left(\sum_j Y_{ij}^n - 1\right)\right\} \qquad (6.8)$$

Note that despite the fact that we are relaxing an equality constraint (and would therefore generally expect that the Lagrange multipliers could be unrestricted in sign), we can restrict the Lagrange multipliers to nonnegative values as long as all demands, h_i, and all distances, d_{ij}, are nonnegative. Doing so will improve the values of the lower bounds that we obtain from the Lagrangian objective function (6.3).

When updating the Lagrange multipliers for the second relaxation [in which (6.2d) is relaxed], these equations are modified as follows:

$$t^n = \frac{\alpha^n(\text{UB} - \mathscr{L}^n)}{\sum_i \sum_j \{Y_{ij}^n - X_j^n\}^2} \qquad (6.9)$$

where all notation is as defined above and

X_j^n = the optimal value of the location variable, X_j, on the nth iteration

Finally, the Lagrange multipliers themselves are updated as follows:

$$\lambda_i^{n+1} = \max\left\{0, \lambda_i^n + t^n(Y_{ij}^n - X_j^n)\right\} \qquad (6.10)$$

With either relaxation, we may initialize the Lagrange multipliers in a variety of ways. One simple approach is to initialize all Lagrange multipliers to some constant value. This is the approach adopted below in the solution of an example problem.

6.5.2 Numerical Example

To illustrate the P-median formulation and the relaxations we will outline, consider the example network shown in Figure 6.18. (To simplify the computations and to reduce the number of numerical values that need to be shown, we have reduced the size of the sample network in this section. Note that this network consists of nodes A, E, G, and H of the network shown in Figure 6.11, with the nodes sequentially labeled in Figure 6.18.)

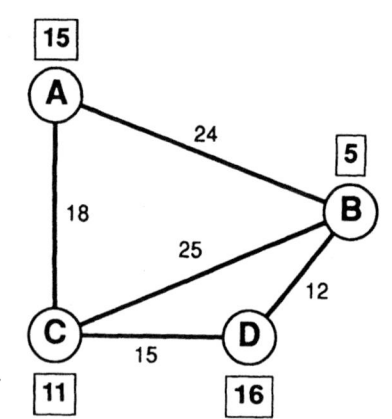

Figure 6.18. Sample network for Lagrangian relaxation example.

For completeness, we begin by stating the P-median formulation for this example problem (for $P = 2$):

MINIMIZE

$$0Y_{AA} + 360Y_{AB} + 270Y_{AC} + 495Y_{AD}$$
$$+ 120Y_{BA} + 0Y_{BB} + 125Y_{BC} + 60Y_{BD}$$
$$+ 198Y_{CA} + 275Y_{CB} + 0Y_{CC} + 165Y_{CD}$$
$$+ 528Y_{DA} + 192Y_{DB} + 240Y_{DC} + 0Y_{DD} \quad (6.2a)$$

SUBJECT TO:

$$Y_{AA} + Y_{AB} + Y_{AC} + Y_{AD} = 1$$
$$Y_{BA} + Y_{BB} + Y_{BC} + Y_{BD} = 1$$
$$Y_{CA} + Y_{CB} + Y_{CC} + Y_{CD} = 1$$
$$Y_{DA} + Y_{DB} + Y_{DC} + Y_{DD} = 1 \quad (6.2b)$$
$$X_A + X_B + X_C + X_D = 2 \quad (6.2c)$$

$$\begin{array}{llll} Y_{AA} \leq X_A & Y_{BA} \leq X_A & Y_{CA} \leq X_A & Y_{DA} \leq X_A \\ Y_{AB} \leq X_B & Y_{BB} \leq X_B & Y_{CB} \leq X_B & Y_{DB} \leq X_B \\ Y_{AC} \leq X_C & Y_{BC} \leq X_C & Y_{CC} \leq X_C & Y_{DC} \leq X_C \\ Y_{AD} \leq X_D & Y_{BD} \leq X_D & Y_{CD} \leq X_D & Y_{DD} \leq X_D \end{array}$$

$$(6.2d)$$

$$X_A, X_B, X_C, X_D = 0, 1 \quad (6.2e)$$
$$Y_{AA}, Y_{AB}, Y_{AC}, Y_{AD}, Y_{BA}, Y_{BB}, Y_{BC}, Y_{BD},$$
$$Y_{CA}, Y_{CB}, Y_{CC}, Y_{CD}, Y_{DA}, Y_{DB}, Y_{DC}, Y_{DD} = 0, 1 \quad (6.2f)$$

An Optimization-Based Lagrangian Algorithm for the P-Median Problem

If we relax constraint (6.2b), we obtain the first relaxation discussed above:

$$\begin{array}{c}\text{MAX}\\\lambda\end{array}\begin{array}{c}\text{MIN}\\X,Y\end{array} (0 - \lambda_A)Y_{AA} + (360 - \lambda_A)Y_{AB} + (270 - \lambda_A)Y_{AC} + (495 - \lambda_A)Y_{AD}$$

$$+ (120 - \lambda_B)Y_{BA} + (0 - \lambda_B)Y_{BB} + (125 - \lambda_B)Y_{BC} + (60 - \lambda_B)Y_{BD}$$

$$+ (198 - \lambda_C)Y_{CA} + (275 - \lambda_C)Y_{CB} + (0 - \lambda_C)Y_{CC} + (165 - \lambda_C)Y_{CD}$$

$$+ (528 - \lambda_D)Y_{DA} + (192 - \lambda_D)Y_{DB} + (240 - \lambda_D)Y_{DC} + (0 - \lambda_D)Y_{DD}$$

$$+ \lambda_A + \lambda_B + \lambda_C + \lambda_D \qquad (6.3)$$

SUBJECT TO:

Constraints (6.2c), (6.2d), (6.2e), and (6.2f)

For illustrative purposes, let us begin with all Lagrange multipliers set to 150. We then obtain $V_A = -180$, since we would set both Y_{AA} and Y_{BA} to 1 if we had a facility located at node A; $V_B = -150$, since we would only set Y_{BB} to 1 if we had a facility at B; $V_C = -150$ for similar reasons; and $V_D = -240$, since we would set both Y_{BD} and Y_{DD} to 1 if we had a facility at node D. Since we want to select two facilities, we choose to locate at nodes A and D (the two nodes with the smallest V values). The Lagrangian objective function is then $600 - 180 - 240 = 180$ ($\lambda_A + \lambda_B + \lambda_C + \lambda_D + V_A + V_D$). This is the lower bound on the objective function at this iteration. If we locate at nodes A and D, the 2-median objective function is 225 with both nodes B and C being assigned to the facility at node D. Thus, the upper bound on the objective function is 225. Since the lower and upper bounds are not the same, we compute the Y_{ij} values and the violations of the relaxed constraints to evaluate new Lagrange multipliers.

With facilities located at nodes A and D, we would set $Y_{AA} = Y_{BA} = Y_{BD} = Y_{DD} = 1$ and all other Y_{ij} values to 0 *in the Lagrangian problem*. This means that node B would be assigned to two facilities and node C would not be assigned to any facility. Constraints (6.2b) are therefore violated for these two nodes. The sum of the squared violations is 2. Using (6.7), we obtain a stepsize of

$$45 = \frac{\alpha^1(\text{UB} - \mathscr{L}^1)}{\sum_i \left\{ \sum_j Y_{ij}^1 - 1 \right\}^2} = \frac{2(225 - 180)}{2}$$

Next, we reduce λ_B by 45 to obtain 105 and increase λ_C by 45 to obtain 105. This ends the first iteration.

At the second iteration, we have $V_A = -150$ (assigning only node A to a facility at A if we locate there), $V_B = -105$ (assigning only node B to a facility at B if we locate there), $V_c = -195$ (assigning only node C to a facility at C if we locate there), and $V_D = -225$ (assigning nodes B, C, and D to a facility at node D if we locate there). Again picking the two smallest values, we set $X_C = X_D = 1$ (the other two X_j values are set to 0) and we obtain an objective function for the Lagrangian problem of $600 - 195 - 225 = 180$ ($\lambda_A + \lambda_B + \lambda_C + \lambda_D + V_C + V_D$). Since this is not better than the previous lower bound, the lower bound remains at 180. In this example, we will halve the constant α whenever we fail to improve the lower bound, so we will have $\alpha^2 = 1$. Locating at nodes C and D, the 2-median objective function is 330 (with node A assigned to the facility at C and node B assigned to the facility at D). Since this is worse than the previous upper bound, the best upper bound remains at 225.

Now with facilities located at nodes C and D, we would set $Y_{CC} = Y_{BD} = Y_{CD} = Y_{DD} = 1$ and all other Y_{ij} values to 0 *in the Lagrangian problem*. This means that node C would be assigned to two facilities and node A would not be assigned to any facility. Constraints (6.2b) are now violated for these two nodes. The sum of the squared violations is again 2. Using (6.7), we obtain a stepsize of

$$22.5 = \frac{\alpha^2(\text{UB} - \mathscr{L}^2)}{\sum_i \left\{ \sum_j Y_{ij}^1 - 1 \right\}^2} = \frac{1(225 - 180)}{2}$$

Next, we reduce λ_C by 22.5 to obtain 172.5 and increase λ_A by 22.5 to obtain 172.5. This ends the second iteration.

At the third iteration, we have $V_A = -172.5$ (assigning only node A to a facility at A if we locate there), $V_B = -105$ (assigning only node B to a facility at B if we locate there); $V_C = -172.5$ (assigning only node C to a facility at C if we locate there), and $V_D = -202.5$ (assigning nodes B, C, and D to a facility at node D if we locate there). Again picking the two smallest values (and breaking the tie between nodes A and C arbitrarily), we set $X_A = X_D = 1$ (the other two X_j values are set to 0) and we obtain an objective function for the Lagrangian problem of $600 - 172.5 - 202.5 = 225$ ($\lambda_A + \lambda_B + \lambda_C + \lambda_D + V_A + V_D$). Since this lower bound equals the best upper bound that we have found so far (corresponding to locating at nodes A and D in iteration 1), we stop. In this case, we have proven that the solution we obtained from the Lagrangian procedure is optimal.

An Optimization-Based Lagrangian Algorithm for the P-Median Problem

Now let us turn to the relaxation of constraint (6.2d). The Lagrangian problem that we now need to solve is

$$\begin{aligned}
\text{MAX} & \text{ MIN} \\
\lambda & \ X,Y \\
& (0 + \lambda_{AA})Y_{AA} + (360 + \lambda_{AB})Y_{AB} + (270 + \lambda_{AC})Y_{AC} + (495 + \lambda_{AD})Y_{AD} \\
& + (120 + \lambda_{BA})Y_{BA} + (0 + \lambda_{BB})Y_{BB} + (125 + \lambda_{BC})Y_{BC} + (60 + \lambda_{BD})Y_{BD} \\
& + (198 + \lambda_{CA})Y_{CA} + (275 + \lambda_{CB})Y_{CB} + (0 + \lambda_{CC})Y_{CC} + (165 + \lambda_{CD})Y_{CD} \\
& + (528 + \lambda_{DA})Y_{DA} + (192 + \lambda_{DB})Y_{DB} + (240 + \lambda_{DC})Y_{DC} + (0 + \lambda_{DD})Y_{DD} \\
& - (\lambda_{AA} + \lambda_{BA} + \lambda_{CA} + \lambda_{DA})X_A \\
& - (\lambda_{AB} + \lambda_{BB} + \lambda_{CB} + \lambda_{DB})X_B \\
& - (\lambda_{AC} + \lambda_{BC} + \lambda_{CC} + \lambda_{DC})X_C \\
& - (\lambda_{AD} + \lambda_{BD} + \lambda_{CD} + \lambda_{DD})X_D
\end{aligned} \quad (6.4)$$

SUBJECT TO: Constraints (6.2b), (6.2c), (6.2e), and (6.2f)

Consider the following initialization of the Lagrange multipliers:

$$\lambda_{ij} = \begin{cases} h_i \cdot \min_{j, j \neq i} \{d_{ij}\} - \epsilon & \text{if } i = j \\ 0 & \text{if } i \neq j \end{cases}$$

In other words, if $i = j$, then the Lagrange multiplier is set equal to the demand at node i times the distance between node i and the closest node other than node i minus some very small number. If $i \neq j$, then the Lagrange multiplier is set equal to 0. This will cause us to locate (on the first iteration) at the P nodes with the largest product of the demand and the second closest node. These are the P nodes which individually would contribute the most to the objective function if we did not locate at the node but could serve all nodes from the second closest facility. The initial allocation variables will be

$$Y_{ij} = \begin{cases} 1 & \text{if } i = j \\ 0 & \text{if not} \end{cases}$$

This will cause the P nonzero λ_{ij} values associated with the selected sites to cancel with the coefficients of the corresponding allocation variables. The Lagrangian objective function will then have a value equal to the sum over all demand nodes at which we do not locate a facility of the demand at the node times the distance to the closest other node. (These statements need to be interpreted somewhat loosely in light of the ϵ term which is introduced to avoid ties in the objective function on the initial iteration. In principle, ties are not a problem since they can be broken arbitrarily.)

Table 6.7 Initial Lagrange Multipliers for Second Relaxation

Node i	Node j			
	A	B	C	D
A	269.9	0.0	0.0	0.0
B	0.0	59.9	0.0	0.0
C	0.0	0.0	164.9	0.0
D	0.0	0.0	0.0	191.9

In this case, we initialize the Lagrange multipliers as shown in Table 6.7. With these Lagrange multipliers, we locate at nodes A and D. The Lagrangian objective function is 224.8 (which is the sum of the λ_{BB} and λ_{CC} terms—for the sites at which we do not locate facilities—as predicted above). This is the initial lower bound on the objective function. Locating at nodes A and D gives an objective function for the 2-median problem of 225. This is the initial upper bound on the objective function. Since all demands and all distances are integer valued in this example, we know that the objective function must be an integer. Therefore, since the lower and upper bounds differ by less than 1, we can stop, knowing that we have obtained an optimal solution to the problem.

In this example, we obtained a provably optimal solution on the first iteration. This will rarely be the case. For example, Figure 6.19 plots the difference between the lower and upper bounds as a percentage of the lower bound for the case $P = 1$. To generate this figure, we started with the same

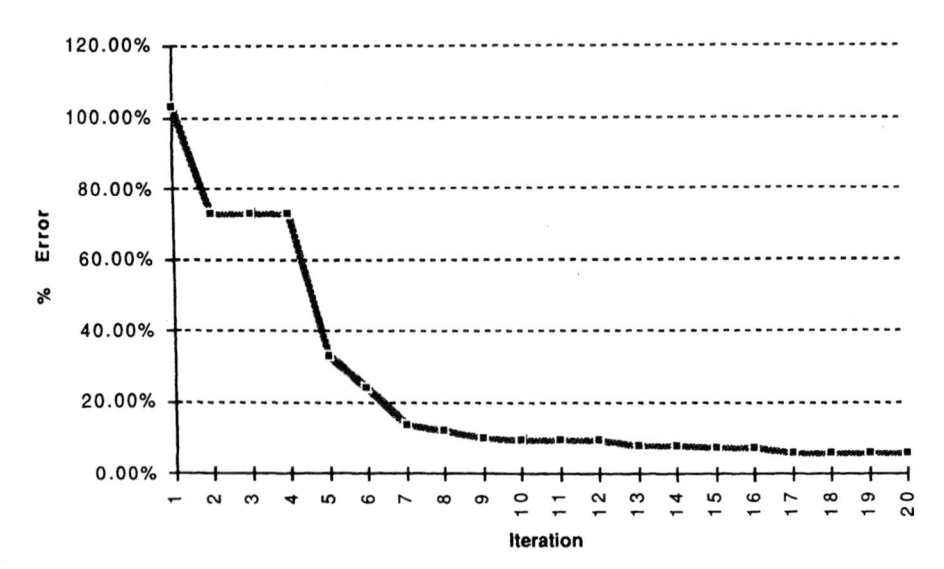

Figure 6.19. Percentage difference between lower and upper bounds using second relaxation for 1-median problem on the network of Figure 6.18.

initial Lagrange multipliers and halved α whenever the lower bound did not improve in three consecutive iterations. Even after 20 iterations, the difference between the lower and upper bounds still exceeds 6 percent.

To further convince yourself that the formula for the initial multipliers may not always work well, consider again the problem of locating two medians on the network of Figure 6.11. It should be clear that with networks larger than the one shown in Figure 6.18 and with a small number of facilities, this initial choice for the Lagrange multipliers is not likely to work well since many demands will have to be assigned to facilities at nodes further than the next closest node. For $P = 2$, the initial lower bound for this problem with Lagrange multipliers generated using the equation above is 2004.88, while the true optimal value can be shown to be 3145, or over a 56 percent difference between the initial lower bound and the true optimal solution. Note that the 2-median objective computed using the first iteration locations for this problem is 4415, or over twice the lower bound and over 40 percent above the optimal value. In short, the fact that we obtained an optimal solution using the initial Lagrange multipliers shown in Table 6.7 is fortuitous and is largely due to the fact that we are using a very small network.

Christofides and Beasley (1982) compared the two relaxations using a slightly different formulation of the P-median problem. In their formulation, they aggregated constraints (6.2d) to obtain

$$\sum_i Y_{ij} \leq nX_j \quad \forall j \qquad (6.2d')$$

Aggregating the constraints in this manner reduces the number of Lagrange multipliers needed, since only one Lagrange multiplier is needed for each candidate location instead of one for each demand node/candidate location pair. However, it is well known that the bound obtained from the linear programming relaxation that results from aggregating the constraints in this way and then relaxing the integrality constraints (6.2e) and (6.2f) is weaker than is the linear programming bound using constraints (6.2d). Christofides and Beasley found that the first relaxation (when applied to the aggregate formulation) is superior.

The fact that the second relaxation results in having to do more iterations than are required using the first relaxation should not be surprising since there are far more constraints being relaxed when (6.2d) is relaxed than when (6.2b) is relaxed. Also, the second relaxation decouples the location variables (X_j) from the allocation variables (Y_{ij}), which would intuitively give a weaker relaxation. For all of these reasons, SITATION uses the first relaxation.

6.5.3 Extensions and Enhancements to the Lagrangian Procedures

In some cases, the Lagrange multipliers tend to oscillate between two sets of values. Crowder (1976) suggested an approach that tends to damp out such fluctuations. This capability is incorporated into the SITATION computer program.

In addition to incorporating Crowder's damping factor, other enhancements to the procedure outlined in Section 6.5.1 are possible. For example, the procedure takes the initial best upper bound to be the value of the (primal) solution obtained at the first iteration. In other words, the initial upper bound is the total demand-weighted distance associated with the sites selected on the initial iteration. An improved initial upper bound can often be found by applying either the myopic algorithm outlined in Section 6.4 or the myopic algorithm coupled with either of the improvement algorithms (the neighborhood search or exchange algorithm). Also, either at each iteration or at each iteration at which the upper bound improves, the algorithm can again apply an improvement algorithm to attempt to reduce the best upper bound even further. SITATION incorporates all of these extensions. Results from applying SITATION to the network shown in Figure 6.11 as well as to the 88-node network presented in Appendix G are given in Section 6.6 below.

6.6 COMPUTATIONAL RESULTS USING THE HEURISTIC ALGORITHMS AND THE LAGRANGIAN RELAXATION ALGORITHM

In this section we outline computational results for the heuristic and Lagrangian-based algorithms described in Sections 6.4 and 6.5. We begin with results for the 12-node problem shown in Figure 6.11.

Table 6.8 gives the objective function values for: (1) the myopic algorithm, (2) the exchange heuristic with substitution applied after each additional facility is added and with substitution applied only after all facilities (in the

Table 6.8 Results of Using Each Algorithm on the Network of Figure 6.11

Number of Facilities	Myopic	Exchange Heuristic Each Iteration	Exchange Heuristic Last Iteration	Neighborhood Search Each Iteration	Neighborhood Search Last Iteration	Lagrangian Algorithm Value	Lagrangian Algorithm Iterations
1	25.795	25.795	25.795	25.795	25.795	25.795	
2	17.000	17.000	17.000	17.000	17.000	17.000	15
3	14.276	13.503	13.503	14.276	14.276	13.178	44
4	11.659	10.719	10.184	11.659	11.659	10.184	27
5	9.227	8.341	7.805	8.497	8.497	7.805	48
6	7.173	6.389	5.854	6.443	6.443	5.854	110
7	5.222	4.038	4.038	4.492	4.038	4.038	18
8	3.600	2.870	2.870	3.195	3.600	2.870	19
9	2.303	1.978	1.978	2.027	2.303	1.978	10
10	1.135	1.135	1.135	1.135	1.135	1.135	8
11	0.324	0.324	0.324	0.324	0.324	0.324	5
12	0.000	0.000	0.000	0.000	0.000	0.000	1

Table 6.9 Lagrangian Parameter Settings for Testing 12-Node Network

Improvement algorithm used at end	Yes
Terminal improvement algorithm	Exchange
Critical percentage difference	0.0001
Maximum number of iterations	400
Minimum α value allowed	0.00005
Number of failures before changing α	4
Restart failure count on improved bound	Yes
Crowder damping term	0.3
Neighborhood search on improved solution	Yes
Initial Lagrange multiplier	10 * average demand

row in question) have been added, (3) the neighborhood search with the algorithm applied after each additional facility is added and with the algorithm applied only after all facilities (in the row in question) have been added, and (4) the Lagrangian procedure using the first relaxation. In computing the Lagrangian objective function, the parameter settings shown in Table 6.9 were used. Table 6.8 also shows the number of iterations needed for the Lagrangian procedure to confirm that the lower and upper bounds were within 0.0001 percent of the lower bound. Table 6.10 gives the difference between the objective function for each heuristic algorithm and the Lagrangian solution as a percentage of the Lagrangian objective function. The same information is presented in Figure 6.20.

Table 6.10 Percentage Differences Between Heuristic Objective Functions and Lagrangian Values

Number of Facilities	Myopic	Exchange Heuristic Each Iteration	Exchange Heuristic Last Iteration	Neighborhood Search Each Iteration	Neighborhood Search Last Iteration
1	0.0%	0.0%	0.0%	0.0%	0.0%
2	0.0%	0.0%	0.0%	0.0%	0.0%
3	8.3%	2.5%	2.5%	8.3%	8.3%
4	14.5%	5.3%	0.0%	14.5%	14.5%
5	18.2%	6.9%	0.0%	8.9%	8.9%
6	22.5%	9.1%	0.0%	10.1%	10.1%
7	29.3%	0.0%	0.0%	11.2%	0.0%
8	25.4%	0.0%	0.0%	11.3%	25.4%
9	16.4%	0.0%	0.0%	2.5%	16.4%
10	0.0%	0.0%	0.0%	0.0%	0.0%
11	0.0%	0.0%	0.0%	0.0%	0.0%
12	0.0%	0.0%	0.0%	0.0%	0.0%
Average (2–11)	13.5%	2.4%	0.2%	6.7%	8.4%

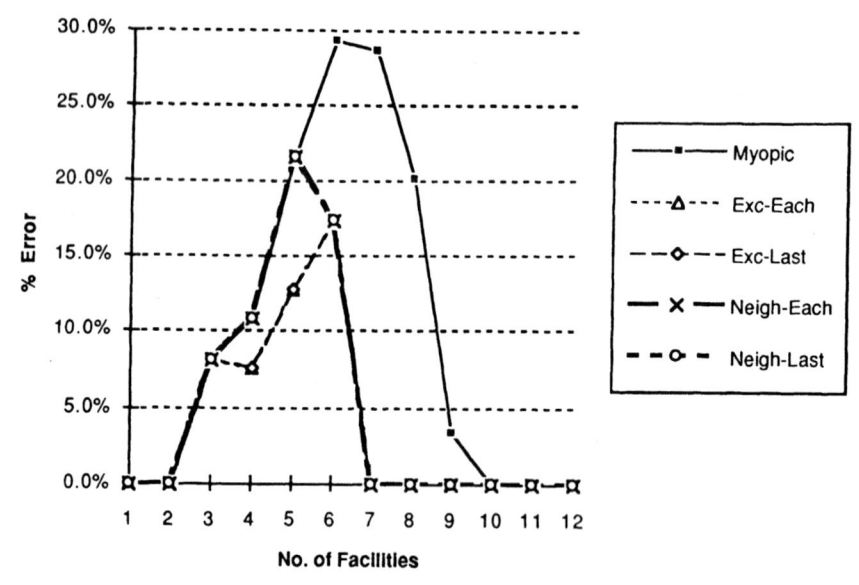

Figure 6.20. Percentage differences between heuristic objective functions and Lagrangian values.

Several points are worth noting. First, the solution obtained from the myopic algorithm can be as much as 30 percent off of the Lagrangian (optimal) solution. Second, for any given number of facilities, the exchange heuristic results are always no worse than the neighborhood search algorithm results. Third, sometimes it is better to perform the neighborhood search algorithm after each additional facility is added and sometimes it is better to use the algorithm only after all facilities have been added. Fourth, note that the cases which are in some sense the most difficult to solve (as measured either by the number of Lagrangian iterations needed or by the percentage error between the heuristic solution and the Lagrangian solution) are those with an intermediate number of facilities. When the number of facilities is small, all algorithms seem to do well. Similarly, when the number of facilities is large, all algorithms do well. When the number of facilities being located is neither large nor small, the problems seem to be more difficult.

Finally, the last row of Table 6.10 shows the average percentage error for each of the different heuristics averaged over the $P = 2$ through $P = 11$ cases. The 1-median case was excluded from the average since all algorithms simply use the myopic search procedure which is optimal when $P = 1$. The $P = 12$ case was also excluded from the average since the solution in this case is trivial: locate a facility at each node. Figure 6.21 shows the average, minimum, and maximum percentage error for the $P = 2$ through $P = 11$ cases for all of the heuristic algorithms. From the figure it is clear that the exchange algorithm, with substitution performed after all nodes have been added using the myopic algorithm, is the best for this network.

Computational Results Using Heuristic and Lagrangian Algorithms 235

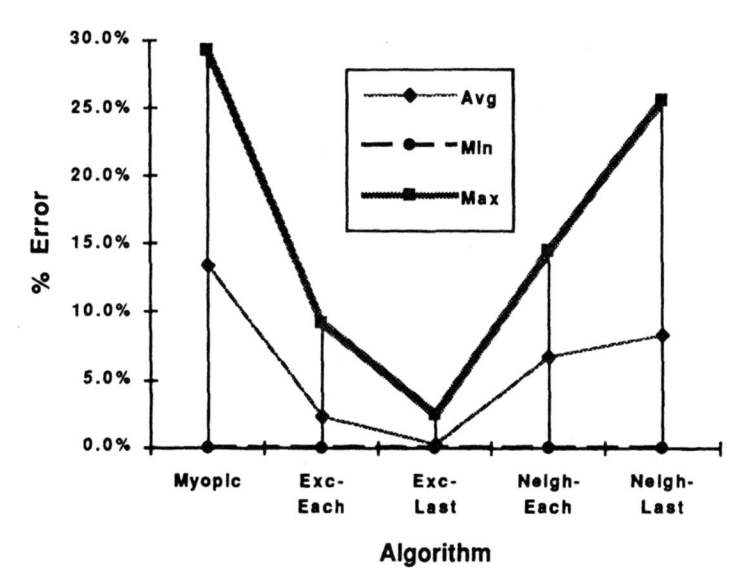

Figure 6.21. Average, minimum, and maximum percentage error by algorithm for the network of Figure 6.11.

We now turn the 88-node network described in Appendix G. All 88 cases were solved using: (1) the myopic algorithm, (2) the exchange algorithm with substitution attempted after each facility is added, (3) the neighborhood search algorithm with the search performed after each facility is added, and (4) the first Lagrangian relaxation approach. The parameters of the Lagrangian procedure were identical to those shown in Table 6.9 except that the percentage difference between the upper and lower bounds was set to 0.01 instead of 0.0001. Also, the maximum number of iterations allowed was equal to the maximum of 400 and 10 times the number of medians being located.

In 84 of the 88 cases, the Lagrangian algorithm converged so that the upper bound was within 0.01 percent of the lower bound. For the 12-, 19-, 34-, and 46-median cases, the bounds at the end of the Lagrangian procedure differed by more than 0.01 percent. Decreasing the minimum allowable value of α to 0.0000001, increasing the number of failures before α is halved to 12 (from 4), and increasing the maximum number of iterations to 4000 allows the algorithm to find solutions to the 19-median and 34-median problems with lower and upper bounds that are within 0.01 percent of each other. In general, decreasing the minimum value of α and increasing the number of failures before halving α, while simultaneously increasing the maximum number of iterations, may allow the algorithm to converge more tightly.

Table 6.11 presents summary statistics comparing the myopic, neighborhood, and exchange algorithms with the Lagrangian lower and upper bounds.

Table 6.11 Summary Statistics on the Performance of the Heuristic Algorithms for the 88-Node Problem

(a) Percentage Differences Between Heuristic Algorithm and Lagrangian Lower Bound			
Measure	Myopic	Exchange	Neighborhood
Maximum	12.428%	3.631%	3.631%
Minimum	0.000%	0.000%	0.000%
Average	4.208%	0.225%	0.181%

(b) Percentage Differences Between Heuristic Algorithm and Lagrangian Upper Bound			
Measure	Myopic	Exchange	Neighborhood
Maximum	12.417%	3.626%	3.626%
Minimum	0.000%	0.000%	0.000%
Average	4.199%	0.216%	0.172%

The myopic algorithm performed least well in terms of the maximum percentage deviation and average percentage deviation from the Lagrangian bounds. The exchange algorithm was the best of the three, though it was not dramatically better than the neighborhood search algorithm. Again, we note that the computational burden associated with the exchange algorithm is considerably greater than is that associated with the neighborhood search algorithm. In no case did any of these three algorithms find a solution that was better than the Lagrangian upper bound, through such a result is not at all impossible since the Lagrangian algorithm is also heuristic.

Finally, the results confirm the general superiority of the Lagrangian relaxation algorithm (in terms of the quality of the solution obtained) when compared to any of the three heuristics outlined in Section 6.4. However, the Lagrangian algorithm generally takes longer to solve any particular problem than do the other heuristics. Also, when solving a P-median problem with any of the other heuristics, we automatically obtain solutions for the 1- through $P - 1$-median problems. Using the Lagrangian algorithm, when solving a P-median problem, we do not obtain results for any other problem. Thus, the computational advantages of the heuristics are greater than they might, at first glance, appear.

6.7 SUMMARY

In many cases, we want to locate facilities so that the total demand-weighted average distance between each demand node and the nearest facility site is minimized. For example, suppose a city wants to locate libraries. To find good locations for the libraries, the city might divide the city into census

Summary

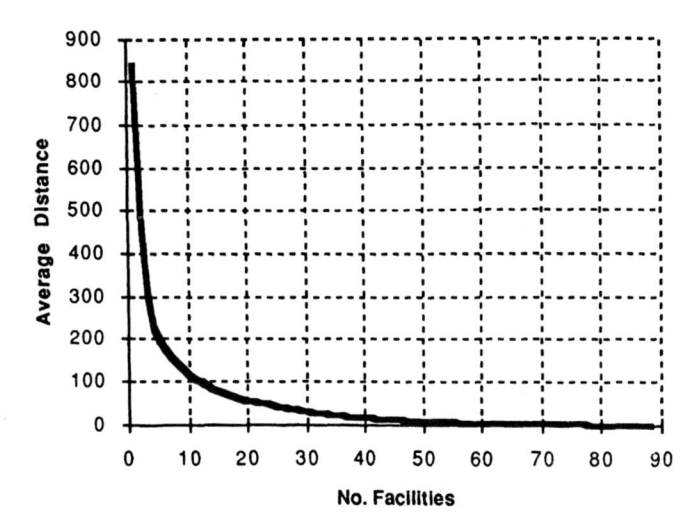

Figure 6.22. Average distance versus number of facilities for the 88-node problem.

blocks. The city might than use the population in each block as a measure of the demand in that block which would be assumed to be concentrated at the centroid of the block. By minimizing the demand-weighted distance between the centroids and the nearest of P facilities—solving a P-median problem—the city would be identifying P library locations that minimize the average distance citizens would have to travel between their homes and the nearest library. Since the construction and acquisition costs will be paid by the city government, but the travel costs will be borne by each individual library user, combining the two costs may be inappropriate.

For the P-median problem, as the number of facilities increases, the demand-weighted distance decreases. This phenomenon is shown in Figure 6.22 which plots the average demand-weighted distance versus the number of facilities that are located for the 88-node problem described in Appendix G.

One of the key properties of the P-median problem that distinguished it from either the covering or the center problems discussed in Chapters 4 and 5, respectively, is that we can show that at least one optimal solution to the P-median problem consists of locating only on the demand nodes. This allows us to narrow the search for optimal facility locations.

The P-median problem was formulated as an integer programming problem. On a tree, polynomial time algorithms exist for locating any number of medians. A linear time algorithm $[O(n)]$ was presented for the 1-median problem on a tree.

For a general network, the problem is NP-complete (for variable P). This suggests the need for heuristic algorithms. A simple myopic construction heuristic was outlined. In addition, two improvement algorithms, an exchange algorithm and a neighborhood search algorithm, were presented. Finally, two

Lagrangian relaxations of the integer programming formulation of the *P*-median problem were outlined. Computational results suggest that for small problems the heuristic algorithms (myopic search, exchange, and neighborhood search) might do very badly. In fact, with a simple 12-node network, cases were found in which the value of the solution found using the myopic algorithm was almost 30 percent greater than the optimal solution. For a larger problem with 88 nodes, the myopic algorithm was always within 13 percent of the optimal solution. The exchange and neighborhood search algorithms were always within 4 percent of the optimal solution for the larger network.

One of the two Lagrangian relaxations results in very good solutions. For the 12-node network, the Lagrangian algorithm was always able to obtain results that were within 0.0001 percent of the lower bound. For the 88-node problem, in all but four cases, the algorithm converged to within 0.01 percent of the lower bound using the default Lagrangian parameters built into the SITATION computer program.

EXERCISES

6.1 For the following tree, with demand weights as shown in the boxes beside each node on the tree, solve for the demand-weighted 1-median. Show the order in which you process nodes, the location of the 1-median, and compute the objective function for the optimal location.

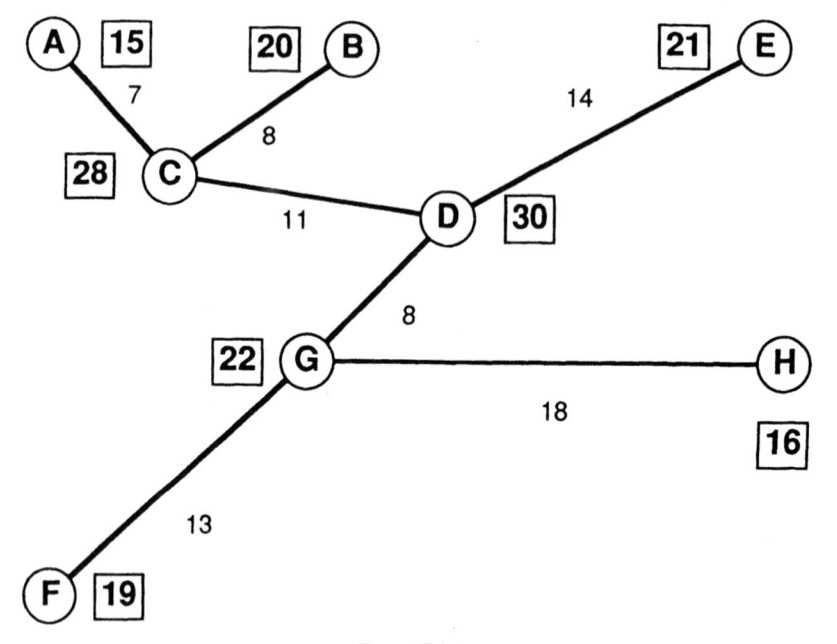

Figure E6.1

Exercises 239

6.2 (a) For the following tree, find the 1-median. Clearly indicate where the solution is.

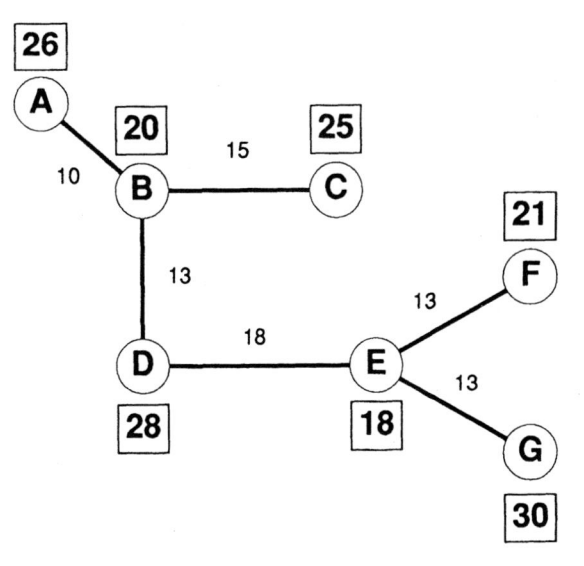

Figure E6.2

(b) What is the objective function for the 1-median problem?

6.3 (a) For the following tree, find the 1-median. Clearly indicate where the solution is.

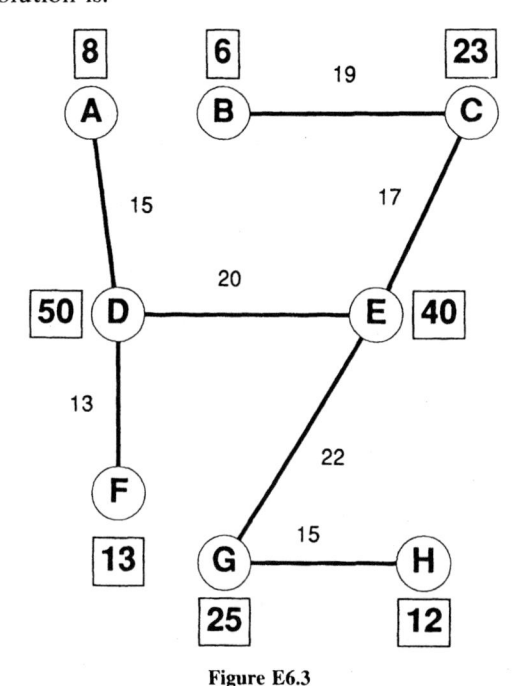

Figure E6.3

(b) What is the objective function for the 1-median problem?

(c) By how much would the demand at node F have to increase for the location of the 1-median to change? Where would the new 1-median be located?

6.4 (a) For the following tree, find the 1-median. Clearly indicate where the solution is.

(b) What is the objective function for the 1-median problem?

(c) How would the optimal location change if the distance between nodes H and I doubled? Briefly justify your answer.

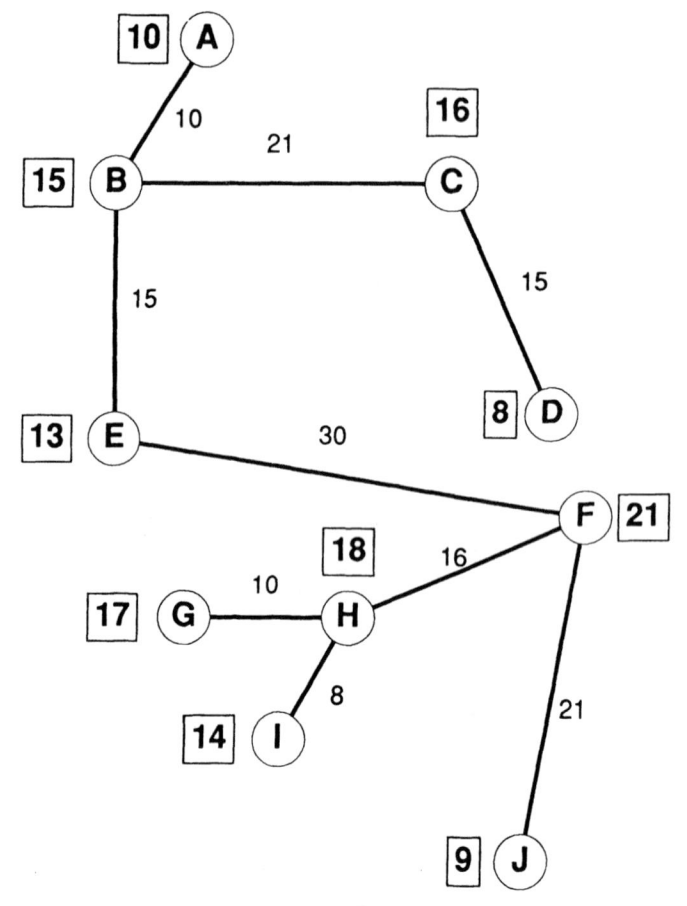

Figure E6.4

6.5 Consider the tree shown in Figure E6.5

(a) Find the *absolute unweighted* 1-*center* of the tree. What is the value of the objective function for this problem at that point?

Exercises

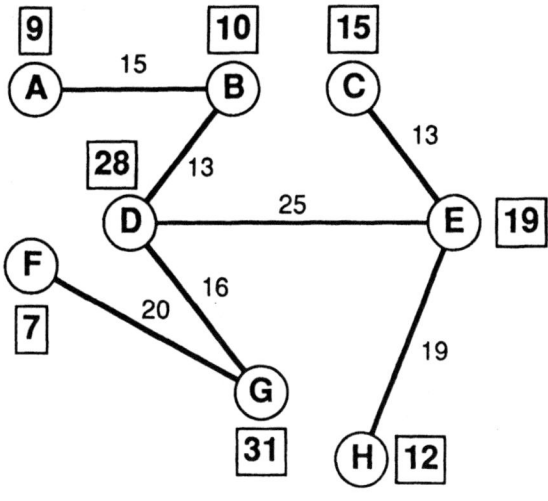

Figure E6.5

(b) Find the *vertex unweighted 1-center* of the tree. What is the value of the objective function for this problem at that point?

(c) Find the *absolute unweighted 2-center* of the tree. What is the value of the objective function for this problem at that point?

(d) Find the *absolute weighted 1-center* of the tree. What is the value of the objective function for this problem at that point?

(e) Find the 1-*median of the tree*. What is the value of the objective function for this problem at that point?

6.6 You are charged with locating an express mail drop box somewhere on the network shown in Figure E6.6. (This is like a mailbox except that only packages that are going out by express mail may be placed in the box.) You have identified the locations of the largest firms in the area (all law firms). These firms are identified by circled numbers on the following map (e.g., ② for the firm of Tort and Retort). Each firm is located at the intersection of two streets as indicated in the following table. The number of packages per day is also given in the following table.

NOTE:

North-South (Numbered) Avenues are separated by 800 feet

East-West (Lettered) Streets are separated by 600 feet

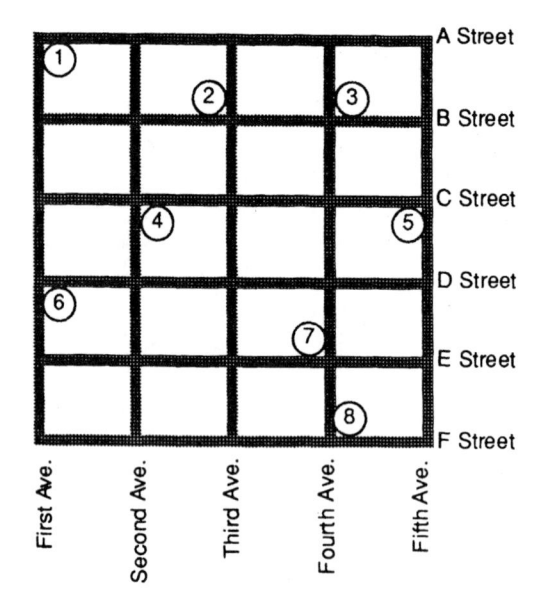

Figure E6.6

Number	Firm	E–W Street	N–S Avenue	Packages/Day
1	Sue Em Ltd.	A Street	First Avenue	30
2	Tort and Retort	B Street	Third Avenue	50
3	Bank and Rupt	B Street	Fourth Avenue	25
4	Jail Em Fast Ltd.	C Street	Second Avenue	45
5	Hang Em Inc.	C Street	Fifth Avenue	60
6	Trial By Jury Inc.	D Street	First Avenue	35
7	Never Guilty Inc.	E Street	Fourth Avenue	70
8	Mob Law and Sons	F Street	Fourth Avenue	20

(a) Assuming you can only walk along north–south avenues and east–west streets, find the location of the drop box that minimizes the demand-weighted distance that people must walk from the office to the drop box. Demand is weighted by the number of packages per day. Give the *intersection* at which the drop box should be located. What is the value of the *objective function* at this point (assuming that the law firms are at the intersections)?

(b) The demands listed above are the *current* demands. However, the criminal law firm of Jail Em Fast Ltd. is growing very rapidly. (Note that a recent study showed that the United States has the largest number of imprisoned people per 100,000 in the population of the countries in the study, ahead of South Africa by 50 percent,

Exercises

2.5 times Venezuela, and over 4 times both Canada and China![1]) How large must the number of package per day delivered from this firm become for it to be optimal to change the location of the drop box? Where would the new drop box be located?

6.7 Consider the following demand and location data:

Node	X Coordinate	Y Coordinate	Demand
1	25	1	86
2	38	22	49
3	6	29	128
4	23	8	101
5	17	25	84
6	17	37	125
7	14	32	12
8	36	16	12

(a) Find the location of the 1-median for this problem using Manhattan distances. Give the value of the optimal objective function.

(b) Plot demands and the optimal facility location on an X–Y plot.

(c) Plot the contribution to the objective function of transportation (movement) in the X direction as we move the facility location in the X direction from 0 to 40.

(d) *Qualitatively* and *briefly* discuss how the *objective function* and the *optimal location* would change if the coordinates of point 6 change from $(17, 37)$ to $(17, 100)$.

(e) We know that an optimal solution to this problem has an X coordinate equal to one of the X coordinates of the demand points. Call this demand point node α for the sake of argument. Similarly, we know that the optimal location has a Y coordinate equal to one of the Y coordinates of the demand points. Call this demand point node β. Note that nodes α and β need not be the same.

Now consider the same problem except that we are now restricted to selecting one of the demand nodes for the facility site. Consider the following conjecture:

Conjecture: An optimal solution to the 1-median problem using Manhattan distances must be at either node α or node β where node α is the node at which the demand-weighted distance in the X direction is minimized and node β is the node at which the demand-weighted distance in the Y direction is minimized.

[1]"Study: U.S. biggest user of prisons," *Chicago Tribune*, February 11, 1992, p. 3, section 1.

Either prove that this conjecture is true or show by a counterexample that it is false.

6.8 (a) Using the SITATION software, find the tradeoff curve between the number of facilities located and the demand-weighted average distance using the **Neighborhood** algorithm for the first demand data set (representing the state population) in the 49-node problem described in Appendix H. This data set is called SORTCAP.GRT.

(b) Identify at least six cases in which the heuristic results are likely to be suboptimal. Briefly indicate why you believe they are suboptimal.

(c) For each of the cases identified in part (b), use the Lagrangian relaxation approach to try to find better solutions.

6.9 (a) Using the SITATION software, find the tradeoff curve between the number of facilities located and the demand-weighted average distance using the **Exchange** algorithm for the first demand data set (representing the state population) in the 49-node problem described in Appendix H. This data set is called SORTCAP.GRT.

(b) Identify at least three cases in which the heuristic results are likely to be suboptimal. Briefly indicate why you believe they are suboptimal.

(c) For each of the cases identified in part (b), use the Lagrangian relaxation approach to try to find better solutions.

(d) Use the Lagrangian relaxation approach to find the true tradeoff curve (of average distance versus the number of facilities located) for this problem. Solve each problem so that the difference between the lower and upper bounds is less than 0.01 percent. Plot the resulting tradeoff curve using a spreadsheet program.

6.10 The Lagrangian algorithm failed to converge to within 0.01 percent for the 12-median problem using the CITY1990.GRT data set as discussed in the Section 6.6. Change the parameters on the Lagrangian procedure in the SITATION program to allow the algorithm to perform more iterations [e.g., change (1) the minimum allowable value of α, (2) the maximum allowable number of iterations, and/or (3) the number of consecutive failures to improve the lower bound before α is halved]. By forcing nodes in and out of the solution, you should be able to obtain a provably optimal solution, or one in which the difference between the lower and upper bounds is 0 percent. What are the parameter settings that you used? What is the optimal objective function value?

Exercises

6.11 Consider the following formulation of the P-median problem:

Inputs

h_i = demand at node i
d_{ij} = distance between demand node i and node j

Decision Variables

Y_{ij} = fraction of demand at node i that is served by a facility at node j

$$X_j = \begin{cases} 1 & \text{if we locate at node } j \\ 0 & \text{if not} \end{cases}$$

Formulation

MINIMIZE $\quad \sum_i \sum_j h_i d_{ij} Y_{ij}$

SUBJECT TO: $\quad \sum_j Y_{ij} = 1 \quad \forall i$

$$\sum_j X_j = P$$

$$X_j - Y_{ij} \geq 0 \quad \forall i, j$$

$$X_j = 0, 1 \quad \forall j$$

$$Y_{ij} \geq 0 \quad \forall i, j$$

In many cases, we not only want to minimize the sum over all demand nodes of the demand-weighted average distance between a demand node and the nearest facility, but we also want to be sure that each demand node has a facility located within at most D_c distance units of it.

(a) Define any additional notation that you need and show how this can be handled by *adding a constraint to the formulation shown above*.

(b) Briefly state how you can deal with this problem by *changing certain data inputs to the problem*.

(c) Briefly discuss how you can solve the problem by *eliminating certain variables from the formulation before you begin to solve the problem*.

(d) Will the new problem always have a feasible solution? If not, give an example of a case in which the new problem will not have a feasible solution. If so, prove that it will always have a feasible solution.

6.12 Seers Ltd., a firm that supplies crystal balls to fortune tellers, has expanded its client base in recent years. Because of this expansion, it now needs to establish a set of warehouses from which it will resupply its customers. Since demand for crystal balls is unpredictable (you never know when a crystal ball will become clouded) and time sensitive (when a crystal ball fails, you need a new one NOW) and since Seers is not the only firm in this business, Seers feels a need to provide timely deliveries to its customers. In particular, Seers would like to guarantee all of its customers deliveries within 48 hours and to maximize the number of customers that are served within 24 hours. Seers estimates that any customer within 250 miles of the nearest warehouse can be served within one day, while those within 500 miles can be served within two days.

(a) Formulate the problem of locating the minimum number of warehouses needed to satisfy all demands within 500 miles and (from among the solutions that serve all demands within 500 miles) maximizing the number of demands within 250 miles of the nearest warehouse as an integer linear programming problem.

(b) Using the first demand set in the 49-node data set described in Appendix H (SORTCAP.GRT) as a proxy for demand and assuming that facilities can only be located at one of these 49 cities, solve the problem using the SITATION program.

Note that you may need to modify the distance matrix in some way to solve this problem using the SITATION software. Also, note that you may need to modify the default Lagrangian options rather significantly to obtain a solution in which the lower and upper bounds are adequately close to each other.

7

Fixed Charge Facility Location Problems

7.1 INTRODUCTION

In most of the models that we have considered so far, the number of facilities to be located was an input into the model. For example, in the P-median and P-center problems, we try to locate P facilities to minimize the total demand-weighted distance or the maximum demand-weighted distance. Similarly, in the maximum covering location problem, we try to locate a given number of facilities to maximize the number of demands that are within some specified distance of the nearest facility. One notable exception was the set covering location model in which we try to minimize the number of facilities needed to cover all demands within a specified distance. In that case, the number of facilities was determined endogenously.

By optimizing an objective function subject to a constraint that we locate a fixed number of facilities, we are implicitly separating operating costs or benefits as captured by the objective function from the construction costs for which the number of facilities is taken as a proxy. Such a separation is often important and necessary in public sector problems in which different agencies or actors bear the costs and receive the benefits or in which the benefits and costs are incommensurable. In the private sector, these problems are less acute. Costs are generally borne and benefits are realized by the same organization. Furthermore, costs and benefits can typically be measured in monetary units.

By using the number of facilities as a proxy for the construction costs, we implicitly assumed that the cost of constructing a facility was the same at each candidate facility site. Clearly, this is not always a valid assumption. In this chapter we consider models which incorporate an explicit cost, f_j, of locating at candidate location j. Such models are typically used for private sector location problems in which a single actor or firm pays the costs and realizes the benefits of the facilities being located and in which costs and

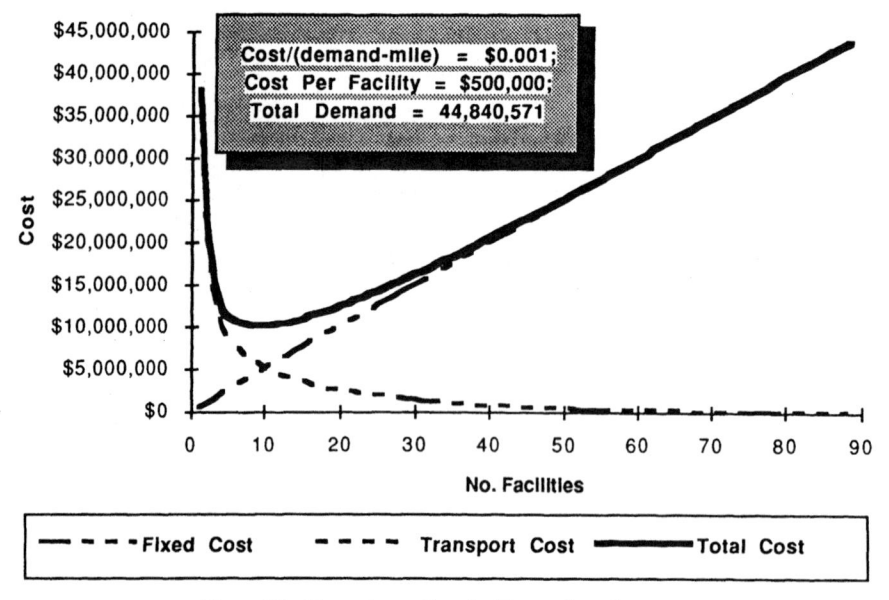

Figure 7.1. Example routing, facility, and total costs.

benefits can be measured in commensurable units.

To see the importance of accounting for fixed location costs, consider again Figure 6.22. The figure shows the average distance as a function of the number of facilities that are located. The figure suggests that we should locate as many facilities as possible, thereby reducing the average distance between a demand node and the nearest facility as much as possible.

If we incorporate the fixed cost of locating facilities, a different picture emerges. Figure 7.1 replicates Figure 6.22, but adds a constant fixed cost of $500,000 for each facility that is located. In addition, the average distance is multiplied by the total demand[1] of 44,840,571 units in the 88-node problem and then by $0.001/mile to convert the average distance into a routing cost. The fixed facility cost is added to the routing cost to obtain the total cost for each number of facilities.

As indicated in Fig. 7.1, the total cost initially declines as the reduction in routing cost that results from the addition of more facilities more than offsets the additional facility location costs. At some point, the cost of additional

---

[1] We need not be overly concerned about the units that are used to measure demand in this case since the units will immediately cancel with those in the cost per mile term. Thus, if demand is measured in tons, the cost per mile term would be measured in dollars per ton-mile. Therefore, we simply refer to demand in a generic sense at this point. This is clarified in the discussion below.

Introduction

facilities exceeds the savings in routing costs and the total cost increases as we add more facilities. For the values used in Figure 7.1, the optimal number of facilities is 10. This results in a total cost of $10,126,720 of which $5,000,000 (or about 50 percent) is the fixed facility costs and the remainder is associated with routing or transport costs.

It is important to note that the cost per mile that is used in the models discussed in this chapter and used in the SITATION program are actually a *cost per mile per unit demand*. Thus, to obtain the routing cost of $37,640,812 if we locate one facility, we multiply the average distance of 839.436491 by the total demand of 44,840,571 and then by the cost per mile of $0.001. Changes in the cost per mile allow us to capture changes in the relative importance of fixed costs and routing costs. Higher values of the cost per mile may reflect: (1) the movement of higher-valued goods, in which case the higher inventory costs associated with goods in transit need to be incorporated in the routing costs; (2) more expensive transport services, such as the use of an air freight service instead of a truck carrier; and/or (3) a greater fraction or frequency of use of the service being modeled by the demand units. For example, in the 88-node problem, the cost per mile of $0.001 used above might represent an actual transportation plus inventory cost of $1.00 per mile but only one in one thousand of the demand units actually shipping or demanding service. If the demand rate doubled, the cost per mile would also double to $0.002.

Table 7.1 shows the sensitivity of the solution to changes in the cost per mile for the 88-node problem with all other parameters held at the values used in Figure 7.1. For low values of the cost per mile (low demand rates, low-valued goods, and/or inexpensive transport services), it is optimal to use few facilities. As the cost per mile increases (higher demand rates, higher-valued goods, more expensive transport services, and/or relatively lower fixed facility costs), additional facilities are added.

The remainder of this chapter develops models and solution algorithms for such fixed charge facility location problems. Section 7.2 is devoted to uncapacitated problems, while capacitated facilities are covered in Section 7.3. Section 7.4 summarizes the chapter.

Table 7.1 Sensitivity of Solution to Cost per Mile in the 88-Node Problem

Cost per Mile	Number of Facilities	Average Distance	Routing Cost	Facility Cost	Total Cost
$0.00001	1	839.436	$376,408	$500,000	$876,408
$0.0001	3	305.457	$1,369,688	$1,500,000	$2,569,688
$0.001	10	114.332	$5,126,720	$5,000,000	$10,126,720
$0.01	38	17.377	$7,791,962	$19,000,000	$26,791,962
$0.1	72	0.964	$4,321,231	$36,000,000	$40,321,231

7.2 UNCAPACITATED FIXED CHARGE FACILITY LOCATION PROBLEMS

To formalize the problem outlined in Section 7.1, we consider the problem of minimizing the sum of the facility location and routing or transportation costs. To do so, we define the following notation:

Inputs

f_j = fixed cost of locating at candidate site j
h_i = demand at node i
d_{ij} = distance from demand node i to candidate location j
α = cost per unit distance per unit demand

Decision Variables

$$X_j = \begin{cases} 1 & \text{if we locate at candidate site } j \\ 0 & \text{if not} \end{cases}$$

Y_{ij} = fraction of demand at node i that is served by a facility at node j

With this notation, we can formulate *the uncapacitated fixed charge facility location problem* as follows:

MINIMIZE $\quad \sum_j f_j X_j + \alpha \sum_i \sum_j h_i d_{ij} Y_{ij} \quad$ (7.1a)

SUBJECT TO: $\quad \sum_j Y_{ij} = 1 \quad \forall\, i \quad$ (7.1b)

$\quad Y_{ij} \leq X_j \quad \forall\, i, j \quad$ (7.1c)

$\quad X_j = 0, 1 \quad \forall\, j \quad$ (7.1d)

$\quad Y_{ij} \geq 0 \quad \forall\, i, j \quad$ (7.1e)

The objective function (7.1a) minimizes the total cost which is the sum of the fixed facility costs and the total demand-weighted distance multiplied by the cost per unit distance per unit demand. Constraint (7.1b) stipulates that each demand node i be served. Constraint (7.1c) says that demands at node i cannot be assigned to a facility at candidate site j unless we locate a facility at node j. Constraints (7.1d) and (7.1e) are the integrality and nonnegativity constraints, respectively. As in previous formulations, since the facilities are uncapacitated, all demand at node i will be assigned to the nearest open facility. Thus, the assignment variables, Y_{ij}, will naturally assume integer values.

We note that the key differences between this model and the *P*-median problem discussed in Chapter 6 are (1) the inclusion of the fixed facility

Uncapacitated Fixed Charge Facility Location Problems

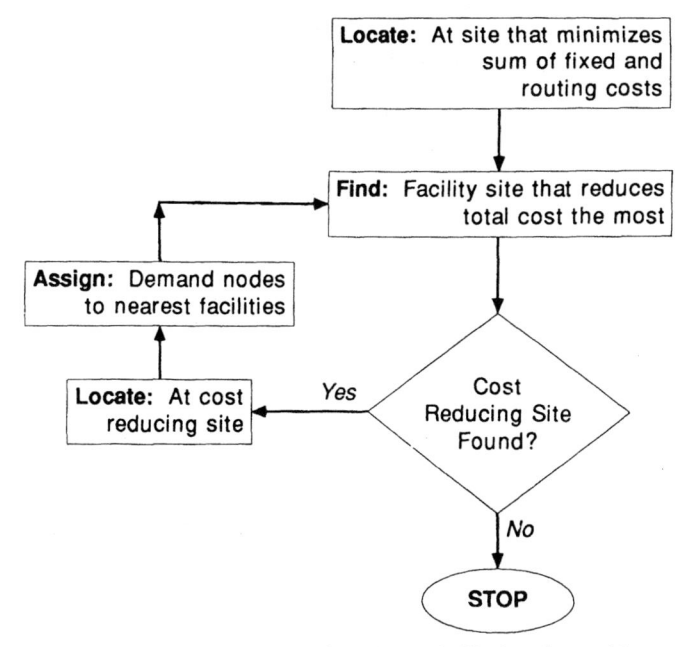

Figure 7.2. ADD algorithm for fixed charge facility location problem.

location costs and (2) the absence of the constraint on the number of facilities to be located. Upon reflection, it should be clear that these are relatively small differences. As such, the algorithms used to solve this model bear a strong resemblance to those used for the P-median problem.

7.2.1 Heuristic Construction Algorithms

A variety of heuristic algorithms have been devised for solving the uncapacitated fixed charge facility location problem. The *ADD algorithm* and the *DROP algorithm* are construction algorithms that take advantage of the general shape of the total cost curve shown in Figure 7.1. Specifically, noting that the total cost generally decreases as facilities are initially added to the solution, the ADD algorithm greedily adds facilities to the solution until the algorithm fails to find a facility whose addition will result in a decrease in the total cost. "Greedily" means that each node that is added to the solution reduces the cost as much as possible, holding the previously selected sites fixed in the solution. Figure 7.2 is a flowchart of the ADD algorithm.

The DROP algorithm works in a similar manner except from the other side of the total cost curve shown in Figure 7.1. In particular, the DROP algorithm begins with a facility at each candidate site. Based on the observation that the total cost generally declines when facilities are removed from a solution in which a facility is located at every candidate site, the DROP

Locate: At all eligible candidate facility sites

Find: Facility site whose removal reduces total cost the most

Assign: Demand nodes to nearest facilities

Remove: Facility from cost reducing site

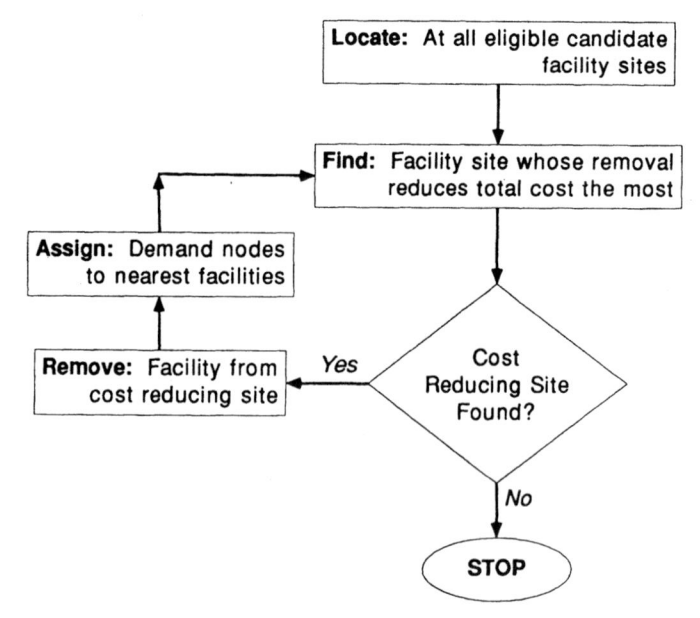

Cost Reducing Site Found? — Yes → Remove; No → STOP

Figure 7.3. Drop algorithm for fixed charge facility location problems.

algorithm then proceeds to greedily remove facilities from the solution until the algorithm can no longer find a facility whose removal will result in a decrease in the total cost. "Greedily" now means that each node that is removed from the solution reduces the cost as much as possible, without changing the status (in the solution or out of the solution) of any other nodes. Figure 7.3 is a flowchart of the DROP algorithm.

To illustrate the use of these algorithms, we again consider the 12-node network of Figure 6.11. Figure 7.4 replicates this figure, adding fixed location costs at each of the candidate nodes. These costs are given by the numbers following the dollar signs. Thus, it would cost $100 to build a facility at node A and $200 to build a facility at node B.

Table 7.2 gives the computations for the addition of the first facility using the ADD algorithm when the cost per mile is $0.35. In each cell of the main part of the table, we give the demand-weighted distance of serving the node in the row in question from a facility located at the node in the associated column. Thus, if we serve node A from a facility at node D, the demand-weighted distance that this pair contributes to the objective function is 825. The last row gives the total demand-weighted distance associated with serving all nodes from the facility in the given column ($\sum_i h_i d_{ij}$) multiplied by the cost per mile ($\alpha = 0.35$) plus the fixed cost (f_j). We locate at the node with the smallest column total. In this case, this is node I. The total cost is $1835.

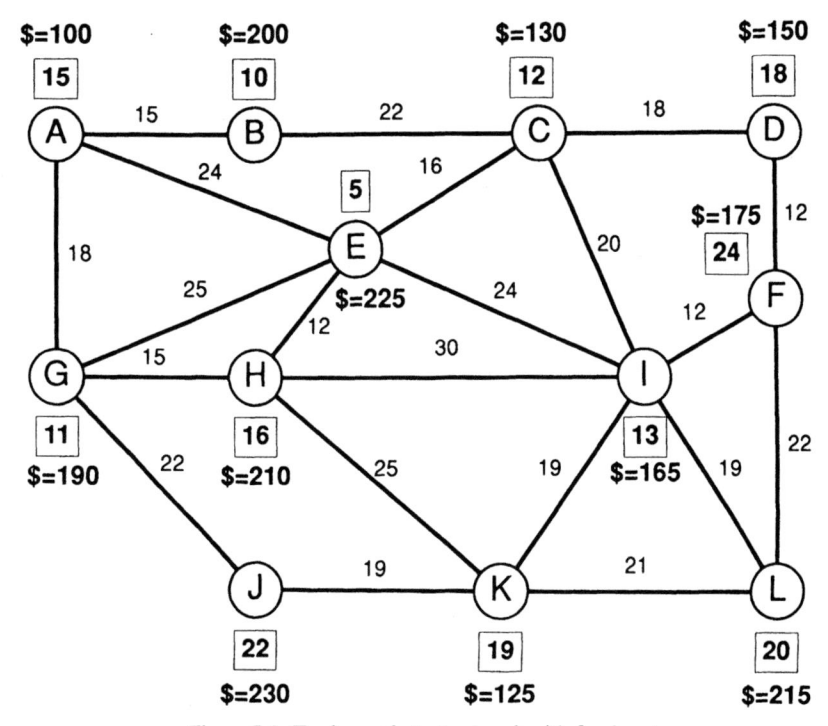

Uncapacitated Fixed Charge Facility Location Problems 253

Figure 7.4. Twelve-node test network with fixed costs.

Table 7.2 Computations for First Facility Location in ADD Algorithm for the Network of Figure 7.4

	A	B	C	D	E	F	G	H	I	J	K	L
A	0	225	555	825	360	900	270	495	720	600	870	1005
B	150	0	220	400	380	520	330	480	420	550	610	610
C	444	264	0	216	192	360	492	336	240	696	468	468
D	990	720	324	0	612	216	1062	828	432	1116	774	612
E	120	190	80	170	0	180	125	60	120	235	185	215
F	1440	1248	720	288	864	0	1368	1008	288	1200	744	528
G	198	363	451	649	275	627	0	165	495	242	440	671
H	528	768	448	736	192	672	240	0	480	592	400	736
I	624	546	260	312	312	156	585	390	0	494	247	247
J	880	1210	1276	1364	1034	1100	484	814	836	0	418	880
K	1102	1159	741	817	703	589	760	475	361	361	0	399
L	1340	1220	780	680	860	440	1220	920	380	800	420	0
Cost	2836	2970	2179	2410	2249	2191	2618	2300	1835	2640	2077	2445

Table 7.3 Computations for the Addition of the Second Facility to the Network of Figure 7.4

	A	B	C	D	E	F	G	H	I	J	K	L
A	0	225	555	720	360	720	270	495	720	600	720	720
B	150	0	220	400	380	420	330	420	420	420	420	420
C	240	240	0	216	192	240	240	240	240	240	240	240
D	432	432	324	0	432	216	432	432	432	432	432	432
E	120	120	80	120	0	120	120	60	120	120	120	120
F	288	288	288	288	288	0	288	288	288	288	288	288
G	198	363	451	495	275	495	0	165	495	242	440	495
H	480	480	448	480	192	480	240	0	480	480	400	480
I	0	0	0	0	0	0	0	0	0	0	0	0
J	836	836	836	836	836	836	484	814	836	0	418	836
K	361	361	361	361	361	361	361	361	361	361	0	361
L	380	380	380	380	380	380	380	380	380	380	380	0
Cost	1485	1669	1675	1819	1684	1834	1456	1654	1835	1642	1640	1917

Table 7.3 gives the computations associated with the addition of the second facility, *given that the first facility is located at node I*. Now the main part of the table gives the demand-weighted distance between the demand node associated with the row in question and the nearer of (1) the facility associated with the column in question and (2) node *I* (at which we have already located a facility). Note that all elements in row *I* are 0, since the demand-weighted distance between node *I* and the nearer of (1) any other site and (2) node *I* is always 0. The final row gives the total demand-weighted distance if we locate the second facility at each of the indicated columns, multiplied by the cost per mile, plus the fixed facility location cost. We locate the second facility at node *G*. The total cost is $1456.

Proceeding in this manner, we locate facilities at node *A* (reducing the cost to $1398), node *K* (reducing the cost to $1374), and, finally, at node *D* (reducing the cost to $1364). If we try to locate another facility, we find that node *C* is the best, but locating there *increases* the total cost to $1405. Thus, the solution from the ADD algorithm is to locate at nodes *A*, *D*, *G*, *I*, and *K* at a total cost of $1364. These results are summarized in Table 7.4.

As facilities are added, the average distance decreases and the total fixed cost increases. As long as the decrease in average distance multiplied by the cost per mile exceeds the increase in fixed cost, it is advantageous to add facilities. Note that once the first five facilities are located at nodes *A*, *D*, *G*, *I*, and *K*, the addition of a sixth facility at node *C* decreases the average distance by 256 units, for a savings of $89.6. This is not sufficient to offset the increase of $130 in the fixed costs. Hence, we stop after the addition of the fifth facility at node *D*.

Turning now to the DROP heuristic, if we locate a facility at each node, we incur only fixed costs (since the distance from any node to the same node

Uncapacitated Fixed Charge Facility Location Problems

Table 7.4 Summary of ADD Heuristic Iterations

Iteration	Add Facility at Node	Demand-Weighted Distance	Total Fixed Cost	Total Cost
1	I	4772	165	1835
2	G	3145	355	1456
3	A	2695	455	1398
4	K	2268	580	1374
5	D	1812	730	1364
6	C	1556	860	1405

Solution: Locate at nodes A, D, G, I, and K for a total cost of $1364

is assumed to be 0). The sum of all fixed costs for the network of Figure 7.4 is $2115. Table 7.5 gives the computations for the first iteration of the DROP heuristic applied to this network, using a cost per mile of $0.35. The cells in the main part of the table now give the demand-weighted distance between the node associated with the row in question and the nearest facility assuming facilities are located at each node *except* the one associated with the column in question. Notice that on the first iteration, all demand-weighted distances are 0 except those associated with the diagonal of the matrix. Thus, the demand-weighted distance for the (A, A) cell is 225. If a facility is not located at node A, the nearest facility would be located at node B, 15 distance units away. Since the demand at node A is 15, the demand-weighted distance from node A to the nearest facility (other than one located at node A) is 225.

Table 7.5 Computations for First Iteration of the DROP Heuristic Applied to the Network of Figure 7.4

	A	B	C	D	E	F	G	H	I	J	K	L
A	225	0	0	0	0	0	0	0	0	0	0	0
B	0	150	0	0	0	0	0	0	0	0	0	0
C	0	0	192	0	0	0	0	0	0	0	0	0
D	0	0	0	216	0	0	0	0	0	0	0	0
E	0	0	0	0	60	0	0	0	0	0	0	0
F	0	0	0	0	0	288	0	0	0	0	0	0
G	0	0	0	0	0	0	165	0	0	0	0	0
H	0	0	0	0	0	0	0	192	0	0	0	0
I	0	0	0	0	0	0	0	0	156	0	0	0
J	0	0	0	0	0	0	0	0	0	418	0	0
K	0	0	0	0	0	0	0	0	0	0	361	0
L	0	0	0	0	0	0	0	0	0	0	0	380
Cost	2094	1968	2052	2041	1911	2041	1983	1972	2005	2031	2116	2033

The final row of Table 7.5 gives the total cost associated with removing a facility from each node. This cost is the sum of all of the fixed costs, except the one associated with the facility at the column in question, plus the total demand-weighted distance given in the column multiplied by the cost per mile. The smallest element in the final row is $1911 corresponding to the removal of the facility at node E. Since this is less than the cost associated with locating at all nodes, we remove node E from the solution. Removing node E from the solution makes sense. Node E has the second highest fixed cost. Only node J has a higher fixed cost. It also has the smallest demand, so removing the facility at node E will inconvenience relatively few people. Finally, there is a facility very close to node E, at node H, only 12 distance units away. Note that 12 is the smallest internodal distance in the network.

Table 7.6 gives the computations associated with the second iteration of the DROP heuristic. Now the main body of the table gives the demand-weighted distance between demands at the node associated with the row in question and the nearest facility assuming facilities are at all nodes except node E (from which we have already removed the facility) and the node associated with the column in question. Note that there is a nonzero element in each column for the row associated with node E. In most cases, demands at node E would be served by the facility at node H, resulting in a demand-weighted distance of 60 (5 demands times 12 distance units). However, if node H is removed, the demand-weighted distance associated with serving demands at node E increases to 80 (5 demands times 16 distance units). Other changes also occur. Thus, if node C is removed, the demand-weighted distance associated with serving demands at that node increases from 192 (shown in Table 7.5) to 216. This occurs because the 12 demands at

Table 7.6 Computations for Second Iteration of the DROP Heuristic Applied to the Network of Figure 7.4

	A	B	C	D	E	F	G	H	I	J	K	L
A	225	0	0	0	0	0	0	0	0	0	0	0
B	0	150	0	0	0	0	0	0	0	0	0	0
C	0	0	216	0	0	0	0	0	0	0	0	0
D	0	0	0	216	0	0	0	0	0	0	0	0
E	60	60	60	60	60	60	60	80	60	60	60	60
F	0	0	0	0	0	288	0	0	0	0	0	0
G	0	0	0	0	0	0	165	0	0	0	0	0
H	0	0	0	0	0	0	0	240	0	0	0	0
I	0	0	0	0	0	0	0	0	156	0	0	0
J	0	0	0	0	0	0	0	0	0	418	0	0
K	0	0	0	0	0	0	0	0	0	0	361	0
L	0	0	0	0	0	0	0	0	0	0	0	380
Cost	1790	1764	1857	1837	1911	1837	1779	1792	1801	1827	1912	1829

Table 7.7 Summary of DROP Heuristic Iterations

Iteration	Remove Facility at Node	Demand-Weighted Distance	Total Fixed Cost	Total Cost
0	—	0	2115	2115
1	E	60	1890	1911
2	B	210	1690	1764
3	G	375	1500	1631
4	I	531	1335	1521
5	J	949	1105	1437
6	D	1165	955	1362
7	L	1585	740	1295
8	H	2038	530	1243
9	C	2438	400	1253

Solution: Remove facilities at nodes B, D, E, G, H, I, J, and L
Locate at nodes A, C, F, and K for a total cost of \$1243

node C can no longer be served by a facility at node E (since that facility was removed). These demands must now be served by the facility at node D which is 2 distance units farther from node C than is node E. Thus, the demand-weighted distance increases by $24 = 216 - 192$ which is equal to the 12 demands at node C multiplied by the additional 2 distance units. The final row again gives the total cost assuming we have facilities at each node except node E and the node associated with the column in question. The smallest value is \$1764 associated with the removal of the facility at node B. Again, this makes sense, since the fixed cost at node B is relatively large, the demand there is relatively small, and there is still a nearby facility. (Node A is only 15 units away from node B.)

Proceeding in this manner, we obtain the results shown in Table 7.7. In all, eight facilities are removed from nodes E, B, G, I, J, D, L, and H, in that order. As facilities are removed, the demand-weighted distance increases and the total fixed cost decreases. As before, as long as the increase in the demand-weighted distance multiplied by the cost per mile is less than the change in the fixed cost, it is desirable to remove facilities. Once eight facilities have been removed, however, the removal of the ninth facility at node C increases the demand-weighted distance by 400 units at a cost of \$140. Since this exceeds the \$130 savings in fixed costs, we stop with the removal of the eighth facility at node H. The DROP heuristic suggests we locate at nodes A, C, F, and K at a total cost of \$1243. Note that in this case we found a solution that is different from that found by using the ADD heuristic. In particular, we now locate fewer facilities at a smaller total cost. The total cost of the solution found using the DROP heuristic is almost 10 percent less than that found using the ADD heuristic in this case.

7.2.2 Heuristic Improvement Algorithms

As in the case of the *P*-median problem, we can develop an exchange (or substitution) heuristic and a neighborhood search heuristic to improve on the solution found using either construction algorithm.

Figure 7.5 is a flowchart of one possible implementation of an exchange algorithm for the uncapacitated fixed charge facility location problem. As in previous exchange algorithms, we begin with any set of facility sites (e.g., those from one of the construction algorithms). In this implementation, we search for the *best* possible substitution before making any site exchanges. In other words, we consider *every* possible combination of a node in the current or incumbent solution and a node that is not in the current solution. After all combinations have been evaluated, we find the best combination. If that combination reduces the total cost, we may exchange a node in the incumbent solution for a node that is not in the solution in the best combination. If the best combination does not result in a total cost savings, we stop. Note that in evaluating each combination, we need to assess the impact of the exchange on both the total demand-weighted distance [the second term in

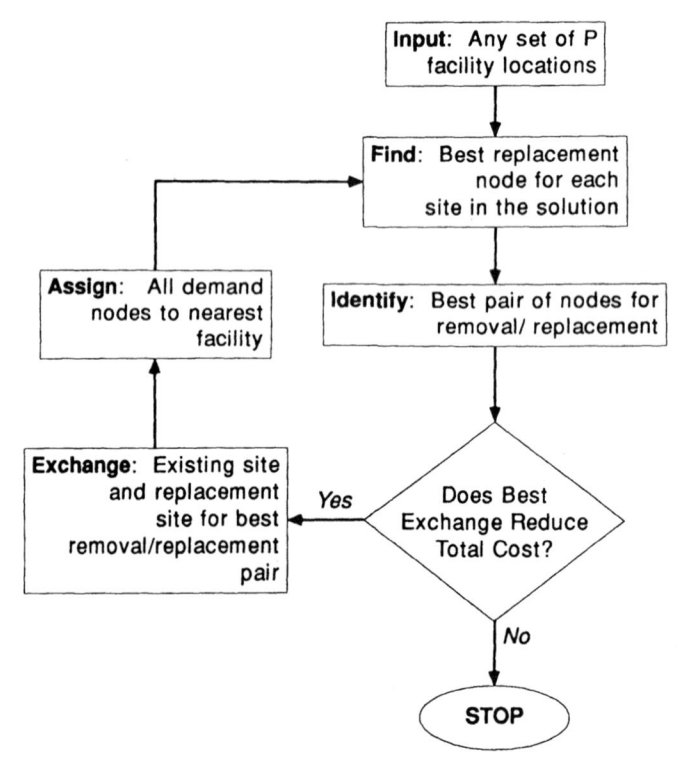

Figure 7.5. Flowchart of exchange algorithm for uncapacitated fixed charge facility location problem.

Table 7.8 Key differences Between Exchange or Substitution Algorithms Discussed to Date

Algorithm of Chapter	Substitute After Finding	Objective Function
4	*First combination* of a node in the solution and a node not in the solution that improves the objective function	Total coverage
6	*Best substitute* node for first node in the solution whose removal will improve the objective function	Average or total demand-weighted distance
7	*Best combination* of a node in the solution and a node not in the solution	Total fixed plus distance related cost

the objective function of (7.1a)] and the total fixed cost [the first term in (7.1a)].

The key differences between this exchange heuristic and the substitution and exchange heuristics outlined in Chapters 4 and 6 for the maximum covering and P-median problems, respectively, are summarized in Table 7.8. The algorithm presented in Chapter 4 effects the first substitution that is found to improve the objective function. The algorithm in Chapter 6 effects the best substitution for the first node in the solution whose removal is found to improve the solution. Thus, if there are multiple improving replacement nodes for the node being removed, the algorithm in Chapter 4 will substitute the first improving node it finds, while that of Chapter 6 will substitute the best improving node. The algorithm of Figure 7.5 effects the best substitution over all possible substitutions. The key tradeoff is that the algorithm in Chapter 4 will tend to find substitutions faster than will the algorithm in this chapter at each iteration. However, the substitutions made by the algorithm in Chapter 4 will be less effective than will those of the algorithm in this chapter. Thus, the algorithm of this chapter will tend to iterate fewer times. The algorithm of Chapter 6 lies between these two extremes. *Note that any one of these substitution or exchange algorithms can be used with any of the three objective functions.* Beginning with the same initial solution and using the same objective function, however, each approach may result in the identification of a different set of facility sites with differing objective function values.

Finally, we note that for the approaches discussed in Chapters 4 and 6, implementations of the substitution algorithms will differ depending on how the indices for the search for the next substitution are initiated and the order over which these indices are searched. Without going into further detail, suffice it to say that there are a large number of possible implementations of a "substitution" heuristic depending on which substitutions are adopted (as

Table 7.9 Summary of Changes Made to ADD Heuristic Solution by Exchange Heuristic

(a) Changes at Each Iteration

Solution	Original Locations	Remove Node	Insert Node	Objective Function
ADD	A, D, G, I, K			1364.20
First substitution	A, D, G, I, K	G	H	1336.95
Second substitution	A, D, H, I, K	I	F	1314.75
Third substitution	A, D, F, H, K	D	C	1294.75

(b) Final Solution Sites and Costs	
Sites:	A, C, F, H, and K
Fixed costs:	$740.00
Demand-weighted distance:	1585.00
Total cost:	$1294.75 = 740 + 0.35(1585)$

outlined in Table 7.8) and how the search for subsequent substitutions is carried out if the substitution procedure of either Chapter 4 or Chapter 6 is adopted. Any approach is valid. Unfortunately, most technical papers are not sufficiently specific about the approach that is adopted to allow for the accurate reproduction of the results.

The neighborhood search heuristic for the uncapacitated fixed charge facility location problem is essentially identical to the neighborhood search algorithm for the P-median problem (whose flowchart is shown in Figure 6.13) except that we are now minimizing the total cost instead of the total demand-weighted distance.

To illustrate these two algorithms, we begin with the solution obtained by the ADD algorithm for the network of Figure 7.4. Table 7.9 summarizes the changes to this solution that are made by the exchange heuristic. Three substitutions are made saving $27.25, $22.20, and $20.00, respectively, for a total savings of $69.45. The total cost of the solution found using the exchange heuristic is $1294.75. Note that this is not as good as the solution found using the DROP heuristic without any improvement algorithm. Also, note that using the exchange algorithm on the solution found by the DROP algorithm does not improve the solution at all.

Table 7.10 summarizes the changes made as a result of applying the neighborhood search algorithm to the ADD heuristic solution to the problem of Figure 7.4. Two substitution/reassignment iterations are performed. The table separates the savings due to substitution from those due to reassignment. Initially, node H replaces node G, saving $6.25. Nodes are then reassigned to the nearest facility and the cost is reduced an additional $21.00. Next, node F replaces node I, saving $15.20. Reassignment of all nodes to

Table 7.10 Summary of Changes Made to ADD Heuristic Solution by Neighborhood Search Heuristic

(a) Changes at Each Iteration				
Solution	Original Locations	Remove Node	Insert Node	Objective Function
ADD	A, D, G, I, K			1364.20
First substitution	A, D, G, I, K	G	H	1357.95
First reassignment	A, D, H, I, K			1336.95
Second substitution	A, D, H, I, K	I	F	1321.75
Second reassassignment	A, D, F, H, K			1314.75

(b) Final Solution Sites and Costs	
Sites:	A, D, F, H, and K
Fixed Costs:	$760.00
Demand-weighted distance:	1585.00
Total cost:	$1314.75 = 760 + 0.35(1585)$

the nearest facility saves an additional $7.00. No additional changes can be identified (since searches are localized to the neighborhood of each facility). The neighborhood search heuristic results in a solution with a cost of $1314.75 and facilities at nodes A, D, F, H, and K. This is not as good as the solution found using the exchange heuristic.

As a final note, we expect that the improvement algorithms will be less effective when applied to the fixed charge facility location problem than they were when applied to the P-median problem. The reason for this is that the improvement heuristics take the number of facilities as given by the initial solution. If the number of facilities is suboptimal, the solution obtained by the improvement heuristic will be doomed to suboptimality. In the P-median problem, this could not happen because the number of facilities to locate was an input to the problem and not an output. To attack this problem, we might consider heuristic algorithms that alternate between construction and improvement procedures. Figure 7.6 is a flowchart of one such hybrid approach in which we begin with the solution obtained using an ADD algorithm. While such an algorithm would undoubtedly do at least as well as the ADD algorithm followed by the exchange heuristic (since the total cost would always be decreasing), there would still be no guarantee that it would find the optimal solution. In addition, as is the case with all of the heuristics we have discussed, there would be no way of knowing how close to, or far from, an optimal solution the heuristic solution is. This leads us to consideration of optimization-based algorithms. In the next two sections, we discuss a Lagrangian-based approach and a dual ascent algorithm for solving the uncapacitated fixed charge facility location problem.

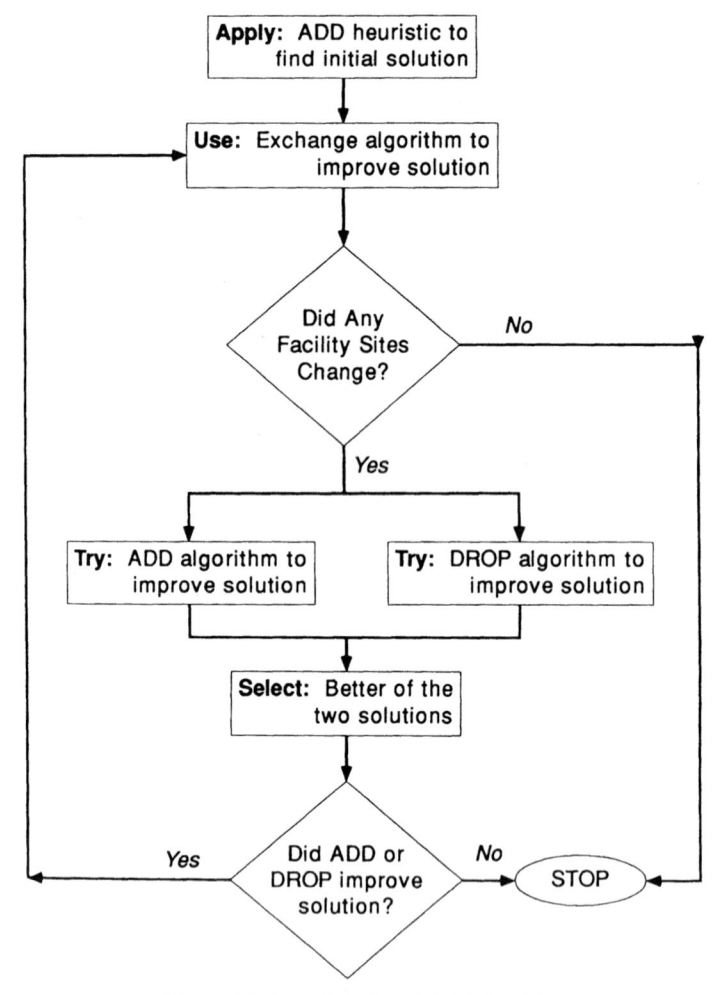

[Flowchart: Apply: ADD heuristic to find initial solution → Use: Exchange algorithm to improve solution → Did Any Facility Sites Change? — No → STOP; Yes → Try: ADD algorithm to improve solution / Try: DROP algorithm to improve solution → Select: Better of the two solutions → Did ADD or DROP improve solution? — Yes (loop back to Exchange algorithm) / No → STOP]

Figure 7.6. Flowchart for a hybrid algorithm.

7.2.3 A Lagrangian Relaxation Approach

The strong similarity between the P-median problem and the uncapacitated fixed charge location problem coupled with the generally outstanding performance of the Lagrangian relaxation approach in solving the P-median problem suggests that we consider using this approach for the uncapacitated fixed charge location problem as well. In particular, we note that the constraints for the uncapacitated fixed charge location problem are identical to those of the P-median problem except that we no longer have an explicit constraint on the number of facilities to be located. Similarly, the second term of the objective function of the uncapacitated fixed charge location problem is identical to the objective function of the P-median problem.

Uncapacitated Fixed Charge Facility Location Problems

As in the case of the P-median problem, we can develop a number of relaxations of the uncapacitated fixed charge facility location problem. For the sake of simplicity, we present only one relaxation below. Specifically, we consider relaxing constraint (7.1b) to obtain the following problem:

$$\text{MAX}_{\lambda} \text{ MIN}_{X,Y}$$

$$\sum_j f_j X_j + \alpha \sum_i \sum_j h_i d_{ij} Y_{ij} + \sum_i \lambda_i \left[1 - \sum_j Y_{ij}\right]$$

$$= \sum_j f_j X_j + \sum_i \sum_j (\alpha h_i d_{ij} - \lambda_i) Y_{ij} + \sum_i \lambda_i \quad (7.2a)$$

SUBJECT TO:

$$Y_{ij} \leq X_j \quad \forall\, i, j \quad (7.1c)$$

$$X_j = 0, 1 \quad \forall\, j \quad (7.1d)$$

$$Y_{ij} \geq 0 \quad \forall\, i, j \quad (7.1e)$$

As before, we are faced with three tasks: (1) solving the relaxed problem in an efficient manner for given values of the Lagrange multipliers, λ_i, (2) converting the relaxed solution into a primal feasible solution, and (3) updating the Lagrange multipliers. The last task will be accomplished using the subgradient optimization procedure identical to that discussed for the P-median problem. The remainder of this section will therefore focus on the first two tasks and on computational results using the approach.

We begin by attacking the first task, that of solving the relaxed problem for fixed values of the Lagrange multipliers. For fixed values of the Lagrange multipliers, we want to minimize expression (7.2a). Consider first the problem involving the allocation variables, Y_{ij}. Clearly, if $\alpha h_i d_{ij} - \lambda_i \geq 0$, we can set $Y_{ij} = 0$. If $\alpha h_i d_{ij} - \lambda_i < 0$, we would like to set Y_{ij} to as large a positive number as possible. But Y_{ij} is constrained to be less than X_j. Thus, we are again led to the notion of computing a value, V_j, associated with locating a facility at node j. Specifically, we compute $V_j = f_j + \sum_i \min(0, \alpha h_i d_{ij} - \lambda_i)$. If we set $X_j = 1$, the Lagrangian objective function will change by this amount. Note that the fixed costs are generally positive, while the summation will always be nonpositive. If $V_j < 0$, it is advantageous to set $X_j = 1$; otherwise, we set $X_j = 0$.

Thus, given values of the Lagrange multipliers, λ_i, we can find optimal values for the location and allocation variables, X_j and Y_{ij}, by using the following two-step procedure:

Step 1: For each candidate site j, compute $V_j = f_j + \sum_i \min(0, \alpha h_i d_{ij} - \lambda_i)$. Set

$$X_j = \begin{cases} 1 & \text{if } V_j < 0 \\ 0 & \text{if not} \end{cases}$$

Step 2: Set

$$Y_{ij} = \begin{cases} 1 & \text{if } X_j = 1 \text{ and } \alpha h_i d_{ij} - \lambda_i < 0 \\ 0 & \text{if not} \end{cases}$$

For any values of the Lagrange multipliers, λ_i, evaluating expression (7.2a) using the location and allocation variables determined using this two-step procedure will provide a *lower bound* on the objective function of the uncapacitated fixed charge facility location problem. As before, the job of the subgradient optimization procedure is to find the values of the Lagrange multipliers that maximize this bound.

The solution obtained using the two-step algorithm is likely to violate some of the relaxed constraints. In particular, it is likely that some demand nodes will not be assigned to any facility (i.e., $\sum_j Y_{ij} = 0$) and others will be assigned to two or more facilities (i.e., $\sum_j Y_{ij} \geq 2$). However, we can construct a primal feasible solution by simply locating facilities at the nodes for which $X_j = 1$ and assigning demands to the nearest facility. The primal objective function (7.1a) evaluated with this set of locations and demand allocations will provide us with an upper bound on the solution. Clearly, the smallest such value over all Lagrangian iterations is the best solution to use. The only problem we may encounter is that it is possible for some values of the Lagrange multipliers to have solutions for which no facilities are located (i.e., $\sum_j X_j = 0$). For iterations of the Lagrangian procedure on which this occurs, we simply do not compute an upper bound on the primal solution (or, equivalently, we set the bound from those iterations to infinity).

To illustrate the Lagrangian procedure, we again use the network of Figure 7.4. Beginning with all Lagrange multipliers set equal to the default SITATION values of 10 times the average demand plus 10 times the average fixed cost ($10 * 15.416667 + 10 * 176.25 = 1916.66667$), the Lagrangian procedure converges in 37 iterations to a solution with a total cost of $1234.95. Only three facilities are sited at nodes A, D, and K. The total fixed cost is 375 and the total demand-weighted distance is 2457.

The lower bound obtained from the Lagrangian procedures is 1234.912. The upper bound is only 0.003 percent greater than this value. Note also that since all fixed costs, demands, and distances are integer, the objective function must increase or decrease in increments of 0.35, which is the smallest possible difference in demand times the smallest possible difference in distance times the cost per mile. (Had this exceeded 1, which is the smallest possible difference in integer fixed costs, the objective function would have to change in increments of at least 1.) Since the difference between the lower and upper bounds is less than this value, we can conclude that the solution is optimal.

Table 7.11 summarizes the solutions we have obtained using the heuristic and Lagrangian procedures. In this example, the ADD heuristic solution is

Table 7.11 Summary of Solutions to the Network of Figure 7.4

Algorithm	Facility Locations	Fixed Cost	Average Distance	Total Cost	Percentage Deviation from Optimal
ADD	A, D, G, I, K	730	1812	1364.20	10.47
DROP	A, C, F, K	530	2038	1243.30	0.68
ADD/substitution	A, C, F, H, K	740	1585	1294.75	4.84
ADD/neighborhood	A, D, F, H, K	760	1585	1314.75	6.46
DROP/substitution	A, C, F, K	530	2038	1243.30	0.68
DROP/neighborhood	A, C, F, K	530	2038	1243.30	0.68
Lagrangian	A, D, K	375	2457	1234.95	0.00

over 10 percent more costly than is the optimal solution. The DROP heuristic solution is within 1 percent of the optimal solution in this case. Note that not only do the objective function values differ when different algorithms are used, but the number of facilities sited differs as well. Only the Lagrangian procedure located the optimal number of facilities.

7.2.4 A Dual-Based Approach

Lagrangian relaxation is not the only optimization-based procedure that enables us to obtain both lower and upper bounds on the solution. In some cases, we can work directly with the original optimization problem, or at least with the linear programming relaxation of it. The following dual-based procedure [originally proposed by Erlenkotter (1978)] is based on this concept. The procedure has been implemented by Erlenkotter in code known as DUALOC. The basic approach that we will adopt in attacking the problem in this section is shown in Figure 7.7.

To facilitate the discussion of this approach, we restate the uncapacitated fixed charge facility location problem, rewriting constraint (7.1c) to facilitate the construction of the dual of the linear programming relaxation of the problem. In addition, we replace the coefficient $\alpha h_i d_{ij}$ of Y_{ij} by the simpler notation c_{ij}, since this term will appear often in the development below:

MINIMIZE $\quad \sum_j f_j X_j + \sum_i \sum_j c_{ij} Y_{ij}$ $\hfill (7.1a)$

SUBJECT TO: $\quad \sum_j Y_{ij} = 1 \quad \forall\, i$ $\hfill (7.1b)$

$\quad X_j - Y_{ij} \geq 0 \quad \forall\, i, j$ $\hfill (7.1c')$

$\quad X_j = 0, 1 \quad \forall\, j$ $\hfill (7.1d)$

$\quad Y_{ij} \geq 0 \quad \forall\, i, j$ $\hfill (7.1e)$

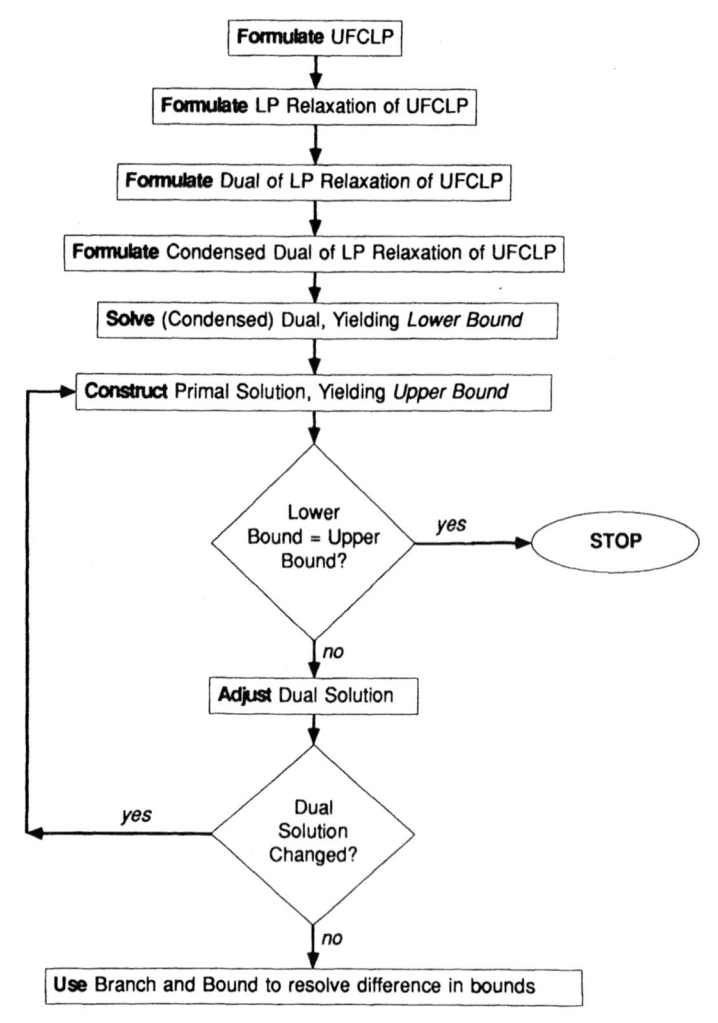

Figure 7.7. Dual-based solution approach for uncapacitated fixed charge location problem.

To develop the linear programming dual, we begin by relaxing the integrality constraint on the location variables, X_j, and rewrite constraint (7.1d) as follows:

$$X_j \geq 0 \quad \forall\, i \tag{7.1d'}$$

Next, we associate dual variables V_i with constraint (7.1b) which stipulate that each demand node j be fully served. We also associate dual variables W_{ij} with constraint (7.1c') which link the location and allocation variables. With

Uncapacitated Fixed Charge Facility Location Problems

the definition of these variables, the associated dual problem is

$$\text{MAXIMIZE} \quad \sum_i V_i \tag{7.3a}$$

$$\text{SUBJECT TO:} \quad \sum_i W_{ij} \leq f_j \quad \forall\, j \tag{7.3b}$$

$$V_i - W_{ij} \leq c_{ij} \quad \forall\, i, j \tag{7.3c}$$

$$V_i \text{ unrestricted} \quad \forall\, i \tag{7.3d}$$

$$W_{ij} \geq 0 \quad \forall\, i, j \tag{7.3e}$$

We begin by showing that we can eliminate the variables W_{ij} from the formulation. For any feasible values of the dual variables V_i, it should be clear that we can pick values of W_{ij} to be as small as possible. W_{ij} will then be constrained by (7.3c) and (7.3e). Specifically, for fixed feasible values of V_i, we will have to pick W_{ij} to satisfy

$$W_{ij} \geq V_i - c_{ij} \quad \forall\, i, j \tag{7.3c'}$$

and

$$W_{ij} \geq 0 \quad \forall\, i, j \tag{7.3e}$$

or, equivalently,

$$W_{ij} = \max\{0, V_i - c_{ij}\} \quad \forall\, i, j \tag{7.4}$$

We can now replace W_{ij} in the dual formulation by (7.4) and, in the process, remove constraint (7.3c) which will automatically be satisfied by this selection of values for W_{ij}. In doing so, we obtain the following condensed dual formulation:

$$\text{MIXIMIZE} \quad \sum_i V_i \tag{7.5a}$$

$$\text{SUBJECT TO:} \quad \sum_i \max\{0, V_i - c_{ij}\} \leq f_j \quad \forall\, j \tag{7.5b}$$

$$V_i \text{ unrestricted} \quad \forall\, i \tag{7.5c}$$

We now turn to the problem of solving the condensed dual problem. The only decision variables that remain are the V_i, whose sum is to be maximized. Since we can fairly assume that all fixed costs, f_j, and all allocation costs, c_{ij}, are nonnegative, a feasible dual solution is $V_i = 0$ for all i. The approach we will adopt is that of successively increasing each V_i by a suitably small amount until no V_i can be increased further. It should be clear that as long as

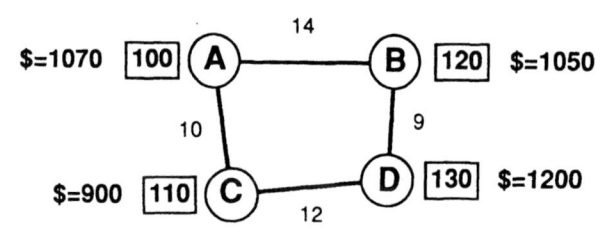

Figure 7.8. Sample network for dual-based approach.

V_i is less than c_{ij}, V_i can be increased without affecting the left-hand side of the jth constraint of the condensed dual. As soon as $V_i = c_{ij}$, further increases in V_i will increase the left-hand side of the jth constraint which must be less than f_j, the right-hand side. Thus, as V_i is increased, its value will affect an increasing number of constraints.

To illustrate the approach, consider the small network shown in Figure 7.8. For $\alpha = 0.7$, the condensed dual problem associated with this network is

MAXIMIZE $\quad V_A + V_B + V_C + V_D$

SUBJECT TO:

(A) $\quad \max\{0, V_A - 0\} + \max\{0, V_B - 1176\} + \max\{0, V_C - 770\}$

$\quad + \max\{0, V_D - 2002\} \leq 1070$

(B) $\quad \max\{0, V_A - 980\} + \max\{0, V_B - 0\} + \max\{0, V_C - 1617\}$

$\quad + \max\{0, V_D - 819\} \leq 1050$

(C) $\quad \max\{0, V_A - 700\} + \max\{0, V_B - 2520\} + \max\{0, V_C - 0\}$

$\quad + \max\{0, V_D - 1092\} \leq 900$

(D) $\quad \max\{0, V_A - 1540\} + \max\{0, V_B - 756\} + \max\{0, V_C - 924\}$

$\quad + \max\{0, V_D - 0\} \leq 1200$

We will define S_j as the slack in the jth dual constraint. Thus, S_A is the slack in the first constraint above. We will also define $\mathbf{M}_i$ to be the set of indices of the constraints j whose slack will be reduced by an increase in the current value of V_i. As indicated above, we will begin with $V_i = 0$ for all values of i. Thus, initially, $\mathbf{M}_i = \{i\}$ (at least for problems in which there are the same

Uncapacitated Fixed Charge Facility Location Problems

Table 7.12 Iterations of the Dual Ascent Algorithm for the Network of Figure 7.8 with $\alpha = 0.7$

Variable	\multicolumn{8}{c}{Key Values at Iteration}
	0	1	2	3	4	5	6	7
V_A	0	*__700__*	700	700	700	700	700	700
V_B	0	0	*__756__*	756	756	756	756	*__777__*
V_C	0	0	0	*__770__*	770	770	*__900__*	900
V_D	0	0	0	0	*__819__*	*__1092__*	1092	1092
S_A	1070	*__370__*	370	370	370	370	*__240__*	240
S_B	1050	1050	*__294__*	294	294	*__21__*	21	*__0__*
S_C	900	900	900	*__130__*	130	130	*__0__*	0
S_D	1200	1200	1200	1200	*__381__*	*__108__*	108	*__97__*
M_A	A	A,C	A,C	A,C	A,C	A,C	A,C	A,C
M_B	B	B	B,D	B,D	B,D	B,D	B,D	B,D
M_C	C	C	C	A,C	A,C	A,C	A,C	A,C
M_D	D	D	D	D	B,D	B,C,D	B,C,D	B,C,D

Values that change at any iteration are shown in **_bold italics_**.

number of candidate locations as there are demand nodes and $d_{ii} = 0$ for all values of i and $d_{ij} > 0$ when $i \neq j$). In other words, initially, an increase in V_i will only affect the slack in the ith constraint. As the value of V_i increases, additional elements will be added to the set M_i.

Table 7.12 shows the values of the dual decision variables, V_i, the dual slack variables, S_j, and the sets, M_i, for each iteration of the dual ascent algorithm. All values of V_i are initially set to 0. Each V_i is examined in turn and is increased until either (1) one more constraints must be added to the set M_i or (2) the slack associated with one of the constraints whose index is already in M_i goes to 0. In other words, at each iteration, one more V_i is increased to the smaller of: (1) the next larger value of c_{ij}, at which point the index j is added to the set M_i; or (2) $\min_{j \in M_i} \{S_j\}$. We alternate between processing the V_i values in ascending and descending order, as suggested by Erlenkotter (1978).

During the first iteration in this example, the increase in all values of V_i is limited by the first condition above. Each V_i increases to the next larger value of c_{ij} and j is added to the set M_i. Thus, V_A increases to 700 at which point further increases in V_A will affect the slack in the third constraint and so C is added to M_A.

For the second pass through the V_i values, we begin by processing V_D. This value can be increased from 819 to 1092, at which point the slack in the third constraint will be affected. The slack in the two constraints in M_D (B and D) is reduced by 273 (1092 − 819). We then add C to M_D. Next, we examine V_C. The difference between the current value of V_C and the next larger value of c_{Cj} is 154(924 − 770). Constraints A and C will be affected by increases in V_C. The slack in constraint A is 370, but the slack in constraint C is only 130. Since this is less than 154, we can only increase V_C by 130. No additional constraint indices need to be added to the set $\mathbf{M}_C$. Finally, we examine V_B. Analogous reasoning indicates that V_B can only be increased by 21 units to 777, at which point constraint B is binding. At this point, constraints B and C are binding. Since either B or C is contained in each set $\mathbf{M}_i$, no further increases in the values of V_i are possible (unless some V_i value is reduced, as we will see below). The value of the dual objective function is 3469, the sum of the V_i values. Since these values are feasible for the dual problem and since this is the dual of a relaxation of the original problem, this value is a *lower bound* on the value of the total cost associated with the uncapacitated fixed charge facility location problem.

Having obtained a solution (which may or may not be optimal for the dual problem), we now try to construct a primal solution related to the dual solution. In particular, we will construct a solution that satisfies all of the primal feasibility constraints for the original integer programming problem and that tries to satisfy all of the complementary slackness conditions. If we are successful in finding a solution that satisfies all of the complementary slackness conditions, we will have an optimal solution. If not, further processing may be required.

The complementary slackness conditions for the linear programming relaxation are

$$[X_j - Y_{ij}]W_{ij} = [X_j - Y_{ij}][\max\{0, V_i - c_{ij}\}] = 0 \qquad \forall\, i,j \quad (7.6a)$$

$$X_j\left[f_j - \sum_i W_{ij}\right] = X_j\left[f_j - \sum_i \max\{0, V_i - c_{ij}\}\right] = 0 \qquad \forall\, j \quad (7.6b)$$

$$Y_{ij}[c_{ij} - V_i + W_{ij}] = Y_{ij}[c_{ij} - V_i + \max\{0, V_i - c_{ij}\}] = 0 \qquad \forall\, i,j \quad (7.6c)$$

$$\left[1 - \sum_j Y_{ij}\right]V_i = 0 \qquad \forall\, i \quad (7.6d)$$

Conditions (7.6d) will automatically be satisfied by any primal feasible solution, since the term in brackets will be 0. Therefore, we can ignore these conditions, since the solution we construct will be primal feasible.

If $V_i \geq c_{ij}$, the term in brackets in conditions (7.6c) will be 0. Since we will only set $Y_{ij} = 1$ if $V_i \geq c_{ij}$, conditions (7.6c) will also be satisfied for any solution we construct.

Conditions (7.6b) can be used to define a set of candidate locations. If the term in brackets (which represents the slack in the jth constraint of the condensed dual problem) is 0, locating at candidate site j will not violate this complementary slackness condition. Adapting the notation used by Erlenkotter (1978) to the problem as defined above, we denote the set of

Uncapacitated Fixed Charge Facility Location Problems

candidate sites for which there is no slack in the jth condensed dual constraint by $\mathbf{J}^*$.

We now examine conditions (7.6a). If there are two or more locations $j \in \mathbf{J}^*$ for which we have set $X_j = 1$ and for which $V_i > c_{ij}$, conditions (7.6a) will have to be violated by any primal feasible solution since we can only assign each demand node i to one facility j. In other words, there will be at least one term $X_j - Y_{ij}$ which will equal 1 for which $\max\{0, V_i - c_{ij}\} > 0$, creating a violation of conditions (7.6a). We can reduce the likelihood of this occurring by only setting $X_j = 1$ when facility j is needed to serve some demand node i.

Thus, for each demand node i, we determine the number of locations $j \in \mathbf{J}^*$ for which $V_i \geq c_{ij}$. If there is only one such location, then we set the corresponding $X_j = 1$. Such facilities are termed essential facilities, since there is a demand node i for which this facility is the only facility that will allow condition (7.6c) to be satisfied.

After we have examined each demand node to determine whether or not there is only one facility location $j \in \mathbf{J}^*$ for which $V_i \geq c_{ij}$, we again examine each demand node. This time, we ask whether or not there is at least one candidate facility site j for which $X_j = 1$ and $V_i \geq c_{ij}$. If there is, we go on to the next demand node. If not, we find the candidate site $j \in \mathbf{J}^*$ for which c_{ij} is a minimum and assign the corresponding $X_j = 1$. Ties may be broken arbitrarily. The set of selected facility sites is denoted by $\mathbf{J}^+$. Demand nodes are assigned to the facility $j \in \mathbf{J}^+$ for which c_{ij} is a minimum. In other words, we set

$$Y_{ij} = \begin{cases} 1 & c_{ij} \leq c_{ik} \text{ for all } k \text{ and } j \in \mathbf{J}^+ \\ 0 & \text{if not} \end{cases}$$

(In the event that there are multiple facilities j which satisfy the condition $c_{ij} \leq c_{ik}$ for all k and $j \in \mathbf{J}^+$, we arbitrarily pick one of them and assign the corresponding $Y_{ij} = 1$ and set all other $Y_{ik} = 0$.)

Applying this procedure to the solution obtained in Table 7.12, we find that $\mathbf{J}^* = \{B, C\}$. For node A, we have $V_A < c_{AB}$ (700 < 980) and $V_A = c_{AC}$ (700 = 700). Therefore, candidate site C is essential for demand node A. We therefore set $X_C = 1$. For node B, we have $V_B > c_{BB}$ (777 > 0) and $V_B < c_{BC}$ (777 < 1764), so candidate site B is essential for demand node B. We therefore set $X_B = 1$. At this point, all facility sites $j \in \mathbf{J}^* = \{B, C\}$ have been selected and we can stop. We have $\mathbf{J}^+ = \{B, C\}$. Demand nodes are now assigned to the closest of these two facilities. Thus, demand nodes B and D are assigned to the facility at node B at a cost of 819 (0 + 819) and demand nodes A and C are assigned to the facility at node C at a cost of 700 (700 + 0). The total transportation cost associated with this solution is therefore 1519. The total fixed cost is 1950 (1050 + 900). The total cost of this solution is 3469 (1950 + 1519). Since this is primal feasible solution, this

Table 7.13 Iterations of the Dual Ascent Algorithm for the Network of Figure 7.8 with $\alpha = 0.5$

	\multicolumn{11}{c}{Key Values at Iteration}
Variable	0	1	2	3	4	5	6	7	8	9	10
V_A	0	*500*	500	500	500	500	500	500	*700*	*715*	715
V_B	0	0	*540*	540	540	540	540	*840*	840	840	840
V_C	0	0	0	*550*	550	570	*660*	660	660	660	*685*
V_D	0	0	0	0	*585*	*780*	780	780	780	780	780
S_A	1070	*570*	570	570	570	570	*460*	460	*260*	*245*	*220*
S_B	1050	1050	*510*	510	510	*315*	315	*15*	15	*0*	0
S_C	900	900	900	*350*	350	350	*240*	240	*40*	*25*	*0*
S_D	1200	1200	1200	1200	*615*	*420*	420	*120*	120	120	*95*
M_A	A	A,C	A,C	A,C	A,C	A,C	A,C	A,C	A,B,C	A,B,C	A,B,C
M_B	B	B	B,D	B,D	B,D	B,D	B,D	A,B,D	A,B,D	A,B,D	A,B,D
M_C	C	C	C	A,C	A,C	A,C	A,C,D	A,C,D	A,C,D	A,C,D	A,C,D
M_D	D	D	D	D	B,D	B,C,D	B,C,D	B,C,D	B,C,D	B,C,D	B,C,D

Values that change at any iteration are shown in **bold italics**.

value represents an upper bound on the solution. Since the lower and upper bounds are equal, we can stop; we have an optimal solution in this case.

The dual ascent procedure coupled with the algorithm outlined above for constructing a related primal feasible solution will not always yield a solution that is provably optimal. To illustrate this point, consider the same example, changing α to 0.5. In this case, we have

$$c_{ij} = \alpha h_i d_{ij} = \begin{bmatrix} 0 & 700 & 500 & 1100 \\ 840 & 0 & 1260 & 540 \\ 550 & 1155 & 0 & 660 \\ 1430 & 585 & 780 & 0 \end{bmatrix}$$

Table 7.13 gives the values of the key variables for each iteration of the dual ascent procedure for this case. After 10 iterations, we have $(V_A, V_B, V_C, V_D) = (715, 840, 685, 780)$. The resulting dual objective function value of $\Sigma_i V_i = 3020$. As before, this is a lower bound on the solution.

Again there is no slack in the constraints associated with candidate locations B and C, so, again, $\mathbf{J}^* = \{B, C\}$. As before, we examine each demand node in turn and find that candidate location B is essential for demand node B and candidate location C is essential for demand node C. Thus, again, we have $\mathbf{J}^+ = \{B, C\}$. Assigning each demand node to the least expensive facility, we assign nodes B and D to the facility at node B and nodes A and C to the facility at node C as before. The total cost of this solution is 3035 (1950 in fixed costs and 1085 in transportation costs). This, again, is an upper bound on the solution. Now, however, the lower and upper bounds are not equal. We now enter the dual adjustment phase shown in Figure 7.7.

Uncapacitated Fixed Charge Facility Location Problems

Notice that $V_A = 715 \geq c_{AB} = 700$ and $V_A = 715 \geq c_{AC} = 500$. Thus, there are two cases of condition (7.6a) in which the term in brackets is positive and for which $X_j = 1$. Since we only have $Y_{ij} = 1$ for one location j for each demand node i, we must have a violation of the complementary slackness conditions. In fact, that violation occurs for the demand node A/facility B combination. Observed that $V_A - c_{AB} = 715 - 700 = 15$ and the difference between the primal and dual objective functions is also 15. In fact, we can show that the difference between the primal and dual objective functions will be

$$\sum_i \sum_{\substack{j \in \mathbf{J}^+ \\ j \neq j^1(i)}} \max\{0, V_i - c_{ij}\}$$

where $j^1(i)$ is the index of the candidate location j to which demand node i is assigned. In other words, $j^1(i)$ is the first closest open facility to demand node i. The reader interested in the derivation of this result is referred to Erlenkotter (1978).

If we reduce the value of one dual variable, we may be able to increase the value of one or more other dual variables. In particular, in this case, if we reduce V_A by 15, we will eliminate the violation of the complementary slackness condition and may be able to increase other dual variables. If the total increase in the other dual variables exceeds 15, we will have succeeded not only in eliminating the violation of the complementary slackness condition but in increasing the dual objective function or the lower bound on the objective function.

In general when we find a violation of complementary slackness condition (7.6a) for some demand node i, we will decrease the value of the corresponding dual variable, V_i, to the next smaller value of c_{ij}. We will then apply the dual ascent procedure beginning with the values of the dual variables that we already have computed. In particular, we will execute the dual ascent procedure *three* times. The first two times, we will allow changes in only a subset of the dual variables. In the *first* pass through the dual ascent procedure, we will allow changes only in those dual variables that are part of the set $\mathbf{I}^+ = \mathbf{I}^+_{j^1(i)} \cup \mathbf{I}^+_{j^2(i)}$, where $\mathbf{I}^+_j$ is the set of demand nodes i which can only be assigned to candidate location j without violating complementary slackness condition (7.6a). $j^1(i)$ and $j^2(i)$ are the first and second closest open facilities to demand node i. Thus, $\mathbf{I}^+$ is the set of demand nodes which forced either the closest or the second closest open facility to demand node i to be in the solution. These are nodes for which it is likely that we will be able to increase the dual variable without being blocked by one or more of the condensed dual constraints. (Note that either or both $\mathbf{I}^+_{j^1(i)}$ or $\mathbf{I}^+_{j^2(i)}$ may be empty sets. If both are, we do not attempt to change the value of V_i.) In the *second* pass through the dual ascent procedure, we augment the set $\mathbf{I}^+$ by $\{i\}$. Finally, in the *third* pass, we allow changes in all of the dual variables.

Table 7.14 Iterations of the Dual Adjustment / Dual Ascent Algorithm for the Network of Figure 7.8 with $\alpha = 0.5$

	Key Values at Iteration			
Variable	0	1	2	3
V_A	715	*700*	700	700
V_B	840	840	*855*	855
V_C	685	685	685	*700*
V_D	780	780	780	780
S_A	220	*235*	*220*	*205*
S_B	0	*15*	*0*	0
S_C	0	*15*	15	*0*
S_D	95	95	*80*	*65*
M_A	A, B, C	A, B, C	A, B, C	A, B, C
M_B	A, B, D	A, B, D	A, B, D	A, B, D
M_C	A, C, D	A, C, D	A, C, D	A, C, D
M_D	B, C, D	B, C, D	B, C, D	B, C, D

Values that change at any iteration are shown in *bold italics*.

To illustrate the dual adjustment procedure, we consider the dual solution obtained in Table 7.13. Since we have a violation of complementary slackness condition (7.6a) for demand node A, we reduce V_A to 700 (from 715). The result of doing so is shown as iteration 1 in Table 7.14. This creates slack in the constraints associated with nodes B and C (and increases the slack associated with node A). The closest open node to demand node A is node C and the second closest in node B. The facility at node C is required for demand node C if we are not to violate condition (7.6a). Similarly, the facility at node B is required for demand node B if we are not to violate the associated condition (7.6a). Thus, in the first pass through the dual ascent procedure, we allow changes only in V_B and V_C. The results of these changes are shown in the columns for iterations 2 and 3 of Table 7.14. Both V_B and V_C increase by 15. The new dual objective function (lower bound) is 3035, which is exactly the value of the primal objective function (or upper bound). We therefore have been able to prove once again that the solution comprised of locating at nodes B and C and assigning demand nodes A and C to the facility at C and nodes B and D to the facility at node D is optimal.

Finally, we note that there are cases in which even the dual adjustment procedure will not allow us to find a provably optimal solution. The reader interested in such a case should consider this example with $\alpha = 0.35$. In these cases, branch and bound is needed to resolve differences between the lower and upper bounds. For a more formal treatment of the dual-based algorithm, the interested reader is referred to the seminal paper on the topic by Erlenkotter (1978). Cornuejols, Nemhauser, and Wolsey (1990) provide an excellent review of the theory underlying uncapacitated facility location problems and algorithms for the solution of these problems.

7.3 CAPACITATED FIXED CHARGE FACILITY LOCATION PROBLEMS

Many of the models we have discussed so far have dealt with problems in which the facilities were uncapacitated. One qualitative exception to this is the maximum expected covering location model in which facilities (often ambulances) were allowed to be busy. In essence, each facility had a capacity of one at any point in time. In this section we formulate a more general capacitated fixed charge location model.

Capacities are important in many facility location problems. For example, a typical automobile assembly plant can assemble approximated 500 vehicles during an 8-hour shift. Operating two shifts, as is commonly done, will double the number of vehicles that can be produced in a day. However, even if the plant operated 24 hours each day, it could only produce approximately 1500 vehicles per day. Similarly, a warehouse has only a fixed number of square feet of storage space. Schools, parking lots and structures, ports, hospitals, and clinics are other facilities that are subject to capacity limitations.

In many cases, fixed capacities are less important than are practical capacities. For example, while an automated teller machine (ATM) might be able to process 1 customer per minute, we would not expect to serve 1440 customers per day. Customers simply do not arrive uniformly over a 24-hour day. Instead, customers for ATM machines (at least those near office complexes) are most likely to arrive just before working hours, just after working hours, and during lunch breaks. Excessive delays will reduce patronage. Thus, the practical capacity of an ATM machine will be significantly less than the theoretical throughput of the machine. Computing such practical capacities requires a knowledge of queuing theory in conjunction with an understanding of customer behavior. Facility location models that incorporate queuing components are beyond the scope of this text. The reader interested in such models is referred to some of the key papers on the topic including Berman, Larson, and Chiu (1985) and Chiu, Berman, and Larson (1985). Larson (1987) discusses the psychology of queuing, a problem that must be understood if the practical capacity of the facilities being located depends on the behavior of customers and their willingness to wait for service.

Given some exogenously specified capacities, the problem we consider in this section is that of locating facilities to minimize the sum of the facility location costs and the travel costs of customers to the facilities subject to constraints that stipulate that all demands must be served, facility capacities must not be exceeded, and customers can only be served from open facilities. In addition to the notation we have already used, we define the following input quantity:

k_j = capacity of a facility at candidate site j if a facility is located there

With this additional definition, we can formulate the capacitated fixed charge facility location problem as follows:

MINIMIZE $\quad \sum_j f_j X_j + \alpha \sum_i \sum_j h_i d_{ij} Y_{ij}$ $\quad$ (7.7a)

SUBJECT TO: $\quad \sum_j Y_{ij} = 1 \quad \forall\, i$ $\quad$ (7.7b)

$\quad\quad\quad\quad\quad\quad Y_{ij} \leq X_j \quad \forall\, i, j$ $\quad$ (7.7c)

$\quad\quad\quad\quad\quad\quad \sum_i h_i Y_{ij} \leq k_j X_j \quad \forall\, j$ $\quad$ (7.7d)

$\quad\quad\quad\quad\quad\quad X_j = 0, 1 \quad \forall\, j$ $\quad$ (7.7e)

$\quad\quad\quad\quad\quad\quad Y_{ij} \geq 0 \quad \forall\, i, j$ $\quad$ (7.7f)

This formation is identical to the uncapacitated fixed charge problem (7.1a)–(7.6a) defined in Section 7.2 above except that we have now included a capacity constraint (7.7d). Note that constraint (7.7c) is not needed in the integer programming formulation of this problem since constraint (7.7d) will ensure that demands at node i are not assigned to a facility at candidate location j if we have not selected candidate location j. However, inclusion of constraint (7.7c) strengthens the linear programming relaxation of the problem considerably. This constraint is also useful in many of the solution algorithms that we will consider. Therefore, we have chosen to include the constraint in the formulation above.

Before discussing solution algorithms, we note that if we are given a set of facility locations that are feasible in the sense that the total capacity of the facilities exceeds the total demand, the problem of assigning the demands to the facilities becomes a transportation problem of the sort discussed in Chapter 2. Specifically, if we are given values $\hat{X}_j$ for the location variables and $\sum_j k_j \hat{X}_j \geq \sum_i h_i$, then the optimal assignment of demands to facilities can be found by solving the following linear programming problem:

MINIMIZE $\quad \alpha \sum_i \sum_j h_i d_{ij} Y_{ij}$ $\quad$ (7.8a)

SUBJECT TO: $\quad \sum_j Y_{ij} = 1 \quad \forall\, i$ $\quad$ (7.8b)

$\quad\quad\quad\quad\quad\quad \sum_i h_i Y_{ij} \leq k_j \hat{X}_j = \begin{cases} k_j & \text{if } \hat{X}_j = 1 \\ 0 & \text{if } \hat{X}_j = 0 \end{cases} \quad \forall\, j$ $\quad$ (7.8c)

$\quad\quad\quad\quad\quad\quad Y_{ij} \geq 0 \quad \forall\, i, j$ $\quad$ (7.8d)

The objective function minimizes the total transportation cost. This is identical to the objective function of a transportation problem. Constraint (7.8b) serves the function of a demand constraint in the transportation problem as it stipulates that all of the demand at node i must be assigned to a facility. Constraint (7.8c) is a supply constraint. Note that the right-hand side of constraint (7.8c) is a constant since the values of $\hat{X}_j$ are given and are either 0 or 1. If a facility is located at candidate site j ($\hat{X}_j = 1$), the right-hand side of (7.8c) is just the capacity of a facility at node j (k_j). If no facility is located at candidate site j ($\hat{X}_j = 0$), the right-hand side of constraint (7.8c) is 0 for that value of j. In this case, we can remove all assignment variables Y_{ij} associated with candidate site j since we know they all must be 0. Finally, constraint (7.8d) is a simple nonnegativity condition. Thus, (7.8a)–(7.8d) defines a transportation problem.

The fact that the capacitated fixed charge location problem reduces to a transportation problem when the location variables $\hat{X}_j$ are known will be important in all of the optimization-based solution approaches discussed below. Note that if the resulting problem is not feasible because the total capacity of the selected facilities is less than the total demand, the primal transportation problem will be infeasible and the dual of the corresponding transportation problem will be unbounded. We can avoid this problem by ensuring that the total capacity of the selected facilities is greater than or equal to the total demand.

7.3.1 Lagrangian Relaxation Approaches

The structure of the capacitated fixed charge location problem lends itself to a number of relaxations. Clearly, if we relax the capacity constraints (7.7d), we will be left with an uncapacitated fixed charge location problem (UFCLP). The excellent computational results that can be attained for this problem using either a Lagrangian relaxation approach as outlined in Section 7.2.3 or the dual-based algorithm discussed in Section 7.2.4 makes this an attractive option.

7.3.1.1 Relaxing the Capacity Constraints (7.7d) Specifically, if we relax constraint (7.7d), we obtain the following optimization problem:

$$\underset{\mu}{\text{MAX}} \underset{X,Y}{\text{MIN}}$$

$$\sum_j f_j X_j + \alpha \sum_i \sum_j h_i d_{ij} Y_{ij} + \sum_j \mu_j \left\{ \sum_i h_i Y_{ij} - k_j X_j \right\}$$

$$= \sum_j (f_j - \mu_j k_j) X_j + \sum_i \sum_j (\alpha h_i d_{ij} + \mu_j h_i) Y_{ij} \qquad (7.9a)$$

SUBJECT TO:

$$\sum_j Y_{ij} = 1 \quad \forall\, i \tag{7.7b}$$

$$Y_{ij} \leq X_j \quad \forall\, i, j \tag{7.7c}$$

$$X_j = 0, 1 \quad \forall\, j \tag{7.7e}$$

$$Y_{ij} \geq 0 \quad \forall\, i, j \tag{7.7f}$$

$$\mu_j \geq 0 \quad \forall\, j \tag{7.9b}$$

For fixed values of the Lagrange multipliers μ_j, we want to minimize objective function (7.9). Substituting $\hat{f}_j = f_j - \mu_j k_j$ and $\hat{c}_{ij} = \alpha h_i d_{ij} + \mu_j h_i$, we obtain the following objective function for the fixed values of μ_j:

MINIMIZE
$$\sum_j \hat{f}_j X_j + \sum_i \sum_j \hat{c}_{ij} Y_{ij} \tag{7.9a'}$$

This objective function is to be minimized subject to constraints (7.7b), (7.7c), (7.7e), and (7.7f). However, this is exactly the structure of the uncapacitated fixed charge location problem discussed in Section 7.2.2.

Figure 7.9 outlines the solution of the capacitated fixed charge facility location problem when we relax the supply constraint. After relaxing the capacity constraint, we initially fix the Lagrange multipliers associated with the relaxed constraint. As indicated above, this results in an uncapacitated fixed charge facility location problem. We solve this using either the Lagrangian procedure (Section 7.2.3) or the dual-based procedure (Section 7.2.4). In either case, the lower bound for the resulting UFCLP is also a lower bound on the capacitated fixed charge facility location problem. If the total capacity of the facility sites identified in the solution to the UFCLP exceeds the demand, we solve the transportation problem (7.8a)–(7.8d) above (with the $\hat{X}_j$ set equal to the value of the location variables from the UFCLP). This allows us to compute and potentially update the upper bound on the capacitated fixed charge location problem (equal to the objective function from the transportation problem plus the sum of the fixed costs of all selected facilities). If the lower and upper bounds are sufficiently close to each other, we stop. If not, we determine whether or not we want to stop the Lagrangian procedure for revising the Lagrange multipliers associated with the capacity constraints. If so, we use branch and bound to close the gap between the lower and upper bounds; if not, we revise the Lagrange multipliers associated with the capacity constraints in an appropriate manner.

Several facets of this algorithm need further explanation. First, we note that, assuming a feasible solution to the capacitated fixed charge facility location problem exists (i.e., $\sum_j k_j \geq \sum_i h_i$), and upper bound can always be obtained by fixing all candidate sites open ($\hat{X}_j = 1 \;\forall\, j$) and solving the resulting transportation problem. The sum of the objective function from the transportation problem plus the sum of all candidate site fixed costs is an

Capacitated Fixed Charge Facility Location Problems

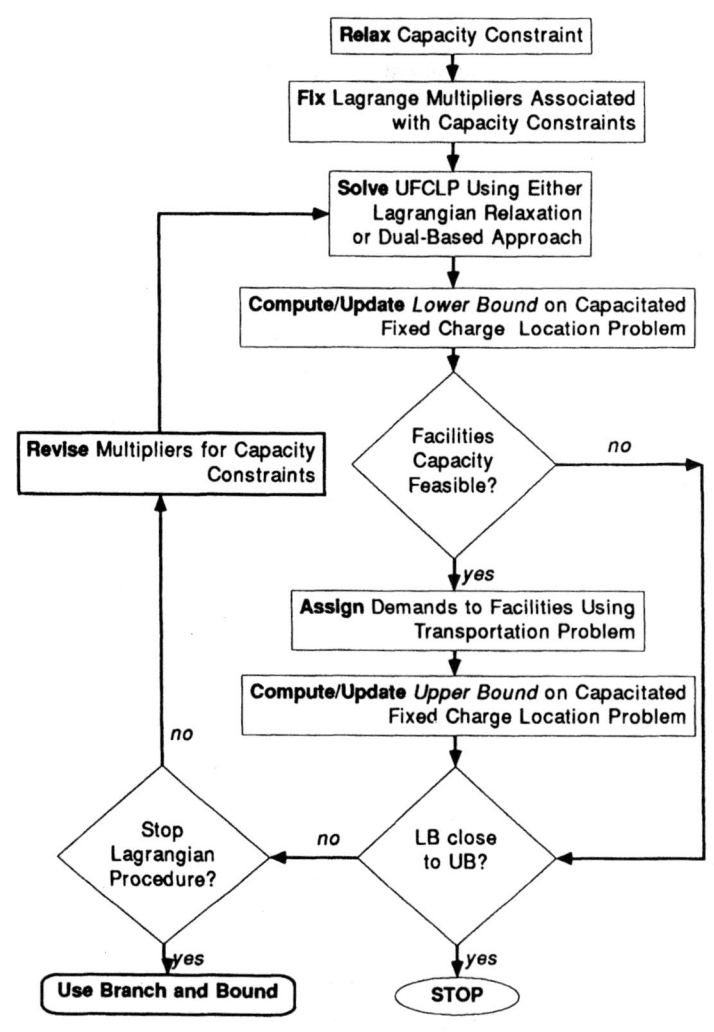

Figure 7.9. Flowchart of an algorithm for the capacitated fixed charge problem with capacity constraint relaxed.

upper bound on the capacitated fixed charge facility location problem. Second, a number of criteria can be used in determining whether or not the lower and upper bounds are close enough to each other. One standard measure is the difference between the two as a percentage of the lower bound. If the fixed costs, transport costs ($\alpha h_i d_{ij}$), capacities, and demands are all integers, the objective function will have to be integer valued. In this case, if the difference between the bounds is less than 1, we can be assured of having the optimal solution.

Third, revising the multipliers on the capacity constraints can be done using standard subgradient optimization procedures. Fourth, if the lower and upper bounds are not sufficiently close to each other, whether or not we stop the Lagrangian procedure should be based on stopping criteria including: (i) a limit on the number of iterations or (ii) a limit on the size of the constant used in computing the stepsize if a subgradient optimization procedure is used to revise the Lagrange multipliers on the capacity constraints. Finally, in using branch and bound to close any gap between the lower and upper bounds, the algorithm shown in Figure 7.9 can be used with the selection of candidate sites in the UFCLP limited by the additional constraints introduced in the branch-and-bound algorithm.

To illustrate this approach, consider the four-supply/eight-demand problem shown in Figure 7.10. Candidate site capacities and fixed costs as well as demands at the demand nodes are shown beside the nodes. Unit shipping cost (αd_{ij}) between the supply (candidate facility) nodes and the demand nodes are shown in the table at the bottom of the figure.

Table 7.15 presents the results of the first 13 major iterations through the algorithm shown in Figure 7.9 when applied to the network of Figure 7.10.

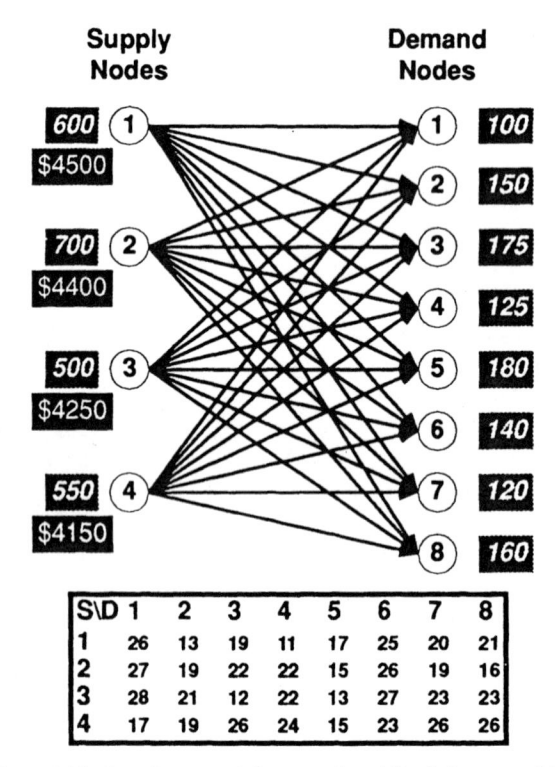

S\D	1	2	3	4	5	6	7	8
1	26	13	19	11	17	25	20	21
2	27	19	22	22	15	26	19	16
3	28	21	12	22	13	27	23	23
4	17	19	26	24	15	23	26	26

Figure 7.10. Sample network for capacitated fixed charge problems.

Capacitated Fixed Charge Facility Location Problems

Table 7.15 Example Results for the Capacitated Fixed Charge Problem Relaxing Constraint (7.7)

Iteration	μ_1	μ_2	μ_3	μ_4	LB	Best UB	LB Loc	UB Loc	Percentage Error This Iteration
1	0.000	0.000	0.000	0.000	26,070.0	34,825.0	1	1, 2, 3, 4	33.58%
2	7.500	0.000	0.000	0.000	27,098.0	34,825.0	1, 3	1, 2, 3, 4	28.52%
3	0.000	0.000	8.500	0.000	26,070.0	34,825.0	1, 3*	1, 2, 3, 4	33.58%
4	7.500	0.000	0.577	0.000	27,293.0	29,280.0	1, 2	1, 2	7.28%
5	4.651	1.950	0.577	0.000	27,890.5	29,280.0	1, 3†	1, 2	4.98%
6	2.817	1.950	2.693	0.000	27,619.9	29,280.0	1	1, 2	6.01%
7	4.326	1.950	2.693	0.000	28,450.3	29,280.0	1	1, 2	2.92%
8	5.080	1.950	2.693	0.000	28,607.3	29,280.0	2	1, 2	2.35%
9	5.080	2.697	2.693	0.000	28,668.9	29,280.0	1, 2	1, 2	2.13%
10	3.043	2.697	2.693	0.000	27,743.9	29,280.0	1	1, 2	5.54%
11	3.742	2.697	2.693	0.000	28,129.0	29,280.0	1	1, 2	4.09%
12	4.003	2.697	2.693	0.000	28,271.0	29,280.0	1	1, 2	3.57%
13	4.118	2.697	2.693	0.000	28,336.2	29,280.0	1	1, 2	3.33%
Best					28,668.9	29,280.0		1, 2	2.13%

*$\hat{f}_3 = 0$, so node 3 is included in the solution.
†The lower bound did not equal the upper bound in the dual ascent approach. Dual adjustment was not performed. The lower bound from the dual ascent algorithm is the value reported in the column for the lower bound.

The table shows the values of the Lagrange multipliers (μ_1 through μ_4) associated with the four relaxed capacity constraints. The lower bound obtained *from that iteration* (as opposed to the best lower bound found over all iterations) is shown in the column labeled LB. The next column gives the *best* upper bound that has been found to date. Initially, the best upper bound is found by allowing a facility to be located at each of the four locations, solving the resulting transportation problem as discussed above, and adding the sum of all of the candidate site fixed costs to the transportation problem objective function. In subsequent iterations, if the set of locations selected by the UFCLP had sufficient capacity to serve all of the demands (i.e., if the total selected capacity exceeded 1150, the total demand), we obtained a new trial upper bound by solving the transportation problem for the selected sites, adding the fixed costs of the selected sites to the transportation problem objective function. If the trial upper bound was better than the best upper bound found so far (as occurred in iteration 4), the new value became the best upper bound.

The next two columns (labeled LB Loc, and UB Loc, respectively) give the locations associated with the current iteration's lower bound (the locations found by solving the UFCLP) and the locations associated with the best upper bound. Note that in many iterations the locations identified by the UFCLP violate the relaxed capacity constraints. For example, in iteration 1, the solution to the UFCLP is to locate only one facility at node A. Node A has a capacity of 600 while the total demand is 1150. As expected, when the demand assigned to candidate site j exceeds the capacity of site j on iteration n, the Lagrange multiplier associated with site j increases at the beginning of iteration $n + 1$, indicating that capacity at that site is deemed to be more valuable. This will decrease the value of the modified fixed cost associated with the node ($\hat{f}_j$), making it more likely that this site will be included in the solution at iteration $n + 1$ all else equal. It will also increase the modified transport costs into that node ($\hat{c}_{ij}$), thereby tending to reduce the demand that is assigned to that node on iteration $n + 1$ all else equal. As a final note, when the value of the Lagrange multiplier was increased, we constrained the value of the Lagrange multiplier to be sure that the resulting modified fixed cost ($\hat{f}_j$) did not become negative. In other words, the following equation was used to update the Lagrange multipliers:

$$\hat{\mu}_j^{n+1} = \max\left\{0, \min\left[(f_j/k_j), (\hat{\mu}_j^n + t^n \delta_j^n)\right]\right\}$$

where

$\hat{\mu}_j^n$ = the value of the Lagrange multiplier for the jth candidate site on iteration n

t^n = the stepsize on iteration n

δ_j^n = the difference between the demand that is assigned to candidate site j on iteration n in the solution to the UFCLP and the capacity of candidate site j

The inner minimization ensures that the Lagrange multiplier does not become so large that the revised fixed cost on iteration $n + 1$ becomes negative. The outer maximization simply ensures that the Lagrange multiplier remains positive since we are relaxing an inequality constraint.

The last column in Table 7.15 gives the difference between the best upper bound and the current iteration's lower bound as a percentage of the current iteration's lower bound. At the end of 13 iterations, the difference between the best lower and upper bounds is slightly more than 2.1 percent. At this point, we stopped the Lagrangian algorithm. Further improvements in the difference between the lower and upper bounds may be obtained using branch and bound as indicated in Figure 7.9.

Within each iteration, the UFCLP was solved using the dual ascent approach discussed in Section 7.2.4. In iteration 3, the fixed charge used in the UFCLP for candidate site 3, $\hat{f}_3$, was 0. Therefore, this site was included in the dual-based solution even though no demands were assigned to this location at this iteration (because $\hat{c}_{13} \geq \hat{c}_{i1}$ for all demand nodes i at this

Table 7.16 Additional Results from Relaxing Constraint (7.7) in the Capacitated Fixed Charge Location Problem

Iteration	$(\delta_1^n)^2$	$(\delta_2^n)^2$	$(\delta_3^n)^2$	$(\delta_4^n)^2$	$\sum_j (\delta_j^n)^2$	α	Stepsize
1	302,500	0	0	0	302,500	2.00000	0.05788
2	105,625	0	140,625	0	246,250	2.00000	0.06276
3	302,500	0	250,000	0	552,500	1.00000	0.01585
4	225,625	105,625	0	0	331,250	1.00000	0.00600
5	105,625	0	140,625	0	246,250	1.00000	0.00564
6	302,500	0	0	0	302,500	0.50000	0.00274
7	302,500	0	0	0	302,500	0.50000	0.00137
8	0	202,500	0	0	202,500	0.50000	0.00166
9	22,500	0	0	0	22,500	0.50000	0.01358
10	302,500	0	0	0	302,500	0.25000	0.00127
11	302,500	0	0	0	302,500	0.12500	0.00048
12	302,500	0	0	0	302,500	0.06250	0.00021
13	302,500	0	0	0	302,500	0.03125	0.00010

iteration). Also, on iteration 5, the lower and upper bounds were not equal at the end of the dual ascent algorithm. We did not employ the dual adjustment procedure discussed in Section 7.2.4. Instead, the lower bound obtained from the dual ascent algorithm is reported as the lower bound at this iteration.

Table 7.16 gives additional information about each major iteration of the Lagrangian approach. The first column simply gives the iteration number. The next four columns, labeled $(\delta_1^n)^2$ through $(\delta_4^n)^2$, give the square of the difference between the demand assigned to a candidate site by the UFCLP and the capacity of the location. [Note that if the facility site is not opened, then the corresponding difference will necessarily be equal to 0. Thus, since candidate site 4 is never selected in the dual ascent algorithm for the UFCLP, $(\delta_4^n)^2$ equals 0 in all iterations shown in the table.] The sixth column gives the sum of these values. This number is used as the denominator in computing the stepsize at each iteration. The next to last column gives the value of the constant α which is multiplied by the difference between the lower and upper bounds to obtain the numerator for the stepsize. The value of α was halved whenever the current iteration's lower bound failed to improve on the best known lower bound. Finally, the last column gives the stepsize. Note that the final stepsize is less than one five-hundredth the size of the initial stepsize. At this point, only minor changes in the values of the Lagrange multipliers are being made (as also indicated in Table 7.15).

7.3.1.2 *Relaxing the Demand Constraints (7.7b)*

An alternative to relaxing the capacity constraints (7.7d) is to relax the demand constraints (7.7b). This leads to a very different set of subproblems. Specifically, when we relax constraints (7.7b), we obtain the following problem:

$$\text{MAX MIN} \atop \lambda \quad X,Y$$

$$\sum_j f_j X_j + \alpha \sum_i \sum_j h_i d_{ij} Y_{ij} + \sum_i \lambda_i \left(1 - \sum_j Y_{ij}\right)$$

$$= \sum_j f_j X_j + \sum_j \sum_i (\alpha h_i d_{ij} - \lambda_i) Y_{ij} + \sum_i \lambda_i \qquad (7.10a)$$

SUBJECT TO:

$$Y_{ij} \leq X_j \qquad \forall\, i, j \qquad (7.10b)$$

$$\sum_i h_i Y_{ij} \leq k_j X_j \qquad \forall\, j \qquad (7.10c)$$

$$\sum_j k_j X_j \geq \sum_i h_i \qquad (7.10d)$$

$$X_j = 0, 1 \qquad \forall\, j \qquad (7.10e)$$

$$0 \leq Y_{ij} \leq 1 \qquad \forall\, i, j \qquad (7.10f)$$

Constraints (7.10b), (7.10c), and (7.10e) correspond to constraints (7.7c), (7.7d), and (7.7e), respectively. Constraint (7.10d) has been added to ensure that the total amount of capacity that is selected is sufficient to serve all of the demands. Constraint (7.10f) is a slight modification of constraint (7.7f) in that we now constrain Y_{ij} to be less than or equal to 1.

As before, we want to minimize the Lagrangian function (7.10a) over the primal decision variables, X_j and Y_{ij}, and to maximize the function over the Lagrange multipliers, λ_i. The second task, that of maximizing the function over the Lagrange multipliers, λ_i, can be done using subgradient optimization as discussed in earlier sections on Lagrangian relaxation. Therefore, we focus on solving the problem for fixed values of the Lagrange multipliers.

For fixed values of the Lagrange multipliers, λ_i, this problem may be solved as follows. It should be clear from constraint (7.10c) that if we do not locate at candidate site j ($X_j = 0$), then no demands can be assigned to that candidate location ($Y_{ij} = 0$ for all demand nodes i and that candidate location j). Thus, the problem is to determine which candidate sites to open to minimize the total cost of the Lagrangian function (7.10a).

If we do not locate at candidate site j, then that site contributes 0 to the Lagrangian bojective function. However, if we do locate at candidate site j, the contribution of that site is determined by the following optimization problem:

MINIMIZE $$V_j = \sum_i (\alpha h_i d_{ij} - \lambda_i) Y_{ij} \tag{7.11a}$$

SUBJECT TO: $$\sum_i h_i Y_{ij} \leq k_j \tag{7.11b}$$

$$0 \leq Y_{ij} \leq 1 \quad \forall\, i \tag{7.11c}$$

(Note that we have one such problem for every candidate location j. Thus, in the objective function and constraints, we do not need to sum over the index j. We also do not need to indicate that the constraints apply to all values of j, since they only apply to the specific candidate location j whose subproblem is being solved at the time. Also, note that V_j is the value of locating at candidate site j. This quantity will always be nonpositive.)

For any candidate location j, this optimization problem is a continuous knapsack problem whose solution may be found greedily. We would like to select sites with large negative coefficients in the objective function and with small demands so that they consume relatively little of the capacity in constraint (7.11b). Therefore, we construct the following ratio for each demand node

$$r_{ij} = \frac{(\alpha h_i d_{ij} - \lambda_i)}{h_i}$$

The quantity r_{ij} is the contribution of demand node i to the objective function divided by the demand at node i (or the amount of capacity consumed by demand node i). Thus, r_{ij} is the rate at which demand node i contributes to the objective function per unit consumption of the capacity. Since the assignment variables, Y_{ij}, can be fractional, we can solve problem (7.11a)–(7.11c) using the following algorithm:

Algorithm for Solving (7.11a)–(7.11c) for a Particular Candidate Site j—a Continuous Knapsack Problem

Step 1: Compute values of r_{ij} for all demand nodes i.

Step 2: Sort the r_{ij} values in increasing order. Let $[i]$ be the index of the demand node with the ith smallest r_{ij} value.

Step 3: Set $\hat{k}_j = k_j$. $\hat{k}_j$ will denote the remaining capacity at candidate site j. Set $m = 1$. m will be an index of the demand nodes.

Step 4:

    a. Set $Y_{[m]j} = \min\{1, \hat{k}_j/h_{[m]}\}$. Note that $Y_{[m]j}$ is the assignment variable for the demand node with the mth smallest value of r_{ij}. Similarly, $h_{[m]}$ is the demand at the node with the mth smallest value of r_{ij}.

    b. Reduce the remaining capacity at candidate site j, $\hat{k}_j$, by $h_{[m]}Y_{[m]j}$.

    c. Increment m by 1.

Step 5:

    a. If m is less than or equal to the number of demand nodes and $\hat{k}_j > 0$, go to step 4.

    b. If m is less than or equal to the number of demand nodes and $\hat{k}_j = 0$, set $Y_{[m]j} = 0$ for all values of m equal to the current value through the number of demand nodes and stop.

    c. If m is greater than the number of demand nodes, stop.

To illustrate this algorithm, suppose at some point we had the following subproblem for candidate node 1:

MINIMIZE

$$V_1 = -200Y_{11} - 1200Y_{21} - 1225Y_{31} - 1625Y_{41} - 0Y_{51}$$
$$- 280Y_{61} - 720Y_{71} - 800Y_{81} \quad (7.12a)$$

SUBJECT TO:

$$100Y_{11} + 150Y_{21} + 175Y_{31} + 125Y_{41} + 180Y_{51} + 140Y_{61}$$
$$+ 120Y_{71} + 160Y_{81} \leq 600 \quad (7.12b)$$

$$0 \leq Y_{i1} \leq 1 \quad \forall\, i \quad (7.12c)$$

Table 7.17 gives the values of r_{i1} for this problem. Sorting the r_{i1} values in ascending order results in the set of pointers $[m]$ given in Table 7.18. Table 7.19 gives the values of the assignment variables and the remaining capacity as we iterate through step 4 of the algorithm given above.

The value of the objective function for the problem given above can readily be computed using the assignment variable values shown in Table

Table 7.17 Values of r_{i1} for the Sample Problem

Demand node, i	r_{i1}
1	−2.00
2	−8.00
3	−7.00
4	−13.00
5	0.00
6	−2.00
7	−6.00
8	−5.00

Capacitated Fixed Charge Facility Location Problems

Table 7.18 Sorted r_{i1} Values and Associated Pointers

Index, m	Points to Demand Node	$r_{[m]j}$
1	4	−13.00
2	2	−8.00
3	3	−7.00
4	7	−6.00
5	8	−5.00
6	1	−2.00
7	6	−2.00
8	5	0.00

7.19. The value of V_1 is -4920. Again, this gives the value of including candidate site 1 in the solution to the Lagrangian problem (7.10a)–(7.10f). Similar values would need to be computed for each other candidate site. Once all of the V_j values have been computed, we can find values for the location variables, X_j, in the Lagrangian problem by solving the following optimization problem:

MINIMIZE $$\sum_j (f_j + V_j) X_j \qquad (7.13a)$$

SUBJECT TO: $$\sum_j k_j X_j \geq \sum_i h_i \qquad (7.13b)$$

$$X_j = 0, 1 \qquad \forall j \qquad (7.13c)$$

Table 7.19 Application of the Greedy Algorithm to Solving the Sample Problem

(a) Iterations

Index, m	Points to Demand Node	$r_{[m]j}$	Demand	$Y_{[m]j}$	Remaining Capacity
0					600.0*
1	4	−13.00	125	1	475.0
2	2	−8.00	150	1	325.0
3	3	−7.00	175	1	150.0
4	7	−6.00	120	1	30.0
5	8	−5.00	160	3/16	0.0
6	1	−2.00	100	0	0.0
7	6	−2.00	140	0	0.0
8	5	0.00	180	0	0.0

Final Assignment Variable Values

$Y_{11} = 0, Y_{21} = 1, Y_{31} = 1, Y_{41} = 1$
$Y_{51} = 0, Y_{61} = 0, Y_{71} = 1, Y_{81} = 0.1875$

*Value of remaining capacity after step 3 and before starting step 4.

The objective function (7.13a) states that the contribution of candidate site j to the Lagrangian objective function will be equal to the fixed cost of locating at candidate site j plus the value of the assignment variables if we locate at candidate site j. The latter term is captured by the value of V_j. Constraint (7.13b) ensures that the capacity of the selected facilities is greater than or equal to the total demand. Constraints (7.13c) are integrality constraints.

Clearly, if the coefficient, $(f_j + V_j)$, of X_j is not positive, we can set X_j equal to 1 in this problem. If, after setting $X_j = 1$ whenever the coefficient of X_j is not positive, the total capacity of all selected sites exceeds the total demand, then we have solved the problem. In general, however, this will not be the case. In other words, setting these X_j values to 1 will not result in our selecting enough facility sites to satisfy the demand constraint (7.13b). This results in our having to solve the remaining integer programming problem.

If we relax the integrality constraint (7.13c) and replace it by the constraint $0 \leq X_j \leq 1$, we can solve the problem using a variant of the algorithm outlined above for problem (7.11a)–(7.11c). We leave the details to the reader noting only that we would like to select locations with small object function coefficients and large capacities (to contribute to the demand constraint). The value of the objective function found using this procedure can be shown to be a lower bound on the objective function for the integer programming problem (7.13a)–(7.13c). We can readily show that at most one location variable X_j will have a fractional value in the solution to the relaxed problem. If all X_j variables are integer, the solution is optimal for the integer programming problem. If one value is fractional, we can obtain an upper bound on the value of the objective function for the integer programming problem by rounding this X_j value up to 1 and evaluating the objective function (7.13a). To use the Lagrangian problem (7.10a)–(7.10f) to obtain valid lower bounds on the problem, however, we must solve (7.13a)–(7.13c) optimally. This can be done using branch and bound as well as other more sophisticated techniques.

To illustrate the solution to this problem, suppose we obtain the following problem after computing all of the V_j values:

MINIMIZE

$$-420X_1 + 770X_2 + 645X_3 + 1870X_4 \quad\quad (7.14a)$$

SUBJECT TO:

$$600X_1 + 700X_2 + 500X_3 + 550X_4 \geq 1150 \quad\quad (7.14b)$$

$$X_j = 0, 1 \quad\quad \forall j \quad (7.14c)$$

We can immediately set $X_1 = 1$ since the coefficient of X_1 is negative. This leaves the following problem:

MINIMIZE $\quad\quad 770X_2 + 645X_3 + 1870X_4$
SUBJECT TO: $\quad\quad 700X_2 + 500X_3 + 550X_4 \geq 550$
$\quad\quad\quad\quad\quad\quad\quad X_j = 0, 1 \quad\quad\quad\quad\quad\quad \forall j$

Capacitated Fixed Charge Facility Location Problems

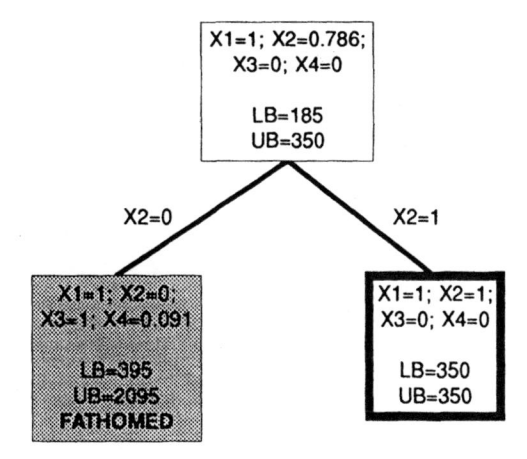

```
         ┌─────────────────┐
         │ X1=1; X2=0.786; │
         │   X3=0; X4=0    │
         │                 │
         │    LB=185       │
         │    UB=350       │
         └─────────────────┘
         X2=0           X2=1
      ╱                      ╲
┌─────────────────┐    ┌─────────────────┐
│  X1=1; X2=0;    │    │  X1=1; X2=1;    │
│  X3=1; X4=0.091 │    │   X3=0; X4=0    │
│                 │    │                 │
│    LB=395       │    │    LB=350       │
│    UB=2095      │    │    UB=350       │
│   FATHOMED      │    │                 │
└─────────────────┘    └─────────────────┘
```

Figure 7.11. Branch and bound tree for solving (7.14a)–(7.14c).

If we solve the linear programming relaxation of this problem (using a variant of the greedy algorithm outlined above), we obtain the following solution:

$$X_2 = 550/700 = 0.78571429$$
$$X_3 = 0$$
$$X_3 = 0$$

Combining these values with $X_1 = 1$ and evaluating (7.14a), we obtain a lower bound of 185. Rounding X_2 up to 1 gives an upper bound of 350. To obtain an all-integer solution, we can use branch and bound. Figure 7.11 gives the branch-and-bound tree for this problem. The optimal solution is to locate at candidate sites 1 and 2 with an objective function (7.14a) of 350.

Figure 7.12 summarizes the entire solution procedure for the capacitated fixed charge location problem when we relax constraint (7.7b). We begin by relaxing constraint (7.7b) and fixing the Lagrange multipliers associated with the constraint at some initial value. We then solve (7.11a)–(7.11c) for each candidate location j to obtain values for V_j. Next, we solve (7.13a)–(7.13c) as indicated above. With these values for the location variables, X_j, we can readily find the values of the assignment variables, Y_{ij}. In particular, if $X_j = 0$, then as indicated above $Y_{ij} = 0$ for all demand nodes i. If $X_j = 1$, we set the Y_{ij} equal to their values from the solution (7.11a)–(7.11c) for that candidate site j. With these values, we can evaluate (7.10a) to obtain a lower bound on the objective function. As always, the largest of these lower bounds over all iterations of the algorithm is the best bound.

We can also obtain an upper bound on the solution by solving the transportation problem (7.8a)–(7.8d) and adding the value of the fixed costs of the selected sites to the value of the transportation problem objective function. The smallest of the upper bounds over all iterations is the best

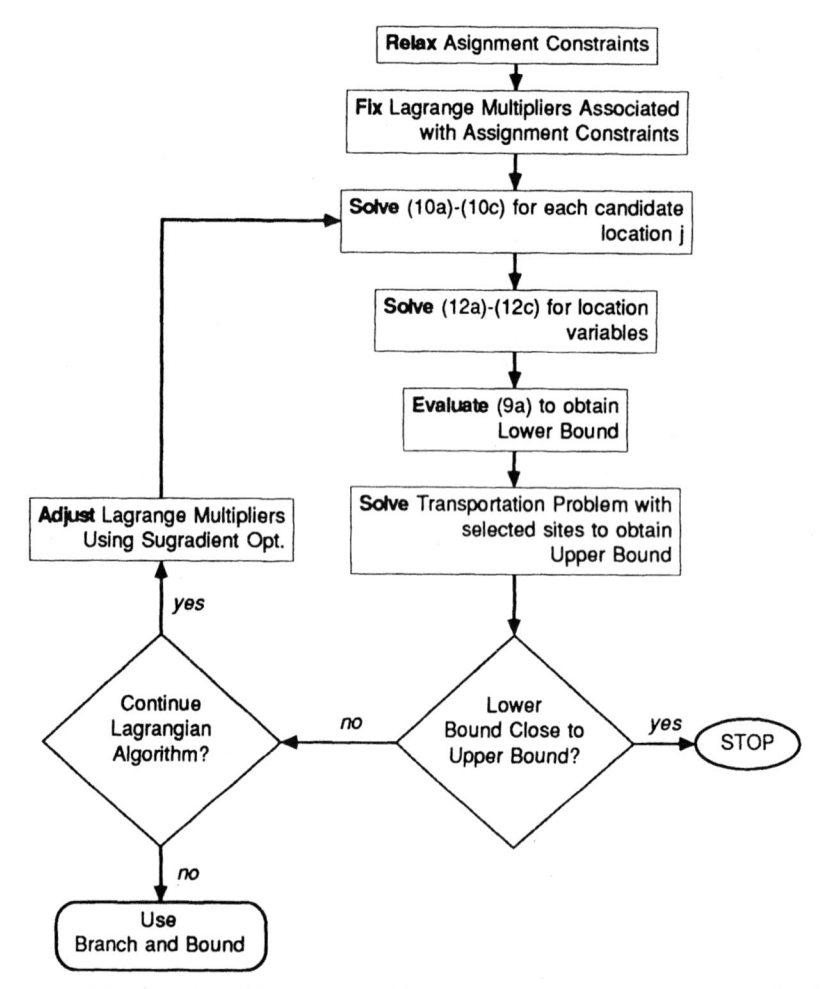

Figure 7.12. Flowchart for solving capacitated fixed charge problem relaxing constraint (7.7b).

value to use when comparing the lower and upper bounds. If the lower and upper bounds are sufficiently close to each other, we can stop. Otherwise, we can either (i) adjust the Lagrange multipliers using a subgradient optimization procedure and continue with the Lagrangian algorithm or (ii) stop the Lagrangian algorithm and employ branch and bound to help close any gap between the lower and upper bounds. [Note that the bounds that we are discussing here are those for the entire capacitated fixed charge location problem and not merely those of problem (7.13a)–(7.13c) as discussed above.]

To illustrate this approach, consider again the problem shown in Figure 7.10. We set the initial Lagrange multipliers for each demand node, λ_i^0, equal

Table 7.20 Sample Initial Lagrange Multipliers for the Network of Figure 7.10

Node	Initial Lagrange Multiplier
1	2800.0
2	3150.0
3	4550.0
4	3000.0
5	3060.0
6	3780.0
7	3120.0
8	4160.0

Table 7.21 Results of Solving Problems (7.11a)–(7.11c) with Initial Lagrange Multipliers of Table 7.20

Candidate Site j	Y_{1j}	Y_{2j}	Y_{3j}	Y_{4j}	Y_{5j}	Y_{6j}	Y_{7j}	Y_{8j}	V_j
1	0.0	1.0	1.0	1.0	0.0	0.0	1.0	0.19	−4920.0
2	0.0	1.0	1.0	0.76	0.0	0.0	1.0	1.0	−3630.0
3	0.0	0.0	1.0	0.0	1.0	0.0	1.0	0.16	−3605.0
4	1.0	1.0	0.0	0.0	0.89	1.0	0.0	0.0	−2280.0

to $\max_j\{\alpha h_i d_{ij}\}$ as shown in Table 7.20. With these values, we must solve problem (7.12a)–(7.12c) to obtain V_1. Similar problems must be solved to obtain V_2, V_3, and V_4. Table 7.21 summarizes the results of solving these four problems. When we add the fixed charges to the values of V_j shown in Table 7.21, we obtain the coefficients of X_j for problem (7.13a)–(7.13c) shown in Table 7.22. Note that these coefficients are exactly the values of the coefficients in problem (7.14a)–(7.14c) whose solution was found using the branch-and-bound tree shown in Figure 7.11. Adding the optimal value of the objective function (7.14a) to the sum of the Lagrange multipliers yields a lower bound of 27,970. Solving the transportation problem (7.8a)–(7.8d) with facilities located at nodes 1 and 2 yields a transportation problem objective

Table 7.22 Coefficients of X_j for Initial Iteration of Lagrangian Algorithm for the Network of Figure 7.10

Candidate Site j	V_j	f_j	Coefficient
1	−4920.0	4500.0	−420.0
2	−3630.0	4400.0	770.0
3	−3605.0	4250.0	645.0
4	−2280.0	4150.0	1870.0

Table 7.23 Results of Using Lagrangian Algorithm Relaxing Constraint (7.7b) on the Network of Figure 7.10

Iteration	Lower Bound	Best Upper Bound	Sites	Percentage Error
1	$27,970	$29,280	1, 2	4.684%
2	$28,444	$29,280	2, 3	2.940%
3	$27,184	$29,280	1, 2	2.940%
4	$28,675	$29,280	1, 4	2.111%
5	$28,709	$29,280	1, 2	1.988%
6	$28,868	$29,280	1, 4	1.428%
7	$28,901	$29,280	1, 2	1.312%
8	$28,893	$29,280	1, 2	1.312%
9	$28,954	$29,280	1, 2	1.127%
10	$28,951	$29,280	1, 4	1.127%
11	$29,020	$29,280	1, 4	0.894%
12	$28,962	$29,280	1, 2	0.894%
13	$29,002	$29,280	1, 2	0.894%
14	$29,019	$29,280	1, 2	0.894%
15	$29,023	$29,280	1, 2	0.887%
16	$29,026	$29,280	1, 4	0.876%
17	$29,023	$29,280	1, 2	0.876%
18	$29,027	$29,280	1, 2	0.872%
19	$29,026	$29,280	1, 4	0.872%
20	$29,026	$29,280	1, 4	0.872%
Best	$29,027	$29,280	1, 2	0.872%

function of 20,380. Adding the fixed costs of nodes 1 and 2 (4500 and 4400, respectively) results in an upper bound of 29,280 and an initial gap of 4.684 percent. Table 7.23 gives the lower bounds and best upper bounds for the first 20 iterations of the algorithm. Successive values of the Lagrange multipliers were found using a standard subgradient optimization procedure. The value of the multiplier α used in the Lagrangian procedure was halved whenever the lower bound failed to improve on the best known lower bound.

The difference between the best upper bound and the best lower bound (as a percentage of the best lower bound) is plotted in Figure 7.13. After 11 iterations, the difference between the lower and upper bounds is less than 1 percent of the lower bound. Further iterations reduce the gap only marginally. To reduce the gap further, we would need to employ branch and bound.

7.3.2 Bender's Decomposition

In this section we outline a different solution approach. This approach is based on decomposing the capacitated fixed charge facility location problem —a mixed integer linear programming problem—into two problems. The first problem is a (nearly) pure integer programming problem with only one

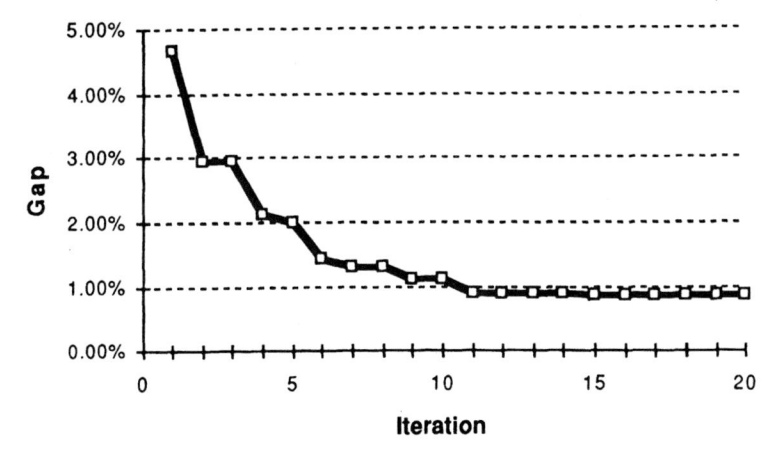

Figure 7.13. Lagrangian relaxation convergence.

continuous variable. This will be referred to as the *master problem* as it will give us tentative facility location sites. This problem incorporates the fixed facility costs explicitly and the transportation costs implicitly. The second problem is a pure linear programming problem. The pure linear programming problem will be the transportation problem (7.8a)–(7.8d), which can be solved very easily. By solving this problem, known as the *subproblem*, we will be able to compute the actual transportation costs for a set of facility sites suggested by the master problem.

To facilitate the discussion of this approach, we reformulate the capacitated fixed charge facility location problem in terms of flow variables, Z_{ij}. Specifically, we define the following new notation:

Inputs

$c_{ij} = \alpha d_{ij}$
 = the unit shipping cost between candidate location j and demand node i

Decision Variables

Z_{ij} = the amount of demand at node i that is satisfied by a facility at location j

The key difference between this decision variable and the allocation variable, Y_{ij}, defined earlier is that Y_{ij} gives the *fraction* of the demand at node i that is satisfied by a facility at location j, while Z_{ij} gives the *amount* of demand at node i that is satisfied by a facility at location j. With this notation, we can reformulate the capacitated fixed charge facility location problem as follows:

MINIMIZE $\quad \sum_j f_j X_j + \sum_i \sum_j c_{ij} Z_{ij}$ (7.15a)

SUBJECT TO: $\quad \sum_j Z_{ij} = h_i \quad \forall\, i$ (7.15b)

$\quad\quad\quad\quad\quad\quad Z_{ij} \leq k_j X_j \quad \forall\, i, j$ (7.15c)

$\quad\quad\quad\quad\quad\quad \sum_i Z_{ij} \leq k_j X_j \quad \forall\, j$ (7.15d)

$\quad\quad\quad\quad\quad\quad X_j = 0, 1 \quad \forall\, j$ (7.15e)

$\quad\quad\quad\quad\quad\quad Z_{ij} \geq 0 \quad \forall\, i, j$ (7.15f)

As before, the objective function (7.15a) minimizes the sum of the fixed facility location costs and transportation costs. The only differences between this function and (7.7a) are that in (7.15a) we have collapsed the distance (d_{ij}) and cost per unit mile (α) into a single-unit cost parameter (c_{ij}). In addition, (7.15a) is stated in terms of flow variables instead of allocation variables. Thus, the demand at node i (h_i) is captured in the flow variable (Z_{ij}). Constraint (7.15b) serves the function of (7.7b) and ensures that all demand at node i is satisfied. As before, constraint (7.15c), which is analogous to (7.7c), is not needed. Since it will not be used in this approach, we will drop it from here on. Constraint (7.15d) is a capacity constraint. Finally, constraints (7.15e) and (7.15f) are integrality and nonnegativity constraints, respectively.

Before describing the application of Bender's decomposition to this problem, it is worth noting that by converting from allocation variables (Y_{ij}) to flow variables (Z_{ij}), we gain one additional advantage that is not apparent in formulation (7.15a)–(7.15f). Specifically, it is now easier to model nonlinear cost functions. The objective function (7.15a) can, in principle, be rewritten as

$$\sum_j f_j X_j + \sum_i \sum_j g_{ij}(Z_{ij}) \quad (7.15a')$$

where $g_{ij}(Z_{ij})$ can be any nonlinear function of the flow between demand node i and a facility at candidate site j. In particular, $g_{ij}(Z_{ij})$ can be a concave function representing volume shipping discounts or it can be a step function representing a cost structure that remains constant for a range of shipment volumes. For example, the U.S. Postal Service uses such a pricing scheme. The functional form and the specific values of the functions $g_{ij}(Z_{ij})$ can depend on the demand node/facility pair (i, j). Thus, longer-distance

Capacitated Fixed Charge Facility Location Problems

shipments might cost more than shorter-distance shipments of equal volume though the relationship between distance and cost need not be linear.

For fixed values of the location variables (X_j), problem (7.15a)–(7.15f) reduces to a transportation problem. Specifically, if we fix the location variables at some values $\hat{X}_j$, we obtain the following problem:

MINIMIZE
Z
$$T(\mathbf{Z}|\hat{\mathbf{X}}) = \sum_i \sum_j c_{ij} Z_{ij} \qquad (7.16a)$$

SUBJECT TO:
$$\sum_j Z_{ij} = h_i \qquad \forall\, i \qquad (7.16b)$$

$$\sum_i Z_{ij} \leq k_j \hat{X}_j = \begin{cases} k_j & \text{if } \hat{X}_j = 1 \\ 0 & \text{if } \hat{X}_j = 0 \end{cases} \qquad \forall\, j \qquad (7.16c)$$

$$Z_{ij} \geq 0 \qquad \forall\, i, j \qquad (7.16d)$$

This formulation is identical to that of (7.8a)–(7.8d) except that it is stated in terms of flow variables instead of allocation variables. We have identified the objective function as $T(\mathbf{Z}|\hat{\mathbf{X}})$ to emphasize that the cost being optimized is a function of the flow variables Z_{ij} (which we represent in vector form by $\mathbf{Z}$) and that the cost is *conditional* on the choice of facility sites as given by the variables $\hat{X}_j$ (which we represent in vector form by $\hat{\mathbf{X}}$). The facility sites represented by $\hat{\mathbf{X}}$ will be generated as the solution to the master problem (7.21a)–(7.21e) as discussed below. With this formulation in mind, we can restate the capacitated fixed charge facility location problem as follows:

MINIMIZE
X
$$\sum_j f_j X_j + \left\{ \underset{\mathbf{Z}}{\text{MINIMIZE}}\, [T(\mathbf{Z}|\mathbf{X})] \right\} \qquad (7.17a)$$

SUBJECT TO:
$$\sum_j k_j X_j \geq \sum_i h_i \qquad (7.17b)$$

$$X_j = 0, 1 \qquad \forall\, j \qquad (7.17c)$$

This minimization is explicitly over the location variables (X_j). The term $T(\mathbf{Z}|\mathbf{X})$ in the objective function implicitly captures the entire transportation problem formulation (7.16a)–(7.16d) including the flow variables (Z_{ij}). In this formulation, we have included constraint (7.17b) which guarantees that the capacity of all selected facilities be greater than or equal to the total demand. This is clearly a necessary condition for any solution to the capacitated fixed charge facility location problem to be feasible. It is also a necessary condition for the transportation problem (7.16a)–(7.16d) to have a feasible solution.

Let us now consider the dual of the transportation problem (7.16a)–(7.16d). We define the following dual variables:

Dual Decision Variables

U_i = dual variable associated with the demand constraint (7.16b)
W_j = dual variable associated with the supply constraint (7.16c)

The dual of the transportation problem (7.16a)–(7.16d) may now be formulated as follows:

MAXIMIZE $\quad D(\mathbf{U}, \mathbf{W}|\hat{\mathbf{X}}) = \sum_i h_i U_i - \sum_j k_j \hat{X}_j W_j \quad$ (7.18a)

SUBJECT TO: $\quad U_i - W_j \leq c_{ij} \quad \forall\, i, j \quad$ (7.18b)

$\quad U_i$ unrestricted $\quad \forall\, i \quad$ (7.18c)

$\quad W_j$ unrestricted $\quad \forall\, j \quad$ (7.18d)

Note that the objective function is linear in the decision variables, since $\hat{X}_j$ is an input and not a decision variable in this formulation. Since the optimal value of the objective function for the transportation problem (7.16a)–(7.16d) must be the same as the optimal value of the dual of the transportation problem (7.18a)–(7.18d), we can rewrite the capacitated fixed charge location problem in terms of the dual of the transportation problem as follows:

MINIMIZE $\quad \sum_j f_j X_j + \left\{ \underset{\mathbf{U},\mathbf{W}}{\text{MAXIMIZE}} \left[D(\mathbf{U}, \mathbf{W}|\mathbf{X}) \right] \right\} \quad$ (7.19a)
$\mathbf{X}$

SUBJECT TO: $\quad \sum_j k_j X_j \geq \sum_i h_i \quad$ (7.19b)

$\quad X_j = 0, 1 \quad \forall\, j \quad$ (7.19c)

As before, we have implicitly captured the entire dual problem (7.18a)–(7.18d) by including the maximization of $D(\mathbf{U},\mathbf{W}|\mathbf{X})$ over the dual decision variables U_i and W_j in the objective function (7.19a).

Let us now consider all combinations $(\mathbf{U}, \mathbf{W})$ of the dual decision variables that correspond to an extreme point of the dual problem (7.18a)–(7.18d). Let us denote the tth such combination by $(\mathbf{U}^t, \mathbf{W}^t)$ in vector notation with elements U_i^t and W_j^t, respectively. Let D^t be the objective function of the dual problem corresponding to the tth extreme point of the feasible region of

Capacitated Fixed Charge Facility Location Problems

the dual problem. In other words, we have

$$D^t = \sum_i h_i U_i^t - \sum_j k_j X_j W_j^t \qquad \forall\, t \qquad (7.20)$$

We know from the theory of linear programming that at least one optimal solution to any linear programming problem occurs at an extreme point of the feasible region. Thus, D^* (the optimal value of the dual problem) must be greater than or equal to D^t for all extreme points t. Therefore, we can rewrite the capacitated fixed charge facility location problem one more time as follows:

$$\text{MINIMIZE}_X \quad \sum_j f_j X_j + D \qquad (7.21a)$$

$$\text{SUBJECT TO:} \quad \sum_j k_j X_j \geq \sum_i h_i \qquad (7.21b)$$

$$D \geq \sum_i h_i U_i^t - \sum_j k_j X_j W_j^t \qquad \forall\, t \qquad (7.21c)$$

$$X_j = 0, 1 \qquad \forall\, j \qquad (7.21d)$$

$$D \geq 0 \qquad (7.21e)$$

In this formulation, we have replaced the implicit representation of the dual problem found in (7.19a) by the dual objective function value (D). Constraint (7.21b) is identical to (7.19b). Using the result from linear programming theory mentioned above (that at least one optimal solution lies at an extreme point), in (7.21c) we constrain the dual value to be greater than or equal to the dual objective function evaluated at each extreme point of the dual feasible region. Constraints (7.21d) and (7.21e) are simply the integrality and nonnegativity constraints, respectively. Note that (7.21e)—the nonnegativity constraint on the dual objective function value—is justified as long as we assume that the unit transport costs (c_{ij}) are nonnegative. In general, this will be the case.

The problem with formulation (7.21a)–(7.21e) is that the number of extreme points of the dual problem is potentially very large. Thus, we do not want to enumerate all of the constraints in (7.21c) explicitly. Also, at the optimal solution to (7.21a)–(7.21e), only a very small subset of the constraints (7.21c) are likely to be binding. Thus, even if we could enumerate all of them, many of them would prove to be unnecessary. On the other hand, if we solve (7.21a)–(7.21e) with only a subset of the constraints in (7.21c), we will obtain a valid *lower bound* on the optimal value of the objective function for the capacitated fixed charge facility location problem. Furthermore, if all of the constraints that are binding in the optimal solution to (7.21a)–(7.21e) happen to be in the subset of constraints that we include, then the value of the objective function will exactly equal the optimal value for the capacitated fixed charge location problem.

We can compute an *upper bound* on the problem by taking any set locations with sufficient capacity, evaluating the optimal solution to the corresponding transportation problem (7.16a)–(7.16d), and adding the fixed costs of the selected facility sites. Note that constraint (7.21b) guarantees that *any* solution to (7.21a), (7.21b), (7.21d), (7.21e), and a *subset* of (7.21c) will have sufficient capacity. Each time we solve the transportation problem (7.16a)–(7.16d) and its dual (7.18a)–(7.18d), we obtain a new extreme point of the dual problem. This will enable us to add another constraint of the form (7.21c), thereby tightening the lower bound obtained by solving (7.21a)–(7.21e) [with a subset of the constraints in (7.21c)].

Figure 7.14 summarizes this approach. We begin by solving the master problem (7.21a)–(7.21e) without constraints (7.21c) which will be generated iteratively. The objective function to this problem is a lower bound on the objective function for the capacitated fixed charge facility location problem. The solution also gives us a feasible candidate set of facility locations. Using these locations, we solve the associated transportation problem (7.16a)–(7.16d) and its dual (7.18a)–(7.18d). From this, we obtain the optimal transportation costs for the candidate sites suggested by the master problem. The sum of these costs and the fixed costs of the sites identified by the master problem constitutes an upper bound on the problem. If the lower and upper bounds are equal, we stop. Otherwise, using the optimal dual variables, we can add another constraint of the form of (7.21c) to the master problem. We again solve the master problem obtaining a new lower bound. Again, if the lower and upper bounds are equal, we stop. In that case, the solution corresponding to the smallest upper bound is the optimal solution. If the bounds are not equal, we again solve the corresponding transportation problem, compute the associated total cost, and update the upper bound if the total cost of the solution is smaller than the current best upper bound value. The process continues until the lower and upper bounds are equal.

To illustrate this approach, we again consider the network of Figure 7.10. We begin by solving the following simple problem [where the letters after the equation numbers correspond to those of problem (7.21a)–(7.21e)]:

MINIMIZE

$$4500X_1 + 4400X_2 + 4250X_3 + 4150X_4 + D \quad (7.22a)$$

SUBJECT TO:

$$600X_1 + 700X_2 + 500X_3 + 550X_4 \geq 1150 \quad (7.22b)$$

$$X_1, X_2, X_3, X_4 = 0, 1 \quad (7.22d)$$

$$D \geq 0 \quad (7.22e)$$

Capacitated Fixed Charge Facility Location Problems

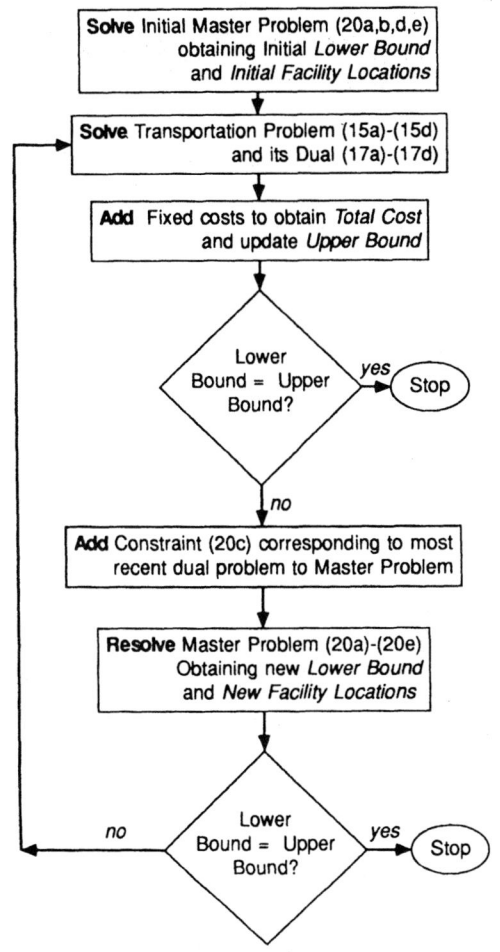

Figure 7.14. Flowchart of Bender's decomposition approach for solving capacitated fixed charge problems.

The solution to this problem is to locate at candidate sites 2 and 4. The objective function value is 8550 which is the initial lower bound on the problem. Solving the transportation problem with these locations, we obtain transport costs of 21,910. The lower bound is therefore 21,910 plus the fixed costs of sites 2 and 4 (4400 and 4150, respectively), or 30,460. Since the lower and upper bounds are not equal, we must add a constraint of the form of (7.21c) to the master problem. The dual variables are

$$\mathbf{U}^1 = [17, 19, 22, 22, 15, 23, 19, 16]$$

Table 7.24 Results of Using Bender's Decomposition on the Problem of Figure 7.10

Iteration	Lower Bound	Locations	Fixed Costs	Transport Costs	Total Cost	Upper Bound
1	8,550.	2, 4	8,550.	21,910.	30,460.	30,460.
2	23,210.	1, 3, 4	12,900.	18,445.	31,345.	30,460.
3	24,210.	1, 2	20,380.	8,900.	29,280.	29,280.
4	25,630.	2, 3	21,220.	8,650.	29,870.	29,280.
5	27,095.	1, 4	20,680.	8,650.	29,330.	29,280.
6	28,360.	1, 2, 4	19,110.	13,050.	32,160.	29,280.
7	29,280.					

and

$$\mathbf{W}^1 = [11, 0, 10, 0]$$

Substituting these values into (7.21c), we obtain the following constraint:

$$D \geq 21{,}910 - 11(600)Y_1 - 10(500)Y_3 \qquad (7.22c')$$

This constraint is added to (7.22a)–(7.22e) and the problem is resolved.

The optimal solution to the revised problem is to locate three facilities at candidate sites 1, 3, and 4. The lower bound is now 23,210 of which 12,900 is fixed costs and the value of D is 10,310. The process continues for a number of additional iterations. Table 7.24 summarizes the results of these iterations giving the lower and upper bounds, the facilities selected by the master problem, and the associated fixed costs and transport costs (obtained from the transportation problem). Note that on some iterations the total cost for the sites suggested by the master problem is not less than the current best upper bound. This happens on iteration 2, for example. Table 7.25 gives the

Table 7.25 Dual Variable Values for Use in Bender's Decomposition for the Problem of Figure 7.10

Iteration	\multicolumn{8}{c}{Demand Locations}	\multicolumn{4}{c}{Candidate Sites}
	1	2	3	4	5	6	7	8	1	2	3	4
1	17	19	22	22	15	23	19	16	11	0	10	0
2	17	13	12	11	13	23	29	21	0	5	0	0
3	27	14	20	12	15	26	19	16	1	0	8	10
4	27	19	12	22	13	26	19	16	11	0	0	10
5	17	18	24	16	15	23	25	26	5	10	12	0
6	17	13	19	11	15	23	19	16	0	0	7	0

Capacitated Fixed Charge Facility Location Problems

Table 7.26 Additional Constraints in Bender's Decomposition for Network of Figure 7.10

Iteration	Constraint Added	Optimal D Value
1	$600Y_1 + 700Y_2 + 500Y_3 + 550Y_4 \geq 1{,}150$	0.
2	$D \geq 21{,}910 - 11(600)Y_1 - 10(500)Y_3$	10,310.
3	$D \geq 18{,}445 - 5(700)Y_2$	15,310.
4	$D \geq 20{,}980 - 1(600)Y_1 - 8(500)Y_3 - 10(550)Y_4$	16,980.
5	$D \geq 21{,}220 - 11(600)Y_1 - 10(550)Y_4$	18,445.
6	$D \geq 23{,}680 - 5(600)Y_1 - 10(600)Y_2 - 12(500)Y_3$	15,310.
7	$D \geq 19{,}110 - 7(500)Y_3$	20,380.

optimal dual variables for the solution of (7.18a)–(7.18d). Finally, Table 7.26 gives the constraint that is added to the problem on each iteration. We have included the capacity constraint for iteration 1.

As indicated in Table 7.24, a total of six solutions need to be tested to solve the problem optimally. There are a total of 15 possible solutions (excluding the solution with no facilities) of which 9 are feasible in terms of the capacity constraint (7.22b). Thus, using Bender's decomposition allowed us to avoid enumerating one-third of the feasible solutions. For larger problems, the savings would be dramatically bigger.

Before ending our discussion of the application of Bender's decomposition to the solution of the capacitated fixed charge facility location problem, we note that constraints (7.21c) readily lend themselves to a straightforward interpretation. Note that if the location variables remain unchanged—that is, they remain the same as the values that were used in obtaining the dual variables for the tth solution to the dual problem—constraint (7.21c) is identical to the dual objective function (7.18a). In other words, the value of D must be greater than or equal to the value of the dual objective function that was computed on the tth iteration. However, the values of the location variables may change in the solution to the master problem. Recall that W_j^t gives the amount by which the optimal value of the transportation problem solution will change for a unit (small) change in the jth right-hand side of the supply constraint (7.16c) using the tth set of locations suggested by the master problem. Thus, $k_j W_j^t$ is an estimate of the amount of which the transportation cost will change if we add or subtract k_j units from the right-hand side of the jth capacity constraint starting with the values that result from the tth set of locations suggested by the master problem. [In fact, it is likely to be an overestimate of the changes since (i) W_j is only the marginal rate of change in the objective function as the right-hand side of the jth capacity constraint changes and (ii) capacity changes beyond the (small) region in which W_j gives an accurate estimate will result in smaller unit changes in the objective function.] But k_j is just the capacity of the jth candidate location. Thus, if facility j had been selected in the tth set of locations, $k_j W_j^t$ gives a conservative estimate of the amount by which the transportation costs associated with the tth set of locations would *increase* if we removed facility j from the solution. The costs would increase because X_j would change from 1 to 0 and $k_j W_j^t$ enters the right-hand side of constraint (7.21c) with a negative sign. Since such a change amounts to removing capacity from the solution, an increase in the transport costs is to be expected. Similarly, if facility j was not in the tth set of locations, $k_j W_j^t$ gives a conservative estimate of the amount by which the transportation costs associated with the tth set of locations would *decrease* if we added facility j to the solution.[2]

7.4 SUMMARY

In this chapter we have discussed fixed charge facility location problems. We began with a discussion of uncapacitated problems which can be seen as an extension of the P-median problems discussed in Chapter 6. In these problems there is an inherent tradeoff between the fixed costs of the facilities and the transport or operating costs. As additional facilities are added, the fixed costs increase but the operating or transport costs decrease.

Two heuristic construction algorithms were presented for the solution of the uncapacitated fixed charge facility location problem. The ADD heuristic sequentially adds facilities greedily until addition of the next facility would result in a cost increase. The DROP heuristic begins with facilities tentatively located at all of the candidate locations. It then sequentially removes facilities greedily until the removal of the next facility would result in an increase in the total cost. Neighborhood and exchange (or substitution) improvement heuristics were also presented. We also outlined a Lagrangian relaxation algorithm for the solution of this problem. Finally, we presented a dual-based approach first introduced by Erlenkotter (1978) that has been shown to be remarkably effective in solving this problem.

In many cases, it is important to consider capacities associated with facilities. For example, the number of cars that can be garaged in a parking structure is limited to the number of spaces in the garage. In other cases, practical capacities limit the ability of facilities to provide unlimited service.

---

[2] The astute reader will recall from Chapter 2 that the values of the dual variables in the transportation problem, *as discussed in that chapter*, are unique up to an additive constant. Thus, we really can only be concerned about differences between dual variables. The interpretation outlined here relies on the magnitude of the dual variable W_j. This suggests that there need to be additional conditions imposed on the problem for this interpretation to be strictly correct and/or the transportation problem discussed here differs slightly from the form of the problem discussed in Chapter 2. The identification of these conditions is left as an exercise for the reader.

These considerations led us to discuss the capacitated fixed charge facility location problem which is a generalization of its uncapacitated cousin discussed in the first part of the chapter. Two alternative relaxations of the problem were discussed. The first involved relaxing the capacity constraints. This naturally led to an uncapacitated problem which could then be attacked using the algorithms outlined earlier for the uncapacitated fixed charge facility location problem. The second approach involved relaxing the demand constraints. This resulted in our having to solve two problems: a continuous knapsack problem and a second problem closely related to the 0–1 knapsack problem. Finally, we outlined the application of Bender's decomposition to the solution of the capacitated fixed charge facility location problem. This entailed alternately solving a (nearly) pure integer programming problem and a transportation problem. All three approaches were illustrated using a small example problem involving four candidate locations and eight demand nodes. The reader interested in more details regarding these approaches is referred to the excellent review of these algorithms by Magnanti and Wong (1990). Also, Van Roy (1986) proposed an approach that merges Bender's decomposition and Lagrangian relaxation into a particularly effective approach to solving this class of problems. This algorithm is beyond the scope of this text.

In the next chapter we consider a number of extensions of the classical location problems we have explored to date. These extensions allow the analyst to consider increasingly realistic problems. In the remainder of the text, our emphasis will shift from the development of solution algorithms to the formulation, interpretation, and use of richer, more complex models.

EXERCISES

7.1 Consider the results shown in Table 7.1

(a) Summarize the fixed cost, commodity value, demand, and transportation cost conditions that will tend to drive the solution toward having only a single facility.

(b) Identify at least one type of facility for which it would likely be optimal to locate only a single facility to serve the entire country.

(c) Identify at least one type of facility for which it would be optimal to have a very large number of facilities—almost one per major city.

7.2 In Chapter 4 we discussed the set covering problem under the assumption that all facility costs were identical. In that case, minimizing the total cost of the selected facilities becomes identical to minimizing the total number of facilities. In many cases, however, the facility costs are not identical. We indicated that the set covering objective function would then be changed to

MINIMIZE $$\sum_j f_j X_j$$

where

f_j = the fixed cost of locating at candidate site j

$$X_j = \begin{cases} 1 & \text{if candidate site } j \text{ is selected} \\ 0 & \text{if not} \end{cases}$$

(a) Show how this extension of the set covering problem can be formulated as a fixed charge facility location problem.

Hint: Consider changing the distances in an appropriate manner. Clearly describe how to obtain the new (modified) distances.

(b) Do any of the row and column reduction rules discussed in Chapter 4 for the set covering problem with identical facility costs apply to this extended problem? If so, which ones? For those that do not apply, justify why they do not work in this case.

(c) For the 12-node network of Figure 7.4 and a covering distance of 30, solve this extension of the set covering problem using the SITATION program. Use the NET-SPEC program to create a 12-node problem suitable for use in SITATION. This will also create a distance file, MDST12.NET. Then use the MOD-DIST program to change the distances in the MDST12.NET file appropriately before running SITATION.

7.3 Discuss how you can solve the maximum covering location problem as a fixed charge facility location problem if you include a constraint fixing the number of facilities equal to P (i.e., you include the constraint $\sum_j X_j = P$ in the formulation of the fixed charge facility location problem). Specifically, what changes would you want to make to the distance matrix to allow an algorithm that could solve the fixed charge location problem with this added constraint to solve the maximum covering location problem?

7.4 (a) Use the approach outlined in Exercise 7.2 to solve for the set covering solution to the 12-node problem of Figure 7.4 when all facility costs are equal. (Set them equal to 100.) Use a covering distance of 30 as in part (c) of Exercise 7.2.

(b) Why are the lower and upper bounds so far apart in your solution to part (a)?

(c) What does the value of the lower bound suggest to you about the linear programming relaxation of the problem? In particular, can you find a solution to the linear programming relaxation of this

Exercises

instance of the set covering problem that has an objective function equal to the lower bound computed by the Lagrangian procedure? If so, what is that solution?

(d) How can you use the constraint discussed in Exercise 7.3 to help tighten the bound? Using this approach, where does SITATION tell you to locate facilities?

7.5 Consider the following uncapacitated fixed charge facility location problem (see the following network). The numbers in the box beside each node are the demand and the fixed facility location cost associated with the node. Using a cost per unit distance per unit demand of 1, solve the problem using the following approaches:

(a) The ADD algorithm

(b) The DROP algorithm

(c) The ADD algorithm followed by the exchange algorithm

(d) Lagrangian relaxation

Compare the solutions that you found using these approaches in terms of (i) the number of facilities located, (ii) the total cost, (iii) the fixed facility location costs that are incurred, and (iv) the transport costs incurred.

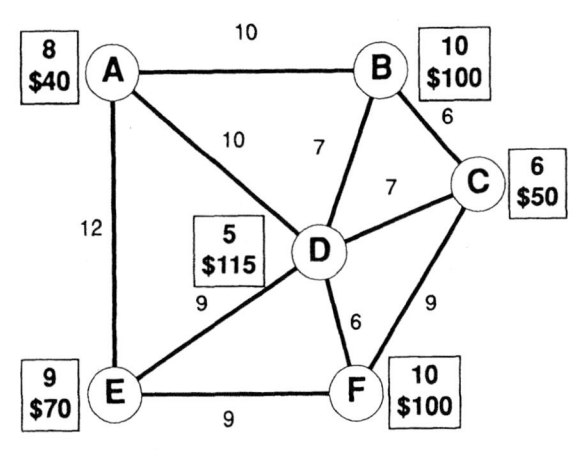

Figure E7.5

7.6 It is often important to understand how the optimal number of facilities and their locations depend on the relative values of the fixed facility location costs and the transport costs. This is particularly important if the cost factors that affect the cost per unit distance (e.g., driver wages, fuel costs, and vehicle acquisition and maintenance costs) change.

(a) For the network of Exercise 7.5, find the cost per mile (accurate to the nearest hundreth of a unit) at which the optimal number of facilities becomes one less than the optimal number found in Exercise 7.5. Where do you locate facilities now? How many do you locate?

(b) Repeat part (a) until you find the cost per mile at which it is optimal to locate only one facility. For each range of the cost per mile in which the number of facilities remains constant, how many facilities do you locate? Where should the facilities be located? Within each range of the cost per mile, do the facility locations change? (Be careful when you consider the case in which you locate only one facility.)

(c) For the cost per mile varying between 0 and 1, plot the following quantities versus the cost per mile:

(i) the total cost
(ii) the fixed facility location costs
(iii) the transport costs

7.7 (a) Use SITATION and the 88-node data set (CITY1990.GRT) to plot the total cost as a function of the number of facilities located for a cost per mile of 0.001. Use the ADD heuristic in the TRADEOFF curve MAIN MENU OPTION.

(b) What is the optimal number of facilities found in part (a)?

(c) What is the range of the number of facilities found such that the total cost is within 2 percent of the optimum?

(d) Again using the ADD heuristic, complete the following table:

Cost per mile	"Optimal" Number of Facilities	Lower limit of 2 Percent Range	Upper limit of 2 Percent Range
0.01			
0.0025			
0.001			
0.00025			
0.0001			

(e) Use a regression model to fit an equation of the following form to the data in the table above:

$$\text{``Optimal'' number} = \alpha(\text{cost per mile})^{\beta}$$

Exercises

In other words, find the least squares estimates of the values of α and β above.

(f) What does the value of β tell you about the sensitivity of the optimal number of facilities to changes in the total demand level?

Hint: Think carefully about the interpretation of the cost per mile.

7.8[3] Warren and Williams (W & W) Warehousing is planning to add a new line of automated warehouses for a specialized product. At present, W & W Warehousing has no such warehouses in its service region and does not handle this product. Since it is not in the business now, it is, in essence, starting with a "clean sheet of paper." W & W Warehousing has identified a number of candidate sites for the new warehouses and, for each site, it has identified a number of alternative sizes or capacities for a warehouse that might be built on the site. Clearly, at most one warehouse can be built at each candidate site. In addition, W & W Warehousing has predicted the demand in each of the markets it hopes to serve as well as the unit shipment cost from each candidate site to each market area. W & W Warehousing is interested in identifying the following quantities:

(a) The optimal number of warehouses

(b) The optimal sites for the warehouses

(c) The optimal sizes for the warehouses

(d) The optimal distribution plan (which warehouses supply which markets)

so that the total investment cost (measured on an equivalent daily payment basis) plus the distribution cost is minimized.

Using the notation below and any additional notation that you clearly define, formulate this problem as an integer linear programming problem. Clearly state the objective function and constraints in both words and mathematical notation. Be sure to indicate clearly any indices of summation and for which values of the indices constraints apply.

Inputs

h_i = demand in market area i

S_{jk} = capacity at candidate site j (or size or supply at site j) if the kth sized warehouse is built there (that is, for site j, there are K alternative sizes of warehouses that can be built which are

---

[3] This problem involves the formulation of a model involving economies of scale. The reader is referred to Osleeb et al. (1986) for a description of such a model applied to investments in ports to be used for exporting coal. Kuby, Ratick, and Osleeb (1991) further describe the application of this model.

indexed by $k = 1, \ldots, K$. This is the size or capacity of the kth one on this list).

F_{jk} = fixed cost (equivalent daily payment of the investment cost) of building the kth sized warehouse on the list of possible warehouse sizes at site j

c_{ij} = unit shipment cost from candidate site j to market area i

$\hat{c}_{jk}$ = unit variable cost of shipping through a warehouse at candidate site j if the kth sized warehouse is built there

Assume that the unit variable costs decrease with the size of the facility (i.e., $c_{jk} < c_{j,k-1}$) but that the fixed costs increase with the size of the facility (i.e., $F_{jk} > F_{j,k-1}$).

Decision Variables

X_{ij} = number of units to ship from a warehouse at site j to market area i

$$Y_{jk} = \begin{cases} 1 & \text{if we build the } k\text{th sized warehouse at candidate site } j \\ 0 & \text{if not} \end{cases}$$

7.9 Knowing that the set covering problem is NP-complete, prove that the uncapacitated fixed charge facility location problem is also NP-complete.

7.10 Knowing that the uncapacitated fixed charge facility location problem is NP-complete (as proven in Exercise 7.9), prove that the capacitated fixed charge facility location problem is also NP-complete.

7.11 At the end of Chapter 7, we provided an interpretation of constraint (7.21c) that is added to the master problem in Bender's decomposition whenever the lower and upper bounds are not equal at the end of an iteration of the algorithm. In a footnote we observed that the interpretation suggested in the text was not entirely consistent with that outlined in Chapter 2. The interpretation in Chapter 7 was based on the dual variables associated with the capacity constraints in the transportation problem. To make the interpretation valid, additional conditions must be imposed on the interpretation and problem context. Identify these additional conditions.

8

Extensions of Location Models

8.1 INTRODUCTION

The covering, center, median, and fixed charge location models discussed in Chapters 4 through 7 are at the *heart* of many of the location models that have been used in siting public and private facilities. To deal with almost any realistic situation, these mathematical models need to be extended in a variety of ways. The focus of this chapter is on a number of the more common extensions discussed in the literature.

8.2 MULTIOBJECTIVE PROBLEMS

Facility location decisions are inherently strategic and long term in nature. As such, there are likely to be many possibly conflicting or competing objectives that need to be considered. For example, landfills (for nonhazardous wastes) are generally considered undesirable facilities. Most people do not want a landfill to be located in their community. Therefore, we would like these facilities far from population centers. On the other hand, we would like to minimize the distance that hauling vehicles need to travel from the waste collection sites to the landfills. However, many of the waste generation sites are population centers. Thus, there is a tradeoff between (i) being far from population centers and (ii) reducing the total number of vehicle miles needed to transport the waste to the landfill. Facility location models can, as we shall see, assist in quantifying this sort of tradeoff.

In locating warehouses, we may want to balance the average distance between customers and the nearest warehouse and the extent of demand coverage. The number of vehicle miles traveled is likely to be related closely to the demand-weighted total distance—the P-median objective. Reducing vehicle miles will therefore reduce the cost of doing business. On the other hand, coverage and the ability to deliver goods to customers in a timely manner are becoming increasingly important as firms become more concerned about customer service. For example, the ability to deliver goods to customers within one day of the time an order is placed may give a firm a significant advantage over its competitors. Thus, there may be a tradeoff between (i) minimizing the total demand-weighted distance (which may leave some customers with inadequate service) and (ii) maximizing the number of customers who receive adequate service.

Before proceeding with a quantitative discussion of the use of facility location models in a multiobjective arena, we need to define a number of concepts associated with multiobjective analysis. Consider Figure 8.1 in which we have two objectives (1 and 2), both of which are to be minimized. Six solutions, labeled A through F, are shown. Solution A minimizes objective 1 but has the largest (worst) value for objective 2. Solutions D and F minimize objective 2. However, solution F has a larger value for objective 1 than does solution D. Therefore, in the absence of other considerations, solution D would always be preferred to solution F. Solution F is therefore *inferior* to solution D or is *dominated* by solution D. In general, *a solution Φ is inferior if there exists some other solution Θ that is as good as Φ in terms of all of the objectives and Θ is strictly better than Φ in terms of at least one objective.* In that case, solution Θ dominates solution Φ which is inferior to or dominated by solution Θ. Solutions A and D are *noninferior* solutions.

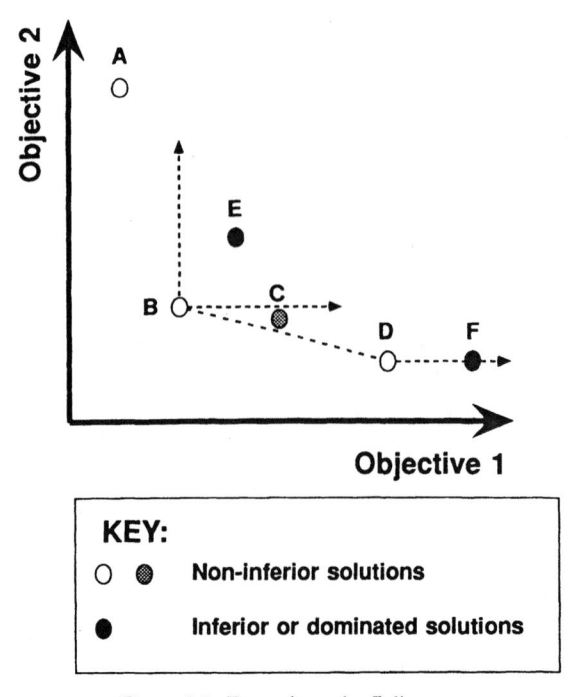

Figure 8.1. Example tradeoff diagram.

Multiobjective Problems

Solutions B, C, and E represent compromise solutions. They are not as good as solution A at achieving objective 1, nor are they as good as solution D at achieving objective 2. However, solution E is not as good at achieving either objective as is solution B. This is evidenced by the fact that it is to the northeast of solution B. Thus, solution E is also inferior or dominated. Finally, solution C is also noninferior as it is better than solution B at achieving objective 2, but not as good at achieving objective 1. Solution C, however, is above the line connecting two other noninferior solutions—solutions B and D.[1] This will make finding solution C rather difficult. In particular, if we try to minimize a weighted (linear) combination of objectives 1 and 2, with appropriate weights we will be able to identify solutions A, B, and D, but not solution C. As the weighting method is one of the easiest methods to use in finding noninferior solutions to multiobjective problems, such solutions are indeed problematic.

Having defined the notions of (a) inferior or dominated and (b) noninferior or nondominated solutions, let us now consider a particular two-objective warehouse location problem. The first objective will be to minimize the total demand-weighted distance between demand or customer nodes and the nearest facility (the P-median objective), and the second objective will be to maximize the number of covered demands (the maximum covering objective). We begin by noting that with fixed total demand, maximizing the number of covered demands is equivalent to minimizing the number of uncovered demands. By making this transformation, we can deal with two minimization problems and can present the analysis in a framework that is similar to that shown in Figure 8.1. Formulation (6.1a)–(6.1e) illustrated this transformation. Below, we show that the same problem can also be structured as a P-median problem if we transform the distance matrix appropriately.[2]

Consider transforming the distance matrix by letting

$$\hat{d}_{ij} = \begin{cases} 0 & \text{if } d_{ij} \leq D_c \\ 1 & \text{if } d_{ij} > D_c \end{cases}$$

In other words, if the original distance between demand node i and candidate location j (d_{ij}) is less than or equal to the coverage distance (D_c), then the revised distance ($\hat{d}_{ij}$) is set equal to 0; otherwise, the revised distance is set equal to 1. The following problem will locate P facilities to minimize the number of uncovered demands. All notation is as defined earlier.

---

[1] Solutions such as solution C are sometimes called convex dominated solutions or duality gap solutions.
[2] The reader is referred to Hillsman (1984) for an excellent discussion of this and similar transformations.

MINIMIZE $$U = \sum_i \sum_j h_i \hat{d}_{ij} Y_{ij} \tag{8.1a}$$

SUBJECT TO: $$\sum_j Y_{ij} = 1 \quad \forall\, i \tag{8.1b}$$

$$\sum_i X_j = P \tag{8.1c}$$

$$Y_{ij} - X_j \leq 0 \quad \forall\, i, j \tag{8.1d}$$

$$X_j = 0, 1 \quad \forall\, j \tag{8.1e}$$

$$Y_{ij} = 0, 1 \quad \forall\, i, j \tag{8.1f}$$

This formulation is exactly that of a P-median problem. The objective function (8.1a) minimizes the total demand-weighted distance. Now, however, the weight on each demand node will be 0 if the nearest facility is within the coverage distance and 1 if the distance to the nearest facility exceeds the coverage distance. Thus, the objective function is equivalent to minimizing the number of uncovered demands. We call this objective function U, to emphasize that it represents the *uncovered* demand. Constraints (8.1b)–(8.1f) are identical to the P-median constraints and should be familiar to the reader by this point.

The traditional P-median objective function is

MINIMIZE $$D = \sum_i \sum_j h_i d_{ij} Y_{ij} \tag{8.2a}$$

This function is minimized subject to constraints (8.1b)–(8.1f) above. We call this objective function D to emphasize that we are minimizing the total demand-weighted *distance*.

Now let us consider minimizing a weighted combination of the objectives (8.1a) and (8.2a) subject to constraints (8.1b)–(8.1f). Specifically, consider the following problem:

MINIMIZE $$\begin{aligned} Z &= kU + D \\ &= k\left\{\sum_i \sum_j h_i \hat{d}_{ij} Y_{ij}\right\} + \sum_i \sum_j h_i d_{ij} Y_{ij} \\ &= \sum_i \sum_j h_i \{k\hat{d}_{ij} + d_{ij}\} Y_{ij} \\ &= \sum_i \sum_j h_i \tilde{d}_{ij} Y_{ij} \end{aligned} \tag{8.3}$$

SUBJECT TO: Constraints (8.1b)–(8.1f)

where $\tilde{d}_{ij} = k\hat{d}_{ij} + d_{ij}$.

In essence, k is the absolute value of the slope of a line through a solution in the uncoverage/distance space as shown in Figure 8.2. If k is very large,

Multiobjective Problems

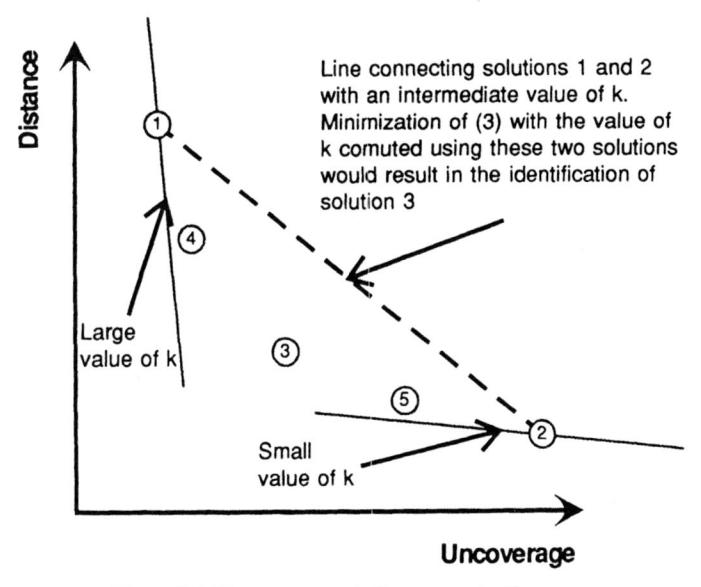

Figure 8.2. Uncoverage and distance tradeoff curve.

(Figure shows Distance on y-axis, Uncoverage on x-axis, with solutions 1–5 plotted. Solution 1 at top-left labeled "Large value of k", solution 2 at bottom-right labeled "Small value of k", solutions 3, 4, 5 intermediate. Dashed line connects solutions 1 and 2 with annotation: "Line connecting solutions 1 and 2 with an intermediate value of k. Minimization of (3) with the value of k comuted using these two solutions would result in the identification of solution 3")

minimizing (8.3) will be equivalent to minimizing U, which is the uncoverage objective (8.1a). In Figure 8.2 this corresponds to identifying solution 1. If k is 0 (or very small), minimizing (8.3) will be equivalent to minimizing D, which is the P-median objective (8.2a). A very small value of k would identify solution 2 in Figure 8.2. Thus, by setting k equal to a very small number and then equal to a very large number, we can identify the noninferior solutions that minimize the total demand-weighted distance and that maximize coverage, respectively. As k varies between these extremes, we should be able to find solutions that represent compromises between minimizing the demand-weighted distance and maximizing coverage (e.g., solutions 3, 4, and 5 in Figure 8.2). It is this observation that motivates the weighting method for solving multiobjective problems.

We begin by trying to find the noninferior solutions that do best in terms of each of the individual objectives. This entails placing a large weight on one objective in question and a very small weight on the other objective. We need a small positive weight on the other objective to ensure that the solutions found are truly noninferior. Thus, returning to Figure 8.1, if we had a weight of 0 on objective 1, we would find either solution D or solution F. To ensure that we find solution D, the noninferior solution, we need to place a small weight on objective 1.[3]

---

[3] In a multiobjective problem, it is sometimes not easy to determine a priori how large or small the weight should be on any one objective to be sure that we first minimize the desired objective and then select the best solution in terms of the other objective from among the alternate optima

(*Continued*)

Once we have found the noninferior solutions that are best in terms of the individual objectives, we take a weighted combination of the two objectives and minimize the weighted combination of the objectives subject to the original constraints on the problem. In particular, suppose we have found two solutions with objective function values $S^1 = (U^1, D^1)$ and $S^2 = (U^2, D^2)$, where $S^n = (U^n, D^n)$ is a vector that contains the objective function values for the nth solution and U^n and D^n are the uncoverage and total demand-weighted distance values associated with the nth solution. The line connecting these two points is given by the following equation:

$$D = D^2 + \left[\frac{D^1 - D^2}{U^1 - U^2}\right] \cdot (U - U^2)$$

or

$$D = \left\{D^2 - \left[\frac{D^1 - D^2}{U^1 - U^2}\right] \cdot U^2\right\} + \left[\frac{D^1 - D^2}{U^1 - U^2}\right] \cdot U$$

Consider finding a new value of U and a new value of D to minimize the function

$$Z = \left[-\frac{D^1 - D^2}{U^1 - U^2}\right] \cdot U + D \qquad (8.4a)$$

subject to constraints (8.1b)–(8.1f). Note that all points on the line connecting solutions 1 and 2 will have the same weighted objective function value. However, if there is a point below the line connecting these two points, minimizing (8.4a) will find a new noninferior point. In Figure 8.2 minimizing (8.4a) would result in our identifying solution 3. Note that $[-(D^1 - D^2)/(U^1 - U^2)]$ is the equivalent of k in (8.3).

In some cases, it is easier and more meaningful to think in terms of the *average* demand-weighted distance instead of the *total* demand-weighted distance. The total demand-weighted distance (D) is equal to the total

---

to the first objective. One approach to this is to solve the problem twice. In the first solution, we eliminate the other objective completely. We record the objective function value for the objective that was optimized. Then we introduce a very small weight on the other objective and solve the weighted objective problem again. If the objective function value for the original objective is the same, the solution to the second problem is then the best possible with respect to the original objective and is noninferior. If the objective function value for the original objective is not the same as it was before we introduced the second objective, the weight on the second objective is too large. In that case we need to resolve the weighted problem with a smaller weight on the second objective.

Multiobjective Problems

demand (H) multiplied by the average demand-weighted distance ($\overline{D}$). Therefore, we can substitute $H\overline{D}$ into (8.4a) to obtain the following equation in terms of uncovered demand (U) and the average demand-weighted distance:

$$Z = \left[-\frac{H(\overline{D}^1 - \overline{D}^2)}{U^1 - U^2} \right] \cdot U + D \qquad (8.4b)$$

Here $[-H(\overline{D}^1 - \overline{D}^2)/(U^1 - U^2)]$ is the equivalent of k in (8.3).

If we find a new noninferior solution as a result of minimizing Z as given in (8.4a) or (8.4b), we can label the new solution $S^3 = (U^3, D^3)$. We could then repeat the process outlined above connecting first solutions 1 and 3 and then solutions 2 and 3 to search for new noninferior solutions. This process would continue until we are unable to find a pair of adjacent noninferior solutions that have not been connected by a line in this manner. (A pair of solutions is considered adjacent in this context if there is no other solution with an uncovered demand value between the uncovered demand values of the two solutions in question.)

To illustrate this approach, consider the 49-node problem given in Appendix H. For this example, we will use demand set 1 in which the 1990 population of each state is assigned to a node at the state capital. Adding Washington, DC to the state capitals of the lower 48 states gives 49 nodes. The total demand for the problem is 247,051,601.

The solution to the maximum covering problem for five facilities with a coverage distance of 250 miles is to locate at nodes 1, 7, 9, 11, and 30 (corresponding to Columbus, OH; Atlanta, GA; Sacramento, CA; Des Moines, IA; and Trenton, NJ). 60,305,344 demands are uncovered (and 186,746,257 are covered). The average demand-weighted distance is 241.772 miles. The 5-median solution is to locate at nodes 1, 3, 4, 6, and 9 (corresponding to Austin, TX; Sacramento, CA; Springfield, IL; Trenton, NJ; and Tallahassee, FL). In this solution, there are 85,734,965 uncovered demands and the average distance is 203.640 miles. Thus, we have $S^1 = (U^1, \overline{D}^1) = (60,305,344, 241.772)$ and $S^2 = (U^2, \overline{D}^2) = (85,734,965, 203.640)$. Substituting these values into $[-H(\overline{D}^1 - \overline{D}^2)/(U^1 - U^2)]$ from (8.4b), given that $H = 247,051,601$, we find that $k = 370.4617$.

Using this value of k in (8.3), we solve the resulting 5-median problem. The solution is to locate at nodes 1, 3, 9, 11, and 14 (Indianapolis, IN; Atlanta, GA; Austin, TX; Sacramento, CA; and Trenton, NJ). Now, 66,328,966 demands are uncovered and the average demand-weighted distance is 207.119. The procedure continues using this new solution to obtain two new lines [and therefore two new values of k in equation (8.3)].

Table 8.1 summarizes the results of this procedure. The $P = 5$ maximum covering solution is labeled solution A; the 5-median solution is labeled solution B. The solution found by using the slope of the line connecting these

Table 8.1 Solutions Obtained in Search for Non-inferior Solutions

Solution	Uncovered Demand	Average Distance	Constant	Facility Locations				
A	60,305,344	241.77225		1	7	9	11	30
B	85,734,965	203.56628		1	3	4	6	9
C	66,328,966	207.11191	370.461680	1	3	9	11	14
D	60,305,344	241.77225	1421.552097	1	7	9	11	30
E	74,692,854	204.58097	44.203219	1	3	4	9	14
F	66,328,966	207.11191	74.758626	1	3	9	11	14
G	85,734,965	203.56628	21.058912	1	3	4	6	9

Solution A is the maximum covering solution; solution B is the P-median solution.

two solutions is labelled C. Solution D is obtained by using a weighted average of the values of solutions A and C. This is identical to solution A so we have not generated a new noninferior solution in this region of the objective space. Next, we consider a weighted average of the objective function values for solutions B and C. This gives us solution E. A weighted average of solutions E and C gives solution F. A weighted average of solutions E and B gives solution G. Finding any remaining points on the tradeoff curve is left as an exercise for the reader.[4]

Figure 8.3 plots the tradeoff curve found using this approach.

To emphasize the importance of considering location problems from a multiobjective perspective, note that solution A that maximizes coverage results in an average distance that is 18.73 percent above the minimum possible value (that of solution B). Similarly, the 5-median solution results in 42.17 percent more uncovered demand than is achievable by the best solution (solution A). Solution C, on the other hand, results in an average distance that is only 1.71 percent greater than the minimum possible value and uncovered demand that is only 9.99 percent greater than the smallest possible value. As such, it represents a good candidate compromise solution.

Finally, we note that the weighting method that we demonstrated above is only one of several different methods that can be used to generate solutions in a multiobjective analysis. One other approach is the constraint method. In this approach, all but one of the objectives are represented as constraints.

---

[4]The modified distances

$$\bar{d}_{ij} = k\hat{d}_{ij} + d_{ij} = \begin{cases} k + d_{ij} & \text{if } d_{ij} > D_c \\ d_{ij} & \text{if } d_{ij} \leq D_c \end{cases}$$

may be computed by beginning with the original distance matrix and setting the G and J parameters in the MOD-DIST program menu for change matrix elements equal to k. The reader is referred to Appendix C for a discussion of MOD-DIST.

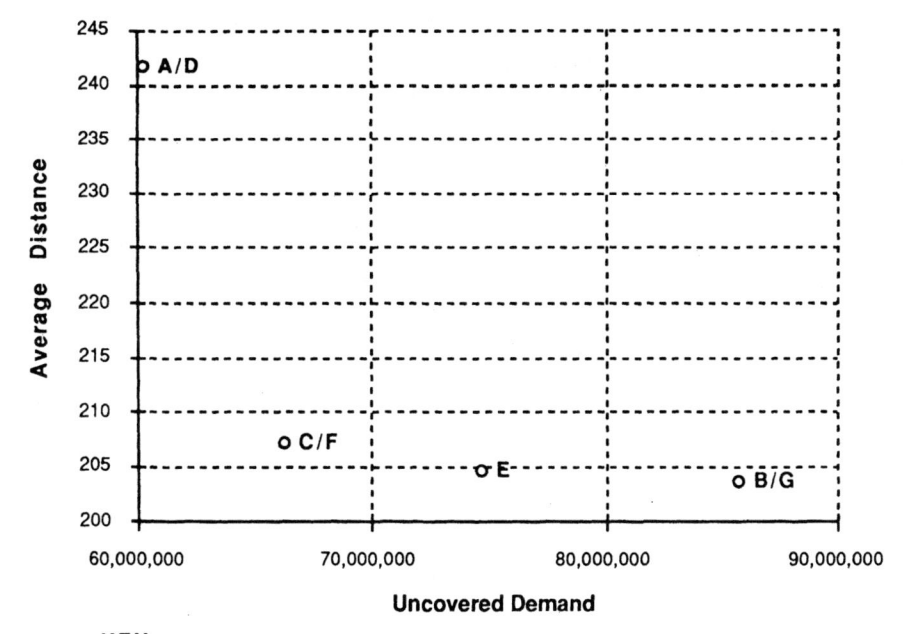

Hierarchical Facility Location Models   317

Figure 8.3. Median and coverage tradeoff curve for 49-city problem.

KEY:
Bold face letters next to open circles are non-inferior solutions

We then minimize (or maximize as appropriate) the remaining objective subject to certain attainment levels of the remaining objectives which are now represented as constraints. By systematically varying the attainment levels of the constrained objectives, we can identify the set of noninferior points. This method allows us to identify convex-dominated or duality gap points such as solution C in Figure 8.1. Cohon (1978) provides an excellent introduction to the field of multiobjective optimization modeling.

8.3 HIERARCHICAL FACILITY LOCATION MODELS

The basic models discussed in Chapters 4 through 7 assumed that there was only one type of facility being located. In many contexts, planners are interested in locating a variety of facilities that interact in one or more of a number of possible ways. In this and the next several sections, we outline a number of models of such interacting facilities.

8.3.1 Basic Notions of Hierarchical Facilities

Often, facilities are hierarchical in terms of the types of services being offered. For example, in designing a location plan for a school system, one would typically have to locate three types of facilities. Primary schools typically include kindergarten through fifth grade.[5] Such schools tend to be small and numerous as parents and communities do not want young children to go too far from home. Middle schools generally include sixth through eighth grade. High schools include ninth through twelfth grade. High schools tend to be large as there are significant economies of scale associated with high school education. For example, it is often difficult to justify a first-rate physics laboratory in a school with only 40 students per grade, many of whom might have little capability or desire to take physics. At first glance, one might imagine that these three facility types can be located independently. However, there are often linkages between the facilities. For example, many school districts like to have all children from a primary school go to the same middle school. Similarly, all students from a particular middle school may be required to attend the same high school. Thus, the facility locations and student allocations at one level are linked to those at other levels. This would be particularly true if the different schools being located faced capacity constraints.

Postal services represent another example of hierarchical facilities. At the lowest level are post boxes at which customers may simply deposit mail. At branch offices, postal patrons may deposit mail, buy stamps, and obtain a limited range of other services. Finally, at central (or main) post offices, customers can access all types of postal services. In addition, a number of "behind the scenes" operations such as sorting mail for an entire city may occur only at central post offices.

Hierarchical facilities are also prevalent in the banking industry. Automatic teller machines (ATMs) allow bank customers to deposit funds in a bank, receive cash, and obtain statements of their current account balance. Branch offices allow patrons to obtain all of these services as well as a variety of other services including maintaining a safety deposit box, applying for residential loans, and purchasing government bonds. Finally, main bank offices typically provide all services available at branch offices as well as others. Thus, applications for large corporate loans may only be handled at the main bank office.

As a final example, consider the location of health facilities in developing countries. Local clinics may provide basic care as well as diagnostic services. Community hospitals will provide basic care and diagnostic services as well as out-patient surgery and limited in-patient services. Regional hospitals may perform out-patient surgery, in-patient surgery, and provide a full range of in-patient services. However, regional hospitals may or may not provide basic care and diagnostic services.

As indicated by the examples above, in hierarchical facility location problems, facilities at different levels of the hierarchy are distinguished by the services they provide. In addition, there must be some sort of link

---

[5]The actual demarcations between primary and middle school and between middle school and high school vary across the country.

between the facilities being located. Often, the link between the facilities is an implicit link that comes in the form of a global budget constraint that limits the amount of money that can be spent on all of the facilities. [Moore and ReVelle (1982) suggest this as an alternative approach to that of explicitly constraining the number of facilities at each hierarchy.] Thus, the problem becomes one of (i) allocating the budget between the different hierarchies, (ii) locating facilities at each hierarchy, and (iii) allocating customers to facilities. In other cases, the link between facilities at different levels of the hierarchy may be more explicit. For example, in a health care context, there may be linkages that identify which local clinics can refer patients to which community hospitals. In postal services, there is a need to delineate which branch offices or main post offices are responsible for the collection of mail from each of the post boxes that are located in the study region.

In thinking about hierarchical facilities, it is often useful to classify systems by the way in which services are offered and by the region to which services are provided by facilities (Tien, El-Tell, and Simons, 1983; Narula, 1986). To facilitate the discussion, we will number the possible services provided by the system from 1 through m. Similarly, we will number the levels in the hierarchy (or the types of facilities) from 1 through m. Level 1 is the lowest order of service or facility (e.g., ATM, primary school, post box) and level m is the highest order. Note that while this scheme of identifying services and facility types using the same numbering scheme is prevalent in the literature, it poses certain important problems as we shall see below. We begin by classifying hierarchical systems by the type of service provided and then extend the classification scheme to encompass the locations to which service is provided by the facilities.

A *successively inclusive facility hierarchy* is one in which a facility at level m (the highest level) offers services 1 through m. Somewhat more generally, we might say that a successively inclusive facility hierarchy is one in which a facility at level k provides all lower-order services offered by a facility at level $k - 1$ (i.e., levels 1 through $k - 1$) plus at least one additional service. This latter definition allows us to begin to dissociate the indexing of the services and the indexing of levels of the facilities in the hierarchy. Note, however, that with the latter definition of a successively inclusive facility hierarchy, those services that are offered by a level k facility that are not offered by a level $k - 1$ facility may be bundled together and called level k services. Postal facilities and banking facilities, at least as discussed above, are examples of successively inclusive facility hierarchies.

At the other extreme, a *successively exclusive facility hierarchy* is one in which a facility at level k offers only service type k. More generally, a successively exclusive facility hierarchy is one in which the set of services offered by a type k facility has no intersection with the set of services offered by a type n facility (for all $k \neq n$). Of the examples discussed above, public schools are an example of a successively exclusive facility hierarchy.

Table 8.2 Summary of Hypothetical Hierarchical Health Care System

Facility	Service Provided			
	Basic Care	Diagnostic Services	Out-Patient Surgery	In-Patient Surgery
Clinic	Yes	Yes		
Community hospital	Yes	Yes	Yes	
Regional hospital			Yes	Yes

Many hierarchical facilities do not fall into either the successively inclusive or successively exclusive categories. For example, a typical health care system is summarized in Table 8.2. Note that community hospitals and clinics constitute a successively inclusive hierarchy since community hospitals provide all services offered by clinics plus at least one additional class of services, namely out-patient surgery. Regional hospitals and clinics, however, illustrate an exclusive hierarchy since the services offered by regional hospitals are not offered by clinics and vice versa. Note, however, that this is not illustrative of a *successively* exclusive hierarchy since the two levels are separated by another level, that of the community hospital. Finally, the relationship between regional hospitals and community hospitals is more complicated and is neither successively inclusive nor successively exclusive.

Having defined successively inclusive and successively exclusive facility hierarchies, we now turn to the similar definitions in terms of the geographic regions to which services are provided. For a successively inclusive facility hierarchy, a *locally inclusive service hierarchy* is one in which a type k facility at location i offers service of types 1 through k to demands at node i, but service of type k only to demands at nodes $j \neq i$. In other words, only service level k is exported to other nodes. To illustrate this notion, the health care example outlined above may be altered so that regional hospitals provide basic care and diagnostic services but only to patients living in the town in which the hospital is located; patients from other towns must receive basic care and diagnostic services at either clinics or community hospitals, but not at the regional hospital in question.

For a successively inclusive facility hierarchy, a *globally inclusive service hierarchy*[6] is one in which a type k facility at location i offers service of types 1 through k to customers at *all* nodes. Finally, for a successively exclusive facility hierarchy, a *successively exclusive service hierarchy* exists when a level

---

[6] Narula (1986) and others refer to this as a *successively inclusive service hierarchy*. We prefer the term *globally inclusive* as this distinguishes this terminology which deals with the region in which service is provided from the terminology which deals with the type of service provided. Also, the term *globally inclusive* contrasts this type of service (in which services are provided to all customers) with *locally inclusive service* (in which some services are only provided to customers in the community in which the facility is located).

Hierarchical Facility Location Models

k facility at location i can provide services of type k to customers at *all* nodes.[7]

8.3.2 Basic Median-Based Hierarchical Location Formulations

We now turn our attention to the problem of formulating hierarchical facility location problems. Most such problems have been posed as variants of *P*-median models. We begin by defining the following notation:

Inputs

h_{ik} = demand for type k services at node i
d_{ij} = distance between node i and candidate location j
P_k = number of type k facilities to locate

Decision Variables

$$X_{jk} = \begin{cases} 1 & \text{if a facility of type } k \text{ is located at candidate site } j \\ 0 & \text{if not} \end{cases}$$

$$Y_{ijk} = \begin{cases} 1 & \text{if demands at node } i \text{ for type } k \text{ services are satisfied by a facility at candidate site } j \\ 0 & \text{if not} \end{cases}$$

With this notation, we can formulate a model for a *successively inclusive facility hierarchy operating under a globally inclusive service hierarchy* as follows:

MINIMIZE $\qquad \sum_i \sum_j \sum_k h_{ik} d_{ij} Y_{ijk} \qquad\qquad (8.5a)$

SUBJECT TO: $\qquad \sum_j Y_{ijk} = 1 \qquad \forall\, i, k \qquad\qquad (8.5b)$

$\qquad\qquad\qquad\quad \sum_j X_{jk} = P_k \qquad \forall\, k \qquad\qquad (8.5c)$

$\qquad\qquad\qquad\quad Y_{ijk} \leq \sum_{h=k}^{m} X_{jh} \qquad \forall\, i, j, k \qquad\qquad (8.5d)$

$\qquad\qquad\qquad\quad X_{jk} = 0, 1 \qquad \forall\, j, k \qquad\qquad (8.5e)$

$\qquad\qquad\qquad\quad Y_{ijk} = 0, 1 \qquad \forall\, i, j, k \qquad\qquad (8.5f)$

---

[7]Naturally, in all such systems a facility will only provide services to those demand nodes that are allocated to the facility. The notions of facility and service hierarchies merely define the set of demand nodes and service requirements which are candidates for assignment or allocation to each facility.

The objective function (8.5a) minimizes the demand-weighted total distance. Constraint (8.5b) stipulates that all demand types at all locations must be assigned to some facility. Constraint (8.5c) limits the total number of type k facilities located to P_k. Constraints (8.5d) are the linkage constraints. They say that demands for type k service that originate at node i cannot be assigned to a facility at node j unless there is a type k or higher-level facility located at node j. Note that h is the maximum number of facility types under consideration. Constraints (8.5e) and (8.5f) are the integrality constraints. Note that the integrality constraints (8.5f) associated with the allocation variables (Y_{ijk}) may be relaxed since demands at node i for service of type k will naturally be assigned to the closest facility which can serve such demands.

In this formulation, constraints (8.5d) link the location (X_{jk}) and the allocation (Y_{ijk}) variables. They also link the different levels of facilities together. Technically, this formulation allows multiple facility types to be located at the same candidate location. However, any optimal solution to (8.5a)–(8.5f) will always have at most one facility type located at each candidate location. The proof of this is left as an exercise for the reader. In some fourmulations, however, this will not be the case. Constraint (8.6) below forces at most one facility type to be located at any candidate location:

$$\sum_k X_{jk} \leq 1 \quad \forall j \tag{8.6}$$

Inclusion of this constraint in formulation (8.5a)–(8.5f) will not alter the optimal solution but may expedite any solution algorithm.

Using the same notation, we can formulate a *successively inclusive facility hierarchy operating under a locally inclusive service hierarchy* as follows:

MINIMIZE $$\sum_i \sum_j \sum_k h_{ik} d_{ij} Y_{ijk} \tag{8.7a}$$

SUBJECT TO: $$\sum_j Y_{ijk} = 1 \quad \forall\, i, k \tag{8.7b}$$

$$\sum_j X_{jk} = P_k \quad \forall\, k \tag{8.7c}$$

$$Y_{ijk} \leq X_{jk} \quad \forall\, j, k; \forall\, i \neq j \tag{8.7d}$$

$$Y_{jjk} \leq \sum_{h=k}^{m} X_{jh} \quad \forall\, j, k \tag{8.7e}$$

$$X_{jk} = 0, 1 \quad \forall\, j, k \tag{8.7f}$$

$$Y_{ijk} = 0, 1 \quad \forall\, i, j, k \tag{8.7g}$$

Hierarchical Facility Location Models

This formulation is identical formulation (8.5a)–(8.5f) except that constraints (8.7d) and (8.7e) replace constraint (8.5d). Constraint (8.7d) states that demands at node i for type k service cannot be served by a facility at node j unless a type k facility is located at node j. Note that this constraint applies to all candidate facility locations and to all facility types. It also applies to all demand nodes, provided the demand node and candidate facility location are *different*. Constraint (8.7e) deals with the case in which the demand node and candidate facility location are the *same*. In this case, demands at location j for service of type k can be assigned to the facility at location j provided there is a level k or higher facility located at node j. Alternatively, we may think of constraints (8.7d) and (8.7e) as follows. Constraint (8.7e) is equivalent to (8.5d) in that service of type k can be provided by a type k or higher facility. However, constraint (8.7e) applies only to demands originating at the node at which the facility is located (i.e., $i = j$). Demands at nodes at which a facility of type k is not located can receive type k service only from a type k facility as indicated by constraint (8.7d).

As before, this formulation allows multiple facility types to be located at the same candidate location. Now, however, it might be advantageous to do so. For example, consider the network shown in Figure 8.4. Suppose this represents the network for a two-level hierarchical system. Suppose further that the demand for level 2 services is 2 at each node and the demand for level 1 (the lower level) services is 1 at each of the five nodes. Finally, suppose we can only locate one of each type of facility (i.e., $P_1 = P_2 = 1$). It should be clear that it is optimal to locate the level 2 facility at the center node. The level 2 demands then contribute 80 to the objective function (2 demands at each node times 10 distance units times 4 tip nodes). The level 2 facility at the center node provides level 1 service only to the local demands at the center node. The tip nodes, however, need a level 1 facility to serve them. If this facility cannot be located at the center node, then locating it at any tip node will add 60 to the objective function. If we can locate the level 1 facility at the center node (where the level 2 facility is already located), the four level 1 demands that cannot be served by the level 2 facility at the center

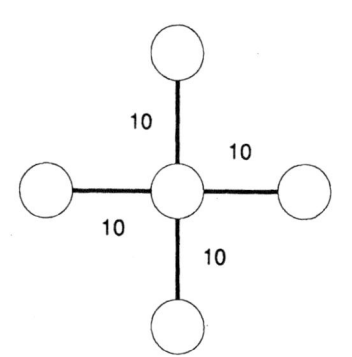

Figure 8.4. Network in which it is optimal to locate multiple facility types at the same node.

node will add only 40 to the objective function. Thus, in this case, it is clearly optimal to locate both facility types at the center node. Again, inclusion of constraint (8.6) will preclude such colocation solutions.

Next, we consider the following formulation of a *successively exclusive facility location* problem:

MINIMIZE
$$\sum_i \sum_j \sum_k h_{ik} d_{ij} Y_{ijk} \tag{8.8a}$$

SUBJECT TO:
$$\sum_j Y_{ijk} = 1 \quad \forall\, i, k \tag{8.8b}$$

$$\sum_j X_{jk} = P_k \quad \forall\, k \tag{8.8c}$$

$$Y_{ijk} \leq X_{jk} \quad \forall\, i, j, k \tag{8.8d}$$

$$X_{jk} = 0, 1 \quad \forall\, j, k \tag{8.8e}$$

$$Y_{ijk} = 0, 1 \quad \forall\, i, j, k \tag{8.8f}$$

Formulation (8.8a)–(8.8f) is identical to formulation (8.7a)–(8.7g) except that constraint (8.7e) is omitted and constraint (8.8d) now applies to all demand and candidate location site pairs, even when the two locations are the same. In other words, type k service can only be provided by a type k facility, not by facilities of types $k+1$ through m.

This formulation may be solved as m separate P-median problems. In other words, in this formulation, the location of level k facilities is independent of the location of level h facilities for $h \neq k$. Thus, this is not truly a hierarchical facility location model. However, if we include constraint (8.6), we preclude colocating facilities of different types or levels. Thus, the m median problems are linked because a decision to locate a level k facility at node j would preclude our locating a level h facility at node j for $h \neq k$.

8.3.3 Coverage-Based Hierarchical Location Formulations

Let us now consider a hierarchical location problem with a coverage objective. Since we will now be defining inputs that specify whether a facility of a given type located at a particular candidate location can cover demands of a particular type at some demand node, we intentionally discard the notions of successively inclusive and exclusive facilities. Specifically, we define the following additional notation:

Inputs

$$a_{ij}^{kq} = \begin{cases} 1 & \text{if a type } q \text{ facility located at candidate site } j \\ & \text{can provide service } k \text{ to demands at node } i \\ 0 & \text{if not} \end{cases}$$

Hierarchical Facility Location Models

Decision Variables

$$X_{jq} = \begin{cases} 1 & \text{if we locate a type } q \text{ facility at candidate site } j \\ 0 & \text{if not} \end{cases}$$

$$Z_{ik} = \begin{cases} 1 & \text{if demands for service } k \text{ at node } i \text{ are covered} \\ 0 & \text{if not} \end{cases}$$

With this additional notation, we can formulate a *hierarchical maximum covering location problem* as follows:

$$\text{MAXIMIZE} \quad \sum_i \sum_k h_{ik} Z_{ik} \tag{8.9a}$$

$$\text{SUBJECT TO:} \quad \sum_j X_{jq} = P_q \quad \forall\, q \tag{8.9b}$$

$$Z_{ik} \leq \sum_j \sum_q a_{ij}^{kq} X_{jq} \quad \forall\, i, k \tag{8.9c}$$

$$0 \leq Z_{ik} \leq 1 \quad \forall\, i, k \tag{8.9d}$$

$$X_{jq} = 0, 1 \quad \forall\, j, q \tag{8.9e}$$

The objective function (8.9a) maximizes the total number of demands of all types that are covered. Constraints (8.9b) stipulate that exactly P_q type q facilities are to be located. Constraints (8.9c) stipulate that demands for service k at node i cannot be counted as being covered unless we locate at least one facility at one or more of the candidate locations which are able to provide service k to demand node i. Constraints (8.9d) and (8.9e) are the nonnegativity and integrality constraints, respectively. Note that the coverage variables (Z_{ik}) need not be explicitly constrained to take on only integer values.

In some situations, demands at each demand node can only be counted as being covered if all of the services that are demanded at the node are covered. To formulate such a model, let us introduce the following additional notation:

Inputs

h_i = the "demand" at node i. This is probably a function of the service-specific demands (h_{ik}) at node i such as: (i) $\sum_k h_{ik}$; or (ii) $\sum_k \alpha_k h_{ik}$, where α_k is a weight associated with demands for service type k; or (iii) $\max_k \{h_{ik}\}$.

$$\delta_{ik} = \begin{cases} 1 & \text{if } h_{ik} > 0 \\ 0 & \text{if not} \end{cases}$$

Decision Variables

$$W_i = \begin{cases} 1 & \text{if demands at node } i \text{ are covered} \\ 0 & \text{if not} \end{cases}$$

With this additional notation, the alternate maximum covering hierarchical coverage location problem can be formulated as follows:

MAXIMIZE $\quad \sum_i h_i W_i \quad\quad\quad\quad\quad\quad\quad\quad\quad\quad (8.10a)$

SUBJECT TO: $\quad \sum_j X_{jq} = P_q \quad\quad\quad \forall\, q \quad\quad\quad (8.10b)$

$$Z_{ik} \leq \sum_j \sum_q a_{ij}^{kq} X_{jq} \quad \forall\, i, k \quad\quad (8.10c)$$

$$W_i \leq Z_{ik} + (1 - \delta_{ik}) \quad \forall\, i, k \quad\quad (8.10d)$$

$$0 \leq Z_{ik} \leq 1 \quad \forall\, i, k \quad\quad (8.10e)$$

$$X_{jq} = 0, 1 \quad \forall\, j, q \quad\quad (8.10f)$$

$$0 \leq W_i \leq 1 \quad \forall\, i \quad\quad (8.10g)$$

The objective function (8.10a) maximizes the number of covered demands. Constraints (8.10b) and (8.10c) are identical to (8.9b) and (8.9c), respectively. Constraint (8.10d) stipulates that demands at node i cannot be covered unless all services that are required at demand node i are covered. If there exists demand for service type k at node i (i.e., $\delta_{ik} = 1$), then (8.10d) stipulates that $W_i \leq Z_{ik}$, meaning that node i cannot be covered ($W_i = 1$) unless demands for service type k at node i are covered ($Z_{ik} = 1$). However, if there is no demand for type k services at node i (i.e., $\delta_{ik} = 0$), then node i can be covered ($W_i = 1$) even if $Z_{ik} = 0$. Finally, constraints (8.10e), (8.10f), and (8.10g) are the nonnegativity and integrality constraints on the problem's decision variables.

8.3.4 Extensions of Hierarchical Location Formulations

All of the models outlined above assume that the total number of facilities of a given type was given exogenously. In many cases, there may be a budget for either facilities of a give type or for all of the facilities. In such cases, the constraints of the form

$$\sum_j X_{jq} = P_q \quad \forall\, q$$

must be replaced by budget constraints of some form or by a fixed charge formulation. Let us define the following additional inputs:

f_{jq} = cost of building a type q facility at node j
B_q = budget for type q facilities
B = overall budget for all types of facilities

The constraints on the total number of facilities of each type can now be replaced by one or both of the following types of constraints:

$$\sum_j f_{jq} X_{jq} \leq B_q \quad \forall q \tag{8.11}$$

$$\sum_j \sum_q f_{jq} X_{jq} \leq B \tag{8.12}$$

Constraint (8.11) is a budget constraint that is specific to each type of facility. This would be appropriate if the funding for the different facilities comes from different sources. For example, the funding for clinics may come from the national government, while the monies for community and regional hospitals may come from an international funding agency. Constraint (8.12) limits the expenditures on all types of facilities to B. This would be appropriate if the funding for the different facility types comes from a single source and funding for one type of facility can readily be substituted for funding for another facility type.

The linear constraints (8.11) and (8.12) assume that the cost of building a facility to provide type k service at candidate location j is independent of whether or not we build any other type of facility at node j. This is clearly not always valid. For example, the cost of adding an elementary school wing to a location at which we are building a high school may be much less than the cost of a standalone elementary school. This is simply because of (i) the economies of scale associated with building large structures and (ii) the ability of the two facilities to share resources (e.g., to share an auditorium, gymnasium, and/or a cafeteria) if they are colocated.

The models outlined in Sections 8.3.2 and 8.3.3 all assume that the facilities are uncapacitated. They can readily be extended to the case of capacitated facilities. In this case, we need to replace the 0/1 allocation variables (Y_{ijk}) and coverage variables (Z_{ik} and W_i) by variables that reflect the *number* of demands for service k at node i that are served by a facility at location j or that are covered. Specifically, we introduce the following additional decision variables:

$\hat{Y}_{ijk}$ = the number of demands at node i that receive service of type k from a facility at candidate site j

$\hat{Z}_{ik}$ = the number of demands at node i that are covered for service of type k

$\hat{W}_i$ = the number of demands at node i that are covered

Note that the capacities of the facilities (to provide different services) may be unrelated or they may be interrelated. As an example of unrelated capacities, we can consider the health care system problem outlined above. In that case, the in-patient surgical capacity of the regional hospital may be unrelated to the clinic's capacity, even if they are located in the same physical building. The postal system provides an example of interrelated capacities. In that case, the number of express mail patrons that a clerk can handle per hour is related (inversely) to the number of requests he or she receives for stamp purchases. The formulation of the extensions of the basic models to include capacitated facilities is left as an exercise for the reader.

8.4 MODELS OF INTERACTING FACILITIES

In all of the models we have discussed so far, the facilities being located operate independently. Often, however, this is not the case; rather, the facilities interact in one or more different ways. In this section we present formulations of two such models. We begin with a system in which there are explicit flows between some of the facilities being located. In the second model, the facilities need to be sufficiently close to each other to cover each other.

8.4.1 Flows Between Facilities

Let us return again to the health care example discussed in Section 8.3. Now, however, let us assume that all patients must first be seen by a physician at a clinic before receiving care at either a regional hospital or a community hospital. Thus, all diagnostic and referral services are located at the clinics. In essence, the clinics act as gatekeepers for the community hospitals and the regional hospitals.[8] Figure 8.5 illustrates the flow of patients in such a system. Note that we are assuming that the diagnostic services are perfect in the sense that there is never a need to refer a patient a second time from a community hospital to a regional hospital. Let us further assume that the system operates as a successively inclusive facility hierarchy with a globally inclusive service hierarchy. Finally, we will assume that each of the facilities is capacitated.

To illustrate how such a system can be modeled, we define the following notation (in addition to the notation that was used in the discussion of hierarchical location models in Section 8.3):

Inputs

S_{jk} = capacity of a facility located at candidate site j for providing service k. As before, service level 1 will be the lowest level which, in this case, is the clinic service at which all diagnostic functions are performed.

d_{ij} = travel time between demand node i and candidate facility site j

$\hat{d}_{jm}$ = travel time between candidate site j and candidate site m. Note that we are allowing the times between facilities to differ from the times

---
[8] Many HMOs in the United States try to operate in this manner, in that you need to be seen by your primary physician before you can be treated by a specialist or by a hospital unless the situation is a true emergency.

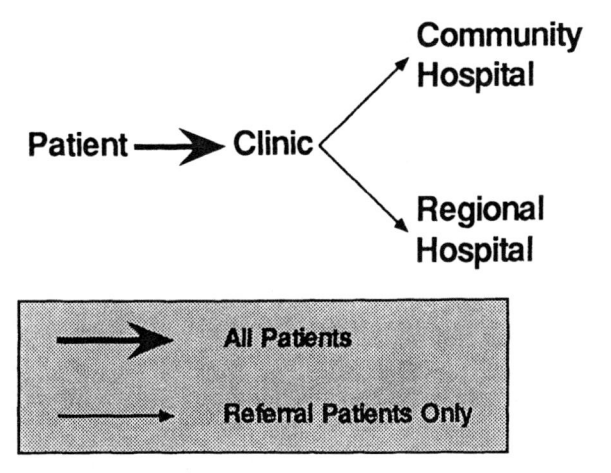

Figure 8.5. Patient flows in hypothetical health care system.

between a demand node and a candidate location even if the distances are the same. The reason for doing so is that the system may employ different modes of travel. For example, in rural settings in developing countries, many patients may need to walk from their village to the nearest clinic. Once there, however, they may be transported by car or ambulance to a community hospital or a regional hospital if a referral is made. Alternatively, we might like there to be a different "time" associated with referral movements to reflect the fact that such patients are likely to be in more serious condition than are other patients. As such, we might like to use a *higher* value of "time" per unit distance for such moves than for the movements from demand nodes to clinics.

Decision Variables

Y_{ijk} = number of patients requiring service level k going from demand node i to a clinic at candidate site j

Z_{jnk} = number of patients at a clinic at candidate site j referred to a facility at candidate site n for service level k

With this notation, we can formulate the following model in which the objective is to minimize the total movement time of the patients:

MINIMIZE

$$\sum_i \sum_j d_{ij} \left\{ \sum_k Y_{ijk} \right\} + \sum_j \sum_n \hat{d}_{jn} \left\{ \sum_{k=2}^{m} Z_{jnk} \right\} \qquad (8.13a)$$

SUBJECT TO:

$$\sum_j Y_{ijk} = h_{ik} \qquad \forall\, i, k \qquad (8.13b)$$

$$\sum_i Y_{ijk} - \sum_n Z_{jnk} = 0 \qquad \forall j; k = 2, \ldots, m \qquad (8.13c)$$

$$\sum_j X_{jk} = P_k \qquad \forall k \qquad (8.13d)$$

$$\sum_k X_{jk} \leq 1 \qquad \forall j \qquad (8.13e)$$

$$\sum_i \sum_k Y_{ijk} \leq S_{j1} \sum_{q=1}^{m} X_{jq} \qquad \forall j \qquad (8.13f)$$

$$\sum_j Z_{jnk} \leq S_{nk} \sum_{q=k}^{m} X_{nq} \qquad \forall n; k = 2, \ldots, m \qquad (8.13g)$$

$$X_{jk} = 0, 1 \qquad \forall j, k \qquad (8.13h)$$

$$Y_{ijk} \geq 0 \qquad \forall i, j, k \qquad (8.13i)$$

$$Z_{jnk} \geq 0 \qquad \forall j, n, k \qquad (8.13j)$$

The objective function (8.13a) minimizes the total patient travel time in the system. The first term in the objective function is the total travel time of patients from demand nodes to clinics. The second term is the total travel time of the referral patients. Note that the summation over services in the second term goes from $k = 2$ to m (the maximum number of services levels) and not from $k = 1$. Constraints (8.13b) ensure that all patients are assigned to some clinic. This essentially ensures that all patients enter the system. Constraints (8.13c) state that all patients who arrive at a clinic at location j and who need service at a level above that which can be provided at the clinic are referred to an appropriate facility. This is a flow conservation constraint. Constraints (8.13d) state that exactly P_k facilities of type k are to be located. Constraints (8.13e), which are identical to constraint (8.6) above, ensure that at most one facility of any type may be located at any candidate site. Constraints (8.13f) and (8.13g) are the capacity constraints on the facilities. The differ in that all patients impinge on the diagnostic capacity of the clinics [in (8.13f)], while only patients who need service level k affect the capacity of a facility to deliver level k service [in (8.13g)]. Finally, constraints (8.13h), (8.13i), and (8.13j) are the integrality and nonnegativity constrains, respectively.

Model (8.13a)–(8.13j) is a variant of the P-median model. Clearly, hierarchical location problems and models of interacting facilities can be formulated as variants of the other basic models discussed in Chapters 4 through 7. For example, Church and Eaton (1987) discuss hierarchical covering-based models.

8.4.2 Facilities with Proximity Constraints

In addition to interacting explicitly through flows, facilities also interact in other ways. One of those ways is by serving as backup facilities to each other. One way of modeling this constraint is to require facilities to cover each other in addition to or as opposed to covering demand nodes. Continuing in the health care context, let us imagine a system in which lower-level facilities must be covered by higher-level facilities. Specifically, let us assume that all community hospitals must have a regional hospital located within D^{hr} distance units. Similarly, all clinics must have a community hospital within D^{ch} distance units *and* a regional hospital within D^{cr} distance units.[9] These requirements may be imposed to ensure that patients requiring a higher level of service can obtain that service in a timely manner, without having to model patient flows explicitly.

In this subsection we present a multiobjective model in which we (i) minimize the total cost of the facilities being located, (ii) maximize the number of level 2 (community hospital) demands that are covered, and (iii) maximize the number of level 3 (regional hospital) demands that are covered. We do so subject to constraints that stipulate that all demands must be covered by a facility providing clinic-type services; all clinics must be covered by both a community hospital and a regional hospital; and all community hospitals must be covered by a regional hospital.[10] To formulate this model, we define the following notation:

Inputs

f_{jc}, f_{jh}, f_{jr} = cost of building a clinic (c), community hospital (h), or a regional hospital (r) at candidate site j

$\phi_{ic}, \phi_{ih}, \phi_{ir}$ = demands for clinic (c), community hospital (h), and regional hospital (r) services arising at demand node i

$$a_{ij}^{kq} = \begin{cases} 1 & \text{if a type } q \text{ facility at node } j \text{ can cover} \\ & \text{demands at node } i \text{ for service request of type } k \\ 0 & \text{if not} \end{cases}$$

$$b_{jn}^{ch} = \begin{cases} 1 & \text{if a hospital (h) at node } n \\ & \text{can cover a clinic (c) at candidate site } j \\ 0 & \text{if not} \end{cases}$$

b_{jn}^{cr}, b_{jn}^{hr} = defined similarly for clinic/regional hospital coverage (b_{jn}^{cr}) and community hospital/regional hospital coverage (b_{jn}^{hr})

---

[9] Throughout this section, the subscripts and superscripts c, h, and r refer to clinic, community hospital, and regional hospital, respectively. They are not indices over which summation is appropriate.

[10] The reader should note that there is a difference in this model between a facility covering a demand node and a higher-level facility covering a lower-level facility. Thus, the fact that all demand nodes must be covered by a facility capable of providing clinic-level services and all clinics must be covered by a community hospital and a regional hospital does not imply that all demand nodes will be covered by a community hospital and a regional hospital.

Decision Variables

$$X_{jc} = \begin{cases} 1 & \text{if a clinic (c) is located at candidate site } j \\ 0 & \text{if not} \end{cases}$$

X_{jh}, X_{jr} = defined similarly for community hospitals (X_{jh}) and regional hospitals (X_{jr})

$$Y_{ih} = \begin{cases} 1 & \text{if level 2 (community hospital) demands at node } i \text{ are covered} \\ 0 & \text{if not} \end{cases}$$

$$Y_{ir} = \begin{cases} 1 & \text{if level 3 (regional hospital) demands at node } i \text{ are covered} \\ 0 & \text{if not} \end{cases}$$

With this notation, we can formulate the following multiobjective problem in which lower-level facilities must be covered by higher-level facilities:

MINIMIZE $\Theta_F = \sum_j f_{jc} X_{jc} + \sum_j f_{jh} X_{jh} + \sum_j f_{jr} X_{jr}$ (8.14a)

MAXIMIZE $\Theta_h = \sum_i \phi_{ih} Y_{ih}$ (8.14b)

MAXIMIZE $\Theta_r = \sum_i \phi_{ir} Y_{ir}$ (8.14c)

SUBJECT TO: $\sum_j a_{ij}^{cc} X_{jc} + \sum_j a_{ij}^{ch} X_{jh} + \sum_j a_{ij}^{cr} X_{jr} \geq 1$ $\forall\, i$ (8.14d)

$Y_{ih} \leq \sum_j a_{ij}^{hh} X_{jh} + \sum_j a_{ij}^{hr} X_{jr}$ $\forall\, i$ (8.14e)

$Y_{ir} \leq \sum_j a_{ij}^{rr} X_{jr}$ $\forall\, i$ (8.14f)

$X_{jc} \leq \sum_n b_{jn}^{ch} X_{nh}$ $\forall\, j$ (8.14g)

$X_{jc} \leq \sum_n b_{jn}^{cr} X_{nr}$ $\forall\, j$ (8.14h)

$X_{jh} \leq \sum_n b_{jn}^{hr} X_{nr}$ $\forall\, j$ (8.14i)

$X_{jc}, X_{jh}, X_{jr} = 0, 1$ $\forall\, j$ (8.14j)

$Y_{ih}, Y_{ir} = 0, 1$ $\forall\, i$ (8.14k)

The first objective function (8.14a) minimizes the total cost of the facilities that are located. The second objective function (8.14b) maximizes the number of level 2 (community hospital) demands that are covered. The third objective function (8.14c) maximizes the number of level 3 (regional hospital) demands that are covered. Constraints (8.14d) stipulate that all demand nodes must be covered by at least one clinic or a hospital (regional or community) able to provide clinic services to the demand node. Constraints (8.14e) state that demand node i cannot be considered covered for community hospital services unless at least one community hospital or one regional hospital is located at one of the nodes able to provide community hospital services to node i. Constraints (8.14f) state that node i cannot be counted as being covered for regional hospital services unless a regional hospital is located at one or more of the nodes able to provide regional hospital services to node i. Constraints (8.14g), (8.14h), and (8.14i) link the facility location decisions. They state that each clinic must be covered by at least one community hospital [(8.14g)] and by at least one regional hospital [(8.14h)]. Constraints (8.14i) ensure that each community hospital is covered by at least one regional hospital. Finally, constraints (8.14j) and (8.14k) are the integrality constraints. Chaudhry, Moon, and McCormick (1987) and Chaudhry (1993) discuss similar models in which the facilities being located must not only cover demands but also cover each other.

8.5 MULTIPRODUCT FLOWS AND PRODUCTION / DISTRIBUTION SYSTEMS

In Sections 8.3 and 8.4 we discussed hierarchical facility location problems and problems with interacting facilities. In both sections we implicitly dealt with cases in which there were multiple types of demands. For example, in the postal example discussed in Section 8.3, there were customers who only needed to deposit mail. They could be served at any of the facilities (post boxes, branch offices, and main post offices) that were being located. Customers who wanted to open P.O. boxes represented another type of customer. They could only be served at branch offices and the main post office. Finally, customers who needed a variety of special services might only be able to receive those services at the main post office. Similarly, in the school location problem, we had three classes of customers: primary school students, middle school students, and high school students.

In this section we outline a similar sort of problem. Now, however, instead of multiple demands, we have multiple products that are being produced in a manufacturing environment. For example, we might have different types of automobiles that are being produced at up to 30 different assembly plants. These cars are to be shipped to thousands of car dealers. A firm may have only one or two (or, in very rare cases, three) plants at which a particular vehicle can be produced, even if the firm has a total of 30 assembly plants. Furthermore, even when there is a choice of plant at which to produce the vehicle, the outbound shipment costs are only one of many factors that need to be considered in assigning orders to plants. Thus, cars are generally not produced at the plant nearest the dealer ordering the car.

Dealers that are within about 500 miles of the assembly plant at which the vehicle is made generally receive their cars by direct truck shipment. Vehicles going to dealers located more than 500 miles from an assembly plant are generally shipped by train to a rail ramp near the dealer and then from there to the dealer by truck. One of the key issues in the auto industry is how many rail ramps a firm should have and where they should be located.

In a more general case, a firm may have relatively few products and a number of plants. Products are shipped from plants either directly to markets or to warehouses and from there to markets (or customers). The key issues we are concerned with are: (i) how many warehouses to have, (ii) where to locate warehouses, and (iii) how the products should flow through the system. Implicit in the product flow decision are other decisions about which products should be produced at which plants for which markets. Figure 8.6 is a schematic of such a system in which three products are produced. Product flows are indicated by the different styles of lines. Three plants are shown. In this example, each plant is dedicated to the production of a single product. In general, this will not necessarily be the case. The products are shipped to two warehouses. Warehouse A deals with the products from plants I and II, while warehouse B deals with the products from plants II and III. Thus, warehouse A ships products from plant I to markets 2, 3, and 6 and products from plant II to markets 1 and 4. Similarly, warehouse B ships products from plant II to markets 5 and 6 and products from plant III to markets 3 and 4.

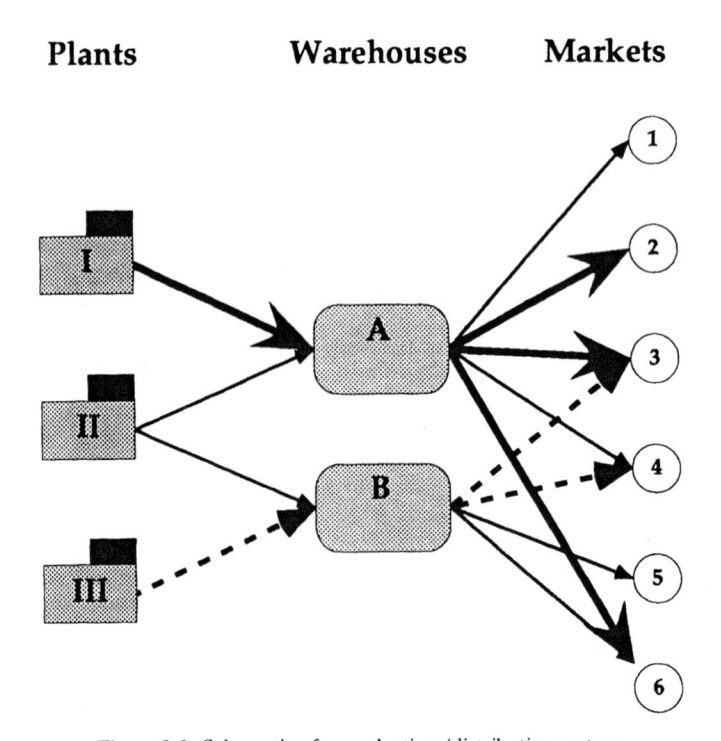

Figure 8.6. Schematic of a production/distribution system.

Multiproduct Flows and Production / Distribution Systems

In this section we formulate a simple model for this class of problems. The problem we will consider will assume that the warehouses are uncapacitated. Extensions to capacitated warehouses are left as an exercise for the reader. To formulate the problem, we define the following notation:

Inputs

h_i^k = demand for product k in market i

f_j = fixed cost of locating a warehouse at candidate site j

c_{ijm}^k = cost of producing one unit of product k at plant m and shipping it to market i via a warehouse at candidate location j

S_m^k = capacity of plant m for production of product k. Note that we are assuming that the production capacities of each plant for different products are independent. In general, this will not be true. Thus, for example, an automobile assembly plant may be able to produce 1000 vehicles per day. If the plant is assigned two types of vehicles (e.g., a Chevrolet model and its Pontiac twin), then the more of one type of vehicle that is produced at the plant (e.g., the Chevrolet model), the fewer of the other type of vehicle (e.g., the Pontiac model) the plant can produce. We make this assumption for the sake of simplicity.

M = a very large number

Decision Variables

Y_{ijm}^k = flow of product k from plant m to market i via warehouse j

$X_j = \begin{cases} 1 & \text{if we locate a warehouse at candidate site } j \\ 0 & \text{if not} \end{cases}$

With this notation, we can formulate a simple production/distribution problem as follows:

MINIMIZE $\quad \sum_j f_j X_j + \sum_i \sum_j \sum_m \sum_k c_{ijm}^k Y_{ijm}^k \quad$ (8.15a)

SUBJECT TO: $\quad \sum_i \sum_m \sum_k Y_{ijm}^k \leq MX_j \quad \forall j \quad$ (8.15b)

$\qquad \sum_j \sum_m Y_{ijm}^k \geq h_i^k \quad \forall i, k \quad$ (8.15c)

$\qquad \sum_i \sum_j Y_{ijm}^k \leq S_m^k \quad \forall m, k \quad$ (8.15d)

$\qquad Y_{ijm}^k \geq 0 \quad \forall i, j, m, k \quad$ (8.15e)

$\qquad X_j = 0, 1 \quad \forall j \quad$ (8.15f)

The objective function (8.15a) minimizes the sum of the fixed warehouse location costs and the variable costs. Note that a large number of cost components can be incorporated into the variable costs (c_{ijm}^k), including the unit production costs at plant m, the unit transport costs between plant m and a warehouse at candidate site j, any unit variable warehousing costs at site j, and the transport costs between site j and market i. All of these unit costs may depend on the product in question. The left-hand side of constraint (8.15b) is the flow through a warehouse at candidate site j. Constraint (8.15b) states that this flow can only be positive if we locate a warehouse at candidate site j. Constraints (8.15c) state that the total amount of product k that is shipped into market i from all plants and all warehouses must be greater than or equal to the demand for product k in market i. This is simply a multiproduct extension of the demand constraints that we used in the transportation problem formulation of Chapter 2. Similarly, constraints (8.15d) are multiproduct extensions of supply constraints used in the transportation problem. Finally, constraints (8.15e) and (8.15f) are the nonnegativity and integrality constraints, respectively.

Whenever production, distribution, and/or warehousing are considered, it is important to explore the inventory implications associated with the plant production schedules and with the shipping plans to and from the warehouses. To illustrate how inventory costs can alter a shipping plan, consider the network shown in Figure 8.7. Two products are sold in market 1. Plant I produces product I; plant II produces product II. The demand for product I in market 1 is 12 units per year (one item every 30 days), while the demand for product II is 120 units per year (one item every 3 days). Warehouse A is closer to plant I and warehouse B is closer to plant II. Material is shipped from the warehouses when 3 units of inventory have been accumulated for shipment to market 1. (In other words, we may wait until a truckload of material has been accumulated and, for the sake of argument, a truckload is defined as 3 units of either product.) The production schedules at the plants are independent of the shipping schedules from the warehouses. [In a large custom production environment (as in common in the automobile industry), coordination of production with the outbound shipping schedules from warehouses may be sufficiently difficult as to be nearly impossible.]

At first glance, it might seem as though the logical shipping plan would be for plant I to ship to warehouse A from which product I would be sent to market 1. Similarly, plant II would ship to warehouse B from which product II would be sent to market 2. However, this shipment plan would involve large inventory costs for items at warehouse A. Only product I goes through warehouse A for market 1. Since warehouses wait to accumulate 3 items of inventory before shipping from the warehouse and since production and shipping schedules from the warehouses are uncoordinated, the average time a unit of product I spends in inventory at warehouse A is 45 days. Thus, the inventory carrying cost of the product at warehouse A is $45\tau_I$ dollars, where τ_I is the daily inventory carrying cost of product I. The total annual shipping

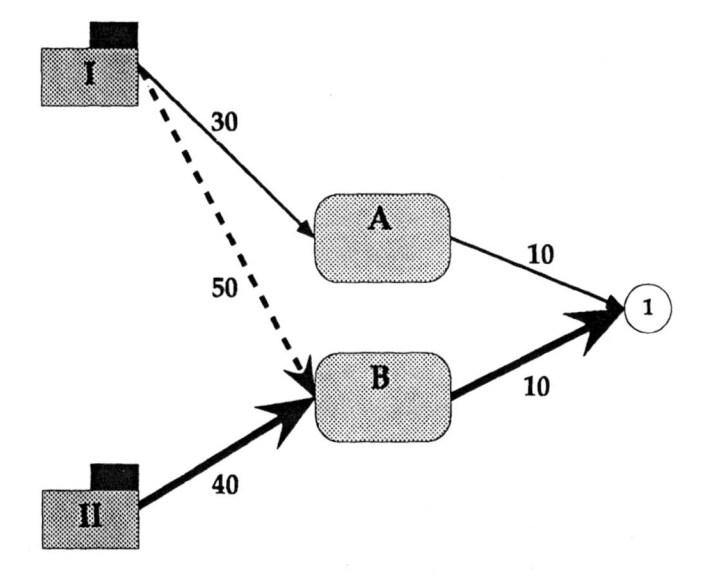

Multiproduct Flows and Production / Distribution Systems 337

**Demand for product from plant I is 12 units per year.
Demand for product from plant II is 120 units per year.**

Unit shipping costs are shown beside each link.

**Material is shipped from warehouses when
3 units have accumulated.**

**Production is independent of the shipping schedule
and is uniformly distributed over the year.**

Figure 8.7. Inventory implications on shipping plans.

and inventory cost of the plan in which each plant ships to market 1 via the warehouse that is closest (in terms of unit shipping cost) to it is

$$(30 + 10) \cdot 12 + 45\tau_I \cdot 12 + (40 + 10) \cdot 120 + 4.5\tau_{II} \cdot 120$$

$$= 6480 + 540\tau_I + 540\tau_{II}$$

where τ_{II} is the daily inventory carrying cost of product II.

On the other hand, if plant I shipped its product to warehouse B, it would incur larger shipping costs, but lower inventory costs, since products I and II could be consolidated at warehouse B. The annual shipping and inventory cost of this plan would be

$$(50 + 10) \cdot 12 + 360 \cdot \left(\frac{3}{132}\right) \cdot 0.5 \cdot \tau_I \cdot 12 + (40 + 10) \cdot 120$$

$$+ 360 \cdot \left(\frac{3}{132}\right) \cdot 0.5 \cdot \tau_{II} \cdot 120$$

$$= 6720 + 49.09\tau_I + 490.91\tau_{II}$$

where we are using 360 days per year in estimating the inventory carrying costs (since 12 and 120 are divisors of 360 and not of 365). Note that the time between shipments is $(3/132)$ years or $360 \cdot (3/132)$ days. Each item waits approximately half of this time. Therefore, the inventory cost per item is this time multiplied by 0.5 and then multiplied by the appropriate daily inventory carrying cost. This must then be multiplied by the number of items of this product that are being shipped.

If the inventory carrying costs of the two products are the same (i.e., $\tau_I = \tau_{II} = \tau$), then this second plan is less expensive if

$$6480 + 1080\tau > 6720 + 540\tau$$

or

$$\tau > \tfrac{4}{9} \approx 0.44$$

This is a highly simplified example. Most such problems would be complicated by the presence of far more products going through each of the warehouses for may other markets. Nevertheless, this example illustrates that incorporation of inventory considerations may lead to the adoption of shipping plans that call for the use of longer, more costly transport lanes in an attempt to realize inventory cost reductions through consolidation.

A large number of other extensions of the basic production/distribution problem can be considered. One important extension is to allow products to be shipped from warehouses on routes that incorporate multiple customers. Section 8.6 discusses this sort of problem. Formulation of other extensions of the basic production/distribution problem discussed above and consideration of the means of solving the problem are left as exercises for the reader. Suffice it to say that this problem has much of the character of the fixed charge facility location problems discussed in Chapter 7. In a classic paper in this field, Geoffrion and Graves (1974) outline the user of Bender's decomposition to solve one variant of this sort of problem.

8.6 LOCATION / ROUTING PROBLEMS

In every model we have examined so far, we have implicitly assumed that customers are served directly (on an out-and-back basis) from the facilities being located. Figure 8.8 illustrates such a routing scheme. Each customer is served on his or her own route. Such a routing scheme is clearly appropriate in a number of contexts. If customers come individually to the facilities, this sort of modeling assumption is likely to be a good approximation of the total travel cost incurred by the customers. One example of this would be the case of patients going to health care clinics in developing countries. Alternatively, if customers are served individually from the facilities, this is a good way to model the transport costs. For example, in locating ambulances in a city, it is fair to assume that each ambulance will handle a single patient at any one time. Thus, this way of approximating the transport costs is appropriate. Finally, if goods are delivered from the facilities to customers in truckload quantities, this is a good modeling approach.

In many cases, however, customers are not served individually from the facilities. Rather, customers are consolidated into routes which may contain many customers. This is common in less-than-truckload (LTL) shipping. Such problems are considerably more difficult than are the simpler facility location problems we have discussed to date. One of the reasons for the added difficulty in solving these problems is that there are far more decisions that need to be made by the model. These decisions include: (i) how many

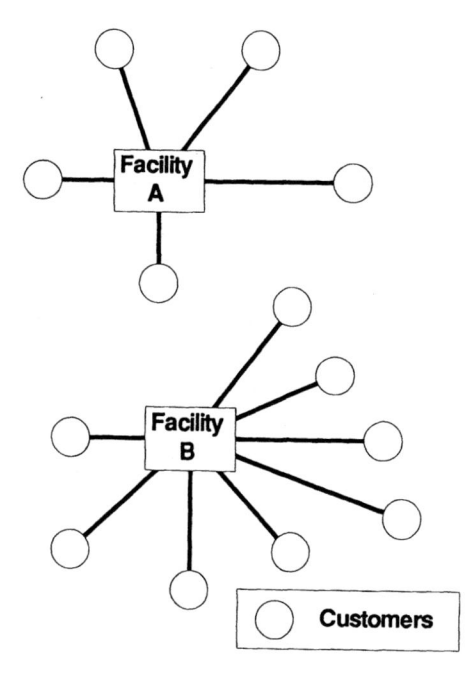

Figure 8.8. Schematic of out-and-back shipping.

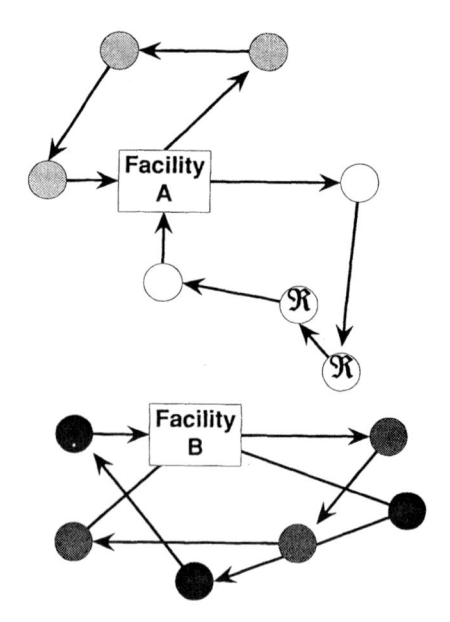

Figure 8.9. Schematic of a distribution system with multicustomer routes.

facilities to locate, (ii) where the facilities should be, (iii) which customers to assign to which warehouses, (iv) which customers to assign to which routes, and (v) in what order customers should be served on each route. To appreciate the difficulty associated with this problem, it is sufficient to realize that the final problem—that of optimally sequencing the customers on a route once the facilities have been located and customers have been assigned to facilities and to routes—is a traveling salesman problem. This is the quintessential NP-complete problem.[11]

Figure 8.9 illustrates this sort of routing scheme for the same set of customers and warehouses that were used in Figure 8.8. Four routes are shown. Note that two customers (those indicated by the ℜ symbol) have been reallocated from warehouse B to warehouse A. Also, note that it is common for different routes to overlap. Overlapping routes result from the imposition of constraints in the routing portion of the problem. Typical routing constraints include vehicle capacity constraints and time windows during which customers either must or must not be visited.

We will begin by considering a relatively simple problem of locating facilities and determining routes from the facilities such that each customer is on a single route and the total cost is minimized. The cost is composed of

---

[11] The reader is referred to *The Travelling Salesman Problem: A Guided Tour of Combinatorial Optimization*, 1985 (E. L. Lawler, J. K. Lenstra, A. H. G. Rinnooy Kan, and D. B. Shmoys, eds.), Wiley, New York.

Location/Routing Problems

three primary components: (i) the fixed cost of locating facilities, (ii) the distance-related travel costs, and (iii) any fixed costs associated with using a vehicle. In dealing with vehicle routing problems of this sort, it is important that we carefully distinguish between demand nodes, candidate warehouse locations, and the union of these two sets of nodes which is the set of all nodes. Therefore, in defining the notation to be used, we begin by defining the relevant sets used in the model:

Sets[12]

$\mathbf{I}$ = set of demand nodes
$\mathbf{J}$ = set of candidate facility sites
$\mathbf{N} = \mathbf{I} \cup \mathbf{J}$ = the set of all nodes
$\mathbf{K}$ = set of all vehicles that can be used

Inputs

h_i = demand at customer node i
f_j = fixed cost of locating a facility at candidate site j
c_{ijk} = cost of traveling between node i and node j using vehicle k
g_k = fixed cost of using vehicle k
u_k = capacity of vehicle k (measured in the same units as the customers demands, h_i)

Decision Variables

$$X_j = \begin{cases} 1 & \text{if we locate at candidate site } j \\ 0 & \text{if not} \end{cases}$$

$$Y_{jk} = \begin{cases} 1 & \text{if vehicle } k \text{ operates out of a warehouse at candidate site } j \\ 0 & \text{if not} \end{cases}$$

$$Z_{ijk} = \begin{cases} 1 & \text{if node } i \text{ immediately precedes node } j \text{ on a route using vehicle } k \\ 0 & \text{if not} \end{cases}$$

---

[12] Note that indices i and j will be used for any of the sets $\mathbf{I}$, $\mathbf{J}$, and $\mathbf{N}$. In all cases, the set over which the summation extends will be explicitly stated. The key point is that notation such as c_{ijk} refers to the cost of traveling between nodes i and j using vehicle k, where the indices i and j apply to the set of *all* nodes (in this case).

With this notation, we can define one variant of a location/routing problem as follows[13]:

MINIMIZE

$$\sum_{j \in J} f_j X_j + \sum_{i \in N} \sum_{j \in N} \sum_{k \in K} c_{ijk} Z_{ijk} + \sum_{j \in J} \sum_{k \in K} g_k Y_{jk}$$

(8.16a)

SUBJECT TO:

$$\sum_{k \in K} \sum_{i \in N} Z_{ijk} = 1 \qquad \forall j \in I \qquad (8.16b)$$

$$\sum_{i \in N} Z_{ijk} - \sum_{i \in N} Z_{jik} = 0 \qquad \forall j \in N; \forall k \in K \qquad (8.16c)$$

$$\sum_{i \in \mathfrak{S}} \sum_{j \in \mathfrak{S}} \sum_{k \in K} Z_{ijk} \leq |\mathfrak{S}| - 1 \qquad 2 \leq |\mathfrak{S}| \leq |I|$$

$$\forall \mathfrak{S} \subseteq I \qquad (8.16d)$$

$$\sum_{i \in I} Z_{ijk} = Y_{jk} \qquad \forall j \in J; \forall k \in K \qquad (8.16e)$$

$$\sum_{i \in I} Z_{jik} = Y_{jk} \qquad \forall j \in J; \forall k \in K \qquad (8.16f)$$

$$Y_{jk} \leq X_j \qquad \forall j \in J; \forall k \in K \qquad (8.16g)$$

$$\sum_{j \in J} Y_{jk} \leq 1 \qquad \forall k \in K \qquad (8.16h)$$

$$\sum_{i \in I} h_i \left\{ \sum_{j \in J} Z_{ijk} \right\} \leq u_k \left\{ \sum_{j \in J} Y_{jk} \right\} \quad \forall k \in K \qquad (8.16i)$$

$$X_j = 0, 1 \qquad \forall j \in J \qquad (8.16j)$$

$$Y_{jk} = 0, 1 \qquad \forall j \in J; \forall k \in K \qquad (8.16k)$$

$$Z_{ijk} = 0, 1 \qquad \forall i, j \in N; \forall k \in K \qquad (8.16l)$$

The objective function (8.16a) minimizes the sum of the fixed facility location costs, the distance-related routing costs, and the fixed costs associ-

---

[13] The user is cautioned that, just as there are many different facility location problems, there are also many different vehicle routing problem statements. In addition, for every statement of a vehicle routing problem, there are several valid ways of formulating the problem. Different formulations often lead to different solution algorithms. The number of *possible* combinations of facility location and vehicle routing problems is *very* large. Thus, the model presented here is but one of many possible ways of viewing combined facility location and vehicle routing problems.

Location / Routing Problems

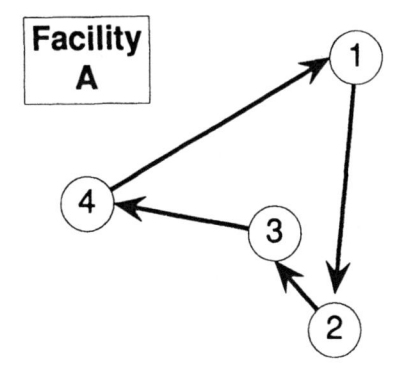

[Figure showing Facility A connected to nodes 1, 2, 3, 4 with a subtour among 1, 2, 3, 4]

This subtour is precluded by the constraint

$Z_{12} + Z_{13} + Z_{14} +$
$Z_{21} + Z_{23} + Z_{24} +$
$Z_{31} + Z_{32} + Z_{34} +$
$Z_{41} + Z_{42} + Z_{43} \leq 3$

where
$\mathcal{S} = \{1, 2, 3, 4\}$

Figure 8.10. Example subtour elimination constraint.

ated with using the vehicles. Constraints (8.16b) state that every customer j must be on exactly one route. They do so by requiring that there be exactly one node (a customer or a facility) served by one vehicle preceding every customer j. Constraints (8.16c) are flow conservation constraints. They state that if vehicle k enters node j (from any node i), then it must depart from node j (to some other node i).

Constraints (8.16d) are a bit tricky to interpret. They are subtour elimination constraints. They prevent a vehicle from being assigned to a set of demand nodes only (and not visiting a facility at all). They do so by requiring that for any subset ($\mathcal{S}$) of demand nodes of cardinality 2 or more,[14] the total number of connections between pairs of nodes in the subset must be less than or equal to the cardinality of the subset minus 1. Note that for every subset of nodes, the number of links in a subtour that visits only the nodes in the subset of nodes would equal the number of nodes in the subset. By requiring that the number of links connecting nodes in the subset be strictly less than the number of nodes in the subset, constraints (8.16d) preclude the formation of such subtours. Figure 8.10 illustrates a subtour and the constraint from among those shown in constraint set (8.16d) that would preclude

---

[14] The cardinality of a set is simply the number of elements in the set.

its formation. For this subtour, $Z_{12} = Z_{23} = Z_{34} = Z_{41} = 1$. Thus, the constraint would be violated since the left-hand side of the constraint would be 4 while the right-hand side would only be 3. (In Figure 8.10 we have dropped the vehicle index, k, for simplicity.)

Constraints (8.16e) and (8.16f) state that if vehicle k is assigned to a route emanating from a facility at site j, then at least one link goes into node j [(8.16e)] and one leaves node j [(8.16f)]. Constraints (8.16g) state that a vehicle can be assigned to a route from site j only if we locate a facility at site j. Constraints (8.16h) state that each vehicle can be assigned to at most one facility. Constraints (8.16i) are the vehicle capacity constraints. Finally, constraints (8.16j), (8.16k), and (8.16l) are the integrality constraints.

One of the key problems with formulation (8.16a)–(8.16l) is that there are a huge number of constraints. Not only are there eight different classes of constraints, but the total number of constraints is very large even for very small problems. Constraints (8.16d) represent most of the total number of constraints, since they apply to all subsets of the demand nodes (except for the empty subset and subsets of cardinality 1). Thus, if there are $|\mathbf{I}|$ demand nodes, the number of constraints in set (8.16d) is $2^{|\mathbf{I}|} - (|\mathbf{I}| + 1)$ which is $O(2^{|\mathbf{I}|})$. Specifically, for a very small problem of 20 demand nodes, there are 1,048,555 constraints of the form of (8.16d). For a problem with 40 demand nodes (which is still a small problem), there are over 10^{12} constraints. Finally, for a problem with 100 nodes, there are over 10^{30} such constraints. In short, the number of constraints is not polynomially bounded as a function of the size of the problem.

To put the problem in a slightly different light, if a computer could enumerate 10^{15} constraints every second,[15] it would take the computer over 40 million years to enumerate all of the constraints for a problem with 100 nodes. After that, the computer would have to solve the problem. Most of us would rather not wait this long for a solution! Thus, in solving such problems, a relaxation of the problem in which constraints (8.16d) are omitted completely and other constraints may be relaxed is solved initially. Then constraints of the form of (8.16d) that are violated by the relaxed solution are identified and added (possibly along with other constraints known as facets). Alternatively, a variety of heuristics could be used in attacking this difficult class of problems (e.g., Perl and Daskin, 1985). A complete discussion of the means of solving this sort of problem is beyond the scope of this text. For an overview of such solution methods, the reader is referred to Laporte (1988), Laporte, Nobert, and Taillefer (1988), and Laporte, Nobert, and Arpin (1986).

Difficult as the formulation is, a number of extensions are worth mentioning. Often, constraints are imposed on the total duration of a route. Such constraints frequently arise from contractual agreements with drivers (or regulatory limits) that restrict the number of consecutive hours they can

---

[15] This would be an exceptionally fast computer by today's standards.

drive. In addition, in many industries, there are windows on the times at which customers may (or may not) be visited by a vehicle. For example, in the fast-food industry, most establishments do not want deliveries during the noon rush hour. Employees of the establishment are frequently enlisted to assist in unloading the vehicle. During the noon peak period, all employees are needed to prepare and serve food. Also, large delivery vehicles take up parking spaces that could be used by patrons.

A second extension involves incorporating plant-to-warehouse flows (as discussed in Section 8.5) and other multiechelon distribution system designs. Finally, in many cases, it is important to include warehouse capacities as well as vehicle capacities.

In the remainder of this section, we describe and formulate a somewhat different location/routing problem that arises in the newspaper delivery business. Newspapers are among the most economically perishable commodities produced for day-to-day life. There is almost no value to yesterday's paper today. In fact, marketing managers within the industry suggest that if home subscription papers are not delivered by early morning (when commuters begin to leave their homes), subscription rates will fall dramatically. In many areas this means that papers need to be delivered by 6:00 A.M. At the other end of the production/distribution process are a set of pressures from the editorial staff to start the press runs as late as possible so that late-breaking news and sports scores can be incorporated in the paper. Thus, newspapers must deal with pressures to start the production process (the presses) as late as possible and to deliver the papers as early as possible. In addition, editorial and advertising departments are lobbying for more and more zoned products, or newspapers tailored to specific regions and communities of a city. This, too, complicates the production and distribution problem because printing different editions of a paper often necessitates lengthy press setups which eat into the already limited window during which production and distribution must take place. To put the magnitude of this problem into context, a major metropolitan newspaper in a large North American city (e.g., Chicago) may produce roughly a dozen editions daily and close to a total of 500,000 papers for home delivery between about midnight and 6:00 A.M. Clearly, this is a major logistical undertaking.

Figure 8.11 is a schematic of a typical distribution system used by a major urban newspaper. As presses are very expensive, newspapers are often produced at a single plant. From there, they are typically distributed in truckload shipments to distribution centers. At the centers, additional processing, such as the insertion of nontimely feature or advertising material, may take place. Papers are then trucked to drop-off points in less-than-truckload shipments. At the drop-off points, local delivery personnel pick up the papers for final distribution to homes and businesses.

One of the key issues associated with this process is that of determining the optimal number and location of the distribution centers. Below, we formulate a model that incorporates (i) the plant-to-distribution-center truck-

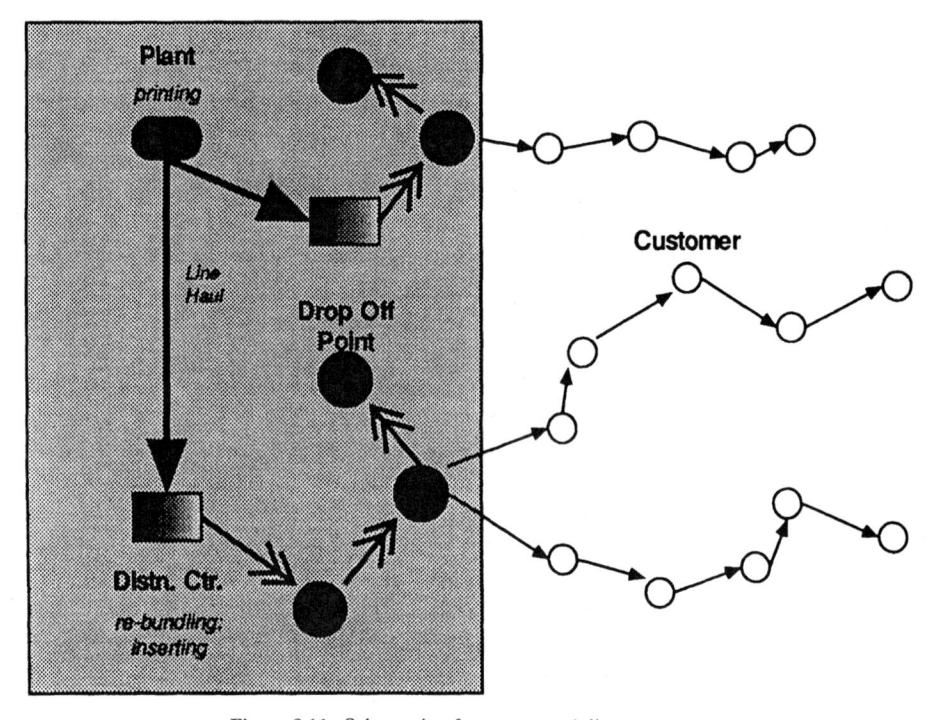

Figure 8.11. Schematic of newspaper delivery system.

load costs, (ii) the fixed costs associated with the selection of particular distribution centers, and (iii) the routing costs (including fixed vehicle costs) associated with the paths connecting distribution centers and drop-off points. (We use the term *path* instead of *route* to emphasize that vehicles depart from a distribution center and visit drop-off points. They do not need to return to the distribution center and we are not concerned with how they are utilized after visiting the final drop-off point.) In addition, we embed within the model a notion of coverage in that the total time from the plant to a drop-off point must not exceed some value. This value may be specific to the drop-off point in question reflecting the fact that some carriers will have longer routes (and so must receive their papers earlier) than others and some communities may be more sensitive to early delivery times than are others.

In formulation (8.16a)–(8.16l), routes were found endogenously. In this formulation, we will adopt a somewhat different approach. Here we will (at least theoretically) enumerate all of the possible paths. A path will consist of a sequence of drop-off points. In particular, the initial drop-off point of a path (as well as the full sequence of drop-off points) will be given for each possible path. We will than select distribution centers and paths and we will assign the selected paths to distribution centers so as to minimize the total cost while ensuring that all of the drop-off sites are covered by exactly one of

Location/Routing Problems

the selected paths. To formulate the model in this way, we define the following notation:

Inputs

f_j = daily equivalent of the fixed cost of locating a distribution center at candidate site j plus the daily cost of the truckload shipment from the plant to a distribution center at candidate site j (including the fixed cost of the vehicle plus the distance-related operating costs)

c_{jk} = cost of assigning path k to a distribution center at candidate site j
= cost of going from candidate site j to the first drop-off site on path k plus the cost of traversing path k, provided all of the drop-off points on path k can be served from a distribution center at candidate site j by their respective deadlines; ∞ otherwise

$a_{ik} = \begin{cases} 1 & \text{if drop-off point } i \text{ is on route } k \\ 0 & \text{if not} \end{cases}$

Decision Variables

$X_j = \begin{cases} 1 & \text{if a distribution center is located at candidate site } j \\ 0 & \text{if not} \end{cases}$

$Y_{jk} = \begin{cases} 1 & \text{if route } k \text{ is assigned to a distribution center at candidate site } j \\ 0 & \text{if not} \end{cases}$

With this notation, one variant of the newspaper distribution center facility location problem may be formulated as follows:

$$\text{MINIMIZE} \quad \sum_j f_j X_j + \sum_j \sum_k c_{jk} Y_{jk} \quad (8.17a)$$

$$\text{SUBJECT TO:} \quad \sum_j \sum_k a_{ik} Y_{jk} = 1 \quad \forall\, i \quad (8.17b)$$

$$Y_{jk} \le X_j \quad \forall\, j, k \quad (8.17c)$$

$$X_j = 0, 1 \quad \forall\, j \quad (8.17d)$$

$$Y_{jk} = 0, 1 \quad \forall\, j, k \quad (8.17e)$$

The objective function (8.17a) minimizes the total cost. As indicated above, a large number of cost components can be embedded in this cost function including the fixed costs of locating a distribution center at a candidate site, all of the plant-to-distribution-center transport-related costs, and the fixed and mileage-related distribution-center-to-drop-off routing costs. Constraints (8.17b) stipulate that each drop-off point i be served by exactly one of the selected paths. This constraint is very similar to a coverage constraint. In addition, it is similar to the constraints found in the P-median and fixed charge facility location models that stipulate that each demand node be assigned to a facility. Constraints (8.17c) state that path k cannot be assigned to distribution center j unless we select distribution center j. Constraints (8.17d) and (8.17e) are standard integrality constraints.

This model is clearly similar in structure to the uncapacitated fixed charge facility location model discussed in Chapter 7, despite the fact that it encompasses a large number of additional issues. These issues include plant-to-distribution-center travel, drop-off points being served on paths from the distribution center instead of being served in an out-and-back style, and constraints on the total time between departure from the plant and arrival at the drop-off point. As before, however, there is a problem with this formulation. In this case, the number of possible paths can potentially be very large. However, the tight time constraints that operate in the newspaper delivery context make it unlikely that long paths (that visit many drop-off points) will be feasible. Such long paths (or the assignment of short paths to remote distribution centers from which it is impossible to satisfy the delivery deadlines) can be eliminated a priori. Variables (and columns) corresponding to these paths and/or path distribution center assignments need not be included in the formulation as they will enter with infinite cost ($c_{jk} = \infty$). This should make this model tractable for moderate-sized problems using column generation techniques.[16] Alternatively, a number of heuristic approaches can be used in attacking this and related problems.[17]

From the discussion of both the general model (8.16a)–(8.16l) and the newspaper production/distribution problem formulation (8.17a)–(8.17e), it should be clear that location/routing problems are exceptionally difficult problems to solve. Before attacking such problems, we need to think carefully about whether the two problems—facility location and vehicle routing—really need to be solved jointly. On the one hand, it should be clear that modeling the routing component of the problem using the more traditional out-and-back structure may introduce significant errors in the estimation of the routing costs. These errors may, in turn, lead to suboptimal facility locations. On the other hand, facility location decisions are generally viewed as strategic decisions that can only be changed over a very long time horizon. Thus, no firm is likely to relocate its warehouses on a monthly basis, let alone a daily basis. At most, a few warehouses will be opened each year while a few others may be closed. Vehicle routing decisions, however, are tactical and often operational decisions that can be changed on relatively little notice.

---

[16] Column generation techniques are beyond the scope of this text. They have been used extensively in such areas as airline crew scheduling and in selected vehicle routing applications.
[17] We note that while the integer programming formulation presented above—model (8.17a)–(8.17e)—is similar in structure to the uncapacitated fixed charge facility location problem, the linear programming relaxation of the problem may be very weak (Berger, 1994; Berger, Coullard, and Daskin, 1995). Among other things, this suggest that additional constraints may be needed to strengthen the linear programming relaxation. This is an area of ongoing research.

8.7 HUB LOCATION PROBLEMS

In all of the models we have discussed so far, service could be performed by or at a facility. Thus, educational services could be provided at a school, emergency medical services could be provided at an emergency scene by ambulances based at different locations, and postal services could be delivered by post offices of different types. In this section we outline problems in which services are of a different type. In this section services involve the movement of people, goods, or information between an origin location and a destination location. *Each origin/destination pair represents a different service* that needs to be provided. Thus, packages moving from Los Angeles to Chicago are not interchangeable with packages moving from Atlanta to New Orleans.

If we have N nodes, each of which can be an origin or a destination, it is clear that we have $N(N-1)$ origin/destination pairs in a fully connected network in which each node is connected directly to each other node.[18] Figure 8.12 illustrates such a network for the $N = 6$ case. For such a network, if we have a transportation service that can serve five origin/destination pairs each day for each vehicle, then with 18 vehicles, we can serve exactly 10 nodes with daily service.

If, on the other hand, we designate one node as a hub and connect all other nodes to that node, then we need only $2(N-1)$ connections to provide service to all origin/destination pairs. Figure 8.13 illustrates such a network for the $N = 6$ case. Again, if each vehicle can serve five origin/destination pairs and if we have 18 vehicles, we can now serve 46 cities with this network topology.[19] In short, with fixed transport resources, we can serve many more cities with a hub-and-spoke network than we can with a fully connected network. Alternatively, we can serve the same number of cities but do so with

---

[18] Note that in this accounting origin/destination pair (i,j) [e.g., (Los Angeles, Chicago)] is counted separately from origin/destination pair (j,i) [e.g., (Chicago, Los Angeles)].

[19] This argument ignores vehicle capacities. It should be clear that the volume of goods or the number of people transported on each link in the hub-and-spoke network will be considerably greater than the number transported on each link of a fully connected network. This also ignores differences in the distances between nodes as it assumes that the number of origin/destination pair trips a vehicle can service is independent of the distances between the nodes in question. This is clearly an approximation that may be valid in cases in which the time required at each node is large compared to the internodal travel time.

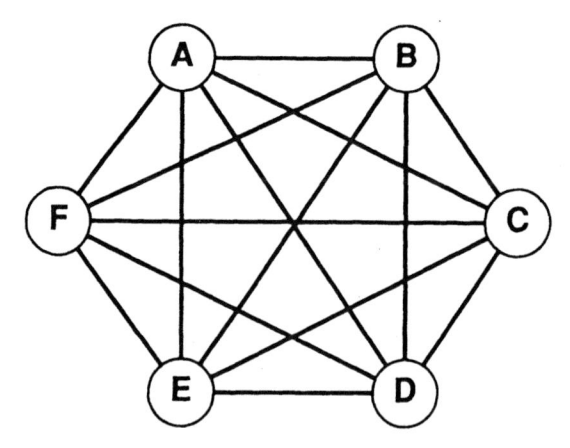

Figure 8.12. A fully connected network with 6 nodes and 30 origin/destination pairs.

a far higher frequency of service using a hub-and-spoke network. It is this sort of reasoning that led many airlines to adopt hub-and-spoke networks following deregulation of the industry in 1978.

The obvious drawback of a hub-and-spoke system is that for all origin/destination pairs except those for which either the origin or the destination is the hub, more than one transportation leg must be traversed to go from the origin to the destination. Thus, while more nodes can be connected at higher frequency using the same fleet resources, more circuitous routes must be used when cities are connected using a hub-and-spoke system than when they are directly connected. O'Kelly (1986b) proposes a simple equation along these lines for comparing the demand-weighted travel costs with the connection costs for hub-and-spoke and fully connected networks.

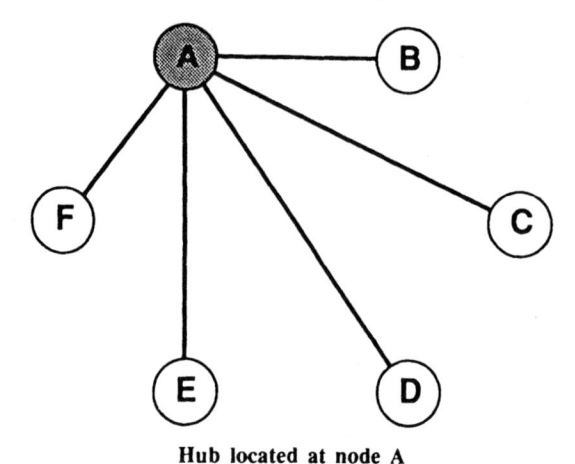

Hub located at node A

Figure 8.13. A hub-and-spoke network with 6 nodes and 30 origin/destination pairs.

Hub Location Problems

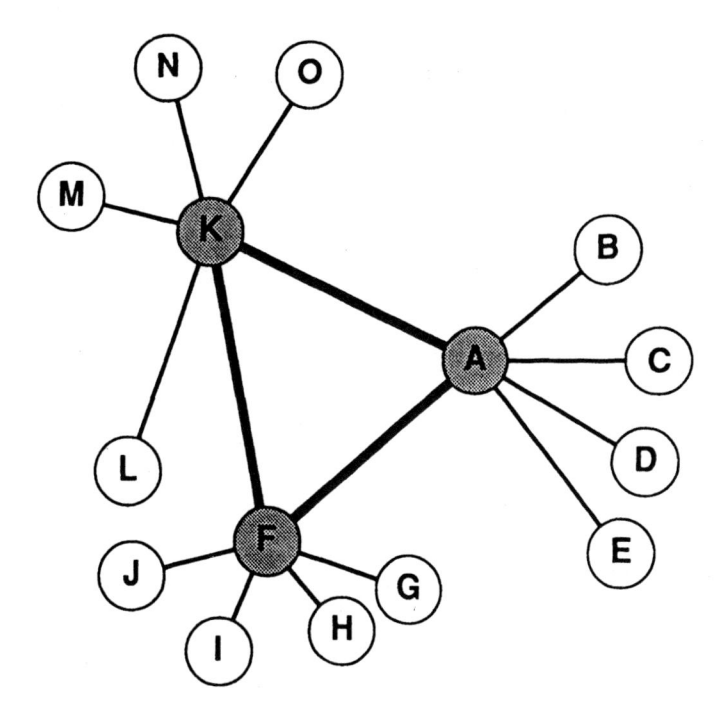

Hubs located at nodes A, F, and K

Figure 8.14. Example three-hub network.

In multiple-hub systems, another advantage of hub-and-spoke systems may come into play. In such networks, we typically assume that all hubs are connected directly to each other and that the spoke cities are connected to a single hub. Figure 8.14 illustrates such a network for a system serving 15 cities of which 3 are hubs. It should be clear that the number of passengers or tons of cargo moving on the hub-to-hub links is likely to be greater than the number of passengers or tons moving on the hub-to-spoke or spoke-to-hub links. In Figure 8.14, if the flow between each origin/destination pair is 10 units (e.g., passengers or tons), then there would be 140 units of flow each way between each spoke city and the hub to which it is connected. There would be 250 units of flow each way between each of the hubs. Note that in a fully connected system for the 15-node network shown in Figure 8.14, there would only be 10 units of flow each way on each link, again assuming that the flow between each origin/destination pair is 10. If there are economies of scale associated with the transport system being used, the operating costs per passenger-mile or per ton-mile may be significantly less when using a hub-and-spoke system than they would be when using a fully connected network. The reason for this is that the spoke/hub connections consolidate all of the flow originating (or terminating) at the spoke node. The hub-to-hub links consolidate all of the flow originating at the origin hub or any of the spoke nodes connected to it and terminating at the destination hub or any of the spoke nodes connected to it.[20]

While there are obvious advantages and disadvantages to hub-and-spoke networks, it should also be clear that the performance of the network—as measured, for example, by the total demand-weighted distance that packages must move—depends critically on the location of the hub or hubs. Having introduced some of the issues associated with hub-and-spoke networks in a qualitative manner, the remainder of this section formulates a number of location problems that arise in the context of such systems.

We begin by formulating a simple model for the location of a single hub. In this model, we minimize the total demand-weighted cost associated with connecting all nodes via a single hub as in Figure 8.13. To formulate this model, we define the following notation:

Inputs

h_{ij} = demand or flow between origin i and destination j

c_{ij} = unit cost of local (non hub to hub) movement between nodes i and j

Decision Variables

$X_j = \begin{cases} 1 & \text{if a hub is located at node } j \\ 0 & \text{if not} \end{cases}$

$Y_{ij} = \begin{cases} 1 & \text{if node } i \text{ is connected to a hub located at node } j \\ 0 & \text{if not} \end{cases}$

With this notation,[21] the single-hub location model may be formulated as follows:

MINIMIZE $\quad \sum_i \sum_j \sum_k h_{ik}(c_{ij} + c_{jk}) Y_{ij} Y_{kj}$ (8.18a)

SUBJECT TO: $\quad \sum_j X_j = 1$ (8.18b)

$\qquad\qquad Y_{ij} - X_j \leq 0 \quad \forall\, i, j$ (8.18c)

$\qquad\qquad X_j = 0, 1 \quad \forall\, j$ (8.18d)

$\qquad\qquad Y_{ij} = 0, 1 \quad \forall\, i, j$ (8.18e)

---

[20] O'Kelly (1986a) outlines this sort of argument in greater detail. He also shows that, as the discount for hub-to-hub transportation increases, or as the extent of economies of scale increases, hubs are likely to spread out further and further.

[21] Note that we do not explicitly need a location variable since we will have $Y_{ij} = 1$ when node j is the hub. For consistency with many of the models formulated in earlier chapters, however, we have introduced such a location variable, X_j. This is clearly also true of many of the earlier models including, for example, the P-median model. By separating the location and allocation variables, however, the model formulation often becomes clearer.

Hub Location Problems

The objective function (8.18a) minimizes the total cost associated with the transport through the hub. The demand or flow from origin node i to destination node j (h_{ij}) is multiplied by the cost of going from node i to a hub at node k and from there to destination node j ($c_{ik} + c_{kj}$). This total demand-weighted cost is counted if there is a hub at location k. Constraint (8.18b) stipulates that we locate only one hub. Constraints (8.18c) state that demand node i cannot be connected to a hub at j unless we locate the hub at j. Finally, constraints (8.18d) and (8.18e) are standard integrality constraints.

Note that the objective function is quadratic since it involves the product of decision variables. However, since there is only one hub, if node i is assigned to a hub at node j, then all nodes k ($k \neq i$) must be assigned to the hub at node j. Thus, we can rewrite the objective function as follows:

$$\sum_i \sum_j \sum_k h_{ik}(c_{ij} + c_{jk}) Y_{ij} Y_{kj}$$

$$= \sum_i \sum_j c_{ij} Y_{ij} \left(\sum_k h_{ik} \right) + \sum_j \sum_i c_{ji} Y_{ij} \left(\sum_k h_{ki} \right)$$

$$= \sum_i \sum_j c_{ij} Y_{ij} (O_i + D_i) \tag{8.19}$$

The first equation results from just this observation. The first term of the second line of (8.19) simply states that if node i is connected to a hub at node j, then the flow that incurs the cost of this connection (c_{ij}) is the total flow out of node i. The second term is similar, but it applies to the total flow into node i if node i is connected to a hub at j. The total flow out of node i is denoted by O_i, the total flow originating at node i (i.e., $O_i = \Sigma_k h_{ik}$). The total flow into node i is denoted by D_i, the total flow destined for node i (i.e., $D_i = \Sigma_k h_{ki}$).

By transforming objective function (8.18a) into (8.19), we can find the optimal 1-hub location with this objective function by total enumeration in $O(N^2)$ time.

The multihub extension of the problem of minimizing the total demand-weighted transport cost may also be formulated using the following additional notation:

Inputs

α = discount factor for line-haul movement between hubs ($0 \leq \alpha < 1$)
P = the number of hubs to locate

This variant of the multihub location problem may now be formulated as follows:

MINIMIZE
$$\sum_i \sum_k c_{ik} Y_{ik} \left(\sum_j h_{ij} \right) + \sum_k \sum_i c_{ki} Y_{ik} \left(\sum_j h_{ji} \right)$$
$$+ \alpha \sum_i \sum_j \sum_k \sum_m h_{ij} c_{km} Y_{ik} Y_{jm} \tag{8.20a}$$

SUBJECT TO: $\quad \sum_j Y_{ij} = 1 \quad \forall \, i \quad\quad\quad (8.20b)$

$$\sum_j X_j = P \quad\quad\quad (8.20c)$$

$$Y_{ij} - X_j \leq 0 \quad \forall \, i, j \quad\quad\quad (8.20d)$$

$$X_j = 0, 1 \quad \forall \, j \quad\quad\quad (8.20e)$$

$$Y_{ij} = 0, 1 \quad \forall \, i, j \quad\quad\quad (8.20f)$$

The objective function (8.20a) minimizes the total cost associated with the P hub locations and the assignment of nodes to the hubs. The first term is the cost of connecting all trips originating at node i to the hub k to which node i is attached. This is analogous to the first term of (8.19). The second term is the cost of connecting all trips destined for node i to the hub k to which node i is attached. This is analogous to the second term of (8.19). The third term is the hub-to-hub connection cost. It counts the total flow from node i to node j assuming nodes i and j are connected to different hubs. (If nodes i and j are connected to the same hub k, then $c_{kk} = 0$ will set the contribution of these nodes i and j to 0 in the third term.) Constraints (8.20b) ensure that each node i is assigned to exactly one hub. Constraints (8.20c) through (8.20f) are identical to constraints (8.18b) through (8.18e) except that we require P hubs to be selected in (8.20c) as opposed to the one hub that is selected in (8.18b). With the exception of the use of the strong form of constraints (8.20d) as opposed to the weaker more aggregate form and the explicit inclusion of the location variables (X_j), this formulation is essentially identical to that proposed by O'Kelly (1987). Constraints (8.20b) through (8.20f) are identical to the constraints of the P-median problem. Since the objective function involves the minimization of the demand-weighted total distance, using the terminology suggested by Campbell (1994), we refer to this model as the *P-hub median problem*.

The assignment of demand nodes to facilities is trivial in the P-median problem; each demand node is assigned to the nearest open facility. In the P-hub median problem, however, the assignment of demands is not as easy. It may not be optimal to assign demand nodes to the nearest hub. To see qualitatively why this is so, consider again Figure 8.14. In this figure, node L is assigned to a hub at node K despite the fact that L is closer to the hub located at F. This may be an optimal assignment if node L interacts more strongly with nodes K, M, N, and O than it does with nodes F through J.[22]

---

[22] This statement also assumes that the interhub discount term (α) is not too small.

Hub Location Problems

In practice, for many problems, the number of hubs to be selected is likely to be relatively small.[23] Recognizing this, O'Kelly (1987) proposed two different heuristics for the solution of this problem. Both heuristics involve enumerating all possible P hub locations from among the candidate hub sites. In HEUR1, O'Kelly assigns each demand node to the nearest hub. Clearly, if $\alpha = 0$ in objective function (8.20a), the nonlinear third term would be 0 and the problem would reduce to a P-median problem. In that case, this heuristic would be optimal. In HEUR2, for each configuration of hubs, O'Kelly evaluates all possible ways of assigning nonhub nodes to their closest and second closest hubs. There are

$$\binom{N}{P} = \frac{N!}{P!(N-P)!}$$

ways of selecting P hubs from N nodes. For each of these, HEUR2 evaluates all of the 2^{N-P} possible ways of assigning the nonhub nodes to the closest and second closest hubs. Clearly, the execution time of HEUR2 grows quickly with the number of nodes. O'Kelly identifies cases in which a second closest assignment does, in fact, reduce the total cost. However, for cases in which there are significant interhub discounts ($\alpha < 0.5$), O'Kelly finds that HEUR1 does quite well.

The P-hub median model may be formulated as an integer *linear* programming problem if we introduce additional decision variables. Campbell (1994) formulates this model in this way. Throughout this section we will use many of the concepts presented in Campbell (1994), though a number of changes have been made in the formulations. We begin by defining the following additional notation:

Inputs

c_{ij}^{km} = unit cost of travel between origin i and destination j when going via hubs at nodes k and m
    $= c_{ik} + \alpha c_{km} + c_{mj}$
f_k = fixed cost of locating a hub at candidate site k

Decision Variables

$$Z_{ij}^{km} = \begin{cases} 1 & \text{if flow from origin } i \text{ to destination } j \\ & \text{uses hubs at candidates sites } k \text{ and } m \\ 0 & \text{if not} \end{cases}$$

---

[23] Most of the major domestic airlines use between one and four major hubs. Truck companies, on the other hand, may use far more hubs.

With this notation, we can formulate the following variant of the P-hub median location problem as follows:[24]

MINIMIZE
$$\sum_i \sum_j \sum_k \sum_m c_{ij}^{km} h_{ij} Z_{ij}^{km} \qquad (8.21a)$$

SUBJECT TO:
$$\sum_k X_k = P \qquad (8.21b)$$

$$\sum_k \sum_m Z_{ij}^{km} = 1 \quad \forall\, i, j \qquad (8.21c)$$

$$Z_{ij}^{km} \leq X_k \quad \forall\, i, j, k, m \qquad (8.21d)$$

$$Z_{ij}^{km} \leq X_m \quad \forall\, i, j, k, m \qquad (8.21e)$$

$$X_k = 0, 1 \quad \forall\, k \qquad (8.21f)$$

$$Z_{ij}^{km} \geq 0 \quad \forall\, i, j, k, m \qquad (8.21g)$$

The objective function (8.21a) minimizes the demand-weighted total travel cost. Constraint (8.21b) stipulates that exactly P hubs should be located. Constraints (8.21c) state that each origin/destination pair (i, j) must be assigned to exactly one hub pair. Note that since k may equal m in constraint (8.21c), we do not preclude the possibility of flow between origin i and destination j going through only a single hub. In that case, $\alpha c_{km} = \alpha c_{kk} = 0$. Constraints (8.21d) and (8.21e) stipulate that flow from origin i to destination j cannot be assigned to a hub at location k [(8.21d)] or m [(8.21e)] unless a hub is located at these candidate sites. Finally, constraints (8.21f) and (8.21g) are the standard integrality and nonnegativity constraints, respectively. As before, the similarity between this model and the P-median model should be clear.

One of the key difficulties associated with hub location models should also be evident from this formulation. The number of assignment variables (Z_{ij}^{km}) can be extremely large. In fact, if every origin or destination node is a candidate hub, then there can be $O(N^4)$ such variables. For a relatively small problem with 32 origins and destinations, we would have over one million such decision variables. In short, the size of these problems grows very quickly with the number of nodes in the problem unless some a priori means of eliminating nodes as candidate hub locations is used. The use of quadratic

---

[24] As discussed below, this formulation differs from that of (8.20a) through (8.20f) in that it allows spoke nodes to be assigned to multiple hubs. The inclusion of constraints (8.23e) which parallel constraints (8.20b) forces each demand node to be assigned to a single hub.

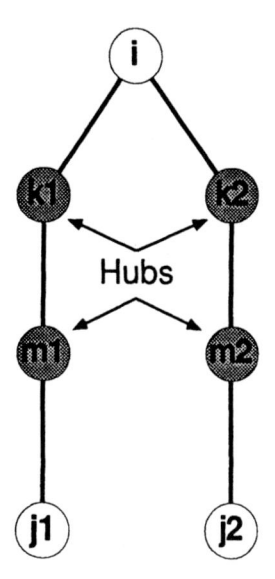

Figure 8.15. Schematic of multiple spoke assignment.

formulations [e.g., objective function (8.20a)] reduces the number of decision variables dramatically but does not make solution of the problem any easier.

A number of important extensions of the P-hub median location problem can be formulated. The fixed costs of hub locations can be modeled by including a fixed cost term in the objective function and by eliminating constraint (8.21b) on the number of hubs to be located. [See O'Kelly (1992).] The new objective function becomes

MINIMIZE $$\sum_i \sum_j \sum_k \sum_m c_{ij}^{km} h_{ij} Z_{ij}^{km} + \beta \sum_k f_k X_k \qquad (8.22)$$

where β is a weight on the capital or fixed costs to allow exploration of the tradeoff between capital costs and transport (or operating) costs. This is important since these costs are sometimes borne by different groups.[25]

Formulation (8.21a)–(8.21g) allows each of the spoke nodes to be assigned to multiple hubs. Thus, for example, the assignment shown in Figure 8.15 is entirely possible as a result of solving this model. In this figure two origin/destination flows are shown: origin i to destination $j1$ and origin i to destination $j2$. Origin i is assigned to two different hubs: $k1$ and $k2$. In many cases, it is desirable (for operational reasons) to have each of the spoke nodes assigned to a single hub. This might result in one or the other of the assignments shown in Figure 8.16. In both of the assignments shown, origin i is assigned to only one hub.

---

[25] The reader is referred to Section 8.8 for additional discussion of cases in which costs are borne by different groups.

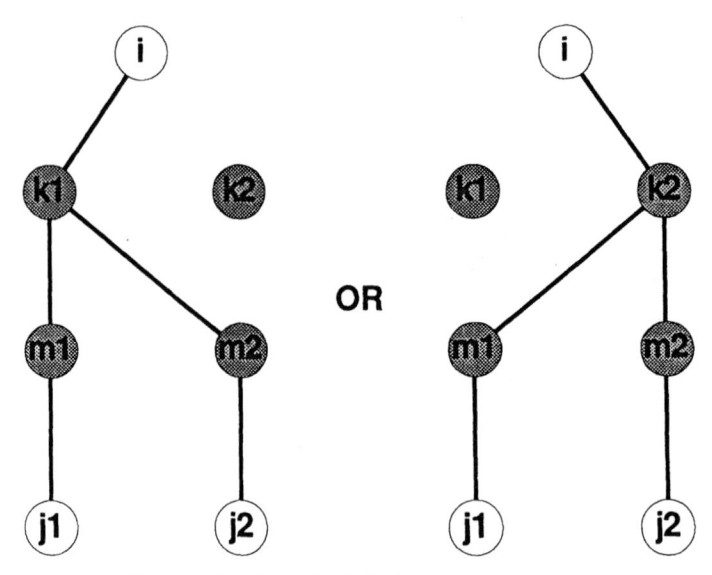

Figure 8.16. Schematic of single spoke assignment.

A single spoke assignment P-hub median location problem, analogous to (8.20a) through (8.20f), can be formulated as follows:

$$\text{MINIMIZE} \quad \sum_i \sum_j \sum_k \sum_m c_{ij}^{km} h_{ij} Z_{ij}^{km} \tag{8.23a}$$

$$\text{SUBJECT TO:} \quad \sum_k X_k = P \tag{8.23b}$$

$$\sum_k \sum_m Z_{ij}^{km} = 1 \quad \forall\, i, j \tag{8.23c}$$

$$Y_{ik} \leq X_k \quad \forall\, i, k \tag{8.23d}$$

$$\sum_k Y_{ik} = 1 \quad \forall\, i \tag{8.23e}$$

$$Y_{ik} + Y_{jm} - 2Z_{ij}^{km} \geq 0 \quad \forall\, i, j, k, m \tag{8.23f}$$

$$X_k = 0, 1 \quad \forall\, k \tag{8.23g}$$

$$Y_{ik} = 0, 1 \quad \forall\, i, k \tag{8.23h}$$

$$Z_{ij}^{km} = 0, 1 \quad \forall\, i, j, k, m \tag{8.23i}$$

The objective function and the first two constraints are identical to those of formulation (8.21a)–(8.21g). Constraints (8.21d) and (8.21e) are replaced by constraints (8.23d), (8.23e), and (8.23f). Constraints (8.23d) state that

Hub Location Problems

spoke i cannot be assigned to a hub at location k unless we locate a hub at location k. Constraints (8.23e) are the key constraints that stipulate that each spoke node i be assigned to exactly one hub. Constraints (8.23f) state that the flow from origin i to destination j cannot be routed through hubs at sites k and m unless spoke node i is assigned to a hub at k and spoke node m is assigned to a hub at j. Constraints (8.23g)–(8.23i) are standard integrality constraints.

Instead of forcing each spoke node to be assigned to a single hub, we may want to stipulate that the flow along any spoke/hub connection exceed some minimum value. Alternatively, we could incorporate a fixed cost for connecting a spoke to a hub. This, too, would discourage, but not preclude, multiple spoke-to-hub assignments. Campbell (1994) shows how these extensions can be formulated. Fixed charges for connecting spoke and hub nodes can be incorporated in the problem by adding the following term to the objective function, where g_{ik} is the fixed cost of connecting spoke node i to a hub at candidate location k:

$$\sum_i \sum_k g_{ik} Y_{ik} \quad (8.24)$$

To ensure that the flow on any spoke/hub connection is at least some minimum value L_{ik}, we need to introduce constraints of the following form:

$$\sum_m \sum_j h_{ij} Z_{ij}^{km} + \sum_p \sum_s h_{pi} Z_{pi}^{sk} \geq L_{ik} Y_{ik} \quad \forall\, i, k \quad (8.25)$$

The first term of (8.25) is the flow from spoke node i to hub k and then from there to any hub/destination pair. The second term is the flow from any origin/hub pair to hub k and then from there to destination i. The sum of these two flows—the flow from spoke i to hub k and the flow from hub k to spoke i—must be greater than or equal to the minimum allowable flow between i and k if spoke node i is connected to hub node k. The complete formulation of this extension of the problem is left as an exercise for the reader. Note that some constraint must be included to ensure that the assignment variables (Z_{ij}^{km}) are 0 if either spoke node i is not assigned to a hub at k or spoke node j is not assigned to a hub at m.

Hub capacities can also be incorporated in the model. If γ_k is the capacity of a hub at candidate site k, then we simply need to add constraints of the following form:

$$\sum_m \sum_i \sum_j h_{ij} Z_{ij}^{km} + \sum_{\substack{s \\ s \neq k}} \sum_i \sum_j h_{ij} Z_{ij}^{sk} \leq \gamma_k X_k \quad \forall\, k \quad (8.26)$$

The first term of (8.26) counts the total number of units of flow that use hub k as the first of the two hubs. The second term counts the total flow that uses hub k as the second hub. To avoid double-counting flow from i to j that might use *only* hub k, the summation over hubs in the second term excludes hub k.

Clearly, the extensions outlined above—fixed hub location costs, single hub/spoke assignments, fixed costs for hub/spoke assignments or lower bounds on hub/spoke flows, and hub capacities—may be combined in a variety of interesting ways to reflect many different problem contexts.

The models formulated above deal with the overall level of service in that they minimize the demand-weighted total travel cost between origin/destination pairs. In short, they are median-like formulations. As is the case in the traditional *P*-median problem, the use of such a formulation may lead to some origin/destination pairs (demands) receiving very poor service. This leads us to consider the adoption of other objective functions. As before, a center-like objective function would minimize the worst service over all origin/destination pairs.[26] In particular, let us define the following additional notation:

Decision Variables

R_{ij} = ratio of the travel cost under a hub-and-spoke system to the cost of a direct connection from origin i to destination j

R_{max} = largest value of R_{ij}
  = $\max_{i,j} \{R_{ij}\}$

With this additional notation, we can formulate a center-like hub location model in which we minimize the largest ratio of the hub/spoke travel cost to the direct travel cost as follows:

MINIMIZE	R_{max}		(8.27a)
SUBJECT TO:	$\sum_k X_k = P$		(8.27b)
	$\sum_k \sum_m Z_{ij}^{km} = 1$	$\forall\, i, j$	(8.27c)
	$Z_{ij}^{km} \leq X_k$	$\forall\, i, j, k, m$	(8.27d)
	$Z_{ij}^{km} \leq X_m$	$\forall\, i, j, k, m$	(8.27e)
	$R_{max} \geq R_{ij}$	$\forall\, i, j$	(8.27f)
	$c_{ij} R_{ij} = \sum_k \sum_m c_{ij}^{km} Z_{ij}^{km}$	$\forall\, i, j$	(8.27g)
	$X_k = 0, 1$	$\forall\, k$	(8.27h)
	$Z_{ij}^{km} \geq 0$	$\forall\, i, j, k, m$	(8.27i)

[26] Campbell (1994) identifies a number of different ways of thinking about center and covering objectives in the context of hub location models. The reader is referred to this paper for more details of these concepts. O'Kelly and Miller (1991) compare a number of solution strategies for the 1-hub center problem in which the hub can be located anywhere in the plane.

Hub Location Problems

The objective function (8.27a) minimizes the maximum ratio of the hub-and-spoke travel cost to the direct travel cost. Constraints (8.27f) define the maximum in terms of the origin/destination specific ratios. Constraints (8.27g) define the origin/destination specific service ratios (R_{ij}) in terms of the assignment variables (Z_{ij}^{km}). Note that $\sum_k \sum_m c_{ij}^{km} Z_{ij}^{km}$—the right-hand side of (8.27g)—is the travel cost between origin i and destination j under the hub-and-spoke system. c_{ij} on the left-hand side of (8.27g) is simply the direct travel cost between i and j. The remaining constraints in this formulation are identical to their counterparts in the P-hub median model, formulation (8.21a)–(8.21g).

Since R_{ij} may be viewed as a proxy for the level of service between origin i and destination j, we can imagine a large number of variants of this basic model. To introduce one final hub location model, we can consider the following covering-based formulation that uses R_{ij} as a level of service variable. An origin/destination pair will not be considered covered unless R_{ij} is less than some critical value R_{ij}^c which may depend on the origin/destination pair. Specifically, we introduce the following additional notation:

Inputs

$$a_{ij}^{km} = \begin{cases} 1 & \text{if hub pair } (k, m) \text{ can cover origin/destination pair } (i, j) \\ 0 & \text{if not} \end{cases}$$

For example, we might have

$$a_{ij}^{km} = \begin{cases} 1 & \text{if } \dfrac{c_{ij}^{km}}{c_{ij}} \leq R_{ij}^c \\ 0 & \text{if not} \end{cases}$$

where R_{ij}^c is the maximum or critical service ratio for origin/destination pair (i, j).

With this notation, we can consider the following covering-like hub location model:

MINIMIZE $\quad \sum_k f_k X_k \quad$ (8.28a)

SUBJECT TO: $\quad \sum_k \sum_m Z_{ij}^{km} = 1 \quad \forall\, i, j \quad$ (8.28b)

$$Z_{ij}^{km} \leq a_{ij}^{km} X_k \quad \forall\, i, j, k, m \quad (8.28c)$$

$$Z_{ij}^{km} \leq a_{ij}^{km} X_m \quad \forall\, i, j, k, m \quad (8.28d)$$

$$X_k = 0, 1 \quad \forall\, k \quad (8.28e)$$

$$Z_{ij}^{km} \geq 0 \quad \forall\, i, j, k, m \quad (8.28f)$$

The objective function (8.28a) minimizes the total cost of all of the hubs that are located. Clearly, if we set all of the fixed hub location costs (f_k) to 1, the objective function can be made to minimize the number of hubs that are located. Constraint (8.28b) requires that each origin/destination pair be assigned to exactly one hub pair. This constraint is identical to constraint (8.21c) in the P-hub median location formulation. Constraints (8.28c) and (8.28d) stipulate that origin/destination pair (i, j) cannot be assigned to hub pair (k, m) unless (i) we locate hubs at candidate sites k [(8.28c)] and m [(8.28d)] and (ii) the level of service to origin/destination pair (i, j) when routed through hub pair (k, m) is better than the critical value for the origin/destination pair (i.e., $a_{ij}^{km} = 1$). Constraints (8.28e) and (8.28f) are integrality and nonnegativity constraints, respectively. This formulation allows nonhub cities to be assigned to multiple hubs. To preclude this, constraints (8.23d), (8.23e), (8.23f), and (8.23h) must be added to this formulation.

Note that we can eliminate any constraints of the form of (8.28c) and (8.28d) whenever $a_{ij}^{km} = 0$. We can also eliminate the corresponding assignment variable (Z_{ij}^{km}) from the formulation, since these constraints require that the variable be equal to 0.

A wide range of important extensions of these models can be formulated. First, fixed fleet and frequency of service planning can be included in hub location models. The lower bounds on the flows between spokes and hubs [constraints (8.25)] and the inclusion of fixed costs associated with spoke/hub connections [expression (8.24)] are proxies for these concerns. Second, the models outlined above assume that flows travel through only a single hub or a pair of hubs. In many applications (e.g., LTL trucking, communication networks, and computer networks), flows travel through more than two hubs. Third, just as we formulated models in which facilities interacted in a variety of ways (Section 8.4), we can extend the basic models outlined above to allow hubs to interact as well. Fourth, we can envision models in which some form of backup service is required. For example, instead of forcing spoke nodes to be assigned to only one hub, we may force each spoke node to be assigned to (or covered by in some sense) more than one hub. Models of this form may be particularly important if the reliability of the system is an issue, as such models allow multiple paths between origin/destination pairs. Again, computer, telecommunication, and power generation and transmission networks are cases in which network redundancy may be particularly important. Finally, spoke nodes need not be connected directly to the hubs. Instead, they may be connected via routes or paths leading to models similar to those outlined in Section 8.6. Both Flynn and Ratick (1988) and Kuby and Gray (1993) discuss models of this sort.

As suggested above, solution of hub location problems is difficult. The large number of decision variables (Z_{ij}^{km}) that enter most, if not all, of the models leads to very large formulations even for instances with relatively few nodes. Very small problems can often be solved using standard integer programming packages. For larger problems, heuristics (such as O'Kelly's HEUR1 and HEUR2 outlined above) must be used.

8.8 DISPERSION MODELS AND MODELS FOR THE LOCATION OF UNDESIRABLE FACILITIES

In all of the models discussed so far, "closer was better." It was always better to locate facilities close to demands or close to each other. In some cases, closer is not necessarily better. In this section we outline two classes of models in which "farther is better," at least in some sense. In the first class of models, known as *dispersion models*, we want to locate a given number of facilities so that the facilities are as far from each other as possible. We are not explicitly concerned with the location of the facilities relative to demand nodes in this class of problems. Such problems arise in a number of contexts. For example, in locating silos for nuclear weapons, it is desirable to have the facilities (silos) as far away from each other as possible so that it is difficult for your enemies to destroy all your weapons with only a few of their own. In a more peaceful context, when locating franchise outlets, it is often desirable to spread them out as much as possible to minimize the extent to which stores of the same company compete with each other.

Models for the location of *noxious* or *obnoxious facilities* represent the second class of problems that we address in this section.[27] In these models, we want to locate facilities to maximize some measure of the distance between demand nodes and the *nearest* facility. Thus, in locating nuclear reactors, hazardous waste dumps, landfills, or solid waste repositories, we would like to find locations that are as far from population centers as possible. Note that now we are concerned with maximizing a function of the distance between demands and the facilities being located, not simply the distance between facilities. The desire to keep such facilities far away from population centers is so pervasive that it is often referred to as the NIMBY (Not In My Back Yard) phenomenon.

8.8.1 Dispersion Models

We begin by discussing the problem of dispersing facilities on a network. We assume that there is a discrete set of candidate locations at which the facilities may be located and that we want to locate P facilities to maximize

---

[27]As indicated by Erkut and Neuman (1989), a *noxious* facility is one that poses a health or welfare threat to individuals. An *obnoxious* facility is one that poses a threat to the lifestyles of those in close proximity to the facility. Following their example, we do not distinguish between these two terms since the elementary models we formulate can be applied to either type of facility. We refer to such facilities as *undesirable*.

the minimum separation between any pair of facilities. To formulate such a model, we define the following notation:

Inputs

d_{ij} = distance between candidate locations i and j
M = a very large number (This must be such that $M \geq \max_{i,j}\{d_{ij}\}$.)
P = the number of facilities to locate

Decision Variables

$$X_j = \begin{cases} 1 & \text{if we locate a facility at candidate site } j \\ 0 & \text{if not} \end{cases}$$

D = minimum separation between facilities.

With this notation, we can formulate a P-dispersion model as follows:

MAXIMIZE $\quad D \quad$ (8.29a)

SUBJECT TO: $\quad \sum_j X_j = P \quad$ (8.29b)

$$D \leq d_{ij} + (M - d_{ij}) \cdot (1 - X_i)$$
$$+ (M - d_{ij}) \cdot (1 - X_j) \quad \forall j; \forall i < j \quad (8.29c)$$

$$X_j = 0, 1 \quad \forall j \quad (8.29d)$$

The objective function (8.29a) maximizes the minimum interfacility distance. Constraint (8.29b) stipulates that we locate exactly P facilities. Constraints (8.29c) define the minimum interfacility distance in terms of the selected facility locations. Constraint (8.29c) may be rearranged to obtain

$$D + (M - d_{ij}) \cdot X_i + (M - d_{ij}) \cdot X_j \leq 2M - d_{ij}$$

If $X_i = X_j = 1$, then constraint (8.29c) reduces to $D \leq d_{ij}$. In other words, if we locate at both candidate site i and candidate site j, then the maximum distance between any pair of facilities cannot exceed the distance between sites i and j. If only one of the two sites is chosen (i.e., $X_i = 0$, $X_j = 1$ or $X_i = 1$, $X_j = 0$), then constraint (8.29c) becomes $D \leq M$. Thus, we want M to be sufficiently large so that this constraint is not binding. Hence, we set $M = \max_{i,j}\{d_{ij}\}$. Finally, if $X_i = X_j = 0$, constraint (8.29c) becomes $D \leq 2M - d_{ij}$. Again, we do not want this constraint to be binding and again

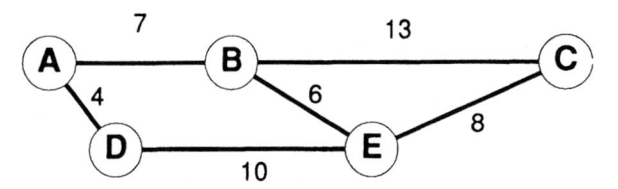

Dispersion Models and Models for the Location of Undesirable Facilities

Figure 8.17. Network for dispersion example.

setting $M = \max_{i,j}\{d_{ij}\}$ will ensure that this constraint does not constrain the value of D. Constraints (8.29d) are simple integrality constraints.

To illustrate the use of this formulation, consider the network shown in Figure 8.17. The minimum path distances for this network are given by

$$[d_{ij}] = \begin{bmatrix} 0 & 7 & 20 & 4 & 13 \\ 7 & 0 & 13 & 11 & 6 \\ 20 & 13 & 0 & 18 & 8 \\ 4 & 11 & 18 & 0 & 10 \\ 13 & 6 & 8 & 10 & 0 \end{bmatrix}$$

For this network, we can use $M = 20$. With this value, formulation (8.29a)–(8.29d) becomes

MINIMIZE D

SUBJECT TO: $X_A + X_B + X_C + X_D + X_E = P$

$D + 13X_A + 13X_B \leq 33$

$D + 0X_A + 0X_C \leq 20$

$D + 16X_A + 16X_D \leq 36$

$D + 7X_A + 7X_E \leq 27$

$D + 7X_B + 7X_C \leq 27$

$D + 9X_B + 9X_D \leq 29$

$D + 14X_B + 14X_E \leq 34$

$D + 2X_C + 2X_D \leq 22$

$D + 12X_C + 12X_E \leq 32$

$D + 10X_D + 10X_E \leq 30$

$X_A, X_B, X_C, X_D, X_E = 0, 1$

Table 8.3 Results of Solving Formulation (8.29a)–(8.29d) for the Network of Figure 8.17

P	Linear Programming Objective Function at Root Node	Iterations	Branches in Tree	Integer Objective Function	Locations
2	20.000000	11	0	20	A, C
3	17.603073	44	8	11	B, C, D
4	12.533320	24	2	6	A, B, C, E

The results of solving this problem using LINDO (Schrage, 1991)—a standard branch-and-bound code—are given in Table 8.3. As can be seen from this simple example, the linear programming bound from this formulation is not very tight. In the case of four facilities, the linear programming bound is over twice the optimal value of the objective function. Nevertheless, relatively few branches needed to be evaluated for each of the solutions. Also, the sequence of optimal facility locations does not exhibit any apparent structure. For example, the optimal solution for $P = 2$ is not a subset of the optimal solution for $P = 3$. Similarly, the optimal solution for $P = 3$ is not a subset of the optimal solution for $P = 4$. This suggests that greedy-type algorithms may not be very effective in solving this class of problems. The reader interested in dispersion models is referred to Kuby (1987) who discusses this model and related dispersion models. This paper also indicates a number of important relationships between the P-dispersion problem and the $(P - 1)$-center problem. Erkut and Neuman (1989) provide a comprehensive review and taxonomy of location models for undesirable facilities. More recently, Erkut and Verter (1994) present an annotated bibliography of hazardous materials logistics including models used in siting facilities handling undesirable materials.

8.8.2 A MAXISUM Model for the Location of Undesirable Facilities

In the case of undesirable facilities, one of the common goals that arises in siting facilities is that of maximizing some function of the distance between demand nodes and the nearest facilities. In this section we outline the formulation of a model that *maximizes* the total demand-weighted distance between demand nodes and the *nearest* of the chosen facility sites. Formulation of such models at first seems straightforward since the P-median problem is identical to this problem statement except that in the P-median problem we want to *minimize* the total demand-weighted distance. This suggests simply changing the objective function from a minimization to a maximization in the P-median formulation. We refer to this formulation as a *maxisum model* since we are maximizing the sum of the demand-weighted distances.

Dispersion Models and Models for the Location of Undesirable Facilities

Adopting this approach, we define the following notation (which is identical to that used in the P-median problem):

Inputs

h_i = demand at node i
d_{ij} = distance between demand node i and candidate site j
P = number of facilities to locate

Decision Variables

$X_j = \begin{cases} 1 & \text{if we locate at candidate site } j \\ 0 & \text{if not} \end{cases}$

$Y_{ij} = \begin{cases} 1 & \text{if demands at node } i \text{ are served by a facility at node } j \\ 0 & \text{if not} \end{cases}$

With this notation, we can *begin* to formulate a maxisum model as follows:[28]

$$\text{MAXIMIZE} \quad \sum_i \sum_j h_i d_{ij} Y_{ij} \tag{8.30a}$$

$$\text{SUBJECT TO:} \quad \sum_j Y_{ij} = 1 \quad \forall \, i \tag{8.30b}$$

$$\sum_j X_j = P \tag{8.30c}$$

$$Y_{ij} \leq X_j \quad \forall \, i, j \tag{8.30d}$$

$$X_j = 0, 1 \quad \forall \, j \tag{8.30e}$$

$$Y_{ij} = 0, 1 \quad \forall \, i, j \tag{8.30f}$$

This model is simply the P-median formulation with an objective function to be maximized instead of minimized. Unfortunately, this formulation is not valid for the maxisum problem we want to solve because it does not guarantee that demands will be assigned to the nearest facility. In fact, it does just the opposite. For any given set of facility locations, it will assign demand node i to the most remote of the facility locations in computing the demand-weighted distance. To see this, consider the network shown in

---
[28] The reader is cautioned at this point that this naive approach is *not* sufficient. As indicated below, additional constraints are needed to ensure that demand nodes are assigned to the *nearest* facility.

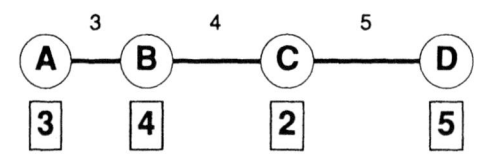

Figure 8.18. Network for maxisum example.

Figure 8.18. The solution to formulation $(8.30a)$–$(8.30f)$ for $P = 2$ is

$$X_A = X_D = 1 \quad X_B = X_C = 0$$

$$Y_{AD} = Y_{BD} = Y_{CA} = X_{DA} = 1$$
All other $Y_{ij} = 0$

Objective function $= 146$

In this solution, demands are assigned to the *farthest* facility, not the closest facility. The true objective function for facilities at nodes A and D can be computed by setting $Y_{AA} = Y_{BA} = Y_{CD} = X_{DD} = 1$ and all other $Y_{ij} = 0$. With these values, the objective function with facilities at A and D is 22. As we shall see, there is a substantially better solution to this problem when $P = 2$.

Before we present the true solution to this particular problem, however, we need to identify additional constraints that will force demand nodes to be assigned to the closest selected facility. Let us begin by considering the addition of the following constraints for node B:

$$-X_B + Y_{BB} \geq 0 \qquad (8.31a)$$

$$-X_A + Y_{BB} + Y_{BA} \geq 0 \qquad (8.31b)$$

$$-X_C + Y_{BB} + Y_{BA} + Y_{BC} \geq 0 \qquad (8.31c)$$

$$-X_D + Y_{BB} + Y_{BA} + Y_{BC} + Y_{BD} \geq 0 \qquad (8.31d)$$

The effect of these four constraints, in conjunction with $(8.30b)$ and $(8.30d)$, is to force node B to be assigned to the nearest facility. Constraint $(8.31a)$ is equivalent to $Y_{BB} - X_B \geq 0$. Thus, if a facility is located at node B ($X_B = 1$), this constraint will force node B to be assigned to that facility ($Y_{BB} = 1$); otherwise (if $X_B = 0$), constraint $(8.30d)$ will force node B not to be assigned to a facility at node B ($Y_{BB} = 0$).

Node A is the next closest site to node B. Constraint $(8.31b)$ forces node B to be assigned to either a facility at node B or to a facility at node A if a

Dispersion Models and Models for the Location of Undesirable Facilities 369

Table 8.4 Indices $[m]_i$ for the Network of Figure 8.18

Demand Node i	$m = 1$	$m = 2$	$m = 3$	$m = 4$
A	A	B	C	D
B	B	A	C	D
C	C	B	D	A
D	D	C	B	A

facility is located at node A. If a facility is also located at node B, constraint (8.31a) will have forced node B to be assigned to that facility ($Y_{BB} = 1$) and constraint (8.31b) will be satisfied. If a facility is not at B, but is at A, constraint (8.31b) will force node B to be assigned to that facility ($Y_{BA} = 1$). Constraints (8.31c) and (8.31d) operate similarly.

The key to constraints (8.31a)–(8.31d) is that we have sorted the candidate locations in terms of increasing distance from the demand node in question, node B. Let $[m]_i$ denote the index of the mth farthest candidate location from demand node i. For the network of Figure 8.18, we have the values shown in Table 8.4.

Having introduced this notation, we can now define the following constraint whose addition to formulation (8.30a)–(8.30f) will ensure that demand nodes are assigned to the closest selected facility:

$$-X_{[m]_i} + \sum_{k=1}^{m} Y_{i[k]_i} \geq 0 \quad \forall\, i;\, m = 1, \ldots, N - 1 \qquad (8.32)$$

where N is the number of nodes. Note that we do not need the Nth constraint for any demand node i, since this simply says

$$-X_{[N]_i} + \sum_{j} Y_{ij} \geq 0$$

But satisfaction of constraint (8.30b) will ensure that this constraint is also always satisfied, no matter what the value of $X_{[N]_i}$.

With the addition of this constraint, the maxisum problem for the network of Figure 8.18 becomes

MAXIMIZE

$$9Y_{AB} + 21Y_{AC} + 36Y_{AD} + 12Y_{BA} + 16Y_{BC} + 36Y_{BD}$$
$$+ 14Y_{CA} + 8Y_{CB} + 10Y_{CD} + 60Y_{DA} + 45Y_{DB} + 25Y_{DC}$$

SUBJECT TO:

$$X_A + X_B + X_C + X_D = P$$

$$Y_{AA} + Y_{AB} + Y_{AC} + Y_{AD} = 1$$

$$Y_{BA} + Y_{BB} + Y_{BC} + Y_{BD} = 1$$

$$Y_{CA} + Y_{CB} + Y_{CC} + Y_{CD} = 1$$

$$Y_{DA} + Y_{DB} + Y_{DC} + Y_{DD} = 1$$

$$Y_{AA} \leq X_A \quad Y_{BA} \leq X_A \quad Y_{CA} \leq X_A \quad Y_{DA} \leq X_A$$

$$Y_{AB} \leq X_B \quad Y_{BB} \leq X_B \quad Y_{CB} \leq X_B \quad Y_{DB} \leq X_B$$

$$Y_{AC} \leq X_C \quad Y_{BC} \leq X_C \quad Y_{CC} \leq X_C \quad Y_{DC} \leq X_C$$

$$Y_{AD} \leq X_D \quad Y_{BD} \leq X_D \quad Y_{CD} \leq X_D \quad Y_{DD} \leq X_D$$

$$-X_A + Y_{AA} \geq 0$$

$$-X_B + Y_{AA} + Y_{AB} \geq 0$$

$$-X_C + Y_{AA} + Y_{AB} + Y_{AC} \geq 0$$

$$-X_B + Y_{BB} \geq 0$$

$$-X_A + Y_{BB} + Y_{BA} \geq 0$$

$$-X_C + Y_{BB} + Y_{BA} + Y_{BC} \geq 0$$

$$-X_C + Y_{CC} \geq 0$$

$$-X_B + Y_{CC} + Y_{CB} \geq 0$$

$$-X_D + Y_{CC} + Y_{CB} + Y_{CD} \geq 0$$

$$-X_D + Y_{DD} \geq 0$$

$$-X_C + Y_{DD} + Y_{DC} \geq 0$$

$$-X_B + Y_{DD} + Y_{DC} + Y_{DB} \geq 0$$

$$X_A, X_B, X_C, X_D = 0, 1$$

All $Y_{ij} \geq 0$

Note that the assignment variables (Y_{ij}) need only be constrained to be nonnegative. We do not need to explicitly force them to be integer valued, since the remainder of the formulation will ensure that this condition is met.

Table 8.5 gives the solution to this problem for the network shown in Figure 8.18 as well as the corresponding P-median solution. It is worth noting how different the solutions are even for this small example. Only when we locate at every candidate site ($P = 4$) are the solutions to the maxisum and P-median model the same in this case.

Table 8.5 Solutions of Maxisum and *P*-median Models for the Network of Figure 8.18

	Maxisum Model		*P*-Median Model	
P	Locations	Objective Function	Locations	Objective Function
1	*A*	86	*B*	62
2	*A, B*	53	*B, D*	17
3	*A, B, C*	25	*A, B, D*	8
4	*A, B, C, D*	0	*A, B, C, D*	0

While we have discussed the location of undesirable facilities in terms of the maxisum model, it is important to note that this is only one of many different approaches that can be adopted in modeling the location of undesirable facilities.[29] In general, such problems are multiobjective. For example, in addition to maximizing the demand-weighted total distance between demands and the nearest facilities, we may also want to maximize the minimum demand-weighted distance between a demand node and the nearest facility. This would be a *maximin objective*. Erkut and Neuman (1989, p. 287) emphasize the need for multiobjective approaches to the siting of undesirable facilities when they state, "The multiple constituency, multiobjective nature of the problem severely limits the usefulness of single objective models." They go on to suggest (p. 289), "Current models can be used to generate a small number of candidate sites, but the final selection of a site is a complex problem and should be approached using multi-objective decision making tools."

While we want many undesirable facilities far from demand centers, it is often these demand centers that are served by the undesirable facilities being sited. For example, while no one wants a landfill in his or her backyard, large population centers generate much of the waste that needs to be transported to landfills. Locating such facilities too far from population centers increases the operating costs associated with waste disposal. Thus, a conflicting objective would be that of minimizing the transport costs. In many cases, this translates into a *P*-median objective. As indicated in Table 8.5, this objective and the maxisum objective are in clear conflict. The optimal locations resulting from the two models differ significantly.

Routing is also an important issue in locating undesirable facilities. Hazardous materials often travel along paths other than the shortest distance or cost path to avoid large concentrations of population. List and Mirchandani (1991) and ReVelle, Cohon, and Shobrys (1991) develop joint location/routing models for hazardous wastes.

---

[29] The reader is referred to Erkut and Neuman (1988) for a survey of models that have been used in locating obnoxious facilities.

In many cases, we need to consider the fixed costs associated with the location of undesirable facilities. These costs can often be very large as special measures are needed to handle and contain the hazardous materials.

Finally, equity is a major concern in the location of undesirable facilities. For example, no single community wants to be the sole recipient of all of a region's waste. If all communities can be shown to be (nearly) equally affected by the facilities being located, local opposition may be muted since no one is being unduly hurt. One way of modeling this sort of concern is through covering constraints. We may want to ensure that all communities have at least one of the undesirable facilities within some "dissatisfaction" distance. While spreading out undesirable facilities in this way may sound counterintuitive and counterproductive since we want these facilities to be located far from population centers, evening out the negative impacts of the location of these facilities *may* be one way of ensuring a solution that can be implemented and that is not likely to be blocked by residents of a few affected communities.

In a widely cited paper, Ratick and White (1988) propose a multiobjective model for the location of undesirable facilities. Three objectives are considered. The first objective minimizes the facility location costs. The second objective minimizes opposition to the siting plan. In their model, the opposition to a facility at a particular site increases with the scale of the facility being located. The opposition is also proportional to the population that is covered by the facility. The coverage distance can depend on the scale of the facility in their model. The total opposition measure is the sum of the individual facility measures. The third objective maximizes equity. Equity is defined in terms of what Ratick and White term a *complementary anticover* model. If a facility is located at candidate site i, the equity index for a candidate site i is defined as the number of other facilities that are sited outside of the (scale-dependent) coverage distance associated with the facility at site i. If no facility is located at candidate site i, the equity index for the candidate site is set to a very large number. The overall equity index for the location plan is the minimum of the equity indices over all candidate facility sites. As such, it is a maximin objective since it maximizes the minimum equity index over all candidate locations. The rationale for this measure is that the location plan is likely to be perceived as being more equitable if other communities are also impacted by other undesirable facilities.[30]

Erkut and Neuman (1992) build on the Ratick and White model to include fixed facility location costs as well as transport costs. In the Erkut and Neuman model, equity and opposition are represented as decreasing functions of the distance between a population center and the facilities being located and as increasing functions of the sizes of the facilities. They solve the problem by enumeration of the feasible facility sizes and locations. They

---

[30]The model formulated by Ratick and White (1988) is a pure location model in that it does not encompass the *allocation* of demand nodes to facilities.

argue that enumeration may not be an unreasonable approach to solving such problems since the number of feasible candidate sites and the number of possible facility sizes may both be small. Finally, Wyman and Kuby (1994) also outline a multiobjective facility location model for undesirable facilities. Their objectives include: fixed and transport cost minimization, risk minimization, and disequity minimization. Wyman and Kuby argue that equity can be modeled by minimizing the maximum demand-weighted distance that material must travel between a generation site and a processing facility.

8.9 SUMMARY

In this chapter we have outlined a number of important extensions of the basic location models discussed in Chapters 4 through 7. Many facility location problems, no matter what the context, involve multiple and sometimes conflicting objectives. In such cases, there will not be a single optimal solution. Instead, there will be a number of noninferior solutions. Our goal in such cases is to identify the noninferior solutions to highlight the tradeoffs that must be made in siting the facilities in question. Section 8.2 dealt with such problems.

One important class of location problems involves the location of a number of different but related facilities. Hierarchical facility location problems were discussed in Section 8.3. Median-based and coverage-based models were formulated for a number of different contexts depending on the way in which facilities at different levels of the hierarchy serve customers. Section 8.4 examined two other types of interactions between facilities: cases in which there is flow between facilities and cases in which the facilities must be located close to each other to provide backup service (or related services) to each other. In Section 8.5 we discussed a special class of interacting facilities, those that arise in production and distribution systems. In many distribution contexts, vehicles visit customers on routes that serve many customers instead of serving each customer directly from one of the facilities. This led to the formulation of location/routing models. A generic model in this category was formulated in Section 8.6. We then outlined a specific problem within this class of problems that arises in the distribution of newspapers. This problem was formulated and shown to be related to the uncapacitated fixed charge facility location problem discussed in Chapter 7. Many facility location problems involve the location of hubs at which flows are collected for movement along major links at lower than usual unit costs. Hub location problems were outlined in Section 8.7.

Finally, Section 8.8 outlined two classes of models for the location of undesirable facilities. Dispersion models attempt to locate a given number of facilities to maximize some function of the distance between the facilities themselves. Such models are useful in locating franchise outlets. The second class of models of undesirable facilities dealt with maximizing a function of the distance between demand nodes and the nearest facility. Such models are useful in locating undesirable facilities such as landfills, hazardous waste disposal sites, and nuclear reactors.

In general, the focus in this chapter has been on formulating models and not on solution methods. In some cases, the problems may be solved using relatively straightforward extensions of the basic models discussed in earlier chapters. In other cases, small instances of the problems may be solvable using standard mixed integer programming software. In many cases, specially devised algorithms are needed to solve the problems. In some cases, the best we can do is to use heuristic algorithms. Construction and improvement algorithms (similar to the greedy adding and substitution algorithm for the maximum covering problem) are effective in some cases. For other problems, more sophisticated techniques including genetic algorithms (Goldberg, 1989) or tabu search (Glover, 1990) must be employed. Many of the problems outlined in this chapter are the subject of ongoing research.

EXERCISES

8.1 Consider the following multiobjective problem:

MINIMIZE	$Z_1 = 100 X_1 + X_2$
MINIMIZE	$Z_2 = X_1 + 125 X_2$
SUBJECT TO:	$X_1 + 0.04 X_2 \geq 3$
	$13 X_1 + X_2 \geq 51$
	$2 X_1 + 0.25 X_2 \geq 9$
	$3 X_1 + 2 X_2 \geq 20$
	$X_1 - 3 X_2 \leq 6$
	$-15 X_1 + X_2 \leq 40$
	$X_1, X_2 \geq 0$ and integer

Note that the decision variables must be nonnegative and integer. They do *not* need to be binary (0/1) variables.

(a) Find the solution that minimizes Z_1 first and then, from among the solutions that do so, minimizes Z_2 second. What weights on Z_1 and Z_2 must you use to find the solution?

(b) Find the solution that minimizes Z_2 first and then, from among the solutions that do so, minimizes Z_1 second. What weights on Z_1 and Z_2 must you use to find the solution?

Exercises

(c) Use the weighting method described in the text to find all the noninferior points you can using this approach. Clearly identify the weights on Z_1 and Z_2 that you must use to find each solution.

(d) Carefully plot the constraints, the feasible region, and the objective functions. Did the weighting approach find all of the noninferior points? If not, identify all noninferior solutions not found in part (c).

8.2 For the 88-node data set (CITY1990.GRT) and the first demand set in the file, use SITATION to find at least five points on the tradeoff curve of uncovered demand versus average distance (as was done in Section 8.2). Use a coverage distance of 300 and 7 facilities. Plot the results carefully using a good spreadsheet program such as EXCEL or QUATRO PRO. Also include a table of the following form:

Solution Number	Facility Locations	Average Distance	Uncovered Demand	Weight on Uncovered Demand
1				Infinite
2				0
3				
4				
5				

To do this, you will need to first find the solution to the maximum covering problem. This should be solution 1 above. Then you will need to find the solution for the P-median problem. This should be solution 2 above. Then you will need to find at least three other solutions. Note that the total demand is 44,840,571 for this data set.

For the next three solutions, you will need to change parameters G and J in the MOD-DIST matrix change mode to generate new distance matrices for solving this problem. You will then need to solve the P-median problem with the revised distance matrix to get the next set of locations. However, to evaluate the actual average distance (and possibly the actual coverage, depending on how you do this), you will need to go back to the original distance matrix. Using the original matrix and the set of sites found using the revised matrix, use SITATION to evaluate the actual average distance and the actual coverage.

Note: You can also evaluate the true average demand-weighted distance from a solution that you obtain after modifying the G and J parameters using MOD-DIST. You need to think about how to do this transformation. When you go back to revise the G and J parameters again, however, be sure to delete the current (modified) MDST88 file and use the original

MDST88 file (which has probably been renamed MDST88.G00 by MODDIST). The original MDST88.GRT file should be 42,116 bytes long and should have a date of 10-19-94 and a time of 12:34:56.

If you have doubts about which MDST88 file is the original one, delete the file named MDST88.GRT and rerun SITATION loading CITY1990.GRT. SITATION will then recreate the correct original MDST88.GRT file (with a different time and date, of course).

8.3 Consider the following problem. Our primary objective is to maximize the number of *covered* demands. As in the maximum covering model, demands will be *covered* if there is at least one facility within D_c distance units of the demand node. Our secondary objective is to minimize the demand-weighted distance between a demand node and the facility to which the demand node is assigned *for those demands that are assigned to a facility more than D_c distance units away*.

Furthermore, to account approximately for busy periods, we want to extend the maximum covering problem in the following three ways:

(i) Each facility being located can have at most H_{max} demands assigned to it;

(ii) Each demand must be fully assigned to a facility or possibly to a number of different facilities (even if the facility or facilities to which it is assigned is/are more than D_c distance units away).

(iii) Each facility must be within D_F distance units of at least one other facility. We will refer to D_F as the backup facility distance.

To formulate this problem, we define the following notation:

Indices

i = index of demand points

j, k = indices used for candidate locations

Inputs

d_{ij} = distance between node i and candidate location j

D_c = coverage distance

$$a_{ij} = \begin{cases} 1 & \text{if } d_{ij} \leq D_c \\ 0 & \text{if not} \end{cases}$$

D_F = backup facility distance

$$b_{jk} = \begin{cases} 1 & \text{if } d_{jk} \leq D_F \text{ and } j \neq k \\ 0 & \text{if not (i.e., } d_{jk} > D_F \text{ or } j = k) \end{cases}$$

(Note that we set $b_{jj} = 0 \; \forall \; j$ so that we prevent a facility from serving as its own backup facility.)

h_i = demand at node i

H_{max} = maximum demand that can be assigned to any facility

M = a very large number (used to weight one of the objectives)

Decision Variables

Y_{ij} = *fraction* of demand at node i that is assigned to a facility at node j

$$X_j = \begin{cases} 1 & \text{if we locate at candidate location } j \\ 0 & \text{if not} \end{cases}$$

With this notation, formulate the weighted objective function and constraints defined below:

MAXIMIZE

The total covered demands *as the primary objective*

MINIMIZE

The demand-weighted distance between demand nodes and the facilities to which they are assigned *for the uncovered demands as the secondary objective*

Note that these should be formulated as *one* objective function.

SUBJECT TO:

Locate exactly P facilities

All of the demand at a node is assigned to some facility

Capacity of the facilities

Demands can only be assigned to open facilities

Each facility must have at least one other facility within D_F distance units

Nonnegativity and Integrality

8.4 In many facility location problems, we are interested in simultaneously locating two types of facilities. For example, in a developing country, we might want to locate regional hospitals and local health care clinics.

Consider the following problem. *We want to locate H hospitals and C clinics* in a developing country. In general, we will have $H \ll C$ (i.e., H is much less than C). For a variety of reasons, *each clinic must be located within a distance D_{hc} (a critical hospital to clinic distance) of at least one of the hospitals*. Some of the possible reasons for this requirement include the need to resupply clinics with supplies and medications from the hospitals and the need for physicians at a hospital to visit clinics on both a routine (inspection) basis and an emergency basis to deal with critically ill patients whose transport to the hospital may be impossible. We are given the populations, h_i, of each of a large number of rural villages whose people will be served by this health care system. *Our objective is to minimize the total* (over all villages) *demand-weighted average distance between a village and the nearest health care facility to the village.* In other words, if one of the few hospitals is closer to a village than any of the clinics, then patients from that village will go to the hospital directly; otherwise, they will go to the nearest clinic.

Using the notation below, *formulate* this problem. Note that in each part of the problem you are given the critical constraints (or objective function) in words. You must formulate them using the notation defined below. Be sure all indices of summation are indicated clearly and that you indicate the indices for which each constraint applies (e.g., for all i).

Indices

i = index of villages (or demand locations)
j = index of candidate hospital sites
k = index of candidate clinic sites

Inputs

H = maximum number of hospitals to be located
C = maximum number of clinics to be located
d_{ij} = distance between village i and candidate hospital j
d_{ik} = distance between village i and candidate clinic k
d_{jk} = distance between candidate hospital j and candidate clinic k
D_{hc} = critical coverage distance between hospitals and clinics
$a_{jk} = \begin{cases} 1 & \text{if candidate clinic } k \text{ is within } D_{hc} \\ & \text{distance units of candidate hospital } k \\ 0 & \text{if not} \end{cases}$
h_i = population in village i

Decision Variables

$X_j = \begin{cases} 1 & \text{if candidate hospital site } j \text{ is selected} \\ 0 & \text{if not} \end{cases}$

$Y_k = \begin{cases} 1 & \text{if candidate clinic site } k \text{ is selected} \\ 0 & \text{if not} \end{cases}$

$W_{ij} = \begin{cases} 1 & \text{if patients at village site } i \\ & \text{go to a hospital at candidate site } j \\ 0 & \text{if not} \end{cases}$

$V_{ik} = \begin{cases} 1 & \text{if patients at village site } i \\ & \text{go to a clinic at candidate site } k \\ 0 & \text{if not} \end{cases}$

The objective function and key constraints of the problem are listed below:

(a) Minimize the total demand-weighted distance between a village and the nearest health care facility.

(b) Locate exactly H hospitals.

(c) Locate exactly C clinics.

(d) Every village is assigned to exactly one facility (either a clinic or a hospital).

(e) Each clinic must be within D_{hc} distance units of the nearest hospital.

(f) Demands at village i can only be assigned to a hospital at candidate site j if we locate a hospital at candidate site j.

(g) Demands at village i can only be assigned to a clinic at candidate site k if we locate a clinic at candidate site k.

(h) Integrality.

8.5 In Chapter 4 we discussed the problem of maximizing coverage subject to a constraint that we locate exactly (or no more than) P facilities. This model implicitly assumes that the cost of locating at each candidate location is approximately the same. This, however, may not be the case. Thus, in many cases, it is important to be able to analyze the tradeoff between the fixed facility location costs and the percentage of the demands that are covered.

(a) Formulate the problem of (i) maximizing the number of covered demands and (ii) minimizing the total fixed costs of the selected facilities as a two-objective problem. (Note that you will not have an explicit constraint limiting the number of facilities being located.) Clearly define all inputs and all decision variables and state the objective functions and constraints in both words and using notation.

(b) Reformulate the two-objective problem outlined in part (a) but now minimize the number of uncovered demands instead of maximizing the number of covered demands.

(c) Using a weight of α for the objective of minimizing the number of uncovered demands and $(1 - \alpha)$ for the weight on the total cost, restructure the problem of part (b) as a weighted objective problem.

(d) Show that the problem you formulated in part (c) can be thought of as an uncapacitated fixed charge facility location problem. What decision variable redefinitions and input transformations must you do to obtain this formulation?

(e) Using SORTCAP.GRT, the first demand data set (representing the state populations) for the 49-node problem described in Appendix H and a coverage distance of 450 miles, use SITATION to find at least four noninferior points on the (approximate) tradeoff curve of uncovered demand versus total facility location cost. Be sure that two of the points correspond to (i) the point that covers all demand at minimum total fixed cost and (ii) the point that has the minimum total cost (and, from among all of the solutions with minimum cost, is the one that minimizes the uncovered demand).

Clearly discuss how you obtain the inputs for this analysis.

For each solution, give the value of α that you used. Also give the total fixed cost, the number of uncovered demands, and the number of facilities that are located, and the locations of the facilities.

Hint: It is probably a good idea to set all of the Lagrange multipliers to 0 at the beginning of the Lagrangian algorithm for the fixed charge problem.

(f) How does the solution that covers all demand at minimum cost compare with the traditional maximum cover solution in terms of (i) the total cost of the two solutions and (ii) the number of facilities located?

8.6 Discuss why constraints (8.6) will automatically be satisfied by any optimal solution to formulation (8.5*a*)–(8.5*f*).

8.7 Consider the *P*-median-based successively inclusive facility model operating under a globally inclusive service hierarchy [formulation (8.5*a*)–(8.5*f*)]. Reformulate this model to include facility capacities. Clearly define all new notation (inputs and decision variables) that you use. As always, clearly state the objective function and constraints in both words and notation.

8.8 Consider the coverage-based hierarchical facility location model [formulation (8.10*a*)–(8.10*g*)]. Reformulate this model to include facility capacities. Clearly define all new notation (inputs and decision variables) that you use. As always, clearly state the objective function and constraints in both words and notation.

8.9 (a) In formulation (8.13*a*)–(8.13*j*) would all patients always go to the nearest clinic? If so, why? If not, why not?

(b) Why is or is this not a desirable property for the model to have?

(c) Given the locations of the facilities of different types, how can you solve the patient flow portion of the problem?

8.10 In the production/distribution model given in the text—formulation (8.15*a*)–(8.15*f*)—we used flow variables, Y_{ijm}^k. One of the problems

Exercises

with this formulation is that it leads to the use of rather weak linkage constraints (8.15b) between the flow variables and the warehouse location variables, X_j. Modify this formulation so that it uses variables that represent the *fraction* of the demand for product k in market i that is produced at plant m and goes via warehouse j.

8.11 (a) Suppose we (temporarily) eliminate the production capacity constraint (8.15d) from the formulation of the production/distribution problem (8.15a)–(8.15f). How can you solve this problem?

Hint: What well-known problem does this reduce to in this case? How can you reformulate the problem so that it looks exactly like this well-known problem?

(b) Use this notion to outline a solution algorithm for the production/distribution problem (8.15a)–(8.15f).

8.12 (a) Modify the formulation of the production/distribution problem (8.15a)–(8.15f) to include capacities for each of the candidate warehouses. Clearly define all notation. Clearly state both the objective function and the constraints in words and in notation.

(b) Again, ignoring the production capacity constraint for this problem (the capacitated warehouse location problem), how can you solve this problem? Again, you should think about what well-known model this relaxation of the problem reduces to.

8.13 The production/distribution model given in the text—formulation (8.15a)–(8.15f)—assumed that there were no economies or diseconomies of scale in production. (Economies of scale imply that over the relevant range of production the unit cost decreases with increasing output. Diseconomies of scale mean that the unit cost of production increases with increasing levels of output over the relevant range of production.)

(a) If the production process exhibits diseconomies of scale, modify the formulation to account for the increasing unit costs of production.

(b) Would the approach you outlined in part (a) work if the production process exhibits economies of scale? Briefly justify your answer.

8.14 The production/distribution model given in the text—formulation (8.15a)–(8.15f)—seems to assume that all shipments must go through a warehouse (as shown in Figure 8.5).

(a) Can the formulation given in the text allow for direct plant-to-market shipments? If so, how? If not, modify the formulation to allow for this possibility, clearly defining any new notation that you use.

(b) Direct shipping from a plant to a market may *seem* to always dominate the more indirect method of shipping through a warehouse in terms of unit costs. Briefly discuss at least three reasons why this might not be so and why, even if it is so, we might want to use warehouses.

8.15 Using the notation defined in Section 8.7 and constraint (8.25), formulate a complete model in which the two-way flow on each hub i/spoke k connection must be greater than or equal to some lower bound L_{ik} if the hub is connected to the spoke.

8.16 Using the notation defined in Section 8.7, formulate a P-hub maximum covering problem in which coverage is defined in terms of the origin/destination specific level of service ratios, R_{ij}. Clearly define all notation. Also, state the objective function and all constraints in words and using notation.

8.17 Consider a P-hub location problem in which each of the nonhub nodes is connected to *every* hub. Such a network configuration would allow at most one stop service between every pair of cities. The hub through which service would be delivered for any origin/destination pair would be the hub which provides the least cost (or best) service. [Sasaki and Drezner (1994) outlined such a hub-and-spoke model, though they did not formulate it as a P-median problem as suggested below.]

(a) Show that this problem can be formulated as a variant of a P-median problem.

(b) What are the "demand" nodes in the P-median problem?

8.18 The National Association for the Advancement of Colored People (NAACP) has alleged that there is "environmental racism" associated with the location of such undesirable facilities as landfills, incinerators, and hazardous waste sites. They charge that such facilities "are disproportionately located near minority communities" (White, 1994, p. 1).

(a) What data would you want to collect to test whether or not this charge is valid?

(b) What social and economic forces are likely to have caused undesirable facilities to be located near minority communities, if, in fact, the charges are valid?

(c) Assuming the NAACP is correct in its charge of environmental racism, what factors inherent in traditional location models are likely to have resulted in this outcome?

(d) Again, assuming the charges can be supported by data, how can you modify a traditional location model, such as the capacitated fixed charge facility location model, to incorporate factors that will result in a more equitable siting plan?

9

Location Modeling in Perspective

9.1 INTRODUCTION

This book has focused on classical facility location models including covering, center, median, and fixed charge problems, and on extensions of these basic models. We have identified the properties of those models and have developed solution algorithms for the models. As such, much of the emphasis has been on the formulation and solution of optimization models for facility location.

In this chapter we will briefly outline one paradigm of the broader planning and problem-solving process in which mathematical facility location models typically arise. In so doing, we will be able to emphasize a number of key points that tend to be glossed over when the focus is only on the analysis and solution of mathematical models.

Throughout the discussion in this chapter, we refer to three different groups of individuals involved in the planning process: analysts, planners, and decision makers. *Analysts* typically have a strong technical background and are capable of formulating and solving complex mathematical location problems. *Planners* often have a somewhat broader perspective, be they in public sector agencies or private organizations. Their concerns typically span those of the technical analysis but encompass additional facets of the problem that are often difficult to model explicitly. Finally, *decision makers* are those individuals associated with the planning process who are charged with making decisions and implementing courses of action to resolve the problem(s) at hand. They are the clients of the technical analysis and planning process. Often, they are not technically oriented.

9.2 THE PLANNING PROCESS FOR FACILITY LOCATION

Figure 9.1 outlines one paradigm of the planning process in which facility location modeling is embedded. Four broad steps are outlined: (i) problem

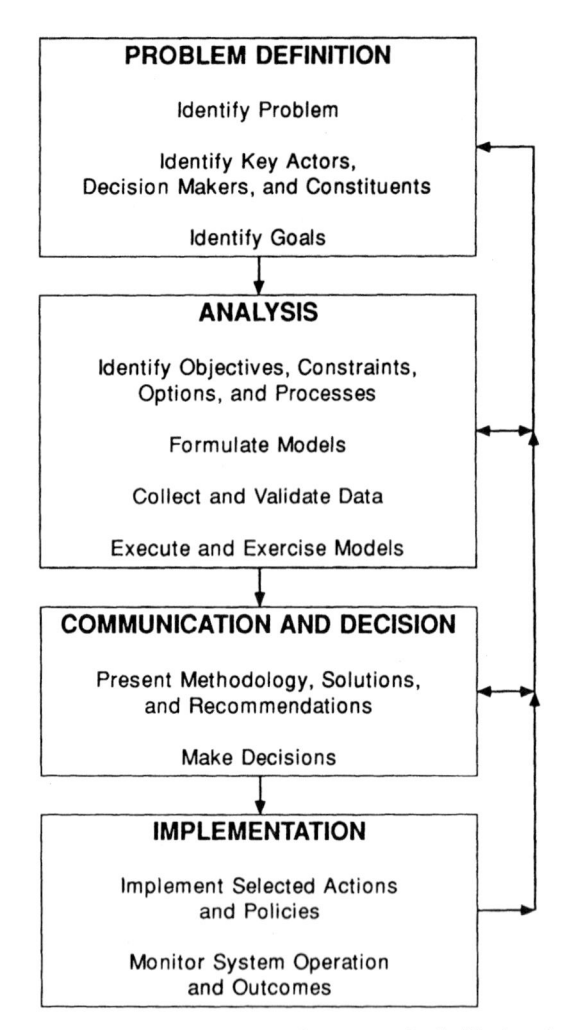

```
┌─────────────────────────────────┐
│      PROBLEM DEFINITION         │
│                                 │
│        Identify Problem         │
│                                 │
│      Identify Key Actors,       │
│ Decision Makers, and Constituents│
│                                 │
│        Identify Goals           │
└─────────────────────────────────┘
                │
                ▼
┌─────────────────────────────────┐
│           ANALYSIS              │
│                                 │
│  Identify Objectives, Constraints,│
│      Options, and Processes     │
│                                 │
│       Formulate Models          │
│                                 │
│    Collect and Validate Data    │
│                                 │
│    Execute and Exercise Models  │
└─────────────────────────────────┘
                │
                ▼
┌─────────────────────────────────┐
│   COMMUNICATION AND DECISION    │
│                                 │
│  Present Methodology, Solutions,│
│       and Recommendations       │
│                                 │
│        Make Decisions           │
└─────────────────────────────────┘
                │
                ▼
┌─────────────────────────────────┐
│         IMPLEMENTATION          │
│                                 │
│    Implement Selected Actions   │
│         and Policies            │
│                                 │
│     Monitor System Operation    │
│         and Outcomes            │
└─────────────────────────────────┘
```

Figure 9.1. Schematic of the planning process for facility location.

definition, (ii) analysis, (iii) communication and decision, and (iv) implementation. In this section we discuss these major steps and the associated subtopics in detail.

9.2.1 Problem Definition

Careful definition and identification of the problem is critical to the success of any planning process. Facility location problems almost always involve long-term decisions. Location decisions are therefore inherently *strategic* in nature. As such, they are inseparable from the strategic goals of (i) the enterprise whose facilities are being located and (ii) the actors involved in the decision process. Thus, we must understand the strategic goals of the enterprise in question and the key problems the organization faces in realizing those goals. We must then ask ourselves dispassionately whether improved facility locations are likely to resolve the problems and to enhance the ability of the firm or agency in achieving its strategic goals.

For example, when asked to help locate ambulances for a city, we should begin by considering the mission of the emergency medical services in the city. The goals of such services are to save lives and to reduce the severity of diseases and injuries. As such, there are a number of actions that an emergency medical service can undertake to help it achieve these goals. These actions include:

a. Installing a 911 emergency phone service if the system is not already tied into such a service;
b. upgrading the skills of the paramedics through improved training;
c. altering the standing orders of the paramedics to allow them to perform more invasive emergency procedures (after appropriate training);
d. initiating a CPR (cardiopulmonary resuscitation) training program for the public;
e. enhancing the emergency room(s) to which patients are taken;
f. instituting a multitiered service (which generally allows more vehicles to be deployed since not all vehicles and personnel are equipped and trained at the highest level);
g. improving the skills of the dispatchers; and, finally,
h. changing the number and location of the ambulance bases.

Changing the number and location of the ambulance bases is clearly just one of a number of different actions an EMS department can adopt to improve the service it provides. Therefore, in dealing with an EMS department, analysts in general and facility location specialists in particular should consider the range of actions available to the department. They should ask themselves whether improving the ambulance base locations is likely to be the most cost-effective way of improving the emergency medical services in question. The problems faced by an enterprise may have little to do with the locations of the organization's facilities. We must avoid falling into the trap of "viewing every problem as a nail (i.e., location problem) simply because all we have in our toolbox is a hammer (i.e., a location model)."

In addition to understanding the overall goals of the enterprise with which we are dealing, we must also identify the key actors, decision makers, and constituents associated with any problem. For example, there are a number of actors involved in decisions regarding the location of ambulances. The emergency medical services department is perhaps the primary actor. Within the EMS department, two groups of actors are typically involved: administrators and paramedics. In addition to the overall goal of saving lives, administrators are often concerned about budgets and employee morale. Employee morale is affected in part by workloads which depend on the allocation of demands to ambulances and on ambulance locations as well. Paramedics are concerned with saving lives and with their workloads. The EMS department often shares facilities and buildings with the fire department whether or not the EMS department is a separate third city service or a part of the fire department. Thus, the fire department is another party with a vested interest in the location of ambulances. Other key factors in the decision process include the city council which must approve budgets and, of course, the taxpayers and the general populace.

Just as it is important to understand the goals of a public agency, it is also critical that we identify the strategic goals of private firms. For example, in locating distribution centers for a retail firm, if the firm wants to be the low-cost firm in the industry, locating relatively few warehouses to realize the full range of economies of scale associated with the inventory and warehousing operations may be optimal. On the other hand, if the firm wants to be known for its customer service, locating a large number of warehouses may be critical to ensure that the firm can meet customer demands in a timely manner.

We will encounter multiple actors in private sector location problems just as we do in public sector problems. The owner of the warehouses, the users of the warehouses (who may differ from the owner), carriers, and customers are all interested parties in warehouse location decisions, for example. Each may have different goals and perceptions of the problems that motivate the need for locating the warehouses.

9.2.2 Analysis

Having identified the key actors, goals, and problems in the problem definition phase of the planning process, we are in a position to begin the analysis. One of the first tasks is to translate the goals into quantifiable objectives that are tied to facility location decisions. If the goals cannot be translated into objectives that depend in some obvious way on the facility locations to be selected, then it is likely that improving facility locations will not help the organization achieve its strategic goals. In that case, analysts, planners, and decision makers need to return to the problem definition phase of the planning process.

To illustrate the need to convert goals into quantifiable objectives, we again consider the EMS case. It is difficult to measure the number of lives that are saved by an EMS department. However, there is strong evidence that indicates that the long-term prognosis for a patient is enhanced if prompt emergency care is available. This suggests that covering, center, and median objectives are all reasonable proxies for the goal of saving lives. In adopting one or more of these objective functions, however, we must always bear in mind that the travel time between an ambulance base and a medical incident is only one of several components of the total time between the onset of the emergency and the initiation of medical care. Other components include the time needed to recognize that a medical emergency has occurred (which can be reduced through public training), the time needed to contact the EMS dispatcher (which can be reduced through the installation of citywide 911 services), dispatch delay (which can be reduced through enhanced dispatcher training), and delays between receipt of the call by an EMS crew and the time the vehicle begins its journey.[1]

After appropriate quantifiable objectives have been developed, we must identify any constraints that may limit the attainment of the objectives,[2] as well as particular options and processes that need to be modeled. Thus, we may find that there are some locations at which it is infeasible to locate one of the facilities in question. For example, the auto industry will often exclude locations next to chemical plants as locations for rail ramps fearing that the chemicals used at the plant will damage the finish on the automobiles if the chemicals are released to the environment. In other cases, systems may operate under single sourcing constraints that require demand nodes to be assigned uniquely to a facility even in the face of capacity constraints. [Daskin and Jones (1993) outline two approaches to solving such problems when the number of demand nodes greatly exceeds the number of facilities.]

Planners and decision makers may also want studies to be conducted of particular candidate sites even if they are not demand nodes. While we may be able to prove that inclusion of non–demand node candidate locations will not enhance the objective function (as in the case of the P-median model), incorporating these sites into the analysis framework may be critical for three reasons. First, unmodeled objectives may be better attained at one or more of these sites. Second, inclusion of these locations and our ability to demonstrate to the decision makers that locating at these sites does not improve (and often will degrade) the quantifiable objectives is likely to enhance the credibility of our analysis in the eyes of the decision makers. Third, if some demand nodes are precluded as candidates sites, the node optimality property is generally lost. In this case, inclusion of non–demand node candidate sites may improve the solution.

---

[1] At one time in Austin, TX, this time was not insignificant. EMS crews were housed in apartment buildings with vehicles parked outdoors. All drugs and expensive medical supplies were stored in the apartment for security reasons. These drugs and supplies had to be loaded onto the ambulance before it could travel to the scene of an incident. Housing ambulances in garages at dedicated bases was expected to reduce this component of delay significantly by allowing drugs and equipment to be stored in the vehicle at all times.

[2] In thinking about constraints in this context, we are interested in those mathematical constraints over and above the constraints that are generally associated with particular objective functions. For example, in the P-median problem we are referring to constraints over and above the constraints that stipulate that each demand node is served and that we locate P facilities.

Decision makers and planners may ask that the models analyze specific policies of interest to them. For example, in the Austin EMS study, the decision makers were interested in studies of the effects of adopting a two-tiered system and of limiting the candidate locations to either zones in which the city owned land or, more restrictively, to zones in which a fire station existed.

Analysis of key aspects of the process underlying the operation of the facilities being located may also be critical. Almost all of the models we have discussed assume that the nearest facility is always available when needed. In emergency services such as fire departments and ambulance services, this is not likely to be a valid assumption. In such cases, more complicated models that incorporate queuing components may need to be used to account for vehicle and facility busy periods.

After the strategic goals have been translated into quantifiable objectives and key constraints, options, and processes have been identified, we must formulate one or more mathematical location models. Model formulation is often an art. Two principles should guide our attempts to formulate a model. First, the model(s) should capture as much of the problem as realistically as possible. Second, we must be able to solve the model either heuristically or optimally.

We note that there is often a tension between (i) formulating a complicated model that captures the problem well but that is difficult to solve and (ii) adopting a simpler formulation that represents the real world in a more approximate manner but that is easier to solve. Complex formulations that capture much of a problem but that cannot be solved effectively are often of less value than are simpler models that capture only a portion of the problem but that can be solved. This is particularly so if we can communicate the limitations of the model to the decision makers so that exogenous judgment can be applied to the model results.

Data are critical to the success of any facility location analysis. Without good data on customer demands, distances, costs, and other relevant inputs, even the best modeling exercise is of limited value.[3] Good data are essential if we are to develop credible solutions and recommendations.

Data validation is difficult. A number of logical checks can be performed on the data. First, we should ask whether the demand and facility coordinates (e.g., longitudes and latitudes) are consistent with the labeled locations. If a city is supposed to be in Illinois, are the longitude and latitude consistent with the location being in Illinois? Second, the reported or measured dis-

---

[3]Note that we do not argue that the absence of good data makes a modeling exercise useless. Often, the process of identifying goals, actors, objectives, constraints, and options is of considerable value in and of itself, even if no model is solved. This planning process often forces decision makers, analysts, and other key actors to adopt a common terminology and a common set of assumptions about the underlying problems.

The Planning Process for Facility Location

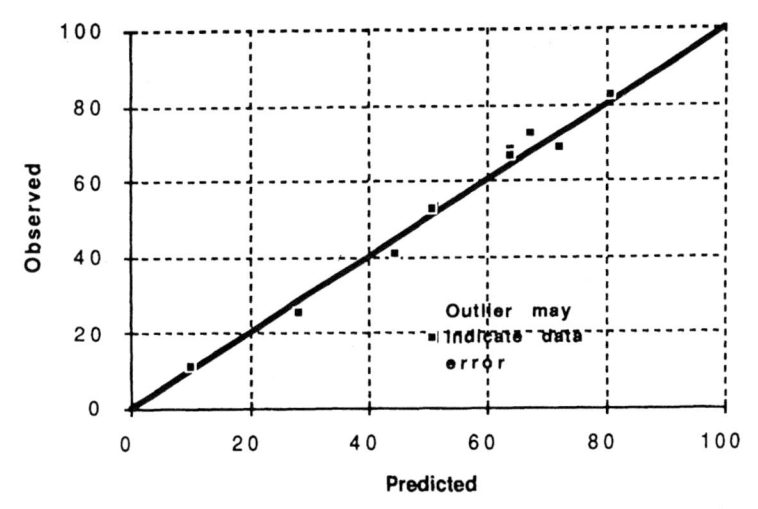

Figure 9.2. Hypothetical measured and predicted data showing potential data error.

tances can be compared to computed or predicted distances to be sure that they are approximately equal. This can be done using a plot similar to that shown in Figure 9.2. In this figure predicted and measured distances are plotted against each other. We would expect most points to fall close to the predicted = measured line. Points that fall either very far above or below this line may indicate the presence of data errors. Brimberg and Love (1992, 1993) have developed a number of models that enable analysts to predict distances using only the coordinates of the demand locations or candidate facility sites. These models are generalizations of the standard l_n norms discussed in Section 1.4.3. Finally, data should pass elementary math tests. For example, the sum of reported mutually exclusive population segments (e.g., by ethnicity or by age) should be equal to (or less than, if the segments are not collectively exhaustive) the total population of a zone. Data that are equal to the ratio of other data elements can also be checked. For example, many data sets report the overall population of a zone, the number of households in a zone, and the average number of people per household. The number of people per household should equal the population divided by the number of households. Unfortunately, this is not always the case. *Just because the data come from a computerized database does not mean they are correct.*

Models and data should also be tested by using them to represent existing conditions. This process, known as calibration, enables us to compare the model predictions with the measured actual performance. For example, we can compare the predicted or modeled average response time for an EMS system with the actual average response time. If the two numbers are close enough we may be willing to accept both the validity of the model and the data.[4] If, on the other hand, the predicted and actual performance differ significantly, we are faced with the difficult task of sorting out whether the source of the discrepancy is in the data or in the model (or both). When the data and model fail this sort of validation test, they may still be usable if we believe that they can accurately predict either the *direction* of changes in the performance measures as a result of location/allocation changes or (even better) the *relative* or *absolute change* in the performance measures.

Often, model calibration and data validation are prerequisites for the use of a model or set of models and the data that feed the models. For example, in using the coal logistics system (COLS) model, Kuby, Ratick, and Osleeb (1991) report that the predicted flows through each port had to be within 15 percent of the actual flows when the model was used with historical data. They note that, if calibration is not done, important errors may slip through the process. On the other hand, suspicions of data tampering may arise in the use of calibrated models. They go on to note that no matter how systematic the calibration process, judgement on the part of the analyst is always needed. In the coal transport study (CTS) model developed for China, (Kuby et al. 1995) model calibration was a major task. The analysts first looked for structural errors in the model. Often, these resulted from overly aggregate representations of supply or demand locations or from having aggregated commodities (coal types) too much. Next, they searched for economic errors or errors in the cost coefficients of the model. They conclude that calibrating a model requires a mixture of art, science and skillful negotiation. By meeting frequently with China's State Planning Commission, the analysts were able to identify cases in which the model was behaving idiosyncratically as well as cases in which the experts recognized that the model was actually providing legitimate results that were superior to those they had expected. This back-and-forth process between the analysts and the decision makers helped build confidence in the model and helped educate the planners about what the model could and could not be expected to do.

The final analysis task is that of executing and exercising the model(s). We use the terms *execute* and *exercise* instead of *solve* to emphasize that in using the model(s) we are not trying to find a single solution, but rather to explore the solution and objective spaces[5] with multiple model runs or scenarios. Doing so involves parametrizing the key inputs, particularly those whose values are subject to uncertainty or doubt, performing sensitivity analyses, and identifying key tradeoffs in multiobjective problems. Multiple model runs can be made in which we systematically vary such key inputs as: the number

---

[4] For obvious reasons *accepting* the validity of both the model and the data is dangerous. Data and modeling errors may interact in a way that allows the model to replicate existing conditions but that makes it a poor predictor of altered conditions. The likelihood of this is fortunately rather small.

[5] The *solution space* is the set of all feasible solutions to the problem. The *objective space* is the space spanned by the objective function values corresponding to each point in the solution space.

of facilities to be located, the cost per mile, demands, and demand growth rates. These scenarios help identify solutions that are robust with respect to variations in the inputs. In some cases, key inputs may be difficult to measure. In other cases, there may be disagreement over the appropriate value(s) to use for certain inputs (e.g., future population values). These difficulties become relatively unimportant if we can show that the model's solution is insensitive to these input values. This is also true of the broader planning process. It is sometimes easier to reach a consensus on a course of action than it is to agree on key assumptions.

Allowing decision makers and planners to exercise the models to test their own solutions and ideas is often a critical element in the planning process.[6] This may involve allowing decision makers to force specific sites (or groups of sites) into or out of the solution. It may also involve allowing them to test specific location configurations. Location modeling software should be designed to facilitate this sort of experimentation.

Giving decision makers and planners a chance to use the model(s) in this way results in a number of benefits. First, in exercising the models, decision makers gain a better understanding of the models and of the modeling process. Second, they gain confidence in the solutions provided by the models. Third, they gain an appreciation for the limitations of the models. In so doing, decision makers and planners are better able to integrate their own judgment with the results obtained from the models. Finally, in using the models, decision makers often provide valuable feedback to the analysts and planners about important unmodeled facets of the problem or objectives and goals that had not previously been articulated. This, in turn, leads to feedback between the model execution phase of the process and earlier phases including model formulation and the broader problem definition phase.

Even problems with a single objective function may in fact be multiobjective in the face of different inputs. For example, a simple maximum covering location model for locating fire stations may be at the heart of a multiobjective analysis in which different objectives include: (i) covering population centers (where people are typically located in the evening and at night), (ii) covering schools, office buildings, and retail centers (where people are often located during daytime hours), (iii) covering hospitals and nursing homes (where particularly vulnerable people are located), and (iv) covering industrial sites (at which flammable materials may be stored and at which dangerous processes are used). In location planning contexts that are either explicitly or implicitly multiobjective, one of the primary outputs of exercising the location models is an understanding of the key tradeoffs between objectives and interest groups.

---

[6]Decision makers need not actually run the computer programs themselves. They should, however, be given a chance to request that certain analyses be conducted.

9.2.3 Communication and Decision

After the analysis phase of the planning process, the results of the analysis must be communicated to the decision makers and a course of action must be adopted. Communicating the results of the analysis involves far more than identifying a single "optimal" solution. Rather the report on the analysis should include:

   a. discussion of the mathematical model(s) that were used in clear nontechnical terms;
   b. identification of the limitations of the modeling process;
   c. identification of nonmodeled impacts;
   d. presentation of alternative solutions in both objective space and solution space; and
   e. recommendations regarding the timing, financing, and implementation of particularly promising solutions.

For planners and decision makers to have confidence in the results of a technical analysis, it is important for them to have at least a rudimentary understanding of the methodology that was employed in the analysis. To that end, a report on a technical analysis should discuss the models that were used in nontechnical terms.[7] This should also include a clear identification of the limitations of the models that were employed as well as the key assumptions that had to be made to use the models. These assumptions may deal with the way in which key processes were represented. For example, if a location model assumes that the nearest fire engine will always be available, such an assumption should clearly be documented. In some cases, we need to make key assumptions related to the data that are used by the models. For example, if demand projections for the use of emergency medical services are not available by analysis zone, we may assume that demands are proportional to the population.[8] While this may be a reasonable first-cut assumption, demand rates will vary as a function of a large number of factors including the socioeconomic composition of the zone, the incidence of crime in the zone, and the number of elderly citizens in the area. If decision makers understand that demand was assumed to be proportional to population, they can filter the model recommendations to account for their own qualitative estimates of where demand rates will be disproportionately larger or smaller than average for a given zonal population.

The technical analysis will often ignore certain impacts either because they are difficult to model (e.g., would require transforming a linear formulation

---

[7]Formal statements of the mathematical details of the models (and the algorithms used to solve the models) should often be relegated to appendices.

[8]In fact, a crude rule of thumb is that an urban area will experience about one call for emergency medical services per day for every 10,000 people in the area.

The Planning Process for Facility Location

into a messy nonlinear model), because they are difficult to quantify (e.g., no quantifiable proxy for the impact can been identified), or because data have not been collected related to these impacts. The report on the analysis should also identify important impacts that were not explicitly modeled. At the very least, the report should list those impacts that were outside the domain of the formal analysis. This again will allow decision makers to account for the unmodeled impacts in a qualitative or judgmental manner. Ideally, the report should estimate these impacts for each of the location configurations presented in the report. For example, if certain impacts have not been included explicitly in the optimization-based models because their inclusion would necessitate the use of a nonlinear model, we can always evaluate the nonlinear impacts of a particular configuration exogenously. Thus, even if we cannot optimize over the impacts which must be modeled in a nonlinear manner, we can evaluate those impacts exogenously.

The results of the technical analysis should be presented both in terms of the objectives attained by different location plans (*objective space*) and in terms of the locations themselves (*solution space*). Tradeoff curves similar to Figure 8.3 are one way of presenting the results in objective space. When more than two objectives are involved, other approaches to presenting the results need to be considered. Erkut and Neuman (1992) display three objectives (cost, equity, and a community opposition measure) using a triangle. Regions of the triangle identify weights on the objectives for which different noninferior solutions are optimal. They note that the method will fail to show any convex dominated points even though such points may be useful compromise solutions. Another approach involves the use of *value path diagrams* (Cohon, 1978).[9] In this approach, one vertical axis is drawn for each objective. Lines are then drawn connecting the values on the different vertical (objective) axes for the same location plan. Scaling the data in a meaningful way is often critical.

To illustrate this approach, consider the data shown in Table 9.1. Figure 9.3 is one (of many) value path diagrams that can be drawn from these data. The data for the three axes were scaled as follows:

Coverage: The coverage attained by each solution was plotted as a percentage of the coverage attained by the best of the three solutions (i.e., solution 3). Note that the values shown are not the percentage of the total demand that is covered, since even the best of these solutions only covers 86.67 percent of the total demand.

---

[9] A variety of other methods of displaying multiobjective problem results have also been developed. One particularly interesting approach is GRADS (Klimberg, 1992; Klimberg and Cohen, 1993), a computer-based graphical method of displaying multiobjective problem results. GRADS plots solutions in terms of two user-selected objectives. As the user points at a solution in the two-dimensional objective space, the computer displays a multidimensional star or polygon to show the attainment of the other objectives by the selected solution.

Table 9.1 Sample Data for Value Path Diagram

Solution	Covered Demands Absolute Value	Covered Demands Scaled Value	Average Distance Absolute Value	Average Distance Scaled Value	Total Cost Absolute Value	Total Cost Scaled Value	Locations
1	34,154,576	87.88	255.75543	85.41	$501,905	100.00	5, 7, 28, 46
2	36,902,335	94.95	223.19279	100.00	$841,802	0.00	1, 2, 3, 8
3	38,863,444	100.00	262.16597	82.54	$778,691	18.57	8, 19, 32, 69

Note: Solution 1 minimizes the total cost of the CITY1990.GRT data set shown in Appendix G with a cost per mile of $0.000025. Solution 2 is the 4-median solution for this data set. Solution 3 maximizes coverage within 400 miles for this data set using four facilities.

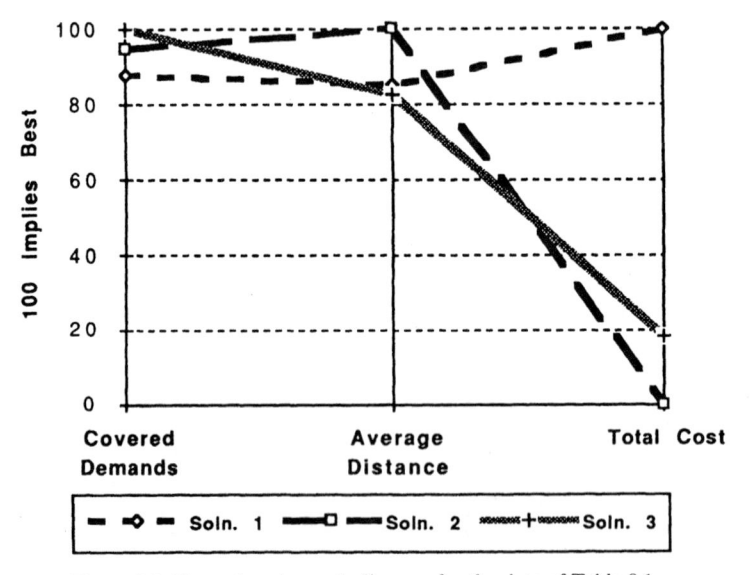

Figure 9.3. Example value path diagram for the data of Table 9.1.

Average Distance: The value shown is the difference between the indicated solution and the best solution (i.e., solution 2) as a percentage of the best solution subtracted from 100. Mathematically, we have

$$\text{Solution } X \text{ score} = \left\{ \frac{2\,(\text{Best distance}) - \text{Solution } X \text{ distance}}{\text{Best distance}} \right\} \cdot 100 \tag{9.1}$$

Therefore, the best solution will have a value of 100, solutions close to

The Planning Process for Facility Location

the best will have scores close to 100, and poor solutions would have low scores.[10]

Total Cost: The difference between the solution's total cost and the largest (worst) total cost (i.e., that of solution 2) is given as a percentage of the difference between the worst and best total cost solutions. Mathematically, we have

$$\text{Solution } X \text{ score} = \left\{ \frac{\text{Worst case} - \text{Solution } X \text{ cost}}{\text{Worst cost} - \text{Best Cost}} \right\} \cdot 100 \quad (9.2)$$

Note that this scoring method ensures that the best solution will have a value equal to 100 and the worst solution will have a value equal to 0. Thus, even small percentage differences between the solutions will be shown on a 0 to 100 scale.

Figure 9.3 displays some of the key tradeoffs between the solutions. Solution 1 minimizes the total cost, is the second best in terms of average distance, and is the worst in terms of coverage. Solution 2 minimizes the average distance, is the second best in terms of coverage, and is the worst in terms of the total cost. Finally, solution 3 maximizes coverage, is the worst in terms of average distance, and is the second best in terms of total cost. Since the differences in coverage *appear* small in this figure, we may be led to the adoption of solution 1 as a reasonable compromise.

However, the reader should note that the method used to scale the objective function values can significantly affect the shape and character of the value path diagram. As indicated above, some scaling systems force the worst solution to have a value of 0 and the best to have a value of 100. Others (like that used for the average distance) do not force the worst solution to have a score of 0. Finally, we can have perfectly logical scales in which all solutions plot strictly between 0 and 100. For example, had we plotted coverage as a percentage of the total demand, the values would have ranged from 76.17 to 86.67. Care should be taken in choosing a scaling system to eliminate biases of the analysts that may implicitly enter the analysis through the way in which solutions are presented.

To highlight this point, consider Figure 9.4 in which all three objectives are plotted using (9.2) above. Now it *appears* as though solution 2 might be the best compromise since each solution is the best in terms of one objective and the worst in terms of another objective. However, when the solutions are ranked in terms of their relative performance on the objective for which they are each second best, solution 2 seems to be the best. The key point is that the choice of scaling procedure can affect the way in which solutions are perceived when using a value path diagram.

---

[10] Note that very bad solutions (those with average distances greater than twice the best value) would have negative values with this scaling system.

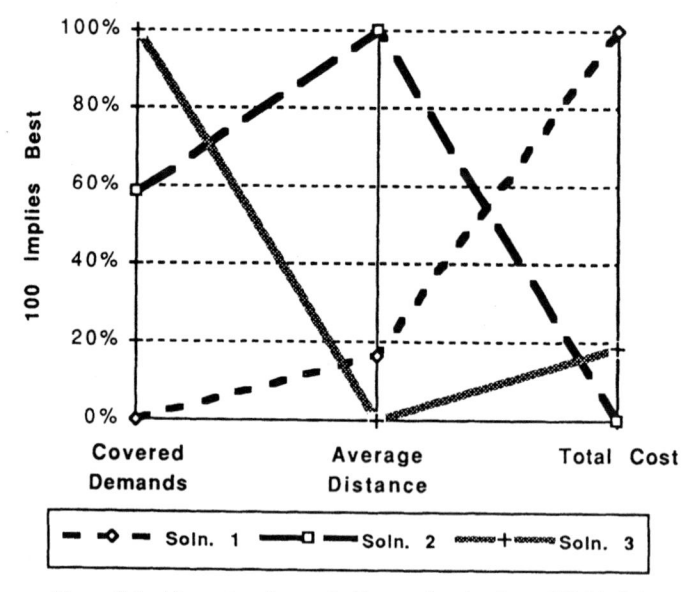

Figure 9.4. Alternate value path diagram for the data of Table 9.1.

In addition to representing solutions in objective space (using either a tradeoff curve or a value path diagram, for example) location plans should also be plotted in the solution space. Doing so typically involves drawing maps showing where facilities are located and the assignment of demand nodes to facilities. Geographic information systems can greatly facilitate the examination of alternative siting plans in the solution space. The reader is cautioned, however, to treat such computerized databases with some degree of caution. The fact that the data have been digitized does not necessarily mean that the data are correct. In one study involving routing school buses in New York City, Braca et al. (1994) found that a very large percentage of the streets that were identified in a database as being one-way streets were incorrectly identified: either they were one-way but in the direction opposite that indicated in the database, or they were two-way streets.

Finally, the technical analysis should include recommendations regarding the timing and implementation of particularly promising solutions. If the location model is dynamic, these recommendations may be the outputs of the modeling process. Otherwise, they need to be examined exogenously.

The technical analysis is an input to the decision-making process. It is clearly an important input, but not necessarily the only one. Numerous assumptions, approximations, and abstractions of the real world will necessarily have been made in the course of conducting the technical analysis. Therefore, the technical analysis will always be supplemented by qualitative judgmental factors brought to bear on the problem by planners and decision makers. In the political arena in which public sector decisions are made, location decisions for one type of facility or service may be tied to compromises being struck on unrelated public projects. In the private sector, such factors as the availability of trained labor or communal amenities may affect location choices. In both cases, a key role of the technical analysis is to indicate the degree to which solutions based on qualitative judgments or political compromises deviate from solutions that are optimal (or noninferior) with respect to the modeled objectives. The result of the decision-making step should ultimately be an agreement on a course of action to be followed. However, it is possible that the initial result will be a call for additional analysis and/or a redefinition of the problem.

9.2.4 Implementation

After a course of action has been adopted, the selected actions and policies need to be implemented. Such actions and policies typically include building new infrastructure and facilities, expanding or closing existing sites, reallocating demands, and/or redefining the operating policies of the system under study. Demands may be reallocated for a number of reasons. Clearly, if new facilities are built or existing facilities are expanded or closed, demands will need to be reallocated to different facilities. Demands may also need to be reallocated if transport costs have changed (as a result of the institution of new services in a region, for example) since the last allocation of demands to facilities. Finally, if demands have changed (particularly in a nonuniform manner) since the last allocation of demands to facilities, a reallocation may be needed.

Operating policies may also be changed as a result of a location study. For example, a location study may suggest that the incremental cost associated with adopting single sourcing constraints is small compared to the savings that may result from simplifying the material handling and ordering processes of the firm.

As solutions are being implemented, the operation of the system and the outcomes of the location study should be monitored. This allows us to change the plan if unforeseen problems arise. In the course of monitoring the implementation and operation of the system, we may identify unanticipated benefits associated with the new location plan. In addition, careful monitoring of the system will allow us to model similar systems better in the future. Finally, this monitoring may lead to further studies of the system being analyzed.

9.2.5 Caveats on the Planning Process

We have outlined a rather detailed planning process that includes problem definition, technical analysis, communication and decision, and implementation. Feedback between the steps of each of these phases and between the phases is likely and often desirable.

The process as described above is the ideal. In practice, compromises must often be made particularly in the technical analysis. The decision process will often not wait for the results of the technical studies. Timely solutions and analyses are critical. An approximate solution to a crude model whose limitations are documented and understood and whose results are delivered on time will be of more value than will the results of an exact solution to a detailed model that are provided after decisions have already been made. Decision makers, planners, and technical analysts share the responsibility for ensuring that there is adequate time for an appropriate analysis.

9.3 SUMMARY

In this chapter we outlined a four-phase planning process. The first phase was problem definition in which we must identify the problem, key actors, and decision makers. We must also clearly understand the goals of the enterprise whose facilities are being located. At the end of the problem definition phase, we must ask ourselves whether improved facility locations are likely[11] to help the enterprise achieve its strategic goals. If not, we should help the organization in ways other than performing a location study.

If improved facility locations are deemed to be potentially valuable to the enterprise, the second phase of the planning process involves conducting the technical analysis. Here we need to convert the organization's strategic goals into quantifiable objectives. We also need to identify key constraints and options that apply to the problem. This leads to our formulating mathematical models. The data that feed the models need to be collected and validated. The models must then be exercised in a number of ways to help analysts, planners, and decision makers understand the options available to them and to gain confidence in the technical analysis.

In the third phase of the process, the results of the technical analysis need to be communicated in clear, nontechnical terms to the decision makers. Location plans should be presented in both objective space (using tradeoff curves, value path diagrams, or other means of highlighting tradeoffs between objectives) and solution space (using maps and figures). The technical analysis should be used as an input into the process by which decisions are made and courses of action are selected.

After decisions have been made, we begin to implement the selected actions and policies. Careful monitoring of the implementation and early operation of the system will enable us to effect midcourse corrections if unanticipated problems arise and to conduct better studies in the future.

---

[11]Clearly, we cannot be certain at the outset, before performing any technical analysis, that improved facility locations will help the organization. However, if we are fairly certain that they will not, we should avoid the temptation of conducting a location study simply because doing so may be convenient or profitable.

EXERCISES

9.1 In discussing the value path diagrams (Figures 9.3 and 9.4), we defined two methods of scaling the data. Can formula (9.1) be used to scale the total cost data shown in Table 9.1? If not, why?

9.2 The We-Rent-Em/U-Dent-Em Inc. truck company rents small trucks to private individuals. Currently, it only has offices in two small midwest cities. However, it has been doing such a "bang up" business that it is ambitiously planning to branch out to a nationwide service over the next five to eight years. The company is concerned about such factors as where to locate facilities in each city it moves into and when to enter each market.

  (a) Identify the key factors that might influence the company's location choices. In particular, what factors would influence its choice of cities in which to locate offices and, within those cities, the locations of the offices themselves.

  (b) What data would you need to help We-Rent-Em/U-Dent-Em Inc. develop an expansion plan?

9.3 **(a)** As discussed in Exercise 1.4, the ever-increasing concern about the environment has brought vehicle emission testing programs under close scrutiny. Clearly identify the key actors associated with such a testing program? For each group of actors, identify the strategic goals that they are likely to have for such a program. Which goals are likely to be affected by the locations of vehicle inspection facilities?

  (b) Identify at least two different quantifiable objectives that are legitimate proxies for the goals of at least some of the key actors you identified in part (a).

  (c) Formulate appropriate mathematical location models for the objectives identified in part (b) and the associated constraints. Discuss any relevant approximations and assumptions that you must make.

  (d) Collect the data needed to exercise the models formulated in part (c). Note that in many cases much of the relevant data can be obtained from publicly available sources of information including census records and commercial geographic information systems.

  (e) Conduct an "exhaustive" study to formulate a location plan for vehicle inspection stations. Present your results in the form of a consulting report for the state environmental protection agency director and the state bureau of motor vehicles.

9.4 Overcrowding in federal, state, and local prisons has led to calls for increased prison space. Repeat the steps outlined in Exercise 9.3 but applied now to the problem of siting and/or expanding prisons in your state. Prepare a final report for the state director of prisons that will help her "arrest" the concerns of citizens in the communities in which the new prisons are to be located.

9.5 See-Better Opticians is planning to license a set of franchises in your city. Describe in detail how you would help "frame" the franchise location problem for the firm's "sight" planners. This is critical since the planners did not provide much "insight" into the problem when the firm last entered a new market. This time around the director of "sight" planning does not want to be embarrassed and does not want to become a "spectacle" when he presents his report to the vice president of strategic planning.

You should discuss the following issues at a minimum:

(a) What objectives should be used in licensing franchise outlets?

(b) How can you model the objectives?

(c) What data should be collected to support the analysis?

Appendix A

SITATION Operations Guide

A.1 WHAT IS SITATION, WHAT DOES IT DO, AND WHAT SYSTEM REQUIREMENTS DOES IT NEED?

SITATION is a program that is designed to solve a number of different facility location problems including:

- Maximum covering location problems;
- P-median problems; and
- Uncapacitated fixed charge location problems.

SITATION allows the user to choose from a variety of heuristics and optimization-based approaches for each of the different objective functions. For example, for the uncapacitated fixed charge problem, the user can choose between: (i) the ADD algorithm, (ii) the DROP algorithm, (iii) an exchange algorithm (initiated by either the ADD or DROP algorithm), (iv) a neighborhood search algorithm (again initiated by either the ADD or DROP algorithm), and (v) a Lagrangian relaxation algorithm.

While SITATION does not employ branch and bound to obtain provably optimal solutions, the Lagrangian approaches available for each of the three objective functions *tend* to do very well for most problems. In those cases in which the lower and upper bounds differ by more than the user would like to declare the solution "optimal," SITATION allows the user to force nodes in and out of the solution. With this capability, the user can walk through a branch-and-bound process manually. This capability is also useful if the user simply wants to determine the effect of requiring a facility to be located at some node, or preventing the program from locating a facility at certain nodes.

In addition, SITATION allows the user to add or delete facilities from a solution manually. It also allows the user to exchange facility locations manually, thereby allowing the user (i) to gain confidence in the solutions determined by the program and (ii) to test his or her own solutions.

SITATION and all other programs that accompany this text are designed to run on an IBM-compatible 286 (*or higher*) *computer under DOS.* The use of a math coprocessor for 286- and 386-level machines is highly recommended. In addition, SITATION must frequently read portions of the program as well as data from the disk. Therefore, to speed up the operation of the program, the user should load the program, related files, and data onto a hard disk as discussed in Section A.2.

The remainder of this appendix is organized as follows. Section A.2 discusses loading SITATION from the distribution disk onto the hard disk. Section A.3 discusses the input files that are needed to run SITATION. Section A.4 outlines the way in which SITATION uses distance files. Section A.5 provides a detailed description of the individual menus and commands that are available using SITATION. Section A.6 summarizes the DOS command line options that can be used to change the way SITATION operates and starts execution. Finally, Section A.7 provides a quick guide to the use of SITATION and walks the user through the keystroke-by-keystroke operation of the program to perform two simple analyses—solving a maximum covering problem and then generating a tradeoff curve showing average distance as a function of the number of facilities located.

A.2   INSTALLING SITATION AND THE OTHER DISTRIBUTRION PROGRAMS AND FILES ON A DISK AND STARTING SITATION

To store SITATION on a hard disk, the user should run the installation program located on the diskette that accompanies this book. The installation requires approximately 900KB of free disk space on your hard drive. To install SITATION and the other software and sample data sets that accompany the text, do the following:

1. Assuming you will be using drive A as the floppy drive for your diskette, at the A:⟩ prompt type **INSTALL**.[1] You may also type **A:INSTALL** at the C:⟩ prompt.
2. Follow the instructions displayed by the installation program. The default choice for the installation directory is LOCATE and the default drive is C. You may change the settings by using the cursor to move to any of the lines, pressing the **ENTER** key, and then either selecting or typing your preferred setting at the prompt. At the end of the process, you will be given the opportunity to review the README.MSD file for more information about the diskette files.

---

[1]Commands to be typed by the user will be shown in **BOLD COURIER** font throughout this appendix and all other appendices dealing with the use of any of the programs on the disk.

Storing SITATION and Other Distribution Programs

The installation program will install the following files on the hard disk:

SITATION.EXE	GOTH.CHR
SITATION.OVR	LITT.CHR
MOD-DIST.EXE	SANS.CHR
MOD-DIST.OVR	TRIP.CHR
NET-SPEC.EXE	CITY1990.GRT
MENU-OKF.EXE	MDST88.GRT
COLORSET.EXE	CAPITALS.ONL
COLORS.COL	SORTCAP.GRT
PRINTER.CNS	MDST49.GRT
PRINTER.ELQ	8NODE.MAN
PRINTER.HP2	MDST8.MAN
PRINTER.HP3	8NODE.EUC
PRINTER.HP4	MDST8.EUC
PRINTER.DJ5	8NODE.NET
ATT.BGI	MDST8.NET
CGA.BGI	FIG74.NET
EGAVGA.BGI	MDST12.NET
HERC.BGI	LOGISTIC.NDE
IBM8514.BGI	LOGISTIC.LNK
PC3270.BGI	README.MSD
VESA.BGI	

The sample SITATION data files included on the diskette are shown in the table below.

Demand and Location Data File	Distance Data File	Number of Nodes	Distance Metric	Brief Description
CITY1990.GRT	MDST88.GRT	88	Great circle	88-node problem on the continental U.S. See Appendix G.
SORTCAP.GRT	MDST49.GRT	49	Great circle	49-node problem on the continental U.S. See Appendix H.
8NODE.MAN	MDST8.MAN	8	Manhattan	8-node example corresponding to Exercise 6 of Chapter 6.
8NODE.EUC	MDST8.EUC	8	Euclidean	8-node example corresponding to Exercise 6 of Chapter 6 but using Euclidean distances.
8NODE.NET	MDST8.NET	8	Network	8-node example corresponding to Exercise 6 of Chapter 6 but using network distances.
FIG74.NET	MDST12.NET	12	Network	12-node network corresponding to Figure 7.4

In addition, **CAPITALS.ONL** is a file that can be used with the **CITY1990.GRT** data set to select only state capitals and Washington, DC as candidate facility sites. **COLORS.COL** is used to change the default color combinations for the programs. It is described in Appendix D. **LOGISTIC.NDE** and **LOGISTIC.LNK** are node and link data files respectively for use with MENU-OKF. The **PRINTER.xxx** files contain printer control codes. The user should copy the appropriate file to the **PRINTER.CNS** file to setup printing functions for SITATION and MENU-OKF. (See Appendix F for a discussion of the **PRINTER.CNS** file and its use.) To do so, type the following command at the DOS prompt.

    COPY PRINTER.xxx PRINTER.CNS

from within the directory in which SITATION was installed where **xxx** is the appropriate file name extension for the printer being used.

SITATION allows the user to print selected graphics screens to a printer. To do so, the user must load the graphics program into memory using the DOS **GRAPHICS** command before SITATION is started. The user should consult his or her DOS manual for additional information on the use of this command.

To start SITATION, the user should change directories to the directory in which the SITATION.EXE and SITATION.OVR files are stored. To do so, the user should type

    CD \LOCATE

at the DOS prompt. The program may then be started by typing

    SITATION

at the DOS prompt. Additional command lined options are outlined in Section A.6.

If the file SITATION.OVR cannot be found, an overlay error will occur. In this case, an error message will be displayed and program execution will stop. If this file can be found, the program will begin be briefly displaying a message indicating whether or not EMS (expanded) memory is being used to hold the overlay program segments. For example, if sufficient EMS memory is available, the following message will be displayed:

    Using Ems for faster OVERLAY swapping

Defining sufficient EMS memory will allow the overlays to be stored in memory and will generally expedite the operation of the program, particularly if a slow hard disk is being used. SITATION will then display an initial screen. This will be followed by a summary of the command line options specified by the user, if any command line options were given. This in turn will be followed by the INITIAL MAIN MENU as described in Section A.5.2.

A.3 DEFINING SITATION INPUTS

SITATION allows the user to define input files that give the locations of the demand nodes and candidate facility sites. This section outlines the format for these input files. From these files, SITATION will compute a distance matrix using one of the following three distance metrics:

- Manhattan or right-angle distances;
- Euclidean or straight-line distances; or
- Great circle distances.

In addition, the user can define his or her own *network distance matrix* for particular problems.

SITATION requires a demand data set as a minimal input. This data set must include the following six fields:

1. *Node ID*. This should generally be a number so that records are numbered consecutively. These values are *not* used by the program, but must be specified as the first field in each record.
2. *X Coordinate or Longitude*. Longitudes should be measured in degrees and fractions of a degree, *not* degrees, minutes, and seconds. If a compatible distance file cannot be found in the directory in which the demand file is located, the X and Y coordinates (or latitudes and longitudes) are used to compute the internodal distances using the distance metric specified by the user.
3. *Y Coordinate or Latitude*. Latitudes should also be measured in degrees and fractions of a degree, *not* degrees, minutes, and seconds.
4. *First Demand Value*. SITATION allows the user to define two different demand values. The first value should be the fourth field in each record of the data set.
5. *Second Demand Value*. Even if there is only one relevant demand value, two demand values must be specified. If only one demand value is relevant, the two values may be the same.
6. *Fixed Cost*. This specifies the fixed cost of locating at each node. To minimize a function of the number of facilities located, all fixed costs

should be set equal to each other. Note that the cost per mile is specified at run time and so the relative value of the fixed and transportation costs can be varied while the program is running.

Finally, the user can include any additional descriptive information about the node following the fixed cost. For example, the data sets described in Appendices G and H both include the city name following the fixed cost field.

This file should be specified as an ASCII file with fields separated by blanks. The data for each demand node must be on a separate record. All demand nodes are automatically considered candidate facility locations, though the user may prevent the program from siting facilities at particular nodes by using the FORCE NODES OPTION MENU described below.

Note that the file containing the demand data is the only file that *must* exist before SITATION is run if great circle distances, Euclidean distances, or Manhattan distances are used. (If network distances are used, a network distance file must also exist. Section A.4 describes the distance files used by SITATION.)

In addition to the demand nodes, the user can also specify an additional file with candidate locations. This should also be an ASCII file with fields separated by blanks. Again, the information for each additional location must be on a separate record. The fields in this data set are:

1. *Node ID*. Again, this is not used by the program, but must be specified. The user should use this as some form of convenient identification.
2. *X Coordinate or Longitude*.
3. *Y Coordinate or Latitude*.
4. *City*. This may be any ASCII field that does *not* contain blanks. It is not used by the program, but must be specified.
5. *State*. This may be any ASCII file that does *not* contain blanks. It, too, is not used by the program, but must be specified.
6. *Fixed Cost*. This is the fixed cost of locating at the additional node.

This file should also be specified as an ASCII file with fields separated by blanks. The data for each additional candidate node must be on a separate record.

The number of demand nodes plus the number of additional candidate facility locations cannot exceed 150.

The program NET-SPEC may be used to create either the demand or the additional candidate location data sets (as outlined in Appendix B) for use with network-based distances. The program MOD-DIST can be used to change the distance data sets in a variety of ways (as outlined in Appendix C).

A.4 SITATION DISTANCE FILES

As indicated above, SITATION allows the user to use Manhattan, Euclidean, great circle, or network distances. After the demand and location (and optionally additional candidate location) data sets are read, the program will search the current directory for a file with the name **MDSTnnn.dis**, where **nnn** is the total number of demand and additional candidate location nodes and **dis** takes on one of the following values: **MAN** (for Manhattan distances), **EUC** (for Euclidean distances), **GRT** (for great circle distances), or **NET** (for network distances). (For example, for a problem with 88 nodes and great circle distances, the program will search for a file named MDST88.GRT.) If such a file is found, it is read. If it is not found, the program will create this file if Manhattan, Euclidean, or great circle distances are being used. If the program cannot find this file and network distances are being used, the program will ask the user if he or she wants to (1) abort the program, (2) provide an alternate name for the distance file, or (3) change to a different distance metric. This information is summarized in the table below.

	DISTANCE METRIC			
	Manhattan	Euclidean	Great Circle	Network
Name of distance file for which SITATION searches	MDSTnnn.MAN	MDSTnnn.EUC	MDSTnnn.GRT	MDSTnnn.NET
SITATION can create distance file if it does not exist	YES	YES	YES	No. If the file does not exists, SITATION allows the user to specify a different distance file name, change the distance metric, or abort the program.
Distance file may be created by	1. SITATION if the file does not exist 2. MOD-DIST 3. User	1. SITATION if the file does not exist 2. MOD-DIST 3. User	1. SITATION if the file does not exist 2. MOD-DIST 3. User	1. NET-SPEC 2. MOD-DIST 3. User

Note that if the user has multiple problems with the same total number of nodes (demand plus additional candidate locations), *extreme care must be taken to be sure that the correct distance file is always available to be read.* For example, if the user initially defines a 50-node problem for cities in New York state (using great circle distances), the program will create a file called

MDST50.GRT. If, at a subsequent time, the user creates a second problem with 50 nodes in California also using great circle distances, SITATION will read the MDST50.GRT distance file that was created for the New York state problem, *unless* this file has either been removed from the disk, renamed, or stored in a different directory from the California data sets.

All distance files use the same format. Each distance is on its own record. Distances are arranged with the row variable changing first, followed by the column variable. In other words, the distances in the first column of the distance matrix are listed first, those in the second column are listed next, and so on. Rows in the distance matrix represent demand nodes and columns represent candidate sites. *Distances specify the distance from the candidate site to the demand node. In all distance computations* (e.g., determining whether or not a candidate site can cover a demand node), *SITATION uses the distance from the candidate site to the demand node.*

To illustrate these concepts, consider the following asymmetric network.

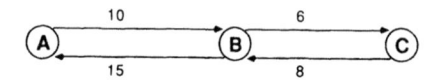

```
        10            6
   (A) ------> (B) ------> (C)
       <------     <------
        15            8
```

The distance matrix for this network would be

Demand Node / Candidate Site	A	B	C
A	0	15	23
B	10	0	8
C	16	6	0

The distance file for this network would be:

0
10
16
15
0
6
23
8
0

If the coverage distance 15, the facility as node B could cover demands at node A, but the facility at node A could not cover demands at node C.

A.5 RUNNING SITATION

A.5.1 General Information

Many times during the execution of the program, the program will temporarily stop running to allow the user to read what is displayed on the screen. The following message will be shown at the bottom of the screen:

```
Holding...Hit any key when ready...
```

After the user has finished reading what is shown on the screen, any key (e.g., the space bar) should be pressed to allow the program to continue execution.

Most menus have a title. The title is surrounded by one, two, or three stars on either side. The MAIN MENU has three stars. Any submenu of the main menu has two stars. Submenus of submenus have one star. In this way the user can know how far from the MAIN MENU he or she is at any time.

In most cases, typing **Q** will return the user to the next higher menu level. For example, if the user is in the DOS FUNCTION MENU, typing **Q** will return the user to the next MAIN MENU.

Section A.7 is a quick guide to the use of SITATION. That section outlines the use of SITATION to do two simple analyses. The user who wants to get up and running quickly might want to jump to that section and to refer to the appropriate portions of Section A.5 as needed.

A.5.2 Initial MAIN MENU

```
        *** MAIN OPTION MENU <v. CMF3.4d> ***

           <Q> Quit
           <D> DOS Directory Functions
           <O> Turn ON Default Mode
           <L> Get Location and Demand Data


           Input the desired option ==>
```

SITATION uses a "smart" menuing system in which the only MAIN MENU options that are displayed are those that the user can safely employ based on the current state of the program as defined by the data that have been read and the additional information that has either been supplied by the user and or computed by the program. The menu shown above is the initial MAIN MENU. The available commands are:

- **Q** This allows the user to stop execution of the program. The user will be asked whether he or she really wants to stop execution of the program. Typing **Y** in response to the prompt will return the user to DOS; typing **N** will return the user to the MAIN MENU shown above.
- **D** This brings up a menu that allows the user to execute a number of DOS functions including changing the directory and displaying a list of the files in the current directory. The next menu to be displayed will be the DOS FUNCTION MENU.

O   This toggles the Default mode on or off. When the MAIN MENU displays the message Turn ON Default Mode, the Default mode is **OFF**; when the MAIN MENU displays the message Turn OFF Default Mode, the Default mode is **ON**. In Default mode, the program does not ask the user selected questions. For example, the user cannot change any of the settings in any of the Lagrangian OPTION SETTING MENUS while in Default mode. Using this mode will expedite the operation of the program. However, *the use of Default mode is not recommended for inexperienced users of the program.*

L   This allows the user to load the demand data. The next menu to be displayed will be the DISTANCE OPTIONS MENU. Generally, this will be the first command selected from the MAIN OPTION MENU. Note that before data are read into the program, a *large* number of data fields are initialized. Thus, to prevent the user from inadvertently losing data, after a demand data set has been read, the program asks the user to verify that he or she wants to reload new data when this command is executed a second or subsequent time.

A.5.3   DOS FUNCTION MENU

```
              ** DOS FUNCTION MENU **

         <Q> Quit--Return to Main Menu

         <G> Get Current Directory
         <C> Change Current Directory
         <M> Make a New Directory
         <R> Remove a Directory

         <D> Display List of Files
         <E> Extended Directory of Files

         <N> ReName a File

         <X> Execute Other DOS Functions

          Input the desired option ==>
```

This menu is entered from the MAIN MENU. It allows the user to execute a number of DOS functions from within SITATION. In particular, the user may need to change directories and/or view a list of (selected) files on the current directory. This menu allows the user to perform these DOS functions. In addition, if sufficient memory is available, the user can temporarily exit to DOS. The commands available are:

Q   This returns the user to the MAIN MENU.

G   This reports the name of the current directory.

C This allows the user to change directories.

M This allows the user to create a new directory.

D This allows the user to display a short list of files on the current directory. The user is prompted for a mask for the files to be viewed. As in DOS, the user can view all files by typing *.* or some subset by using the appropriate combination of characters, ?, and * symbols.

E This allows the user to display an extended list of files on the current directory. Again, the user can select the files to be seen using a DOS-style file mask.

N This allows the user to change the name of a file. The user will be prompted for the name of an existing file whose name is to be changed. The user will then be prompted for a new name for the file. The new name must not be the name of an existing file. This command is useful if the user needs to rename a distance file, for example, to prevent SITATION from using an existing file that does not apply to the case at hand. This command might be needed, for example, to prevent SITATION from reading a distance file that does not apply to the data that are about to be loaded. In that case, the user should rename the current **MDSTnnn.dis** file to some other name (e.g., **MDSTnnn.BAK**). This will force SITATION to generate the correct distance file when great circle, Euclidean, or Manhattan distances are being used.

X This allows the user to exit temporarily from SITATION to DOS to execute any standard DOS function. To return to SITATION, the user simply types **exit** at the DOS prompt. If insufficient memory is available to execute this option, a message to this effect will be displayed, and the user will be returned to the DOS FUNCTION MENU.

A.5.4 DISTANCE OPTIONS MENU

```
            ** DISTANCE OPTIONS MENU **

            <Q> QUIT--Return to Main Menu

            <N> Network Based
            <E> Euclidean
            <G> Great Circle
            <M> Manhattan


            Input the desired option ==>
```

After the user selects **L** from the MAIN MENU to load demand data into the program, the user must select one of the distance metrics shown above.

(If the user is operating in Default mode—default options are on—this menu will be skipped. In that case, the distance will be set to either the distance specified using the command line options discussed in Section A.6 or, if no command line distance options are specified, SITATION will use great circle distances as the default distance metric. When a second or subsequent data set is read, the user must always specify the distance metric using this menu, whether or not the program is operating in Default mode.)

The commands available under this menu are:

Q This allows the user to return to the MAIN MENU. In some cases, when the user is given the option of changing the distance metric because the program could not find an appropriate network distance file, this menu option will not appear.

N This allows the user to select network-based distances. The program will search for a file with the name **MDSTnnn.NET**, where **nnn** is the number of demand and additional candidate nodes in the problem being analyzed.

E This allows the user to select Euclidean distances. The program will search for a file with the name **MDSTnnn.EUC**, where **nnn** is the number of demand and additional candidate nodes in the problem being analyzed. If this file is not found, SITATION will create this file by computing the Euclidean distances between the points using the X and Y coordinates in the demand data file and the file containing the additional candidate nodes.

G This allows the user to select great circle distances. The program will search for a file with the name **MDSTnnn.GRT**, where **nnn** is the number of demand and additional candidate nodes in the problem being analyzed. If this file is not found, SITATION will create this file by computing the great circle distances between the points using the longitudes and latitudes in the demand data file and the file containing the additional candidate nodes.

M This allows the user to select Manhattan or right-angle distances. The program will search for a file with the name **MDSTnnn.MAN**, where **nnn** is the number of demand and additional candidate nodes in the problem being analyzed. If this file is not found, SITATION will create this file by computing the Manhattan distances between the points using the X and Y coordinates in the demand data file and the file containing the additional candidate nodes.

After the user has specified the distance metric to be used, the program will prompt the user for the demand data set to be read with the following prompt:

```
Input the file name to read (e.g. CITY1990.GRT or SORTCAP.GRT) ==>
```

Running SITATION

Note: The file to be read is the file that contains the demand data as described in Section A.3. For example, to use the 88-node data set described in Appendix G, the user would specify file `CITY1990.GRT`. SITATION would then automatically read the **MDST88.GRT** file (assuming great circle distances had been specified using the DISTANCE OPTIONS MENU).

Next the user will have to select one of three demand options using the DEMAND OPTIONS MENU.

A.5.5 DEMAND OPTIONS MENU

```
             ** DEMAND OPTIONS MENU **

             <F>  First Scenario Only
             <S>  Second Scenario Only
             <B>  Both Scenarios Combined

         Input the desired option ==>
```

This menu allows the user to tell SITATION which of the two demand data fields to use. The options are:

- **F** This allows the user to select the first demand data field.
- **S** This allows the user to select the second demand data field.
- **B** This allows the user to have SITATION use the sum of the demands in the two fields as the demand for each node.

After the user has selected a data set, the data are read from the demand file. Next, if the user initiated SITATION with the OTHERS command line option, the program will ask the user if he or she wants to read a file with additional candidate facility locations as follows:

```
Do you want to read a file with
additional candidate locations? (Y/N) ==>
```

If the user answers **Y**, the program will prompt the user for the name of the file containing the additional locations.

After the demand (and optionally the additional candidate location) data have been read, the program will attempt to read the distance file as discussed above. The following sort of message will be displayed:

```
         <<< READING FILE <mdst88.GRT> >>>

              <<< PLEASE WAIT >>>

      <<< SORTING DISTANCES ... PLEASE WAIT >>>
```

Again, if network distances are being used and SITATION cannot find the **MDSTnnn.NET** file that it needs to read, it will ask the user what he or she wants to do using the following message:

```
<<< NETWORK DISTANCE FILE <mdst12.NET> DOES NOT EXIST >>>
 Do you want to ABORT, GIVE another name
 or CHANGE Distance Type? <A / G / C> == >
```

Typing **A** will abort the program and return the user to DOS. Typing **C** will send the user to the DISTANCE OPTIONS MENU discussed in Section A.5.4. Finally, typing **G** will allow the user to specify an alternate name for the network distance file.

To complete the input information, SITATION asks the user for the cost per mile. A brief message will be displayed indicating that the cost per mile is used for evaluation purposes only (to evaluate the total cost of a solution) when either the P-median or the maximum covering model is used. The user will then be asked to specify the cost per mile at the following prompt:

```
Input COST PER MILE == >
```

The value specified here is used in computing the total cost of a solution if the user elects to use either the maximum covering or the P-median objective function. Note that the total cost is simply *reported* in these cases and is not optimized.

A.5.6 Second MAIN MENU

```
            *** MAIN OPTION MENU <v. CMF3.4d> ***

       <Q> Quit
       <D> DOS Directory Functions
       <O> Turn ON Default Mode
       <L> Get Location and Demand Data

       <I> Initial Solution Operations
       <F> Force Specific Sites In / Out of Soln.
       <C> Get Coverage Distance (and Cover List)


             Input the desired option == >
```

Running SITATION

After the demand (and additional candidate location) data have been read, the user is returned to the expanded MAIN MENU shown above. The new commands are:

- **I** This allows the user to read or write an initial solution to the file.
- **F** This allows the user to force specific candidate sites in and out of the solution. In addition, the user can read or write a file containing the status of each node (e.g., forced into the solution, allowed in the solution, part of an initial solution, or excluded from the solution).
- **C** This allows the user to specify the coverage distance. Even if the user is not planning to use the maximum covering objective function, a coverage distance must be specified before any of the algorithms can be used. In these cases, this value is used for *reporting* purposes only.

If the program is not operating in Default mode, a message will be displayed indicating that the coverage distance is used for reporting purposes only when either the P-median or the uncapacitated fixed charge model is used. The user will then be prompted for the coverage distance with the following prompt:

```
Input the critical coverage distance ==>
```

If the program is in Default mode when the user types **C** from the MAIN MENU, the program will prompt the user for the coverage distance (unless the coverage distance was specified on the command line as discussed in Section A.6 and this is the first time during the program run that the coverage distance is being specified). SITATION will then create the cover list.

If a coverage distance has already been specified, typing **C** from the MAIN MENU will result in the user being asked to verify that he or she wants to change the coverage distance. The following prompt will be issued:

```
<<< YOU ARE ABOUT TO CHANGE THE COVER LIST >>>
Are you sure you want to do this? (Y / N) ==>
```

Typing **N** will return the user to the MAIN MENU. Typing **Y** will allow the user to change the coverage distance. This will also remove from memory any current solution.

After the coverage distance has been specified by the user and the cover lists have been successfully generated, the third MAIN OPTION MENU will be displayed.

A.5.7 INITIAL SOLUTION OPTION MENU

```
           ** INITIAL SOLUTION OPTION MENU **

           <Q> Quit--Return to Main Menu
           <S> Show Status of the Nodes
           <R> Read Initial Solution
           <C> Clear Initial Solution
           <W> Write Solution on Disk

           Input the desired option ==>
```

SITATION allows the user to store a solution for subsequent analysis. Initial solutions read in this manner are placed in the solution using some of the heuristic algorithms (e.g., the greedy algorithm for the maximum covering objective or the myopic algorithm for either the P-median problem or the ADD algorithm for the uncapacitated fixed charge location problem). These sites may subsequently be removed from the solution if the user employs one of the exchange heuristics (e.g., the substitution algorithm for the maximum covering objective or either the neighborhood or the exchange heuristics for the P-median problem or the uncapacitated fixed charge location problem) or if the user manually removes or exchanges these sites. The command options include:

- **Q** This returns the user to the MAIN MENU.
- **S** This displays a screen showing the status of each of the nodes. Nodes may be in one of four categories:
    1. *Excluded*. These nodes will not be included in any solution identified by any of the algorithms for any of the objective functions.
    2. *Allowed*. These nodes may be included in a solution identified by SITATION.
    3. *Required*. These nodes are forced into any solution identified by any of the algorithms for any of the objective functions.
    4. *Initial Sol*. These nodes are initially placed into the solution when the user selects one of the algorithms identified above. However, they may subsequently be removed from the solution if one of the improvement algorithms is used and/or if the user manually removes or exchanges the facility site for another location.
- **R** This allows the user to read a file containing an initial solution.
- **C** This option clears any previously read initial solution.
- **W** This option allows the user to store a current solution on the disk for later use as an initial solution or as a set of required locations, as

discussed below. Note that a solution cannot be stored on the disk until the model is run using the options described below.

A.5.8 FORCE NODES OPTION MENU

```
              ** FORCE NODES OPTION MENU **

         <Q> Quit--Return to Main Menu
         <S> Show Status of the Nodes
         <W> Write Status to Disk
         <R> Read Status from Disk

         <D> Disallow All Nodes
         <P> Permit All Nodes

         <F> Force a Site into Soln.
         <U> Unforce Site

         <E> Exclude a Site from Soln.
         <A> Allow a Site into Soln.

         Input the desired option ==>
```

This menu allows the user to force sites in and out of the solution. The commands included in this menu are:

Q This returns the user to the MAIN MENU.

S This displays a screen showing the status of each of the nodes. The screen displayed is identical to that shown using the **S** option of the INITIAL SOLUTION OPTION MENU. It is a good idea to verify the status of the nodes in the problem using this option *before* exiting this menu to be sure that the status of each node is as expected.

W This allows the user to store the status of all of the locations in a file for later use.

R This allows the user to read a previously stored file containing the status of the nodes.

D This allows the user to prohibit (or disallow) all nodes from being included in the solution. This is useful if the user wants to allow only a small number of sites in the solution. In that case, the easiest way to set the status of the nodes is to disallow all nodes and then to allow selected sites back into the solution using the **A** command below.

P This allows the user to permit or allow all nodes to be considered as candidate locations.

F This allows the user to force a site into the solution. This is useful if either a facility already exists at a particular site or if the user wants to walk through a branch-and-bound-type analysis manually.

U This allows the user to change the status of a node that was forced into the solution to simply being allowed in the solution.

E This allows the user to exclude a site from any solution developed by the algorithms for any of the objective functions. This is useful in a variety of contexts including when the user wants to walk through a branch-and-bound-type analysis manually.

A This allows the user to change the status of a node that was forced out of the solution (excluded) to simply being allowed in the solution.

A.5.9 Third MAIN MENU

```
        *** MAIN OPTION MENU <v. CMF3.4d> ***

   <Q> Quit
   <D> DOS Directory Functions
   <O> Turn ON Default Mode
   <L> Get Location and Demand Data

   <I> Initial Solution Operations
   <F> Force Specific Sites In / Out of Soln.
   <C> Get Coverage Distance (and Cover List)

   <R> Run the Model
   <T> Generate Tradeoff Curve

           Input the desired option == >
```

After the user has specified the coverage distance, the user can run any of the algorithms built into SITATION. An expanded MAIN MENU is displayed showing the following two additional options:

R This allows the user to solve a problem using one of the three available objective functions and any of the associated algorithms. The user will be prompted for the desired objective function using the OBJECTIVE FUNCTION MENU described below.

T This allows the user to create a tradeoff curve for any of the three objective functions. Again, the user will be prompted for the desired objective function using the OBJECTIVE FUNCTION MENU described below. One sample tradeoff curve is shown at the end of Section A.7. Note that the Lagrangian algorithms are not available to the user when creating a tradeoff curve. In addition, the neighborhood and exchange algorithms are not available when computing a tradeoff curve for the fixed charge objective function.

The operation of the program for each of these two options is similar (except that certain algorithms may not be used when constructing tradeoff curves as indicated above). After the tradeoff curve has been computed, the values associated with the curve are displayed on the screen. The user may then graph the results on the screen using the DRAWING SELECTIONS menu discussed in Section A.5.14.

A.5.10 OBJECTIVE FUNCTION MENU

```
              ** OBJECTIVE FUNCTION MENU **

         <Q>  QUIT--Return to Main Menu

         <C>  Maximum Covering
         <M>  P-Median
         <F>  Fixed Charge (Uncapacitated)

              Input the desired option ==>
```

After selecting either **R** (to run the program for a single solution) or **T** (to develop a tradeoff curve) from the MAIN MENU, the user must tell SITATION which objective function to employ. This is done through the commands of the OBJECTIVE FUNCTION MENU described below:

- **Q** This allows the user to return to the MAIN OPTION MENU without solving one of the three problems (maximum covering, P-median, or uncapacitated fixed charge).
- **C** This selects the maximum covering objective function. The user will next be asked to specify which algorithm to employ via the covering objective ALGORITHM SELECTION MENU (Section A.5.11).
- **M** This selects the P-median objective function. The user will next be asked to specify which algorithm to employ via the P-median objective ALGORITHM SELECTION MENU (Section A.5.12).
- **F** This selects the fixed charge objective function. The user will next be asked to specify which algorithm to employ via the fixed charge objective ALGORITHM SELECTION MENU (Section A.5.13).

A.5.11 Covering Objective ALGORITHM SELECTION MENU

```
              * ALGORITHM SELECTION MENU *

         <Q>  Quit--Return to Main Menu

         <G>  Greedy Algorithm (Use with initial Soln.)
         <S>  Greedy Algorithm with Substitution

         <L>  Lagrangian Algorithm

              Input the desired option ==>
```

This menu allows the user to select the algorithm to be used in solving maximum covering problems. The available options include:

Q This allows the user to return to the MAIN OPTION MENU without solving a maximum covering problem.

G This allows the user to employ the greedy adding algorithm. If the user has read an initial solution and simply wants to load this solution, he or she should use this algorithm and set the number of facilities to be located equal to the number in the initial solution.

S This allows the user to employ the greedy adding and substitution algorithm. The user will be asked whether or not substitution should be done after each iteration (after each node is added to the solution) or only after a full greedy adding solution has been obtained.

L This allows the user to use Lagrangian relaxation to solve the problem.

After selecting one of the algorithms, the following prompt will ask the user for the number of facilities to locate:

```
Input the number of FACILITIES to locate 1 <= X <= 88 ==>
```

The range of allowable values depends on the number of nodes in the problem and the number of nodes that are forced into the solution, included in the initial solution, and/or excluded from the solution.

After specifying the number of facilities to locate, the user will be asked about a number of other options that are specific to the algorithm that is being used. In all cases, the user will be given the option of automatically excluding dominated nodes. We recommend that this be done in all cases. Dominated nodes are automatically excluded when running in Default mode.

When using the Lagrangian algorithm, the user is also asked if he or she wants to attempt single node substitutions after the Lagrangian algorithm terminates. This often improves the solution. Using this option is also strongly recommended and is done automatically when running in Default mode.

Also, when using the Lagrangian algorithm, the user is presented with the following menu of options that control the operation of the Lagrangian algorithm:

```
                    OPTION SETTING MENU
                                          SUGGEST      CURRENT

<Q> Quit--No More Option Changes
<P> Critical Percentage Difference        0.0100000    0.0100000
<I> Maximum Number of Iterations               400          400
<L> Minimum Alpha Value Allowed           0.0000500    0.0000500
<F> Number of Failures Before Changing Alpha     4            4
<R> Restart Failure Count on Improved Soln.   TRUE         TRUE
<D> Crowder Damping Term                    0.3000      0.30000
<N> Exchange Search on Improved Solution    FALSE        FALSE
<B> Best or Lagrangian LB in Stepsize     LAGRANGE     LAGRANGE

    U(i) = ALPHA*[Avg. Demand] + BETA*[Demand(i) - ALPHA*Avg. Demand]

<C> Constant Factor (ALPHA) for Initial U(i)  1.00000     1.00000
<S> Slope    Factor (BETA ) for Initial U(i)  0.50000     0.50000

Input Desired Option ==>
```

Two columns of values are shown. Suggested values are shown under SUGGEST; the values that will be used are shown under CURRENT. Initially, the current and suggested values are the same. As parameter values are changed, however, the new values are retained, unless a new data set is read into the program (in which case all the current values are again reset to the suggested values). The commands available from this menu include:

- **Q** This tells the program that no additional changes in the Lagrangian algorithm parameters are to be made and that the algorithm is to be executed.
- **P** The Lagrangian algorithm will terminate when any one of a number of conditions is met. One such condition is that the percentage difference between the lower and upper bounds is less than or equal to some value. Smaller values will tend to force the algorithm to execute more iterations but will also tend to result in tighter lower and upper bounds. This option allows the user to change this value.
- **I** The Lagrangian algorithm will terminate when it has performed some maximum number of iterations. Larger values will allow the algorithm to execute more iterations. This option allows the user to change this value.
- **L** The Lagrangian algorithm will also terminate when the constant α falls below some value. Smaller values will allow the algorithm to execute more iterations. This option allows the user to change the minimum allowable value for α.

F The Lagrangian algorithm will halve α whenever the number of failures to improve the upper bound exceeds some value. Larger values will tend to cause the algorithm to execute more iterations. This option allows the user to change the number of failures before α is changed.

R In some cases, the user may want to change α only when the algorithm fails to improve the upper bound after a specified number of *consecutive* iterations. Alternatively, the user may elect to count *all* (including nonconsecutive) failures to improve the upper bound with the current α value when determining when to halve α. If this value is TRUE, the algorithm will only count consecutive failures. If the value is FALSE, the algorithm will count all iterations at the current value of α on which it fails to improve the upper bound. This option *toggles* this value.

D Crowder's damping term allows the user to set the search direction for new Lagrange multipliers equal to a weighted average of all previous search directions. (A value of 0.0 indicates that only the current iterations' direction is to be employed.) Setting the search direction as a weighted average of previous search directions tends to limit the sensitivity of the algorithm to the choice of the initial Lagrange multipliers. This option allows the user to change this term.

N When an improved primal solution (lower bound) is obtained, it is sometimes advantageous to apply a single node exchange algorithm to the new solution in an attempt to improve it further. If this parameter is TRUE, single node exchanges will be attempted when an improved solution is found; if the parameter is FALSE, single node exchanges will not be attempted. This command *toggles* this option.

B The code for the Lagrangian algorithm retains two lower bounds. The first is the value of the best primal feasible solution it has found; generally, this is found by evaluating the coverage obtainable from the facility locations identified at each iteration of the algorithm and ignoring the coverage variables. This is termed the *Best* lower bound. The second is the value of the best primal feasible solution obtained considering both the location and coverage variables. This is termed the *Lagrangian* lower bound. The user has the option of selecting which of these values to use in computing the stepsize for changes in the Lagrange multipliers. Most papers on the subject indicate that the Best values should be used. Computational experience with the code suggests that using the Lagrangian lower-bound values (which will always be less than or equal to the Best values) often results in better solutions. This command *toggles* this option.

Note that this command is only available when the STEPCHANGE command line parameter is specified when starting SITATION from DOS. (See Section A.6.) If this option is not specified, SITATION will only use the Lagrangian bound in computing the stepsize for the maximum covering objective function. For both the *P*-median and fixed charge objective functions, SITATION uses the equivalent of the Best lower bound in computing stepsizes.

c The initial Lagrange multipliers, λ_i^1, are computed using the following formula:

$$\lambda_i^1 = \alpha \bar{h} + \beta(h_i - \bar{h})$$

where α and β are parameters of this equation (not to be confused with the α used in computing the stepsize), $\bar{h}$ is the average demand, and h_i is the demand at node *i*. This command allows the user to change the value of α used in this formula.

s This command allows the user to change the value of β used in the formula for computing the initial Lagrange multipliers.

After the user is through specifying the parameters for the Lagrangian procedure (and the user has pressed **q** to exit from the OPTION SETTING MENU), the Lagrangian algorithm will execute. During execution, the following screen of information will be displayed to indicate the progress of the algorithm:

```
                        LOCATIONS   =           7

                        ITERATION   =         114

Use Main Menu Option    UPPER BOUND = 217275816.  BEST LAG UB = 217175816.6
<S> Show the Results
and Show Results Option LOWER BOUND = 204088310.  LAGRANGE LB =         0.0
<B> Basic Inputs / Outputs
to see final bounds     PERCENT     =       6.462


                        ALPHA       = 0.00003052

                        FAILURES    =           1

                        STEPSIZE    =      927.64  <- +
```

(Only the center and left portions of this information are visible if the STEPCHANGE command line option is not invoked.)

The lower bound that is shown at the end of the Lagrangian algorithm may be improved upon by requesting that single node substitutions be attempted at the end of the Lagrangian procedure. For example, in the model run that generated the results shown above, the single node substitution algorithm resulted in an improved lower bound of 217274909 and a 0.00042 percent difference between the lower and upper bounds as opposed to the 6 percent difference between the bounds that was obtained when the

Lagrangian algorithm terminated.[2] The final lower and upper bounds can be seen by using the BASIC INPUTS command of the SHOW RESULTS MENU described below.

A.5.12 *P*-Median objective ALGORITHM SELECTION MENU

```
              * ALGORITHM SELECTION MENU *

    <Q> QUIT--Return to Main Menu

    <M> Myopic Algorithm (Use with Initial Soln.)
    <E> Exchange Algorithm
    <N> Neighborhood Search Algorithm

    <L> Lagrangian Algorithm


            Input the desired option ==>
```

This menu allows the user to select the algorithm to be used in solving *P*-median problems. The available options include:

- **Q** This allows the user to return to the MAIN OPTION MENU without solving a *P*-median problem.
- **M** This allows the user to employ the myopic algorithm. After this, the user is prompted for the number of facilities to locate. The algorithm then executes and the user is eventually returned to the MAIN MENU. If the user has read an initial solution and simply wants to load this solution, he or she should use this algorithm and set the number of facilities to be located equal to the number in the initial solution.
- **E** This allows the user to employ the exchange algorithm. The user will be asked whether or not substitution should be done after each new

---

[2] These results were obtained using the first demand data in the SORTCAP.GRT data set with a coverage distance of 300 miles, locating 7 facilities. The default Lagrangian parameters were employed. Running the Lagrangian algorithm with the following different parameters will allow the user to confirm the optimality of this solution:

Critical percentage difference	=	0.0000001
Maximum number of iterations	=	4000
Minimum alpha used	=	0.0000001
Number of failures before changing alpha	=	20
Constant factor (ALPHA) for initial $U(i)$	=	0.0
Slope Factor (BETA) for initial $U(i)$	=	0.0

All other parameters should be set to their suggested values.

Running SITATION

facility is added or only after the last facility is added using the myopic algorithm as a starting solution. Next, the user is prompted for the number of facilities to locate. The algorithm then executes and the user is eventually returned to the MAIN MENU.

N This allows the user to employ the neighborhood search algorithm. The user will be asked whether or not the algorithm should be executed after each new facility is added or only after the last facility is added using the myopic algorithm as a starting solution. Next, the user is prompted for the number of facilities to locate. The algorithm then executes and the user is eventually returned to the MAIN MENU.

L This allows the user to employ the Lagrangian relaxation algorithm. The user will be prompted for the number of facilities to locate. Then the program will ask the user if he or she wants to try to improve the solution using an improvement algorithm after the Lagrangian procedure terminates. If so, the user will be able to select either the exchange (or substitution algorithm) or the neighborhood search algorithm. The use of one or the other of these algorithms after the Lagrangian procedure (preferably the exchange algorithm) is highly recommended. This approach is used if the program is in Default mode. After the user tells the program whether or not an improvement algorithm is to be used after the Lagrangian procedure (and if so which one), the following Lagrangian OPTION SETTING MENU is displayed:

```
                    OPTION SETTING MENU

                                              SUGGEST      CURRENT

<Q>  Quit--No More Option Changes
<P>  Critical Percentage Difference           0.0100000    0.0100000
<I>  Maximum Number of Iterations                   400          400
<L>  Minimum Alpha Value Allowed              0.0000500    0.0000500
<F>  Number of Failures Before Changing Alpha         4            4
<R>  Restart Failure Count on Improved Soln.      TRUE         TRUE
<D>  Crowder Damping Term                      0.30000      0.30000
<N>  Neighborhood Search on Improved Solution     TRUE         TRUE

U(i) = ALPHA + BETA*[Dem(i)*Dist(i, [2])] + GAMMA*AvgDem + DELTA*Dem(i)

<C>  Constant Factor (ALPHA) for Initial U(i)      0.0          0.0
<S>  Slope    Factor (BETA)  for Initial U(i)    0.000        0.000
<A>  Average  Factor (GAMMA) for Initial U(i)    10.00        10.00
<Z>  Demand   Factor (DELTA) for Initial U(i)     0.00         0.00
```

This menu is similar to that used for controlling the parameters of the Lagrangian relaxation algorithm for the maximum covering problems (as described in Section A.5.11). Commands **Q**, **P**, **I**, **L**, **F**, **R**, and **D** are identical (except that the term *lower bound* should be used whenever the maximum covering discussion used the term *upper bound* and vice versa since the *P*-median is a minimization problem and the maximum covering is a maximization problem). These commands will not be described again. The other commands are:

N This allows the user to specify whether or not the neighborhood search algorithm should be applied whenever an improved primal (upper bound) solution is found. If the value of this parameter is TRUE, the neighborhood search algorithm will be used in this way; if the value is FALSE, the neighborhood search algorithm will not be used. This command *toggles* this option.

C The initial Lagrange multiplier values, λ_i^1, are set equal to

$$\lambda_i^1 = \alpha + \beta h_i d_{i[2]} + \gamma \bar{h} + \delta h_i$$

where α, β, γ, and δ are parameters of this equation, h_i and $\bar{h}$ are the demand at node i and the average demand, respectively, and $d_{i[2]}$ is the distance from node i to the second closest node. This command allows the user to change the value of α.

S This command allows the user to change the value of β in the equation for the initial Lagrange multipliers.

A This command allows the user to change the value of γ in the equation for the initial Lagrange multipliers.

Z This command allows the user to change the value of δ in the equation for the initial Lagrange multipliers.

After all parameters for the Lagrangian procedure have been specified and the user has quit the OPTION SETTING MENU, the Lagrangian algorithm will be executed. Control will then be returned to the MAIN MENU.

A.5.13 Fixed Charge Objective ALGORITHM SELECTION MENU

```
            * ALGORITHM SELECTION MENU *

        <Q> QUIT--Return to Main Menu

        <A> Add Algorithm (Use with Initial Soln.)
        <D> Drop Algorithm
        <E> Exchange Algorithm
        <N> Neighborhood Search Algorithm

        <L> Lagrangian Algorithm

             Input the desired option ==>
```

Running SITATION

This menu allows the user to select the algorithm to be used in solving uncapacitated fixed charge problems. After the user selects one of the four algorithms, the user will be prompted for the cost per mile to use. Note that even though a cost per mile was specified as part of the initial set of inputs when loading the data, the cost per mile must again be input. While this may initially seem redundant, this feature allows the user to change the cost per mile each time a new fixed charge problem is run. This allows the user to analyze the sensitivity of the solution to changes in the relative importance of the fixed costs and the transport costs. If either set of values is uncertain, such an analysis may be very important.

After the cost per mile is specified, the user will be prompted for the number of facilities to locate. The user may specify either a particular number of facilities to locate [in which case, the algorithm operates as if the additional constraint $\sum_j X_j = P$ has been included in the formulation (where P is the number of facilities to locate)] or the user may ask that the algorithm find the "best" number of facilities to locate by responding with -1 when prompted for the number to locate. In the latter case, the number of facilities located will depend on the algorithm used and on the ability of that algorithm to find good or near optimal solutions. For example, when the ADD algorithm is used, facilities will continue to be added until the addition of any other facility will increase the total cost.

The available options include:

Q This allows the user to return to the MAIN OPTION MENU without solving an uncapacitated fixed charge problem.

A This allows the user to employ the ADD algorithm. After this option is selected, the user is prompted for the number of facilities to locate. After the user indicates the number of facilities to locate, the algorithm executes and the user is eventually returned to the MAIN MENU. If the user has read an initial solution and simply wants to load this solution, he or she should use this algorithm and set the number of facilities to be located equal to the number in the initial solution.

D This allows the user to employ the DROP algorithm. After selecting this algorithm, the user will be prompted for the number of facilities to locate. The algorithm will then load all facilities in the solution and subsequently attempt to remove facilities until either the desired number of facilities have been located or, if the number of facilities to locate was set to -1, until removal of an additional facility will increase the total cost.

E This allows the user to apply an exchange algorithm to the solution constructed using either the ADD algorithm or the DROP algorithm. After selecting this algorithm, the user will be prompted for the number of facilities to locate. The user will then be prompted for the initial algorithm (ADD or DROP) to use in constructing the initial solution to which the exchange algorithm is to be applied.

N This allows the user to apply a neighborhood search algorithm to the solution constructed using either the ADD algorithm or the DROP algorithm. After selecting this algorithm, the user will be prompted for the number of facilities to locate. The user will then be prompted for the initial algorithm (ADD or DROP) to use in constructing the initial solution to which the neighborhood search algorithm is to be applied.

L This allows the user to apply a Lagrangian relaxation algorithm in solving the fixed charge problem. After selecting this algorithm, the user will be prompted for the number of facilities to locate. Next, the user will be asked if he or she wants to use an improvement algorithm after the Lagrangian algorithm terminates. If so, the user will be asked to select either the exchange or the neighborhood search algorithm. As before, the use of one of these algorithms (preferably the exchange algorithm) is highly recommended and is automatically adopted when the program is operating in Default mode. Finally, the user will be permitted to change the parameters that control the operation of the Lagrangian algorithm with the following OPTION SETTING MENU.

```
                        OPTION SETTING MENU

                                                    SUGGEST       CURRENT
<Q> Quit--No More Option Changes
<P> Critical Percentage Difference                  0.0100000     0.0100000
<I> Maximum Number of Iterations                        880           880
<L> Minimum Alpha Value Allowed                     0.0000500     0.0000500
<F> Number of Failures Before Changing Alpha             12            12
<R> Restart Failure Count on Improved Soln.            TRUE          TRUE
<D> Crowder Damping Term                            0.30000       0.30000
<N> Neighborhood Search on Improved Solution           TRUE          TRUE

     U(i) = ALPHA + BETA*[Dem(i)*Dist(i, [2])*($/mile)] + GAMMA*AvgDem
            + DELTA*Dem(i) + EPSILON*AvgFixedCost

<C> Constant   Factor (ALPHA)   for Initial U(i)       0.0           0.0
<S> Slope      Factor (BETA)    for Initial U(i)       0.000         0.000
<A> Average    Factor (GAMMA)   for Initial U(i)      10.00         10.00
<Z> Demand     Factor (DELTA)   for Initial U(i)       0.00          0.00
<E> Fixed $    Factor (EPSILON) for Initial U(i)      10.00         10.00
Input Desired Option ==>
```

The commands, **Q, P, I, L, F, R, D, N, C, S, A**, and **Z** in this menu are identical to those in the *P*-median OPTION SETTING MENU and will not be described again. The other commands available in this menu control the computation of the initial Lagrange multiplier values. The initial Lagrange multipliers are computed using the following formula:

$$\lambda_i^1 = \alpha + \beta h_i d_{i[2]} + \gamma \bar{h} + \delta h_i + \varepsilon \bar{\$}$$

where all values are as defined before, ε is a parameter of the equation, and $\bar{\$}$ is the average fixed cost (averaged over all candidate locations). Menu option **E** allows the user to change the value of the parameter ε in this equation.

After the parameters for the Lagrangian algorithm have been set and the user has quit the OPTION SETTING MENU (using the **Q** command), the algorithm will execute. Control will then be returned to the MAIN MENU.

A.5.14 DRAWING SELECTIONS Menu

```
                    Drawing Selections

       <Q> Quit--Return to Previous Menu

       <M> Medium Resolution Graphics (Print Available)
       <H> High   Resolution Graphics (Print Available)
       <E> EGA    Resolution Graphics (No Printing)


            Input the desired option ==>
```

This menu allows the user to select the resolution of either tradeoff curves or maps that are to be drawn on the screen. The available commands are:

- **Q** This returns the user to the previous menu. Usually, this will be the MAIN MENU, though it may be one of the other menus as well.
- **M** This selects medium resolution graphics (320 × 200). When this option is selected, the user may print the resulting screen by typing **P**, provided the necessary graphics drivers have been loaded into DOS using the DOS GRAPHICS command as discussed above.
- **H** This selects high resolution graphics (640 × 200). When this option is selected, the user may print the resulting screen by typing **P**, provided the necessary graphics drivers have been loaded into DOS using the DOS GRAPHICS command as discussed above.

E   This selects EGA resolution graphics (640 × 350). In this mode, the user cannot print the resulting screen by typing P. However, the standard DOS **Print Screen** keyboard command may be used, provided the necessary graphics drivers have been loaded into DOS using the DOS GRAPHICS command as discussed above.

When this menu appears for the display of a map of the results, the user will be prompted for a title for the map prior to its being displayed. Entering a carriage return without any text will display the map with no title.

A.5.15 Fourth MAIN MENU

```
            *** MAIN OPTION MENU <v. CMF3.4d> ***

       <Q> Quit
       <D> DOS Directory Functions
       <O> Turn ON Default Mode
       <L> Get Location and Demand Data

       <I> Initial Solution Operations
       <F> Force Specific Sites In / Out of Soln.
       <C> Get Coverage Distance (and Cover List)

       <R> Run the Model
       <T> Generate Tradeoff Curve

       <S> Show the Results
       <P> Print the Results
       <E> Exchange Facility Locations
       <M> Map the Results


               Input the desired option ==>
```

After solving a problem by either running the model or generating a tradeoff curve, the user can display, print, or manipulate the results in a variety of ways. An expanded MAIN MENU is displayed with the following additional commands:

S   This command allows the user to display the results on the screen. Control is passed to the SHOW RESULTS MENU described below in Section A.5.16 from which a variety of results may be displayed.

P   This command allows the user to print selected results to a printer. The user will be prompted for a title for the run. Next, the user will be asked if he or she wants a long or short version of the output. The long version of the output prints additional information related to the cover lists. Next, the program attempts to read a file named **PRINTER.CNS** in which printer control parameters are specified. If this file is found, the file will be read and a message to this effect will be shown; if it is not found, SITATION will assume that the printer is compatible with an Epson LQ-850 printer and will use printer control parameters for this printer type. The user will be asked if he or she wants to store this information in a file called **PRINTER.CNS**. (Appendix F provides a brief description of this file.) Next, the user will be reminded to be certain that the printer is on, is online, and that the paper is at the top of the page. When ready, the user may press any key except Q. Pressing Q returns the user to the MAIN MENU. If an error occurs during printing (e.g., a printer is not attached to the computer or the printer is not turned on), SITATION will display a warning message and return to the MAIN MENU.

E This command allows the user to manipulate the results by adding or deleting facilities and by exchanging facility sites. Control is passed to the FACILITY EXCHANGE MENU described in Section A.5.18.

M This command allows the user to draw a map of the solution on the screen. Control is passed to the DRAWING SELECTION menu described in Section A.5.14. After selecting a resolution for the map, the user will be prompted for a title for the map. After the title is input, the map will be drawn on the screen. The map shows uncovered demand nodes, covered demand nodes, and facility sites. Demand nodes are connected to the facility to which they are assigned by a solid line if the demand node is covered and by a dashed line if the node is not covered. At the bottom of the display, SITATION shows a key which includes the following additional information:

    The number of facilities sited;

    The objective function used;

    The cost per mile;

    The total cost;

    The coverage distance;

    The demand-weighted average distance between demand nodes and the facilities to which they are assigned; and

    The percentage of the total demand that is covered.

This option is only available if the problem is defined for (i) great circle distances over the continental United States, (ii) Euclidean distances, or (iii) Manhattan distances. If the inputs do not correspond to one of these three cases, the results cannot be mapped and this menu option will not be displayed.

A sample map is shown at the end of Section A.7.

A.5.16 SHOW RESULTS MENU

```
                  ** SHOW RESULTS MENU **

          <Q> Quit--Return to Main Menu
          <S> Show All Basic Reports

          <B> Basic Inputs / Outputs
          <F> Forced Nodes etc.
          <D> Dominated Nodes

          <T> Tradeoff Curve
          <L> Location Summary
          <E> Extended Summary
          <U> Uncovered Locations Summary

          <A> Assigned Demand Summary
          <C> Coverage Summary
          <N> Number of Times Covered Summary

          <G> Graph the Tradeoff Curve
          <X> Extra SHOW RESULTS Commands

              Input the desired option ==>
```

This menu allows the user to display the results on the screen. The commands available from this menu are:

Q This allows the user to return to the MAIN MENU described in Section A.5.15.

S This allows the user to show many of the reports available from the program. The program will cycle through the following reports before returning to the SHOW RESULTS MENU:

- *Summary of Basic Model Inputs and Results.* This presents a one-screen summary of the inputs and key outputs including the number of nodes, number of facilities located, coverage distance, solution algorithm, cost per mile, and, when using one of the Lagrangian algorithms, the lower and upper bounds on the solution as well as the number of iterations required to obtain the solution.
- *Status of the Nodes.* This screen indicates which demand and additional candidate nodes are excluded from the solution, allowed to be in the solution, required to be in the solution, or are part of an initial solution.
- *List of Dominated Nodes.* This screen displays a list of the demand and candidate facility nodes that are dominated (if the dominated nodes have been identified by the program).

- *Tradeoff Curve.* This screen displays the tradeoff curve information that is relevant to the problem, if a tradeoff curve has been computed. The information that is actually displayed depends on the objective function. For *maximum covering* problems, the program displays a table showing the total and incremental coverage as well as the percentage of the total demand that is covered as a function of the number of facilities that are located. For *P-median* problems, the program shows the average distance and improvement in average distance as a function of the number of facilities that are located. Finally, for *uncapacitated fixed charge* problems, the program shows the total cost, the improvement (which is a positive number), or degradation (which is a negative number) in the total cost as a function of the number of facilities that are located, and the difference between the total cost for the given number of facilities and the best total cost found as a percentage of the best cost.
- *Short Facility Location Summary.* This is a short summary listing the node numbers of the locations at which the program has located facilities.
- *Facility Location and Coverage Table.* This is an extended summary of the solution showing the node number of each location at which a facility is located, the X and Y coordinates (longitude and latitude), and the demand that is *covered* by the facility at that location. In addition, this screen displays the following summary statistics: the total number of covered demands, the percentage of the total demand that is covered, the average distance, and the total fixed cost, transportation cost, and total cost.
- *Facility Location and Assigned Demand Table.* This is also an extended summary of the solution showing the node number of each location at which a facility is located, the X and Y coordinates (longitude and latitude), and the demand that is *assigned* to the facility at that location. In addition, this screen displays the demand-weighted average distance between demand nodes and the nearest facility locations and the total assigned demand (which should equal the sum of the demands at all of the demand nodes).
- *Uncovered Locations and Demands Table.* This table displays the node numbers, X and Y coordinates (longitude and latitude), and the demands of all nodes that are not covered by any facility within the specified coverage distance. In addition, the total number of uncovered demands is shown as is the percentage of the total demand that is not covered.
- *Demand Area Coverage Report.* This screen shows a list of all of the demand node numbers. Each node is highlighted depending on whether or not it is covered.

- *Coverage Summary Table.* This screen shows the number of nodes and total demand that is covered $0, 1, 2, \ldots$ times. In addition, this screen shows the total number of demand nodes, the total demand, and the percentage of the total demand that is covered at least once.

B This allows the user to display the SUMMARY OF THE BASIC MODEL INPUTS AND RESULTS.

F This allows the user to display the STATUS OF THE NODES report.

D This allows the user to display the LIST OF DOMINATED NODES. If no nodes are dominated, or if the dominated nodes have not been computed, a message to this effect is shown.

T This allows the user to *display* the TRADEOFF CURVE information described above. If the tradeoff curve information has not been computed, a message will be displayed indicating that this information is not available. The user should note that when using many of the heuristic algorithms, this information is computed and may be displayed, but the information may be incomplete. For example, if the user employs the greedy adding and substitution algorithm to solve a maximum covering problem locating 5 facilities when a total of 14 are needed, SITATION will show the tradeoff curve for the first 5 facilities (for which it has computed and stored the tradeoff curve information). The information for the remaining 9 facilities will not be shown (as it has not been computed or stored).

L This allows the user to display the SHORT FACILITY LOCATIONS SUMMARY.

E This allows the user to display the FACILITY LOCATION AND COVERAGE TABLE.

U This allows the user to display the UNCOVERED LOCATIONS AND DEMANDS TABLE.

A This allows the user to display the FACILITY LOCATION AND ASSIGNED DEMAND TABLE.

C This allows the user to display the DEMAND AREA COVERAGE REPORT.

N This allows the user to display the COVERAGE SUMMARY REPORT.

G This allows the user to *graph* the tradeoff curve. Control is passed to the DRAWING SELECTIONS menu described in Section A.5.14.

X This allows the user to display a number of other summary statistics and tables. Control is passed to the SHOW EXTRA RESULTS MENU described below in Section A.5.17.

A.5.17 SHOW EXTRA RESULTS MENU

```
             * SHOW EXTRA RESULTS MENU *

        <Q> Quit--Return to Previous Menu

        <L> Location Summary

        <G> Graph the Tradeoff Curve

        <C> Cover Lists
        <U> Utilization of Facility Sites
        <A> Assignment to Sites

        <S> Save Tradeoff Curve

        <F> Fixed Costs


            Input the desired option ==>
```

This menu allows the user to display a number of additional results on the screen. In addition, the user can use the **S** command of this menu to save the tradeoff curve information to a disk file. The commands available with this menu are:

Q This allows the user to return to the SHOW RESULTS MENU described in Section A.5.16.

L This allows the user to display the SHORT FACILITY LOCATIONS SUMMARY.

G This allows the user to *graph* the tradeoff curve. Control is passed to the DRAWING SELECTIONS menu described in Section A.5.14.

C This allows the user to display a list of the nodes that cover or are covered by a user-selected node. The user will be prompted for the node number of the candidate site whose cover list is to be displayed. (Typing **0** allows the user to return to this menu.) The user will then be asked whether he or she wants to display the list of nodes which are covered by this node (**B**) or which cover (**C**) this node. SITATION will then display the appropriate list along with the distance between the selected node and each other node that is either covered by this node or which covers this node. (Note that unless the distance matrix is asymmetric, the two lists should be identical. Also note that SITATION rounds distances to the nearest integer.) After the requested cover list is displayed, SITATION again asks the user for the node number whose cover list is to be displayed. Again, to return to the SHOW EXTRA RESULTS MENU, the user should type **0**.

U This allows the user to display a table showing the demand areas that are assigned to user-selected facility nodes. The user will be prompted for the node number of a site *at which a facility is located*. SITATION will then display a table showing the demand nodes assigned to the facility, the distances between the demand nodes and the facility, the demands of the demand nodes, and whether or not each demand node is covered. In addition, SITATION will show a summary of the (i) total number of demands that are both assigned to the facility location *and* covered by the facility, (ii) total number of demands assigned to the facility, and (iii) total number of demands covered by the facility. (If the user specifies a number of a nonselected node, a message to this effect will be displayed and the user will then again be prompted for a node number of a selected facility facility location site.) If the user specifies node number **0**, control will be returned to the SHOW EXTRA RESULTS MENU.

A This allows the user to display a summary of the facility sites to which each demand node is assigned. Demand nodes are shown in numerical order in a table. For each demand node, the node number of the facility to which the demand node is assigned is displayed as well as the distance between the demand node and the facility to which it is assigned. Covered and uncovered demand nodes are shown in different colors.

S This allows the user to store the tradeoff curve information in an ASCII file on the disk in the current directory. The user will be prompted for a file name. SITATION suggests default names of the form **NnnnDddd.obj**, where **nnn** is the number of nodes in the problem, **ddd** is the coverage distance, and **obj** is equal to TRD for maximum covering problems, MED for *P*-median problems, and FIX for uncapacitated fixed charge location problems. To adopt the suggested name, the user need only type an "at" sign (**@**) when prompted for the file name. Alternatively, the user may specify a name of his or her own choice. Note that the suggested names of the tradeoff curves do not include any encoding of the algorithm used to solve the problem or the cost per mile used in computing the total cost for fixed charge problems.

F This displays a list of the fixed costs associated with each of the nodes. Nodes at which facilities are located are highlighted on the screen.

A.5.18 FACILITY EXCHANGE MENU

```
             ** FACILITY EXCHANGE MENU **

   <Q>  Quit--Return to Main Menu
   <S>  Show Current Solution

   <A>  Add a Node to Solution
   <D>  Delete a Node from Solution
   <E>  Exchange 2 Nodes

   <F>  Find Best Exchange
   <T>  Try and Do All Substitutions
   <L>  Do Limited Substitutions from Current Solution

             Input the desired option == >
```

This menu allows the user to manipulate a solution by adding, deleting, and exchanging facility locations. This is useful in gaining confidence in the solutions provided by SITATION and in testing alternate manually generated solutions. The commands available under this menu include:

- **Q** This allows the user to return to the MAIN MENU described in Section A.5.15.
- **S** This allows the user to display the SHORT FACILITY LOCATIONS SUMMARY. This is useful since the user must often know the sites at which facilities are currently located to use the other options in this menu effectively. For example, a facility cannot be deleted from a site at which no facility is currently located.
- **A** This allows the user to add a facility to the current solution. The user will be prompted for the node number of the candidate site at which a facility is to be added. If a facility exists at the specified site, a warning message will be shown and the user will again be prompted for a node number. Typing **0** as the node number returns the user to the FACILITY EXCHANGE MENU. If a valid node number is given, SITATION displays a message indicating the effect of adding the facility at the specified node on all objective functions. A facility is added at the specified location.
- **D** This allows the user to delete a facility from the current solution. The user will be prompted for the node number of the facility to be removed. If a facility does not exist at the specified site, a warning message will be shown and the user will again be prompted for a node number. Typing **0** as the node number returns the user to the FACILITY EXCHANGE MENU. If a valid node number is given, SITATION displays a message indicating the effect of removing the facility

from the specified node on all objective functions. The facility located at the specified node number is removed from the solution.

E This allows the user to simultaneously add a facility to the current solution and delete a facility from the current solution. The user will first be prompted for a node to remove from the current solution. As before, typing **0** will return the user to the FACILITY EXCHANGE MENU. Next, the user will be prompted for a node number at which a facility is to be added. Again, typing **0** will return the user to the FACILITY EXCHANGE MENU. SITATION will then display a message showing the effect of the exchange on all objective functions. The user will then be asked to confirm that he or she wants to make the indicated exchange. If so, the exchange will be made; if not, no action will be taken by the program. In either case, control is then returned to the FACILITY EXCHANGE MENU.

F This command tells the program to search for the best single node exchange that is possible. "Best" is determined based on the objective function that is currently being used. (*The user should note that exchanges that result in very small improvements will be ignored.*) If such an exchange is found, the effect of the exchange on all objective functions will be indicated and the user will be asked if he or she wants to make the exchange. If so, the exchange will be made; if not, no action will be taken by the program. In either case, control is then returned to the FACILITY EXCHANGE MENU. If no exchange is found that improves the objective function, a message to this effect will be displayed and control will be returned to the FACILITY EXCHANGE MENU.

T This tells the program to find (and implement) all possible exchanges that improve the current objective function. The program will display the effect of each such change as it is made. The user will *not* be asked whether or not he or she wants to make the indicated change. Control will eventually be returned to the FACILITY EXCHANGE MENU. Note that the solution obtained in this manner will not necessarily be the best possible solution for the given number of facilities, since exchange algorithms are heuristics for any of the available objective functions.

L This option allows the user to have SITATION perform a limited number of exchanges from the current solution. This is useful if the user has specified an initial solution (e.g., one consisting of the sites at which facilities are currently located) and he or she wants to know which N facilities to close and which new N facilities to open (where typically N will be much smaller than the total number of sites at which facilities are currently located). The user will be prompted for the number of facilities to exchange. As before, typing **0** returns the user to the FACILITY EXCHANGE MENU. If any other number is given, SITATION will consider single node exchanges from the current solution with the limitation that at most the indicated number of current sites may be exchanged for new sites.

A.6 SITATION COMMAND LINE OPTIONS

This section describes the command line options that the user may employ in starting SITATION from the DOS prompt. SITATION may be started with any combination of the following command line options:

```
%1 %2 %3 sound nologo showmemory stepchange others all help ?
```

Options shown in italics (*%1 %2 %3*) must be specified in the indicated position. Thus, the **%1** options discussed below must immediately follow the program name, SITATION. The **%2** option must be the second option following the program name, SITATION. Other command line options may be in any order and position in the command line.

Command line options may be specified in either all lowercase letters or all uppercase letters (e.g., sound or SOUND). Mixing lower- and uppercase letters within a command line option (e.g., Sound) will cause SITATION to ignore the option.

Each command line option is described below:

%1 This allows the user to specify the distance metric to be used the *first* time a demand data set is loaded into the program. Available options include:

    /g This indicates that great circle distances are to be used.

    /e This indicates that Euclidean distances are to be used.

    /m This indicates that Manhattan or right-angle distances are to be used.

    /n This indicates that network distances are to be used.

    /d This indicates that the built-in default distance metric (great circle) is to be used.

Note: Specifying any **%1** distance metric option causes SITATION to automatically start with the Default mode on. Users who are not very familiar with the implications of having the Default mode on should probably turn this mode off immediately by using the **O** command on the MAIN MENU as soon as the MAIN MENU is displayed.

%2 This allows the user to specify the coverage distance to be used the *first* time a cover list is constructed. In place of **%2** the user would specify a number giving the coverage distance (e.g., **300**).

%3 — This space is reserved for a number of options depending on the implementation of SITATION. For the current version, the only available option is to specify **/s** in the third command line position. This simply allows the user to skip the initial logo screen displayed by SITATION. This has the same effect as specifying **nologo**.

sound — This allows the user to turn on a number of additional warning sounds that SITATION generates. Only very important warning sounds will be used if this option is not specified. Note that the command line parameter **all** does not turn the **sound** option on.

nologo — This allows the user to skip the initial logo screen that is displayed by SITATION. Instead, a simple text screen will be displayed. This expedites the startup of SITATION slightly but makes the startup less fun!

showmemory — SITATION stores many of the large data arrays in dynamic memory. This memory is allocated as needed. If insufficient memory is available, a message will be displayed and the requested action (e.g., loading a large demand node file) will be aborted. When this occurs, the user may want to monitor the amount of available memory. This command line option allows the user to display a screen showing the amount of remaining memory at various stages of the program's execution.

stepchange — This allows the user to change the manner in which the stepsize is computed when using the Lagrangian algorithm for the maximum covering objective. For a further discussion of this, see Section A.5.11.

others — This allows the user to read in additional non–demand node candidate facility locations. If this command line option is not specified, only demand nodes will be considered as candidate locations.

all — This has the combined effect of specifying: **nologo showmemory stepchange others**.

help or **?** — Either of these command line options causes SITATION to display a series of screens that summarize the available command line options. SITATION will *not* be run and the user will eventually be returned to DOS.

Example valid command lines are shown below:

SITATION

This will simply start SITATION in the normal manner. The initial logo

A Quick Guide to the Use of SITATION

screen will be displayed, Default mode will be off, the user will not be permitted to load candidate locations (in addition to the demand nodes), memory information will not be displayed, and the user will not be able to change the way the stepsize is computed when solving maximum covering problems.

SITATION /g 300 /s

This will start SITATION with great circle distances used the first time a demand file is read, with the initial coverage distance set to 300 miles, and skipping the initial logo screen. SITATION will start with the Default mode on.

SITATION nologo sound

This will start SITATION but will skip the initial logo screen and will turn on the additional warning sounds generated by SITATION.

A.7 A QUICK GUIDE TO THE USE OF SITATION

This section provides a quick guide to the use of SITATION. In particular, the section gives the keystroke-by-keystroke commands that are needed to

1. change to the appropriate directory,
2. initiate SITATION,
3. load a data set into SITATION and specify the cost per mile,
4. specify the coverage distance and generate the cover lists,
5. solve a maximum covering problem using the Lagrangian relaxation algorithm,
6. display the results,
7. display a map of the solution,
8. solve for the (heuristic) tradeoff curve between the average demand-weighted distance and the number of facilities located and then display the tradeoff curve, and
9. return to the DOS prompt.

Commands that the user should type are specified in the column headed "User Should Type." Most commands are single letter commands and should not be followed by a carriage return. A carriage return is shown using the symbol ◁. In some cases, the program will display a screen until a key is pressed. In these cases, the table says the user should press **Any Key**. Pressing the **space bar** is always a safe key to press in these cases.

The map generated in Step 7 and the tradeoff curve computed in Step 8 are shown at the end of the section.

Step	To Do	User Should Type	SITATION Displays	See Section
1	Change to the Directory in Which SITATION Is Stored	CD \LOCATE ◁ or See DOS User's Guide if SITATION is stored in a different directory	DOS prompt	DOS User's Guide
2	Begin Running SITATION from DOS Prompt	SITATION ◁	Message indicating whether or not EMS memory is in use	A.2
			Initial Logo screen	
			Initial MAIN OPTION MENU	A.5.2
3	Load the Data	L	DISTANCE OPTIONS MENU	A.5.4
	Select distance type	G	Request: "Input the file name to read (e.g., CITY1990.GRT or SORTCAP.GRT)"	
	Specify name of file to read	SORTCAP.GRT ◁	DEMAND OPTIONS MENU	A.5.5
	Select demand set	F	Messages indicating that the data are being read, the distances are being read and sorted	
			Message indicating how the cost per mile is used	
			Request: "Input COST PER MILE"	
	Specify cost/mile	0.000025 ◁	Second MAIN OPTION MENU	A.5.6
4	Specify Coverage Distance and Generate Cover List	C	Message indicating how the coverage distance is	A.5.6
			Request: "Input the critical coverage distance"	

A Quick Guide to the Use of SITATION 443

			300 ◁	Message indicating that the cover list is being generated and then how many items are in the cover list	
				Third MAIN OPTION MENU	A.5.9
5	Run the Model		R	OBJECTIVE FUNCTION MENU	A.5.10
	Select objective function to use		C	ALGORITHM SELECTION MENU	A.5.11
	Select algorithm to use		L	Request: "Input the number of FACILITIES to locate $1 \Leftarrow X \Leftarrow 49$"	
			5	Request: "Do you want to try SUBSTITUTION at the end? (Y/N)"	
	Turn substitution on at the end of the Lagrangian procedure		Y	Request: "Do you want to exclude dominated nodes? (Y/N)"	
	Exclude dominated nodes		Y	Message showing that dominated nodes are being computed	
				OPTION SETTING MENU	!.5.11
	Use default options for Lagrangian algorithm		Q	Screen showing progress of Lagrangian Algorithm followed by Holding message	
	Clear current screen		Any Key	Message showing that total distance is being computed	
				Fourth MAIN OPTION MENU	A.5.15
6	Show the Results on the Screen		S	SHOW RESULTS MENU	A.5.16
	Show all basic reports		S	The following series of summary tables and screens. To move to the next one hit any key	
				Summary of Basic Model Inputs and Results	

		Any Key	Status of the Nodes	
		Any Key	List of Dominated Nodes	
		Any Key	Facility Locations	
		Any Key	Facility Location and Coverage Table	
		Any Key	Facility Location and Assigned Demand Table	
		Any Key	Uncovered Locations and Demands Table	
		Any Key	Demand Area Coverage Report	
		Any Key	Coverage Summary Table	
		Any Key	SHOW RESULTS MENU	
	Return to MAIN OPTION MENU	Q	Fourth MAIN OPTION MENU	A.5.15
7	Map the Results	M	DRAWING SELECTIONS MENU	A.5.14
	Select EGA graphics	E	Request: "Input the TITLE for the map"	
		Sample map ◁	Map of the solution	
	Clear the map and return to MAIN OPTION MENU	Any Key	Fourth MAIN OPTION MENU	A.5.15
8	Generate Tradeoff Curve	T	OBJECTIVE FUNCTION MENU	A.5.10
	Select median objective	M	ALGORITHM SELECTION MENU	A.5.12
	Select myopic algorithm	M	Messages indicating progress of the algorithm	
			SUMMARY OF BASIC MODEL INPUTS AND RESULTS	
		Any Key	Tradeoff Curve of Facilities Versus Average Distance Table Screens	

A Quick Guide to the Use of SITATION

		Any Key (four times to scroll through displays)	DRAWING SELECTIONS MENU	A.5.114
	Select EGA graphics	**E**	Tradeoff Curve	
	Clear the tradeoff curve and return to MAIN OPTION MENU	**Any Key**	Fourth MAIN OPTION MENU	A.5.15
9	Quit SITATION and return to DOS	**Q**	Request to verify that you want to quit SITATION: "Are you sure you want to do so? (Y/N)"	A.5.2
	Verify that you want to quit SITATION	**Y**	DOS Prompt	

The map of the solution generated in Step 7 is shown below.

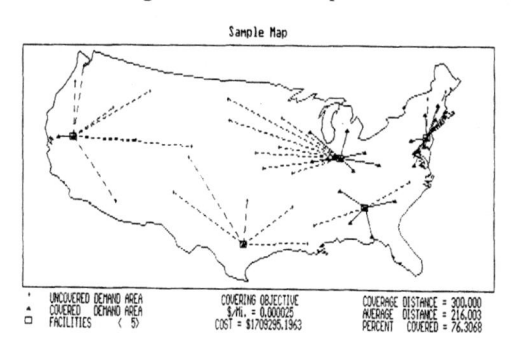

Sample Map

[Map of the United States with facility locations and demand assignments]

- UNCOVERED DEMAND AREA
- COVERED DEMAND AREA
- □ FACILITIES (5)

COVERING OBJECTIVE
$/MI. = 0.000025
COST = $1709295.1963

COVERAGE DISTANCE = 300.000
AVERAGE DISTANCE = 216.003
PERCENT COVERED = 76.3068

The tradeoff curve generated in Step 8 is shown below.

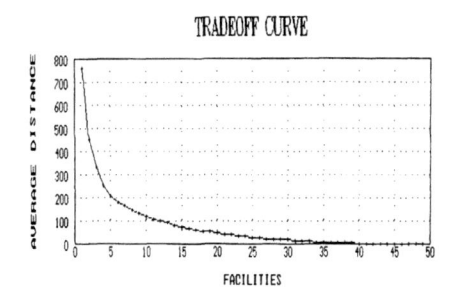

TRADEOFF CURVE

[Graph showing AVERAGE DISTANCE (0–800) on y-axis vs FACILITIES (0–50) on x-axis, with a decreasing curve]

Appendix B

NET-SPEC Operations Guide

B.1 WHAT IS NET-SPEC AND WHAT DOES IT DO?

NET-SPEC is a program that allows the user to specify the inputs for small problems to be used in SITATION. NET-SPEC creates two files: a location/demand data set that specifies the demands and a network distance file with a name **MDSTnnn.NET** where **nnn** is the number of nodes in the data set. NET-SPEC assumes that network distances are being used. In particular, it does not allow the user to specify files that use Euclidean, Manhattan, or great circle distances.[1] The X and Y coordinates of all demand nodes are set to 0.0 by NET-SPEC since network distances are not computed using these coordinates.

NET-SPEC can operate in either of two different modes. In the first mode, the user must specify **all** internodal distances. In the second mode, the user specifies the distances of actual links in the network. Using these link lengths, NET-SPEC computes shortest path distances between all nodes.

When operating in the first mode (i.e., when the user is specifying all internodal distances), NET-SPEC assumes that the distance matrix is symmetric. Thus, to generate a problem with an asymmetric network distance matrix when operating in this mode, the user must first create the location/demand data set using NET-SPEC and then modify the distance file **MDSTnnn.NET** using MOD-DIST as described in Appendix C. In the second mode (i.e., when only actual links are being specified), NET-SPEC allows links to be either two-way or one-way. If one-way links are employed, asymmetric distance matrices can be generated.

---

[1] To set up a file for use with Euclidean, Manhattan, or great circle distances, the user need only construct an ASCII file with the information for each demand node on a separate line (or record). The information needed for each demand node is described in Appendix A, Section A.2. For these distance metrics, SITATION will automatically build the appropriate distance file using the desired distance metric and the node coordinates in the file containing the information on the demand nodes (and optionally on any additional candidate location nodes).

B.2 HOW TO RUN NET-SPEC

To start NET-SPEC, the user need only type **NET-SPEC** at the DOS prompt. Note that the following command line options may be used with NET-SPEC: **sound, nologo, others**, and **all** (which has the combined effect of specifying **nologo others**). The user is referred to Section A.6 of Appendix A for a description of these options.

After the initial logo screen, the user will be asked for the number of nodes in the data set. The maximum number of nodes permitted is 150. (However, for data sets with more than about 10 to 20 nodes, the user is likely to find that writing his or her own program to generate the necessary files is easier than is using NET-SPEC.) To abort the program at this point, the user need only type **0** when prompted for the number of nodes. If the user types any number between 1 and 150 (inclusive) NET-SPEC will check whether or not a distance file with the name **MDSTnnn.NET** already exists, where **nnn** is the number of nodes the user has specified. If such a file exists, NET-SPEC will need to rename the file to proceed. Before renaming the distance file, however, NET-SPEC will ask the user if he or she wants to proceed. The following sort of message will be displayed:

```
<<< FILE [MDST5.NET] ALREADY EXISTS >>>
<<< TO PROCEED WE NEED TO RENAME IT >>>
Type Y to proceed; N to abort program ==>
```

If the user types **N**, the program will stop. Otherwise, NET-SPEC will rename the **MDSTnnn.NET** file with a name **MDSTnnn.N##**. ## number is the first number beginning with 00 such that the file **MDSTnnn.N##** does not already exist. If 100 files with names **MDSTnnn.N00** through **MDSTnnn.N99** exist, NET-SPEC will issue a message indicating that the file **MDSTnnn.NET** cannot be renamed and that the program must abort. In that case, NET-SPEC returns the user to DOS. If NET-SPEC can find a file name in the range **MDSTnnn.N00** to **MDSTnnn.N99** that is not in use, NET-SPEC will rename the **MDSTnnn.NET** file with this name and will display a message telling the user the new name of the **MDSTnnn.NET** file. NET-SPEC will then briefly display information about the amount of memory that is available before and after a large block of memory is allocated for the new distance matrix. (If insufficient memory is available, NET-SPEC will again be forced to abort.)

After NET-SPEC has successfully renamed an existing **MDSTnnn.NET** file and allocated memory for the new distance matrix, NET-SPEC can continue running.

NET-SPEC can be run in two different modes. In the first mode, the user specifies a set of nodes and the links connecting them. NET-SPEC will then compute the shortest path distances between all pairs of nodes and store the resulting distance information in the **MDSTnnn.NET** file. In the second mode, the user must specify all internodal distances for pairs of nodes (i, j), where $i < j$. Thus, after NET-SPEC has successfully renamed an existing **MDSTnnn.NET** file and allocated memory for the new distance matrix, NET-SPEC will ask the user the following question:

```
Do you want to specify existing links only
and have NET-SPEC find all shortest paths?
Y --> existing links only; N --> all distances ==>
```

If the user wants to specify only existing (selected) links and to let NET − SPEC compute the matrix of shortest path distances, he or she should type **Y** in response to this question. If the user wants to specify all internodal distances, he or she should type **N**. If the user elects to specify existing links only, NET-SPEC will ask

```
Are all links two-way links? ==>
```

If the user responds by typing **Y**, NET-SPEC will automatically set the distance from node j to node i equal to the distance from node i to node j when the i-to-j distance is specified by the user. Otherwise, NET-SPEC will expect the user to type in the two distances separately.

After the user indicates the mode in which NET-SPEC is to be run, NET-SPEC will ask if the first and second demand data sets are the same. The following question will be displayed:

```
Is the first demand set the same as the second? ==>
```

If the user types **Y**, NET-SPEC will not prompt the user for information on the second demand data set; otherwise, the user will be asked for two demand values for each node.

Next, the user will be asked for the demand and fixed cost information via the following prompts:

```
Input FIRST demand value for node # ==>

Input SECOND demand value for node # ==>

Input FIXED COST value for node # ==>
```

where # is the node number whose values are to be input.

Next, the user will be asked for the distances between all pairs of nodes. If all distances are being specified, the distance matrix is assumed to be symmetric and so NET-SPEC only asks for distances between a node i and a node j, where i is strictly less than j. (NET-SPEC assumes that the distance between any node i and itself is 0.) The following prompt is used to ask the

How to Run NET-SPEC

user for the distances in this case:

```
Distance from i to j ==>
```

where i and j are numbers.

If the user is specifying the link distances between selected nodes (corresponding to the specification of existing links), NET-SPEC prompts the user for the distances as follows:

```
Input Origin Node Number for Next Link
    (0 stops link input) ==>
Input Destination Node for Next Link ==>
Input Distance from i to j ==>
```

If two-way links are being used, NET-SPEC displays the following message:

```
Distance from j to i also set to ###
```

where ### is the distance specified by the user for the link from i to j. If the links that the user specifies do not allow for a path between all pairs of nodes NET-SPEC displays the following warning message:

```
<<< IMPORTANT WARNING ... NETWORK IS NOT CONNECTED >>>
<<< INFINITE DISTANCES BEING SET TO 32767 >>>
```

After the distances are specified, NET-SPEC will ask the user for the name to use for the location/demand file. NET-SPEC will continue prompting the user for a name until a file name is given that is not currently in use. Finally, NET-SPEC will display the following sort of message indicating that both the location/demand file and the distance file were successfully written:

```
<<< LOCATION / DEMAND FILE [fivenode] WRITTEN >>>
<<< DISTANCE FILE [MDST5.NET] ALSO WRITTEN >>>
```

After these messages are displayed and the user types any key, NET-SPEC will return the user to DOS.

Appendix C

MOD-DIST Operations Guide

C.1 WHAT IS MOD-DIST AND WHAT DOES IT DO?

MOD-DIST is a program that allows the user to read a distance file created by SITATION (or MOD-DIST) and to modify either individual elements of the file or to modify groups of elements of the file. For example, MOD-DIST allows the user to set all distances less than some specified distance to 0. Alternatively, the user can use MOD-DIST to set all distances greater than some value to a large number. This capability is useful when the user wants to run SITATION to solve a variety of problems which can be cast as maximum covering, *P*-median, or uncapacitated fixed charge problems through appropriate modification of the distance matrix. For example, if the user wanted to solve a *P*-median problem in which all demands must be covered by a facility within some specified distance, the user could set all distances greater than this value to a very large number. This would cause the "cost" of any assignment of a demand node to a facility beyond the coverage distance to be prohibitive. Such assignments would not occur in a solution to a *P*-median problem with this modified distance structure.

C.2 HOW TO RUN MOD-DIST

C.2.1 Starting MOD-DIST and Reading the Distance Matrix

To start MOD-DIST, the user need only type **MOD-DIST** at the DOS prompt. Note that the following command line options may be used with MOD-DIST: **sound, nologo, others, showmemory,** and **all** (which has the combined effect of specifying **nologo others showmemory**). In addition, any of the **%1** options may be used with MOD-DIST, though their use is *strongly discouraged* as these options turn on the Default mode. This affects the manner in which distances are read in ways that may not be desirable. The user is referred to Section A.6 of Appendix A for a description of these options.

How to Run MOD-DIST

After the initial logo screen, MOD-DIST will prompt the user for the information needed to read in a data set. Specifically, MOD-DIST begins by displaying the following DISTANCE OPTIONS MENU:

```
           ** DISTANCE OPTIONS MENU **

              <Q>  QUIT--Return to DOS
              <N>  Network Based
              <E>  Euclidean
              <G>  Great Circle
              <M>  Manhattan

           Input the desired option ==>
```

This is a minor variant of the DISTANCE OPTIONS MENU described in Section A.5.4 of Appendix A. The only difference is that typing Q from this menu will return the user to DOS. With this menu, the user must specify the type of distance matrix that is to be read. The options should include: N (for network-based distances), E (for Euclidean distances), G (for great circle distances), and M (for Manhattan distances).

Next, as in SITATION, the following prompt will ask the user for the name of the demand data set:

```
Input the file name to read (e.g. CITY1990.GRT or SORTCAP.GRT) ==>
```

Next, the following DEMAND OPTIONS MENU will be shown:

```
            ** DEMAND OPTIONS MENU **

              <F>  First Scenario Only
              <S>  Second Scenario Only
              <B>  Both Scenarios Combined

           Input the desired option ==>
```

As MOD-DIST does not use the actual demands, any of these options may be selected. Following this selection, MOD-DIST will read the demand data set. If either of the command line options **others** or **all** was specified, MOD-DIST will use the following prompt to ask the user if additional candidate locations are to be read:

```
Do you want to read a file with
additional candidate locations? (Y/N) ==>
```

As in SITATION, if the user responds with **Y**, he or she will be prompted for the name of the file containing the additional candidate locations.

After the demand data (and optionally the additional candidate location data) have been read, MOD-DIST will allocate memory for the distance matrix. If either of the command line options **showmemory** or **all** was specified, MOD-DIST will briefly display a message regarding the amount of memory available before and after this block of memory is allocated. Next, MOD-DIST will read the distance matrix that corresponds to (i) the distance type specified by the user and (ii) the total number of nodes in the demand data file (and optionally in the additional candidate location data file). This distance file must be named **MDSTnnn.dis**, where **nnn** is the total number of nodes and **dis** is **GRT**, **EUC**, **MAN**, or **NET** corresponding to great circle, Euclidean, Manhattan, or network distances.

After the distance matrix has been read, if the total number of nodes is less than or equal to 12, MOD-DIST will ask the user if he or she wants to display the matrix as changes are made. Specifically, the user will have to respond to the following question:

```
Do you want to see the matrix? ==>
```

If the number of nodes exceeds 12, MOD-DIST cannot display the distance matrix.

Next, MOD-DIST will use the following prompt to ask the user if he or she wants to modify individual elements of the matrix or the entire matrix:

```
Change I)ndividual Items or M)atrix Based Coverage Change ==>
```

From this point on, the program operates in one of two different modes. If the user responds by typing **I** to the prompt above, he or she will be allowed to change individual elements in the matrix as described in Section C.2.2. If the user responds by typing **M**, he or she will be asked to specify the parameters of a set of equations that are used to modify each and every element of the distance matrix. This is described in Section C.2.3.

C.2.2 Changing Individual Matrix Elements

In this mode, the user can change individual elements of the distance matrix. This is useful, for example, if the user wants to create an asymmetric distance matrix from a small distance matrix. The user will repeatedly be asked for the *row* (corresponding to the demand node) and *column* (corresponding to the candidate facility node) of the distance matrix element to be changed via

How to Run MOD-DIST

the following prompts:

```
Input ROW Number (DEMAND) of element to change (0 stops) ==>
Input COLUMN Number (FACILITY) of the element to change ==>
```

If the user types **0** in response to the request for the row number of the element to change, MOD-DIST will begin the process of storing the new distance matrix as described in Section C.2.4.

After the user specifies the row and column number of the distance elements to change, MOD-DIST will display the current distance and prompt the user for the new distance as follows:

```
DIST [rrr,ccc]=dist

Input the new distance ==>
```

where `rrr` is the row number specified by the user, `ccc` is the column number specified by the user, and `dist` is the current distance.

After the new distance is specified, the user will again be asked for the row number of the matrix element to change. This process of asking for the row, column, and new distance will continue until the user types **0** when prompted for the row number.

C.2.3 Matrix-Based Changes to the Distance Matrix

Matrix-based changes allow the user to change a distance matrix in a systematic manner. The user will be presented with a screen containing four equations of the form:

$$\text{New Dist} = \alpha + \beta(\text{Old Dist})^{\gamma}$$

where α, β, and γ are parameters to be specified by the user. The initial values of these parameters are 0, 1, and 1, respectively. These initial values result in the new distance being equal to the old distance.

The parameters for this equation can be specified for four different conditions corresponding to all combinations of (i) whether or not the old distance is less than or equal to a user-specified coverage distance and (ii) whether or not the distance matrix element is a diagonal element or not. In short, the user can specify different parameters α, β, and γ for each of the cells in the following table:

Original Distance

		Less Than or Equal to Coverage Distance	Greater Than Coverage Distance
Matrix	Diagonal Element	Parameters A, B, C below	Parameters G, H, I below
Element	Off-Diagonal Element	Parameters D, E, F below	Parameters J, K, L below

After typing **M** when asked whether or not the user wants to change individual matrix elements of the entire matrix, the user will be asked for the coverage distance with the following prompt:

```
Input the COVERAGE DISTANCE 0 <= X <= maxdist ==>
```

where `maxdist` is the largest distance in the old distance matrix. The user should specify the distance corresponding to the cutoff distance between the first and second columns in the table above. Next, the following will be shown:

```
         <<< MENU FOR CHANGING COEFFICIENTS >>>
    COEFFICIENTS FOR D(i,j) <= critdist and i = j
    New Dist = 0.000 + 1.000 * (Old Dist ^ 1.000)
               <A>       <B>                    <C>
    COEFFICIENTS FOR D(i,j) <= critdist and i<>j
    New Dist = 0.000 + 1.000 * (Old Dist ^ 1.000)
               <D>       <E>                    <F>
    COEFFICIENTS FOR D(i,j) > critdist and i = j
    New Dist = 0.000 + 1.000 * (Old Dist ^ 1.000)
               <G>       <H>                    <I>
    COEFFICIENTS FOR D(i,j) > critdist and i<>j
    New Dist = 0.000 + 1.000 * (Old Dist ^ 1.000)
               <J>       <K>                    <L>

    <R> RESET all coefficients to default values
    <Q> QUIT--No more changes to coefficients
```

In the equations above, `critdist` is the coverage distance specified by the user. The user can elect to change any (or all) of the coefficients for any of the four equations using the lettered options **A** through **L**. After selecting

How to Run MOD-DIST

one of the coefficients, the user will be prompted for the new value with one of the following three prompts:

```
Input New Constant      (For options A, D, G, and J)
Input New Slope Term    (For options B, E, H, and K)
Input New Exponent      (For options C, F, I, and L)
```

Selecting option R will reset all of the coefficients to their default values. Again, this will result in the new distance matrix being identical to the old matrix.

When all of the desired changes have been made, the user should press Q.

To avoid very large distances, if the maximum new distance exceeds 10,000, then all of the new distances are scaled so that the maximum new distance is equal to some user-specified value. (If the new maximum distance does not exceed 10,000, no scaling is performed.) After the user has specified all of the parameters for the equations to be used in changing the distances, the user will be prompted for the maximum distance to which all other distances should be scaled if the new maximum distance exceeds 10,000. MOD-DIST will prompt the user as follows:

```
Input the MAXIMUM DISTANCE
(used for scaling if Max{ New Dist } > 10000) == >
```

Thus, if the user specifies a maximum distance of 2000 and if the new (unscaled) distance exceeds, 10,000, all new distances will be computed using the formulae specified by the user and then multiplied by 2000 divided by the largest new (unscaled) distance.

To illustrate this concept, suppose the original distance matrix is given by

Original Distances

Demand Node	Facility Node 1	2	3
1	100	500	900
2	200	600	1000
3	300	700	1100
4	400	800	1200

If the coverage distance is set to 700 and the two equations for distances less than or equal to 700 miles have the β term equal to 0 and the two equations for distances greater than 700 miles have the β term equal to 20 (with all α values equal to 0 and all γ values equal to 1), we would obtain the following matrix of unscaled distances:

Unscaled Distances

Facility Node

Demand Node	1	2	3
1	0	0	18000
2	0	0	20000
3	0	0	22000
4	0	16000	24000

Since the maximum unscaled distance exceeds 10,000, MOD-DIST will scale these distances so that the maximum equals a user-specified value. If that value is 6000, then all unscaled distances will be multiplied by 6000/24,000 or 0.25, resulting in the following matrix of new distances:

New Distances

Facility Node

Demand Node	1	2	3
1	0	0	4500
2	0	0	5000
3	0	0	5500
4	0	4000	6000

C.2.4 Storing the New Distance Matrix

After the user has either changed the individual matrix elements that are to be changed or specified the parameters of the four equations used to change the matrix elements as described in Section C.2.3, MOD-DIST will issue the following sort of warning:

```
<<< FILE [MDST88.GRT] WILL BE OVER-WRITTEN >>>
              <<< WARNING >>>

Type Q to quit now or C to continue ==>
```

This message gives the user one last chance to abort the program before the changes are made to the distance matrix on the disk. Typing **Q** will abort the program; typing **C** will allow the program to continue.

Assuming the user types **C** at the message above, MOD-DIST will rename the **MDSTnnn.dis** file with a name **MDSTnn.D##**. **D** is one of the following letters: **E** for Euclidean distances, **G** for great circle distances, **M** for Manhattan distances, or **N** for network distances. ## is the first number beginning with 00 such that the file **MDSTnnn.D##** does not already exist. If 100 files with names **MDSTnnn.D00** through **MDSTnnn.D99** exist, MOD-DIST will issue a message indicating that the file **MDSTnnn.dis** cannot be renamed and that the program must abort. In that case, MOD-DIST returns the user to DOS. If MOD-DIST can find a file name in the range **MDSTnnn.D00** to **MDSTnnn.D99** that is not in use, MOD-DIST will rename the **MDSTnnn.dis** file with this name and will display the following sort of message telling the user the new name of the **MDSTnnn.dis** file:

```
<<< FILE [MDST88.NET] RENAMED [MDST88.N00] >>>
```

Next, MOD-DIST will indicate whether or not scaling is needed for the distance matrix (if the user is using the change matrix option to revise the distance matrix). Then, MOD-DIST will display the following sort of message indicating that the new file has been written:

```
<<< NEW FILE WRITTEN >>>
<<< FILE IS NAMED >>>
        MDST88.NET
```

Following this message, MOD-DIST returns the user to DOS.

C.3 HOW THE MDSTnnn.dis FILES ARE STORED

In this section we briefly indicate how the **MDSTnnn.dis** files are stored on the disk. There are a number of key points that the user should know:

- All **MDSTnnn.dis** files are for square matrices in which the number of rows equals the number of columns.
- **MDSTnnn.dis** files are stored with the candidate facility node varying first followed by the demand node.
- Distances are stored as nonnegative integers.
- Distances over 9999 should not be used.

Thus, consider a matrix with the following distance elements:

	Candidate Facility Node		
Demand Node	1	2	3
1	0	100	125
2	150	25	160
3	250	180	45
4	210	225	200

This matrix is not a square matrix. Thus, to store this information as an **MDSTnnn.dis** file, the user would need to define an additional dummy candidate node (with very large distances to the demand nodes to preclude assignment of the demands to the dummy facility location). The revised matrix would be

| | \multicolumn{4}{c}{Candidate Facility Node} |
Demand Node	1	2	3	4
1	0	100	125	9999
2	150	25	160	9999
3	250	180	45	9999
4	210	225	200	9999

This matrix would be stored as the following ASCII file:

0
150
250
210
100
25
180
225
125
160
45
200
9999
9999
9999
9999

To specify such a matrix, the user would have to use both NET-SPEC and MOD-DIST.

Appendix D

COLORSET Operations Guide

D.1 WHAT IS COLORSET AND WHAT DOES IT DO?

COLORSET is a program that allows the user to customize most of the display colors used by SITATION, MOD-DIST, NET-SPEC, and MENU-OKF. This is particularly important for users working with LCD screens on which the default color scheme may be hard to read. Other users may simply prefer an alternate color scheme.

COLORSET creates a file called **COLORS.COL** which contains 10 color codes used by each of these programs in displaying information on the screen. A typical **COLORS.COL** file might look like this:

```
     14     Black BACK; Yellow FOR 14
    142     Black BACK; Yellow BLINK FOR 142
     46     Green BACK; Yellow FOR 46
     31     Blue BACK; White FOR 31
    174     Green BACK; Yellow BLINK FOR 174
    112     Gray BACK; Black FOR 112
    240     Gray BACK; Black BLINK FOR 240
    159     Blue BACK; White BLINK FOR 159
     96     Orange BACK; Black FOR 96
    224     Orange BACK; Black BLINK FOR 224
```

The user need not be overly concerned about the format of this file as COLORSET automatically writes the file in a way that can be read by the other programs.

D.2 HOW TO START AND USE COLORSET

To start COLORSET, the user need only type **COLORSET** at the DOS prompt. Note that the command line options **nologo** and **all** may be used with COLORSET. The user is referred to Section A.6 of Appendix A for a description of these options.

After the initial logo screen, COLORSET will display the following screen which allows the user to change the color scheme used with any of the programs which read the **COLORS.COL** file (e.g., SITATION, MOD-DIST, NET-SPEC, and MENU-OKF):

```
  0   1   2   3   4   5   6   7   8   9  10  11  12  13  14  15
 16  17  18  19  20  21  22  23  24  25  26  27  28  29  30  31
 32  33  34  35  36  37  38  39  40  41  42  43  44  45  46  47
 48  49  50  51  52  53  54  55  56  57  58  59  60  61  62  63
 64  65  66  67  68  69  70  71  72  73  74  75  76  77  78  79
 80  81  82  83  84  85  86  87  88  89  90  91  92  93  94  95
 96  97  98  99 100 101 102 103 104 105 106 107 108 109 110 111
112 113 114 115 116 117 118 119 120 121 122 123 124 125 126 127
128 129 130 131 132 133 134 135 136 137 138 139 140 141 142 143
144 145 146 147 148 149 150 151 152 153 154 155 156 157 158 159
160 161 162 163 164 165 166 167 168 169 170 171 172 173 174 175
176 177 178 179 180 181 182 183 184 185 186 187 188 189 190 191
192 193 194 195 196 197 198 199 200 201 202 203 204 205 206 207
208 209 210 211 212 213 214 215 216 217 218 219 220 221 222 223
224 225 226 227 228 229 230 231 232 233 234 235 236 237 238 239
240 241 242 243 244 245 246 247 248 249 250 251 252 253 254 255

    <N>  Blue BACK; Yellow FOR 30      <B>  Gray BACK; Black FOR 112
    <H>  Blue BACK; White FOR 31       <J>  Blue BACK; White BLINK FOR 159
    <R>  Green BACK; Yellow FOR 46     <T>  Purple BACK; Lt Blue FOR 91
    <U>  Red BACK; Yellow FOR 78       <W>  Purple BACK; White FOR 95
    <X>  Red BACK; White FOR 79        <Z>  Cyan BACK; Black FOR 48

    <I> Initial Values; <Q> Quit; <E> End and Save; <S> Save; <D> Defaults
```

The numbers from 0 to 255 will be displayed in different color combinations. The numbers from 0 through 127 will not be blinking while the numbers between 128 and 255 will be blinking. Color schemes 0, 17, 34, 51, 68, 85, 102, 119, and 136 will not be readable as the foreground and background colors are the same. These color combinations should *not* be used.

The bottom half of the screen displays the command options which are described below:

N, H, R, U, X
B, J, T, W, Z These commands allow the user to change the color scheme for the indicated color combinations. For example, in the menu system shown above, typing N would allow the user to change those items that are

normally displayed with yellow letters (foreground) on a blue background (color combination 30) to some other color combination. When one of these commands is issued, the menu options on the bottom of the screen disappear and the user is prompted to provide the new color combination with a prompt of the following form:

```
Input New Color Number for N ==>
```

(Note that N in the message above would be replaced by whatever command option the user had selected.)

After the new color combination is provided, the screen will return to the screen shown above with the color combination that had previously been selected changed to the new color scheme. For example, if the user had typed color code 93 to replace color code 30 for option N, option N would then be displayed as

```
<N> Purple BACK; Pink FOR 93
```

Note that color combinations **N**, **H**, **R**, **U**, and **X** are used to display standard messages and results. As a result, they should probably not be set to any of the blinking color combinations (numbers 128 through 255). On the other hand, color combinations **B**, **J**, **T**, **W**, and **Z** are used to display warning messages. The user may want to set these color combinations to one or more of the blinking combinations so that warning messages stand out.

- **I** This command allows the user to reset all of the color combinations to the values that were present when COLORSET was started.
- **Q** This command allows the user to terminate the execution of COLORSET. A new **COLORS.COL** file will not be written when this command is executed. The user will be returned to DOS and the result will be as if COLORSET had not been run.
- **E** This command allows the user to save the current set of colors in the **COLORS.COL** file and then to terminate the execution of COLORSET. After this command is executed, the user will be returned to DOS. Since a new **COLORS.COL** file is written when this command is executed, all future runs of the programs SITATION, MOD-DIST, NET-SPEC, and MENU-OKF will use the new color combinations. This command has the same effect as that of the **S** command immediately followed by the **Q** command.
- **S** This command allows the user to save the current set of colors in the **COLORS.COL** file. Since a new **COLORS.COL** file is written when this command is executed, all future runs of the programs SITATION,

MOD-DIST, NET-SPEC, and MENU-OKF will use the new color combinations. The user is not returned to DOS after the execution of this command.

D This command resets all of the colors to a default color scheme that is built into COLORSET. This is the color scheme that SITATION, MOD-DIST, NET-SPEC, and MENU-OKF use if the **COLORS.COL** file is absent. In particular, the color scheme is set to the following default values:

```
<N> Blue BACK; Yellow FOR 30    <B> Blue BACK; Yellow BLINK FOR 158
<H> Blue BACK; White FOR 31     <J> Blue BACK; White BLINK FOR 159
<R> Green BACK; Yellow FOR 46   <T> Green BACK; Yellow BLINK FOR 174
<U> Red BACK; Yellow FOR 78     <W> Red BACK; Yellow BLINK FOR 206
<X> Red BACK; White FOR 79      <Z> Red BACK; White BLINK FOR 207
```

Finally, we note that COLORSET and the **COLORS.COL** file will affect about 90 to 95 percent of the color displays. A small number of the color combinations used in each of the programs are not controlled by these color settings.

Appendix E

MENU-OKF Operations Guide

E.1 WHAT IS MENU-OKF AND WHAT DOES IT DO?

MENU-OKF is a program that solves network flow problems using the out-of-kilter flow algorithm. The program readily allows the user to add new links to the problem, delete links, and change link characteristics including the unit cost, minimum allowable flow, and maximum allowable flow on the link. The program is menu driven, hence its name: MENU-OKF.

MENU-OKF uses two paired files to describe any out-of-kilter flow problem. The first file describes the nodes of the problem. Included in this file are the names of each of the nodes. Node names may be up to 10 characters long. MENU-OKF distinguishes between lower- and uppercase letters in node names. Thus, node names (i) supply, (ii) SUPPLY, and (iii) Supply will be treated as three different names. This file must have the file name extension **NDE**. The second file describes the links of the problem. In particular, the origin and destination nodes, unit cost, and lower and upper bounds are supplied in this file for each link of the problem. This file must have the same file name as the associated **NDE** file and must have the file name extension **LNK**. Section E.3 gives example **NDE** and **LNK** files for a small sample problem.

E.2 HOW TO START AND USE MENU-OKF[1]

To start MENU-OKF, the user need only change to the directory in which the **MENU-OKF.EXE** file is installed and then type **MENU-OKF** at the DOS prompt. Note that the command line options **nologo**, **all**, and **sound** may be used with MENU-OKF. The user is referred to Section A.6 of Appendix A for a description of these options.

---

[1]To install MENU-OKF on a hard disk drive, the user is referred to the installation instructions for SITATION contained in Section A.2.

After the initial logo screen, MENU-OKF displays a second screen with a brief set of instructions on the use of the program. After this, the following MAIN MENU is displayed:

```
       <<< MENU-OKF MAIN MENU [3.4MD NLMAA] >>>
   <Q> QUIT--EXIT from MENU-OKF program
   <T> TITLE--name the run for printed output
   <Z> RESET all information--CLEARS workspace
   <O> OPERATING system options

   <R> READ a problem description from disk
   <S> SAVE the data on a disk
   <A> ADD links to the problem
   <C> CHANGE or REMOVE links from the problem

   <X> EXECUTE the OKF algorithm
   <D> DISPLAY the results on the screen or printer

          Select desired option ==>
```

The commands available from the MAIN MENU are described below:

Q   This allows the user to stop execution of the program. The user will be asked whether he or she really wants to stop execution of the program. Typing **Y** in response to the prompt will return the user to DOS; typing **N** will return the user to the MAIN MENU shown above. If changes have been made to the problem that is currently in memory and if those changes have not been saved, the user will be asked if he or she wants to save the changes before leaving MENU-OKF. Similarly, if the problem has been solved but the results have not been either displayed on the screen or printed, the user will be asked if he or she wants to display or print the results before leaving MENU-OKF. The final result of this command is that the user will be returned to the DOS environment.

T   This command allows the user to specify a title for the run. If a run title has been specified, it is displayed at the bottom of the MAIN MENU and is printed on the top of each page when the results are printed on a printer.

Z   This command allows the user to clear the workspace memory. The result of this command is to reset the program to the condition it is in when it is first started (i.e., with no problem in memory). As in the case of an attempt to quit the program, if the current problem has been changed but not saved or if the problem has been solved but the results have not yet been displayed or printed, the user will be asked if he or she really wants to continue with the command to clear the workspace memory. In particular, the following message will be displayed:

```
Data NOT SAVED or Results NOT SHOWN ... WARNING ...
Press <C> to CONTINUE with selected option ...
     <Q> for another chance to SAVE data / SHOW results ... ==>
```

Pressing C will, in this case, clear the workspace. Pressing Q will return the user to the MAIN MENU with the problem that was in the computer's memory intact.

- O This command displays the DOS FUNCTION MENU described in Section A.5.3 of Appendix A. With this menu, the user can perform such DOS functions as change directories, make a new directory, display a list of files, and rename a file. This command is useful in determining the names of the files that can be read by MENU-OKF. Since all node files must have the **NDE** extension, the user can use the DOS FUNCTION MENU to list all such files by using either the D (Display List of Files) or the E (Extended Directory of Files) commands with the *.NDE file mask.
- R This command allows the user to read a problem description from the disk. The following prompt will ask the user for the name of the file to be read:

```
Input the name of the file to READ ==>
<return> with no name goes to MAIN MENU
```

The user should type the name of the file to be read, without any file name extension. For example, if the data for a problem are stored in two files with the names **LOGISTIC.NDE** and **LOGISTIC.LNK**, the user should simply type LOGISTIC. MENU-OKF will add the appropriate file name extensions as the files are read. If the user simply enters return, MENU-OKF will return the user to the MAIN MENU without reading a problem description. At the bottom of the screen, MENU-OKF will display a list of names of files for which both the **NDE** and **LNK** files exist.

- S This command allows the user to store a problem description on the disk. The user will be prompted for the name of a file in which to store the data using the following prompt:

```
Input file name in which to SAVE the data ==>
<return> with no name goes to MAIN MENU
```

The user should specify the name of the **NDE** and **LNK** files in which the data are to be stored. As before, only the file name (and not the file name extensions) should be given. Also, as in the case of reading data from the disk, if the user types return with no name, MENU-OKF will return to the MAIN MENU without storing the data.

If neither the **filename.NDE** nor the **filename.LNK** exists (in the current disk directory), MENU-OKF will store the current problem description on the disk. Files **filename.NDE** and **filename.LNK** (where "filename" is the name specified by the user) will be written to the disk. If either of these files already exists, the following message will be displayed and the user will be asked if he or she wants to erase the existing file:

```
File <filename> already exists ...
Do you want to ERASE it? (Y / N) == >
```

In this message, `filename` is the name given by the user. If the user responds by typing `Y`, MENU-OKF will store the current problem description on the disk creating files named **filename.NDE** and **filename.LNK**. The user will then be returned to the MAIN MENU. If the user responds by typing `N`, MENU-OKF will return the user to the MAIN MENU without storing the current problem description.

> **A** This command allows the user to add links to the network. The user will be prompted for the name of the origin node, the destination node, lower bound, upper bound, and unit cost associated with the link(s) to be added using the following:

```
Input the origin node name...<return> stops input == >
Input the destination node name...== >
Input the lower bound for the link == >
Input the upper bound for the link == >
Input the cost for the link == >
```

Node names may be up to 10 characters long. As the user types in (origin and destination) node names, MENU-OKF will check whether or not the names have already been used in the problem. If they have, MENU-OKF accepts the names. If the user types in a name that has not already been used, MENU-OKF will use the following prompt to ask the user whether or not he or she wants to add the name to the problem description:

```
Node <junk> does not exist...Do you want to ADD it? (Y / N) == >
```

How to Start and Use MENU-OKF

Here `junk` is the node name the user has just entered. If the user responds by typing **Y**, the node name is added to the problem description and MENU-OKF proceeds with the next request for information. If the user responds by typing **N**, MENU-OKF ignores the name and the user is again asked to specify a node name. If the user enters return with no name when asked for the origin node name, MENU-OKF will return to the MAIN MENU.

Lower and upper bounds must be integers between 0 and 32,767. Unit costs must also be integers between $-10,000$ and $10,000$.

- **C** This command allows the user to change the parameters associated with individual links and to remove links from the problem description. The user will first be prompted for the number of the link to change or remove as follows:

```
Input the link number to modify. 0 stops. 0 <= X <= maxlink ==>
```

Here `maxlink` is the largest link number. If the user responds by typing **0**, MENU-OKF will return to the MAIN MENU. If the user specifies any other link number, MENU-OKF will display the following menu through which the user can change the description of the specified link or remove the link from the problem:

```
              LINK 11 INFORMATION
    <O>  ORIGIN                            ST LOUIS
    <D>  DESTINATION                        SEATTLE
    <L>  LOWER BOUND                              0
    <U>  UPPER BOUND                           9999
    <C>  COST                                    17
           <flow>                                 0
           <reduced cost>                        12
           <PI ST LOUIS>                        127
           <PI SEATTLE>                         132
    <R>  REMOVE OR REPLACE THE LINK
    <Q>  QUIT--NO MORE CHANGES ON THIS LINK

              Select desired option ==>
```

The user can change the origin node name (by typing **O**) or the destination node name (by typing **D**). In either case, the user will be prompted for a new name. If the new name has already been used, MENU-OKF changes the old name to the new name. This has the effect of changing the structure of the out-of-kilter flow network. If the new name has not been used as a node name, MENU-OKF will ask the user if he or she wants to add the name to the problem description. If the user responds by typing that he or she does, MENU-OKF will change the old name to the new name. If the user does not want to add the name to the problem description, MENU-OKF will continue asking the user for a new name. (If the user decides not to change the name, he or she should simply type the old name as the new name.) This feature prevents the user from inadvertently typing in new names as is possible since MENU-OKF distinguishes between lower- and uppercase letters throughout the node names.

The user can also change the lower and upper bounds by typing L or U, respectively.

By typing C, the user can change the unit cost associated with the link.

By typing R, the user can remove a link from the problem description or replace a previously removed link. The message <<REMOVED>> will be added next to the menu heading (LINK # INFORMATION) when a link has been removed. The user should note, however, that the information related to the link will not actually be removed until the user returns to the MAIN MENU. In particular, after the user is finished changing the parameters that describe a link, he or she should type Q which will return the user to the prompt for the number of the link to change. Any removed links remain in the problem description until the user responds with link number 0 (to return to the MAIN MENU). This feature allows the user to restore (or replace) links that have been labeled for removal before they are actually removed from the problem description in the computer's memory.

Finally, as indicated above, the user should type Q when he or she has finished making the desired changes to the description of the link in question. This will return the user to the prompt asking for the number of the next link to change. If the user has completed all changes, he or she should type link number 0 which will return the user to the MAIN MENU; otherwise, he or she should type the number of the next link to change.

    X   This MAIN MENU command allows the user to solve the problem by executing the out-of-kilter flow algorithm. The program begins by checking that each node is used at least once as an origin and at least once as a destination. (If this is not the case, the constraints cannot satisfy the circulation flow conditions of the out-of-kilter flow algorithm.) If errors in the connectivity of the network are detected, the following sorts of messages are displayed:

```
Node <Sup> is never used as a destination...FATAL ERROR...
Node <Dem> is never used as an origin...FATAL ERROR...
```

Here `Sup` and `Dem` are node names.

If all nodes are used at least once as origins and as destinations, MENU-OKF will solve the problem. After the problem has been solved, MENU-OKF will display the solution time. If the problem has a feasible (and therefore an optimal) solution, MENU-OKF will return to the MAIN MENU. If no feasible solution exists, the following message will be displayed before MENU-OKF returns to the MAIN MENU:

```
        Inputs yield infeasible solution...

                SEVERE WARNING...
```

D This command allows the user to display the results on the screen and/or to print the results. The user will be asked if he or she wants to display the results for all of the links or only for those with nonzero flows as follows:

```
Do you want to DISPLAY ALL LINKS
or only NON-ZERO FLOW links? (A / N) ==>
```

If the user responds by typing **A**, all links will be displayed. If the user types **N** in response to this question, only those links with nonzero flows will be displayed. This can be particularly useful in problems with many links, only a small number of which might have nonzero flows at the optimal solution.

Next, the user will be prompted for the order in which links are to be displayed as follows:

```
Select a listing order from the options below
<I> Input; <O> Origin Node; <D> Destination Node ==>
```

Links can be listed in either the order they were input (**I**), in order of the origin node name (**O**), or in order of the destination node name (**D**).

Finally, the user will be asked if he or she wants to print the results as follows:

```
Do you want to PRINT the LINK and NODE tables? (Y / N) ==>
```

If the user responds by typing **Y**, MENU-OKF will try to read the **PRINTER.CNS** file. (See Appendix F for a description of this file.) If the file does not exist, MENU-OKF will initialize the printer controls to those of the Epson LQ-850 printer and will then ask if the user wants to store these control codes in a new **PRINTER.CNS** file. After reading (or initializing and possibly storing) the printer control codes, MENU-OKF will display the following message:

```
*** BE SURE PRINTER IS ON, ONLINE AND AT TOP OF PAGE ***
*** <Q> TO ABORT PRINTING AND RETURN TO MAIN MENU      ***
*** <ANY OTHER KEY> TO CONTINUE WITH PRINTING          ***
```

If the user has asked that the results be printed and if an error occurs in printing the results, the following message will be displayed:

```
<<< ------------------------------------ >>>
<<< ERROR ENCOUNTERED IN TRYING TO PRINT >>>
<<< RETURNING TO MAIN MENU NOW ... SORRY >>>
<<< NEXT TIME BE SURE PRINTER IS ON,     >>>
<<< ONLINE, AT TOP OF THE PAGE,          >>>
<<< AND HAS ENOUGH PAPER                 >>>
<<< ------------------------------------ >>>
<<< RESULTS WILL BE SHOWN ON THE SCREEN  >>>
<<< BUT WILL NOT BE PRINTED.             >>>
```

When the results are displayed on the screen or are printed, the following sort of table is used:

LINK	--ORIGIN--	--DESTIN--	LOWER	---FLOW---	UPPER	---COST---	--Red. COST--
1	source	ST LOUIS	0	400	400	125	-2
2	source	BALTIMORE	0	150	350	128	0
3	source	SAN DIEGO	0	200	500	121	0
4	BOSTON	sink	100	100	9999	0	132
5	CHICAGO	sink	150	150	9999	0	130
6	MIAMI	sink	300	300	9999	0	138
7	SEATTLE	sink	200	200	9999	0	132
9	ST LOUIS	CHICAGO	0	150	9999	3	0
10	ST LOUIS	MIAMI	0	250	9999	11	0
12	BALTIMORE	BOSTON	0	100	9999	4	0
14	BALTIMORE	MIAMI	0	50	9999	10	0
19	SAN DIEGO	SEATTLE	0	200	9999	11	0
20	sink	source	0	750	9999	0	0

LINK IN KILTER   * LINK OUT OF KILTER

TOTAL COST = 99700

This table displays the input information about each link (the origin node name, the destination node name, the lower and upper bounds on the flow,

Format of the NDE and LNK Files

and the unit cost) as well as the results (the optimal flow on the link and the reduced cost in the column with the heading Red. COST). The reduced cost indicates the amount by which the optimal cost will change for a small change in the binding bound. Thus, in the example shown above, the reduced cost of -2 for the link from node source to node ST LOUIS indicates that the total cost will go down by 2 units if the upper bound on the link is increased from 400 to 401. As always, the user is cautioned that interpretation of these values requires care as alternate optima may exist.

Links that are out of kilter (in the event that no feasible solution exists) will be highlighted and will be starred.

After all of the links have been displayed, MENU-OKF will show the total cost of the solution. Next, MENU-OKF will show the π values associated with each of the nodes in the following sort of table:

---NODE---	----PI---
source	0
ST LOUIS	127
BALTIMORE	128
SAN DIEGO	121
BOSTON	132
sink	0
CHICAGO	130
MIAMI	138
SEATTLE	132

After this table is displayed, MENU-OKF will return to the MAIN MENU.

E.3 FORMAT OF THE NDE AND LNK FILES

This section provides the user with the formats used for the **NDE** and **LNK** files. This is valuable if the user wants to write his or her own program to set up large problems. Consider the network shown in Figure E.1.

For this network, the **NDE** file would be

```
source
sup1
sup2
sup3
dem1
dem2
sink
```

Figure E.1. Sample network.

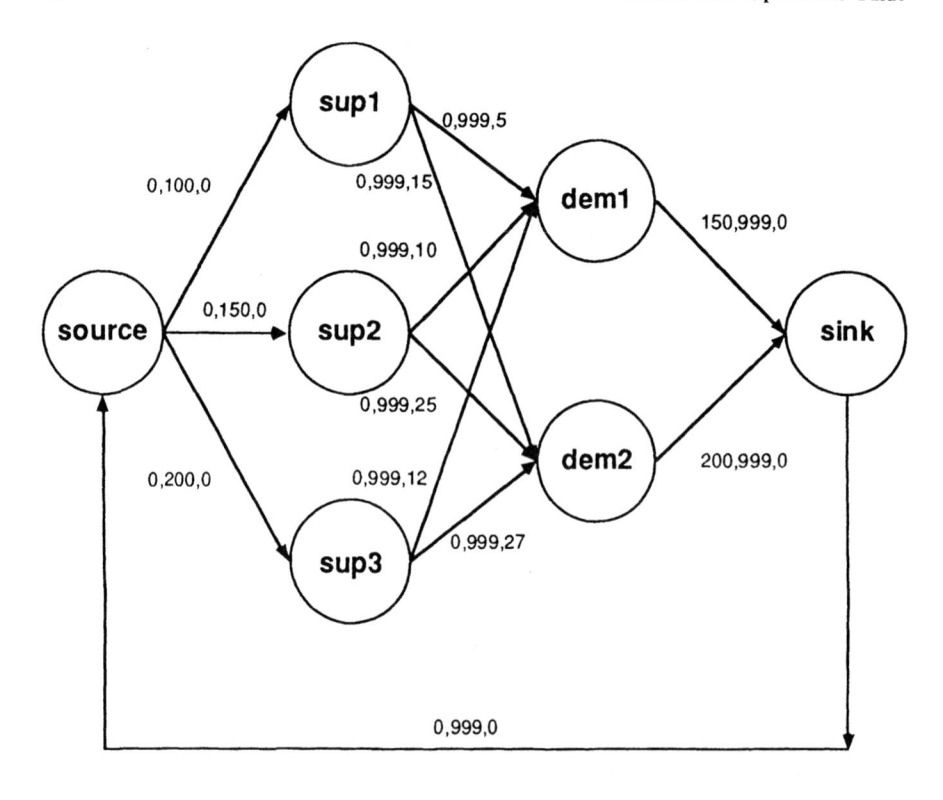

lower bound, upper bound, unit cost

This file contains the names of the nodes. Each record contains the name of a different node. MENU-OKF reads the first 10 characters in each record as the name of the node. Thus, spaces, punctuation, numbers as well as lower- and uppercase characters are interpreted as part of the node names.

The **LNK** file would be

```
12   7   0     0   0
 1   2   0   100   0
 1   3   0   150   0
 1   4   0   200   0
 2   5   0   999   5
 2   6   0   999  15
 3   5   0   999  10
 3   6   0   999  25
 4   5   0   999  12
```

Format of the NDE and LNK Files

4	6	0	999	27
5	7	150	999	0
6	7	200	999	0
7	1	0	999	0

Each record of the **LNK** file contains five fields. The first record of the **LNK** file has the number of links in the first field and the number of nodes in the second field. The remaining three fields should all be set to 0. Following the first record, there will be one record for each link in the network. The fields in these links identify the origin node number, the destination node number, the lower bound, the upper bound, and the unit cost, respectively. Records describing the links may be input in any order (with the obvious exception that the first record must contain the number of links and the number of nodes as described above). Each field in each record must be separated from the following field by one or more blank spaces.

Lower and upper bounds must be between 0 and 32,767. Costs must be between −10,000 and 10,000. All link numbers, lower bounds, upper bounds, and costs must be specified as integers without including a decimal point.

A maximum of 2800 links and 400 nodes may be specified for any MENU-OKF problem.

Appendix F

PRINTER.CNS File Description

F.1 WHAT IS THE PRINTER.CNS FILE AND WHAT DOES IT DO?

The **PRINTER.CNS** file is used by both the SITATION and MENU-OKF programs to obtain printer control characters. When the user attempts to print reports from either of these programs, the program will search the current disk directory for the **PRINTER.CNS** file. If it exists, it will be read. If not, each program is configured to use the control codes for an Epson LQ-850 printer. In addition, if the **PRINTER.CNS** file is not found, each program will ask the user if he or she wants to create a file with the default (Epson LQ-850) control characters for future use.

    PRINTER.CNS is an ASCII file. It may be modified by the user to accommodate other printers. The following table lists the default printer controls built into SITATION and MENU-OKF. These controls will drive an Epson LQ-850 printer.

```
    27    64               <         INIT PRINTER >
    27    69               <              BOLD ON >
    27    70               <             BOLD OFF >
    27    45    49         <        UNDERLINED ON >
    27    45    48         <       UNDERLINED OFF >
    27    52               <           ITALICS ON >
    27    53               <          ITALICS OFF >
    27    71               <     DOUBLE STRIKE ON >
    27    72               <    DOUBLE STRIKE OFF >
    15                     <        COMPRESSED ON >
    18                     <       COMPRESSED OFF >
    12                     <            FORM FEED >
    13                     <      CARRIAGE RETURN >
    10                     <            LINE FEED >
     9                     <        HORIZONTAL TAB >
    55                     <          PAGE LENGTH >
```

What Is the PRINTER.CNS File and What Does It Do?

Note:

1. All of the codes specified above must be included in the **PRINTER.CNS** file.
2. The control codes must be listed in exactly the order shown.
3. Numbers shown are decimal values of the character codes that are equivalent to the indicated command. Thus, to turn bold printing on, the program sends CHR(27) + CHR(69) to the printer.
4. CHR(27) is equivalent to escape.
5. Text shown in ⟨ ⟩ symbols is ignored and is provided only so the user will know what each command does. Note, however, that no embedded numbers should be included inside the ⟨ ⟩ symbols.
6. This file is stored on the distribution disk as both the **PRINTER.CNS** file and the **PRINTER.ELQ** file. The following table shows the other **PRINTER.xxx** files stored on the distribution disk. To use one of these other files, copy the appropriate file as the **PRINTER.CNS** file using the following DOS command:

`COPY PRINTER.xxx PRINTER.CNS`

where `PRINTER.xxx` is the name of the appropriate file.

Table of `PRINTER.xxx` Files on Distribution Disk

File Name	Printer	Comments
PRINTER.ELQ	Epson LQ-850 Printer	
PRINTER.HP2	Hewlett Packard LaserJet II Printer	Italics may not work properly
PRINTER.HP3	Hewlett Packard LaserJet III Printer	
PRINTER.HP4	Hewlett Packard LaserJet IV Printer	
PRINTER.DJ5	Hewlett Packard DeskJet 500 Printer	

Appendix G

Longitudes, Latitudes, Demands, and Fixed Costs for CITY1990.GRT: An 88-Node Problem Defined on the Continental United States

No.	Long.	Lat.	First Demand	Second Demand	Fixed Cost	City	ST
1	73.945	40.671	7,322,564	2,819,401	189,600	New York	NY
2	118.411	34.112	3,485,398	1,217,405	244,500	Los Angeles	CA
3	87.685	41.837	2,783,726	1,025,174	78,700	Chicago	IL
4	95.387	29.769	1,630,553	616,877	58,000	Houston	TX
5	75.135	40.007	1,585,577	603,075	49,400	Philadelphia	PA
6	117.136	32.815	1,110,549	406,096	189,400	San Diego	CA
7	83.102	42.383	1,027,974	374,057	25,600	Detroit	MI
8	96.765	32.794	1,006,877	402,060	78,800	Dallas	TX
9	112.071	33.543	983,403	369,921	77,100	Phoenix	AZ
10	98.505	29.458	935,933	326,761	49,700	San Antonio	TX
11	121.850	37.304	782,248	250,218	259,100	San Jose	CA
12	76.611	39.301	736,014	276,484	54,700	Baltimore	MD
13	86.146	39.776	731,327	291,946	60,800	Indianapolis	IN
14	122.555	37.793	723,959	305,584	298,900	San Francisco	CA

CITY1990.GRT: An 88-Node Problem

15	81.658	30.335	635,230	241,384	62,900	Jacksonville	FL
16	82.987	39.989	632,910	256,996	66,000	Columbus	OH
17	87.967	43.063	628,088	240,540	53,500	Milwaukee	WI
18	90.006	35.106	610,337	229,829	55,700	Memphis	TN
19	77.016	38.905	606,900	249,634	123,900	Washington	DC
20	71.018	42.336	574,283	228,464	161,400	Boston	MA
21	122.350	47.622	516,259	236,702	137,900	Seattle	WA
22	106.438	31.849	515,342	160,545	58,500	El Paso	TX
23	81.679	41.480	505,616	199,787	40,900	Cleveland	OH
24	89.931	30.066	496,938	188,235	69,600	New Orleans	LA
25	86.785	36.172	488,374	198,585	74,400	Nashville-Davidson	TN
26	104.873	39.768	467,610	210,952	79,000	Denver	CO
27	97.751	30.306	465,622	192,148	72,600	Austin	TX
28	97.336	32.754	447,619	168,274	59,900	Fort Worth	TX
29	97.513	35.467	444,719	178,662	54,900	Oklahoma City	OK
30	122.658	45.539	437,319	187,268	59,200	Portland	OR
31	94.555	39.122	435,146	177,607	56,100	Kansas City	MO
32	118.160	33.789	429,433	158,975	222,900	Long Beach	CA
33	110.892	32.196	405,390	162,685	66,800	Tucson	AZ
34	90.244	38.636	396,685	164,931	50,700	St. Louis	MO
35	80.835	35.198	395,934	158,991	81,300	Charlotte	NC
36	84.423	33.763	394,017	155,752	71,200	Atlanta	GA
37	76.044	36.739	393,069	135,566	96,500	Virginia Beach	VA
38	106.625	35.117	384,736	153,818	85,900	Albuquerque	NM
39	122.225	37.772	372,242	144,521	177,400	Oakland	CA
40	79.977	40.439	369,879	153,483	41,200	Pittsburgh	PA
41	121.467	38.567	369,365	144,444	115,800	Sacramento	CA
42	93.267	44.962	368,383	160,682	71,700	Minneapolis	MN
43	95.916	36.128	367,302	155,447	60,500	Tulsa	OK
44	84.506	39.140	364,040	154,342	61,900	Cincinnati	OH
45	80.211	25.776	358,548	130,252	79,200	Miami	FL
46	119.793	36.781	354,202	121,807	80,300	Fresno	CA
47	96.012	41.264	335,795	133,842	54,600	Omaha	NE
48	83.582	41.664	332,943	130,883	48,900	Toledo	OH

49	78.860	42.890	328,123	136,436	46,700	Buffalo	NY
50	97.343	37.687	304,011	123,249	56,700	Wichita	KS
51	93.104	44.948	272,235	110,249	70,900	St. Paul	MN
52	91.126	30.449	219,531	83,340	67,900	Baton Rouge	LA
53	78.659	35.822	207,951	85,822	96,600	Raleigh	NC
54	77.475	37.531	203,056	85,337	66,600	Richmond	VA
55	90.208	32.321	196,637	71,865	54,600	Jackson	MS
56	93.617	41.577	193,187	78,453	49,500	Des Moines	IA
57	96.688	40.816	191,972	75,402	61,700	Lincoln	NE
58	89.388	43.080	191,262	77,361	75,200	Madison	WI
59	86.284	32.354	187,106	69,968	62,200	Montgomery	AL
60	92.354	34.722	175,795	72,573	64,200	Little Rock	AR
61	71.420	41.822	160,728	58,905	113,000	Providence	RI
62	111.930	40.777	159,936	66,657	67,200	Salt Lake City	UT
63	72.684	41.766	139,739	51,464	133,800	Hartford	CT
64	84.554	42.709	127,321	50,635	48,400	Lansing	MI
65	116.226	43.607	125,738	50,852	67,700	Boise City	ID
66	84.281	30.457	124,773	50,442	72,400	Tallahassee	FL
67	95.692	39.038	119,883	49,936	48,800	Topeka	KS
68	123.022	44.925	107,786	40,936	60,300	Salem	OR
69	89.645	39.781	105,227	45,006	59,200	Springfield	IL
70	73.799	42.666	101,082	42,121	101,800	Albany	NY
71	80.886	34.039	98,052	33,919	72,600	Columbia	SC
72	74.764	40.223	88,675	30,744	71,300	Trenton	NJ
73	81.630	38.351	57,287	25,306	66,100	Charleston	WV
74	105.954	35.679	55,859	22,789	99,000	Santa Fe	NM
75	76.885	40.276	52,376	21,520	38,400	Harrisburg	PA
76	104.792	41.145	50,008	20,243	68,700	Cheyenne	WY
77	100.767	46.805	49,256	19,315	67,900	Bismarck	ND
78	119.743	39.148	40,443	15,895	99,300	Carson City	NV
79	71.560	43.232	36,006	14,222	112,400	Concord	NH
80	92.190	38.572	35,481	14,162	61,500	Jefferson City	MO
81	122.894	47.042	33,840	14,951	77,800	Olympia	WA
82	76.503	38.972	33,187	14,061	138,500	Annapolis	MD

83	75.517	39.159	27,630	9,903	88,700	Dover	DE
84	84.865	38.191	25,968	11,037	61,500	Frankfort	KY
85	112.020	46.597	24,569	10,428	63,200	Helena	MT
86	69.730	44.331	21,325	8,856	79,500	Augusta	ME
87	100.322	44.373	12,906	5,063	59,500	Pierre	SD
88	72.572	44.266	8,247	3,514	94,100	Montpelier	VT
			44,840,571	17,224,029		Total	

Cities selected are the 50 most populous cities in the lower 48 states according to the 1990 Population and Housing Census as well as capitals of the lower 48 states.

The **First Demand** is the 1990 population. The **Second Demand** is the number of households in the city in 1990. The **Fixed Cost** is the 1990 median home value in the city.

Cities are sorted by their 1990 population.

All data are from the 1990 Population and Housing Census (U.S. Department of Commerce, 1990a, 1990b).

Appendix H

Longitudes, Latitudes, Demands, and Fixed Costs for SORTCAP.GRT: A 49-Node Problem Defined on the Continental United States

No.	Long.	Lat.	First Demand	Second Demand	Fixed Cost	City	ST
1	121.467	38.567	29,760,021	369,365	115,800	Sacramento	CA
2	73.799	42.666	17,990,455	101,082	101,800	Albany	NY
3	97.751	30.306	16,986,510	465,622	72,600	Austin	TX
4	84.281	30.457	12,937,926	124,773	72,400	Tallahassee	FL
5	76.885	40.276	11,881,643	52,376	38,400	Harrisburg	PA
6	89.645	39.781	11,430,602	105,227	59,200	Springfield	IL
7	82.987	39.989	10,847,115	632,910	66,000	Columbus	OH
8	84.554	42.709	9,295,297	127,321	48,400	Lansing	MI
9	74.764	40.223	7,730,188	88,675	71,300	Trenton	NJ
10	78.659	35.822	6,628,637	207,951	96,600	Raleigh	NC
11	84.423	33.763	6,478,216	394,017	71,200	Atlanta	GA
12	77.475	37.531	6,187,358	203,056	66,600	Richmond	VA
13	71.018	42.336	6,016,425	574,283	161,400	Boston	MA

SORTCAP.GRT: A 49-Node Problem

14	86.146	39.776	5,544,159	731,327	60,800	Indianapolis	IN
15	92.190	38.572	5,117,073	35,481	61,500	Jefferson City	MO
16	89.388	43.080	4,891,769	191,262	75,200	Madison	WI
17	86.785	36.172	4,877,185	488,374	74,400	Nashville-Davidson	TN
18	122.894	47.042	4,866,692	33,840	77,800	Olympia	WA
19	76.503	38.972	4,781,468	33,187	138,500	Annapolis	MD
20	93.104	44.948	4,375,099	272,235	70,900	St. Paul	MD
21	91.126	30.449	4,219,973	219,531	67,900	Baton Rouge	LA
22	86.284	32.354	4,040,587	187,106	62,200	Montgomery	AL
23	84.865	38.191	3,685,296	25,968	61,500	Frankfort	KY
24	112.071	33.543	3,665,228	983,403	77,100	Phoenix	AZ
25	80.886	34.039	3,486,703	98,052	72,600	Columbia	SC
26	104.873	39.768	3,294,394	467,610	79,000	Denver	CO
27	72.684	41.766	3,287,116	139,739	133,800	Hartford	CT
28	97.513	35.467	3,145,585	444,719	54,900	Oklahoma City	OK
29	123.022	44.925	2,842,321	107,786	60,300	Salem	OR
30	93.617	41.577	2,776,755	193,187	49,500	Des Moines	IA
31	90.208	32.321	2,573,216	196,637	54,600	Jackson	MS
32	95.962	39.038	2,477,574	119,883	48,800	Topeka	KS
33	92.354	34.722	2,350,725	175,795	64,200	Little Rock	AR
34	81.630	38.351	1,793,477	57,287	66,100	Charleston	WV
35	111.930	40.777	1,722,850	159,936	67,200	Salt Lake City	UT
36	96.688	40.816	1,578,385	191,972	61,700	Lincoln	NE
37	105.954	35.679	1,515,069	55,859	99,000	Santa Fe	NM
38	69.730	44.331	1,227,928	21,325	79,500	Augusta	ME
39	119.743	39.148	1,201,833	40,443	99,300	Carson City	NV
40	71.560	43.232	1,109,252	36,006	112,400	Concord	NH
41	116.226	43.607	1,006,749	125,738	67,700	Boise City	ID
42	71.420	41.822	1,003,464	160,728	113,000	Providence	RI
43	112.020	46.597	799,065	24,569	63,200	Helena	MT
44	100.322	44.373	696,004	12,906	59,500	Pierre	SD
45	75.517	39.159	666,168	27,630	88,700	Dover	DE
46	100.767	46.805	638,800	49,256	67,900	Bismarck	ND
47	77.016	38.905	606,900	606,900	123,900	Washington	DC

48	72.572	44.266	562,758	8,247	94,100	Montpelier	VT
49	104.792	41.145	453,588	50,008	68,700	Cheyenne	WY
			247,051,601	10,220,590		Total	

Cities selected are the capitals of the lower 48 states as well as Washington, DC.

The **First Demand** is the 1990 state population. The **Second Demand** is the 1990 population of the state's capital. The **Fixed Cost** is the 1990 median home value in the city.

Cities are sorted by their 1990 state population.

All data are from the 1990 Population and Housing Census (U.S. Department of Commerce, 1990a, 1990b).

References

Aho, A. V., Hopcroft, J. E., and J. D. Ullman, 1983, *Data Structures and Algorithms*, Addison-Wesley, Reading, MA.

Ahuja, R. K., T. L. Magnanti, and J. B. Orlin, 1993, *Network Flows: Theory, Algorithms and Applications*, Prentice-Hall, Englewood Cliffs, NJ.

Aly, A. A. and J. A. White, 1978, "Probabilistic Formulation of the Emergency Service Location Problem," *Journal of the Operational Research Society*, 29 1167–1179.

Ballou, R. H., 1968, "Dynamic Warehouse Location Analysis," *Journal of Marketing Research*, 5, 271–276.

Belardo, S., J. Harrald, W. A. Wallace, and J. Ward, 1984, "A Partial Covering Approach to Siting Response Resources for Major Maritime Oil Spills," *Management Science*, 30, 1184–1196.

Benedict, J. M., 1983, "Three Hierarchical Objective Models Which Incorporate the Concept of Excess Coverage to Locate EMS Vehicles or Hospitals," M. S. Thesis, Department of Civil Engineering, Northwestern University, Evanston, IL.

Berger, R., 1994, "Analysis and Solution of a Location–Routing Model," Course Project, IEMS D52 and CE D71-2, Spring Term, Department of Industrial Engineering and Management Science, Northwestern University, Evanston, IL.

Berger, R., C. Coullard, and M. S. Daskin, 1995, "A Joint Location/Routing Model for Newspaper Distribution," to be presented at the TIMS/ORSA Meeting, Los Angeles.

Berkey, D., S. Homer, and A. Kanamori, 1993, "Intractable Problems: P = NP,'" *Bostonia*, Fall, 30–37.

Berman, O., R. C. Larson, and S. S. Chiu, 1985, "Optimal Server Location on a Network Operating as an $M/G/1$ Queue," *Operations Research*, 33(3), 746–771.

Berman, O., R. C. Larson, and N. Fouska, 1992, "Optimal Location of Discretionary Service Facilities," *Transportation Science* 26(3), 201–211.

Bertsekas, D., 1991, *Linear Network Optimization: Algorithm and Codes*, MIT Press, Cambridge, MA.

Braca, J., J. Bramel, B. Posner, and D. Simchi-Levi, 1994, "A Computerized Approach to the New York City School Bus Routing Problem," to appear in *IIE Transactions*.

Brandeau, M. L. and S. S. Chiu, 1989, "Overview of Representative Problems in Location Research," *Management Science*, 35(6), 645–674.

Brimberg, J. and R. F. Love, 1992, "A New Distance Function for Modelling Travel Distances in a Transportation Network," *Transportation Science*, 26, 129–137.

Brimberg, J. and R. F. Love, 1993, "General Considerations on the Use of the Weighted l_p Norm as an Empirical Distance Measure," *Transportation Science*, 27, 341–349.

Broin, M. W. and T. J. Lowe, 1986, "A Dynamic Programming Algorithm for Covering Problems with (Greedy) Totally Balanced Constraint Matrices," *SIAM Journal of Algebraic and Discrete Methods*, 7, 348–357.

Campbell, J. F., 1994, "Integer Programming Formulations of Discrete Hub Location Problems," *European Journal of Operational Research* **72**(2), 387–405.

Carson, Y. M. and R. Batta, 1990, "Locating an Ambulance on the Amherst Campus of the State University of New York at Buffalo," *Interfaces*, **20**(5), 43–49.

Chaudhry, S. S., I. D. Moon, and S. T. McCormick, 1987, "Conditional Covering: Greedy Heuristics and Computational Results," *Computers and Operations Research*, **14**, 11–18.

Chaudhry, S. S., 1993. "New Heuristics for the Conditional Covering Problem," *Opsearch*, **30** 42–47.

Chhajed, D., R. L. Francis, and T. J. Lowe, 1993, "Contributions of Operations Research to Location Analysis," *Location Science*, **1**, 263–287.

Chiu, S. S., O. Berman, and R. C. Larson, 1985, "Locating a Mobile Server Queuing Facility on a Tree Network," *Management Science*, **31**, 764–772.

Christofides, N. and J. E. Beasley, 1982, "A Tree Search Algorithm for the p-Median Problem," *European Journal of Operational Research*, **10**, 196–204.

Church, R. L. and D. J. Eaton, 1987, "Hierarchical Location Analysis Using Covering Objectives," In *Spatial Analysis and Location–Allocation Models* (A. Ghosh and G. Rushton, eds.), Van Nostrand Reinhold, New York, pp. 163–185.

Church, R. L., and R. S. Garfinkel, 1978, "Locating an Obnoxious Facility on a Network," *Transportation Science*, **12**, 107–118.

Church, R. L. and M. Meadows, 1979, "Location Modeling Utilizing Maximum Service Distance Criteria," *Geographical Analysis*, **11**, 358–379.

Church, R. L. and C. ReVelle, 1974, "The Maximal Covering Location Problem," *Papers of the Regional Science Association*, **32**, 101–118.

Cohon, J. L., 1978, *Multiobjective Programming and Planning*, Academic Press, New York.

Cornuejols, G., G. L. Nemhauser, and L. A. Wolsey, 1990, "The Uncapacitated Facility Location Problem," in *Discrete Location Theory* (P. B. Mirchandani and R. L. Francis, eds.), Wiley, New York, Chapter 3, pp. 119–171.

Crowder, H., 1976, "Computational Improvements for Subgradient Optimization," *Symposia Mathematica*, **19**, Academic Press, New York.

Daskin, M. S., 1982, "Application of an Expected Covering Model to EMS System Design," *Decision Sciences*, **13**(3), 416–439.

Daskin, M. S., 1983, "A Maximum Expected Covering Location Model: Formulation, Properties and Heuristic Solution," *Transportation Science*, **17**, 48–70.

Daskin, M. S., 1987, "Location, Dispatching and Routing Models for Emergency Services with Stochastic Travel Times," in *Spatial Analysis and Location-Allocation Models* (A. Ghosh and G. Rushton, eds.), Van Nostrand Reinhold Co., New York, pp. 224–265.

Daskin, M. S. and A. Haghani, 1984, "Multiple Vehicle Routing and Dispatching to an Emergency Scene," *Environment and Planning A*, **16**, 1349–1359.

Daskin, M. S., A. E. Haghani, and C. Malandraki, 1986, "Computational Experiments with Maximum Covering Algorithms," presented at the ORSA/TIMS Conference, Miami.

Daskin, M. S., K. Hogan, and C. ReVelle, 1988, "Integration of Multiple, Excess, Backup, and Expected Covering Models," *Environment and Planning B*, **15**, 15–35.

Daskin, M. S., W. J. Hopp, and B. Medina, 1992, "Forecast Horizons and Dynamic Facility Location," *Annals of Operations Research*, **40** 125–151.

Daskin, M. S. and P. C. Jones, 1993, "A New Approach to Solving Applied Location/Allocation Problems," *Microcomputers in Civil Engineering*, **8**, 409–421.

Daskin, M. S., P. C. Jones, and T. J. Lowe, 1990, "Rationalizing Tool Selection in a Flexible Manufacturing System for Sheet Metal Products," *Operations Research*, **38**, 1104–1151.

References

Daskin, M. S. and E. Stern, 1981, "A Hierarchical Objective Set Covering Model for EMS Vehicle Deployment," *Transportation Science*, **15**, 137–152.

Densham, P. J. and G. Rushton, 1992, "A More Efficient Heuristic for Solving Large P-Median Problems," *Paper in Regional Science*, **71**, 307–239.

Desrochers, M., Y. Dumas, F. Soumis, and P. Trudeau, 1991, "Column Generation Approaches to Airline Crew Scheduling Problems," presented at the 1991 TRISTAN Conference, Montreal.

Eaton, D. and M. S. Daskin, 1980, "A Multiple Model Approach to Planning Emergency Medical Service Vehicle Deployment," in *Proceedings of the International Conference on Systems Science in Health Care* (C. Tilquin, ed.), Pergamon, Montreal, pp. 951–959.

Eaton, D., M. S. Daskin, D. Simmons, B. Bulloch, and G. Jansma, 1985, "Determining Emergency Medical Service Vehicle Deployment in Austin, Texas," *Interfaces*, **15**(1), 96–108.

Ecker, J. G. and M. Kupferschmid, 1988, *Introduction to Operations Research*, Wiley, New York.

Erkut, E. and S. Neuman, 1988, "A Survey of Analytical Models for Locating Undesirable Facilities," Research Report 88-5, Department of Finance and Management Science, Faculty of Business, University of Alberta, Edmonton, Alberta.

Erkut, E. and S. Neuman, 1989, "Analytical Models for Locating Undesirable Facilities," *European Journal of Operational Research*, **40**, 275–291.

Erkut, E. and S. Neuman, 1992, "A Multiobjective Model for Locating Undesirable Facilities," *Annals of Operations Research*, **40**, 209–227.

Erkut, E. and V. Verter, 1994, "Hazardous Materials Logistics: An Annotated Bibliography," Research Report 94-1, Department of Finance and Management Science, Faculty of Business, University of Alberta, Edmonton, Alberta.

Erlenkotter, D., 1978, "A Dual-Based Procedure for Uncapacitated Facility Location," *Operations Research*, **26**, 992–1009.

Fitzsimmons, J. A., 1973, "A Methodology for Emergency Ambulance Deployment," *Management Science*, **19**, 627–636.

Floyd, R. W., 1962, "Algorithm 97, Shortest Path," *Comm. ACM*, **5**, 345.

Flynn, J. and S. Ratick, 1988, "A Multiobjective Hierarchical Covering Model for the Essential Air Services Program," *Transportation Science*, **22**, 139–147.

Francis, R. L., L. F. McGinnis, Jr., and J. A. White, 1992, *Facility Location and Layout: An Analytical Approach*, Prentice-Hall, Englewood Cliffs, NJ.

Fulkerson, D. R., 1961, "The Out-of-Kilter Method for Minimal Cost Flow Problems," *SIAM Journal of Applied Mathematics*, **9**, 18–27.

Garey, M. R. and D. S. Johnson, 1979, *Computers and Intractability: A Guide to the Theory of NP-Completeness*, W. H. Freeman and Co., New York.

Garfinkel, R. S., A. W. Neebe, and M. R. Rao, 1977, "The m-Center Problem: Minimax Facility Location," *Management Science*, **23**, 1133–1142.

Geoffrion, A. M. and G. W. Graves, 1974, "Multicommodity Distribution System Design Using Bender's Decomposition," *Management Science*, **20**, 822–844.

Ghosh, A. and F. Harche, 1993, "Location–Allocation Models in the Private Sector: Progress, Problems, and Prospects," *Location Science*, **1**, 81–106.

Glover, F., 1990, "Tabu Search: A Tutorial," *Interfaces*, **20** (4), 74–94.

Glover, F., Klingman, D., and N. V. Phillips, 1992, *Network Models in Optimization and Their Applications in Practice*, Wiley, New York.

Goldberg, D. E., 1989, *Genetic Algorithms in Search, Optimization and Machine Learning*, Addison-Wesley, Reading, MA.

Goldberg, J., R. Dietrich, J. M. Chen, M. Mitwasi, T. Valenzuela, and E. Criss, 1990a, "A Simulation Model for Evaluating a Set of Emergency Vehicle Base Locations: Development, Validation, and Usage," *Socio-Economic Planning Sciences*, **24**, 125–141.

Goldberg, J., R. Dietrich, J. M. Chen, M. Mitwasi, T. Valenzuela, and E. Criss, 1990b, "Validating and Applying a Model for Locating Emergency Medical Vehicles in Tucson, AZ," *European Journal of Operations Research*, **49**, 308–324.

Goldberg, J. and L. Paz, 1991, "Locating Emergency Vehicle Bases when Service Time Depends on Call Location," *Transportation Science*, **25** 264–280.

Goldberg, J. and F. Szidarovszky, 1991a, "A General Model and Convergence Results for Determining Vehicle Utilization in Emergency Systems," *Stochastic Models*, **7**, 137–160.

Goldberg, J. and F. Szidarovszky, 1991b, "Methods for Solving Nonlinear Equations Used in Evaluating Emergency Vehicle Busy Probabilities," *Operations Research*, **39**, 903–916.

Golden, B., L. Bodin, T. Doyle, and W. Stewart, Jr., 1980, "Approximate Traveling Salesman Algorithms," *Operations Research*, **28**, 694–711.

Goldman, A. J., 1971, "Optimal Center Location in Simple Networks," *Transportation Science*, **5**, 212–221.

Hakimi, S. L., 1964, "Optimum Location of Switching Centers and the Absolute Centers and Medians of a Graph," *Operations Research*, **12**, 450–459.

Hakimi, S. L., 1965, "Optimum Distribution of Switching Centers in a Communication Network and Some Related Graph Theoretic Problems," *Operations Research*, **13**, 462–475.

Handler, G. Y., 1990, "p-Center Problems," in *Discrete Location Theory* (P. B. Mirchandani and R. L. Francis, eds.), Wiley, New York, Chapter 7, pp. 305–347.

Handler, G. Y. and P. B. Mirchandani, 1979, *Location on Networks*, MIT Press, Cambridge, MA.

Hillier, F. S. and G. Lieberman, 1986, *Introduction to Operations Research*, 4th ed., Holden Day, Oakland, CA.

Hillsman, E. L., 1984, "The p-Median Structure as a Unified Linear Model for Location–Allocation Analysis," *Environment and Planning A*, **15**, 305–318.

Hoffman, A. J., A. W. J. Kolen, and M. Sakarovitch, 1985, "Totally Balanced and Greedy Matrices," *SIAM Journal of Algebraic and Discrete Methods*, **6**, 721–730.

Hooker, J. N., 1986, "Karmarkar's Linear Programming Algorithm," *Interfaces*, **16**(4), 75–90.

Hurter, A. P. and J. S. Martinich, 1989, *Facility Location and the Theory of Production*, Kluwer, Boston, MA.

Jacobsen, S. K., 1990, "Multiperiod Capacitated Location Models," in *Discrete Location Theory* (P. B. Mirchandani and R. L. Francis, eds.), Wiley, New York, Chapter 4, pp. 173–208.

Jarvis, J. P., K. A. Stevenson, and T. R. Willemain, 1975, "A Simple Procedure for the Allocation of Ambulances in Semi-Rural Areas," Technical Report 13-75, Operations Research Center, MIT, Cambridge, MA.

Jensen, P. A. and J. W. Barnes, 1980, *Network Flow Programming*, Wiley, New York.

Kariv, O. and S. L. Hakimi, 1979a, "An Algorithmic Approach to Network Location Problems. I: The p-Centers," *SIAM Journal on Applied Mathematics*, **37**, 513–538.

Kariv, O. and S. L. Hakimi, 1979b, "An Algorithmic Approach to Network Location Problems. II: The p-Medians," *SIAM Journal on Applied Mathematics*, **37**, 539–560.

Karp, R. M., 1972, "Reducibility Among Combinatorial Problems," in *Complexity of Computer Computations* (R. E. Miller and J. W. Thatcher, eds.), Plenum, New York, pp. 85–103.

Kennington, J. L. and R. V. Helgason, 1980, *Algorithms for Network Programming*, Wiley, New York.

Klimberg, R., 1992, "GRADS: A New Graphical Display System for Visualizing Multiple Criteria Solutions," *Computers and Operations Research*, **19**, 707–711.

References

Klimberg, R. and R. M. Cohen, 1993, "Experimental Evaluation of a Graphical Display System to Visualizing Multiple Criteria Solutions," Working Paper Series 93-26, Boston University School of Management, Boston.

Kolesar, P. and W. E. Walker, 1974, "An Algorithm for the Dynamic Relocation of Fire Companies," *Operations Research*, 23, 249-274.

Krarup, J. and P. M. Pruzan, 1990, "Ingredients of Locational Analysis," in *Discrete Location Theory* (P. B. Mirchandani and R. L. Francis, eds.), Wiley, New York, Chapter 1, pp. 1-54.

Kuby, M. J., 1987, "Programming Models for Facility Dispersion: The p-Dispersion and Maxisum Dispersion Problems," *Geographical Analysis*, 19, 315-329.

Kuby, M. J., 1989, "A Location-Allocation Model of Lösch's Central Place Theory: Testing on a Uniform Lattice Network," *Geographical Analysis*, 21, 316-337.

Kuby, M. J. and R. G. Gray, 1993, "The Hub Network Design Problem with Stopovers and Feeders: The Case of Federal Express," *Transportation Research*, 27A, 1-12.

Kuby, M., S. Qingqi, T. Watanatada, S. Xufei, X. Zhijun, C. Wei, Z. Chuntai, Z. Dadi, Y. Xiaodong, P. Cook, T. Friesz, S. Neuman, L. Fatang, R. Qiang, W. Xusheng, and G. Shenhaui, 1995, "Planning China's Coal and Electricity Delivery System," Paper Presented as Part of the 1994 Edelman Prize Competition, Institute of Management Sciences, TIMS/ORSA Meeting, Boston, May 1994, and forthcoming in *Interfaces*, January, 1995.

Kuby, M., S. Ratick, and J. Osleeb, 1991, "Modeling U.S. Coal Export Planning Decisions," *Annals of the American Association of Geographers*, 81, 627-649.

Laporte, G., 1988, "Location-Routing Problems," in *Vehicle Routing: Methods and Studies* (B. L. Golden and A. A. Assad, eds.), North-Holland, Amsterdam, pp. 163-197.

Laporte, G., Y. Nobert and D. Arpin, 1986, "An Exact Algorithm for Solving a Capcitated Location-Routing Problem," *Annals of Operations Research*, 6, 293-310.

Laporte, G., Y. Nobert, and S. Taillefer, 1988, "Solving a Family of Multi-Depot Vehicle Routing and Location-Routing Problems," *Transportation Science*, 22, 161-172.

Larson, R. C., 1974, "A Hypercube Queuing Model for Facility Location and Redistricting in Urban Emergency Services," *Computers and Operations Research*, 1, 67-95.

Larson, R. C., 1975, "Approximating the Performance of Urban Emergency Service Systems," *Operations Research*, 23, 845-868.

Larson, R. C., 1987, "Perspectives on Queues: Social Justice and the Psychology of Queueing," *Operations Research*, 35(6), 895-905.

Larson, R. C. and A. R. Odoni, 1981, *Urban Operations Research*, Prentice-Hall, Englewood Cliffs, NJ.

Lawler, E., 1976, *Combinatorial Optimization: Networks and Matroids*, Holt, Reinhart and Winston, New York.

Linder-Dutton, L., R. Batta, and M. H. Karwan, 1991, "Equitable Sequencing of a Given Set of Hazardous Materials Shipments," *Transportation Science*, 25(2), 124-137.

List, G. F. and P. B. Mirchandani, 1991, "An Integrated Network/Planar Multiobjective Model for Routing and Siting for Hazardous Materials and Wastes," *Transportation Science*, 25(2), 146-156.

List, G. F., P. B. Mirchandani, M. A. Turnquist, and K. Zografos, 1991, "Modeling and Analysis for Hazardous Materials Transportation: Risk Analysis, Routing/Scheduling and Facility Location," *Transportation Science*, 25(2), 100-114.

Louveaux, F. V., 1993, "Invited Review—Stochastic Location Analysis," *Location Science*, 1, 127-154.

Love, R. F., J. G. Morris, and G. O. Wesolowsky, 1988, *Facility Location: Models and Methods*, North-Holland, New York.

Magnanti, T. L. and R. T. Wong, 1990, "Decomposition Methods for Facility Location Problems" in *Discrete Location Theory* (P. B. Mirchandani and R. L. Francis, eds.), Wiley, New York, Chapter 5, pp. 209-262.

Maranzana, F. E., 1964, "On the Location of Supply Points to Minimize Transport Costs," *Operational Research Quarterly*, **15**, 261-270.

Marsten, R., R. Subramanian, M. Saltzman, I. Lustig, and D. Shanno, 1990, "Interior Point Methods for Linear Programming: Just Call Newton, Lagrange, and Fiacco and McCormick!" *Interfaces*, **20**(4), 105-116.

Martinich, J. S., 1988, "A Vertex-Closing Approach to the p-Center Problem," *Naval Research Logistics*, **35**, 185-201.

Maze, T. H., S. Khasnabis, K. C. Kapur, and M. Kutsal, 1982, "A Methodology for Locating and Sizing Transit Fixed Facilities and the Detroit Case Study," Final Report, Urban Mass Transportation Administration, Grant Number MI-11-0004, Wayne State University, Detroit.

Maze, T. H., S. Khasnabis, K. Kapur, and M. S. Poola, 1981, "Proposed Approach to Determine Optimal Number, Size, and Location of Bus Garage Additions," *Transportation Research Record*, **781**, Washington, DC.

Megiddo, N., E. Zemel, and S. L. Hakimi, 1983, "The Maximum Coverage Location Problem," *SIAM Journal of Algebraic and Discrete Methods*, **4**, 253-261.

Minieka, E., 1970, "The m-Center Problem," *SIAM Review*, **12**, 138-139.

Minieka, E., 1977, "The Centers and Medians of a Graph," *Operations Research*, **25**, 641-650.

Minieka, E., 1978, *Optimization Algorithms for Networks and Graphs*, Marcel Dekker, New York.

Minieka, E., 1983, "Anticenters and Antimedians of a Network," *Networks*, **13**, 359-365.

Mirchandani, P. B., 1990, "The p-Median Problem and Generalizations," in *Discrete Location Theory* (P. B. Mirchandani and R. L. Francis, eds.), Wiley, New York, Chapter 2, pp. 55-117.

Mirchandani, P. B. and R. L. Francis, 1990, *Discrete Location Theory*, Wiley, New York.

Mirchandani, P. B. and A. R. Odoni, 1979, "Locations on Medians on Stochastic Networks," *Transportation Science*, **13**, 85-97.

Moore, G. C. and C. ReVelle, 1982, "The Hierarchical Service Location Problem," *Management Science*, **28**, 775-780.

Mukundan, S. and M. S. Daskin, 1991, "Joint Location/Sizing Maximum Profit Covering Models," *INFOR*, **29** (2), 139-152.

Narula, S. C., 1986, "Minisum Hierarchical Location-Allocation Problems on a Network: A Survey," in *Location Decisions: Methodology and Applications*, Annals of Operations Research (J. P. Osleeb and S. J. Ratick, eds.), **6**, J. C. Baltzer, pp. 257-272.

Nemhauser, G. L. and L. A. Wolsey, 1988, *Integer and Combinatorial Optimization*, Wiley, NY.

Noble, B., 1969, *Applied Linear Algebra*, Prentice-Hall, Englewood Cliffs, NJ.

O'Kelly, M. E., 1986a, "Activity Levels at Hub Facilities in Interacting Networks," *Geographical Analysis*, **18**, 343-356.

O'Kelly, M. E., 1986b, "The Location of Interacting Hub Facilities," *Transportation Science*, **20**, 92-106.

O'Kelly, M. E., 1987, "A Quadratic Integer Program for the Location of Interacting Hub Facilities," *European Journal of Operational Research*, **32**, 393-404.

O'Kelly, M. E., 1992, "Hub Facility Location with Fixed Costs," *Papers in Regional Science*, **71**, 292-306.

O'Kelly, M. E. and H. J. Miller, 1991, "Solution Strategies for the Single Facility Minimax Hub Location Problem," *Papers in Regional Science*, **70**, 367-380.

Osleeb, J. P. and S. J. Ratick, 1990, "A Dynamic Location-Allocation Model for Evaluating the Spatial Impacts for Just-in-Time Planning," *Geographical Analysis*, **22**, 50-69.

References

Osleeb, J. P., S. J. Ratick, P. Buckley, K. Lee, and M. Kuby, 1986, "Evaluating Dredging and Offshore Loading Locations for U.S. Coal Exports Using the Local Logistics System," in *Location Decisions: Methodology and Applications, Annals of Operations Research* (J. P. Osleeb and S. J. Ratick, eds.), **6**, J. C. Baltzer, pp. 163-180.

Papadimitriou, C. H. and K. Steiglitz, 1982, *Combinatorial Optimization: Algorithms and Complexity*, Prentice-Hall, Englewood Cliffs, NJ.

Perl, J. and M. S. Daskin, 1985, "A Warehouse Location-Routing Model," *Transportation Research*, **19B**, 381-396.

Perl, J. and P.-K. Ho, 1990, "Public Facilities Location under Elastic Demand," *Transportation Science*, **24** (2), 117-136.

Phillips, D. T. and A. Garcia-Diaz, 1981, *Fundamentals of Network Analysis*, Prentice-Hall, Englewood Cliffs, NJ.

Plane, D. R. and T. E. Hendrick, 1977, "Mathematical Programming and the Location of Fire Companies for the Denver Fire Department," *Operations Research*, **25**, 563-578.

Ratick, S. J., J. P. Osleeb, M. Kuby, and K. Lee, 1987, "Interperiod Network Storage Location-Allocation (INSLA) Models," in *Spatial Analysis and Location-Allocation Models* (A. Ghosh and G. Rushton, eds.), Van Nostrand Reinhold, New York, pp. 269-301.

Ratick, S. J. and A. L. White, 1988, "A Risk-Sharing Model for Locating Noxious Facilities," *Environment and Planning B*, **15**, 165-179.

ReVelle, C., J. Cohon, and D. Shobrys, 1991, "Simultaneous Siting and Routing in the Disposal of Hazardous Wastes," *Transportation Science*, **25** (2), 138-145.

ReVelle, C. and K. Hogan, 1989, "The Maximum Availability Location Problem," *Transportation Science*, **23**, 192-200.

ReVelle, C. and V. Marianov, 1991, "A Probabilistic FLEET Model with Individual Vehicle Reliability Requirements," *European Journal of Operations Research*, **53**, 93-105.

ReVelle, C., D. Marks, and J. C. Liebman, 1970, "An Analysis of Private and Public Sector Location Models," *Management Science*, **16**, 692-707.

Rubin, D. S. and H. M. Wagner, 1990, "Shadow Prices: Tips and Traps for Managers and Instructors," *Interfaces*, **20** (4), 150-157.

Sahni, S. and E. Horowitz, 1978, "Combinatorial Problems: Reducibility and Approximation," *Operations Research*, **26** (5), 718-759.

Sasaki, M. and Z. Drezner, 1994, "On the Airline Hub and Spoke System," Paper SD13.4 Presented at the TIMS/ORSA Meeting, Boston, April 24, 1994.

Schilling, D. A., V. Jayaraman, and R. Barkhi, 1993, "A Review of Covering Problems in Facility Location," *Location Science*, **1**, 25-55.

Schrage, L., 1991, *LINDO: An Optimization Modeling System—Text and Software*, The Scientific Press, South San Francisco, CA.

Sweeney, D. J. and R. L. Tatham, 1976, "An Improved Long-Run Model for Multiple Warehouse Location," *Management Science*, **22**, 748-758.

Swoveland, C., D. Uyeno, I. Vertinsky, and R. Vickson, 1973, "Ambulance Location: A Probabilistic Enumeration Approach," *Management Science*, **20**, 686-698.

Teitz, M. B. and P. Bart, 1968, "Heuristic Methods for Estimating Generalized Vertex Median of a Weighted Graph," *Operations Research*, **16**, 955-961.

The Travelling Salesman Problem: A Guided Tour of Combinatorial Optimization, 1985 (E. L. Lawler, J. K. Lenstra, A. H. G. Rinnooy Kan, and D. B. Shmoys, eds.), Wiley, New York.

Tien, J. M., K. El-Tell, and G. R. Simons, 1983, "Improved Formulations to the Hierarchical Health Facility Location-Allocation Problem," *IEEE Transactions on Systems, Man, and Cybernetics*, **SMC-13**, 1128-1132.

Toregas, C., R. Swain, C. ReVelle, and L. Bergman, 1971, "The Location of Emergency Service Facilities," *Operations Research*, **19**, 1363–1373.

U.S. Department of Commerce, 1990a, Bureau of the Census, Data Services Division, Washington, DC 20233, CD ROM 90-1C, 1990 Census of Population and Housing, Summary Tape File 1C, Issued February 1992.

U.S. Department of Commerce, 1990b, Bureau of the Census, Data Services Division, Washington, DC 20233, CD ROM 90-3A-10, 1990 Census of Population and Housing, Summary Tape File 3A, Connecticut, Maine, Rhode Island, Vermont, Issued September 1992.

Van Roy, T. J., 1986, "A Cross Decomposition Algorithm for Capacitated Facility Location," *Operations Research*, **34**, 145–163.

Van Roy, T. J. and D. Erlenkotter, 1982, "A Dual-Based Procedure for Dynamic Facility Location," *Management Science*, **28**, 1091–1105.

Vasudevan, J., E. Malini, and D. J. Victor, 1993, "Fuel Savings in Bus Transit Using Depot–Terminal Bus Allocation Model," *Journal of Transport Management*, **17**, 409–416.

Wagner, H. M., 1969, *Principles of Operations Research*, Prentice-Hall, Englewood Cliffs, NJ.

Walker, W. E., 1974, "Using the Set Covering Problem to Assign Fire Companies to Fire Houses," *Operations Research*, **22**, 275–277.

Weaver, J. R. and R. L. Church, 1983a, "Computational Procedures for Location Problems on Stochastic Networks," *Transportation Science*, **17**, 168–180.

Weaver, J. R. and R. L. Church, 1983b, "A Median Facility Location Problem with Nonclosest Facility Service," Working Paper 83-011, Department of Management Science and Statistics, College of Commerce and Business Administration, University of Alabama, University, AL.

Weaver, J. R. and R. L. Church, 1984, "The VAST Median Facility Location Model," Working Paper 84-012, Department of Management Science and Statistics, College of Commerce and Business Administration, University of Alabama, University, AL.

Wesolowsky, G. O., 1973, "Dynamic Facility Location," *Management Science*, **19**, 1241–1248.

Wesolowsky, G. O. and W. G. Truscott, 1976, "The Multiperiod Location–Allocation Problem with Relocation of Facilities," *Management Science*, **22**, 57–65.

White, B. P., 1994, "NAACP Takes Civil Rights into '90s," *The Chicago Tribune*, July 10, Section 1, pp. 1 and 6.

Wyman, M. M. and M. Kuby, 1994, "Proactive Optimization: General Framework and a Case Study Using a Toxic Waste Location Model with Technology Choice," presented at the 1993 International Symposium on Locational Decisions, ISOLDE VI, Lesvos and Chios, Greece.

Zionts, S., 1974, *Linear and Integer Programming*, Prentice-Hall, Englewood Cliffs, NJ.

Author Index

Page numbers in *italics* are references to authors who are included in the "et al." portion of a citation.

Aho, A. V., 35
Ahuja, R. K., 35, 54, 88, 90
Aly, A. A., 6
Arpin, D., 344

Ballou, R. H., 14
Barkhi, R., 92
Barnes, J. W., 35, 54, 59, 61
Bart, P., 216
Batta, R., 7, 9
Beasley, J. E., 231
Belardo, S., 5
Benedict, J. M., 108
Berger, R., 348
Bergman, L., *4*
Berkey, D., 90
Berman, O., 120, 275
Bertsekas, D., 54
Bodin, L., *208*
Braca, J., 396
Bramel, J., *396*
Brandeau, M. L., 10
Brimberg, J., 389
Broin, M. W., 106
Buckley, P., *307*
Bulloch, B., *5*

Campbell, J. F., 354, 355, 359-360
Carson, Y. M., 7
Chaudhry, S. S., 333
Chen, J. M., *133*
Chhajed, D., 11
Chiu, S. S., 10, 275
Christofides, N., 231
Chuntai, Z., *390*
Church, R. L., 5-6, 9, 94, 110-111, 134, 330
Cohen, R. M., 393

Cohon, J. L., 9, 317, 371, 393
Cook, P., *390*
Cornuejols, G., 274
Coullard, C., 348
Criss, E., *133*
Crowder, H., 231-232

Dadi, Z., *390*
Daskin, M. S., 2, 4-6, 14, 76, 105-106, 108, 122, 127, 130, 200, 344, 348, 387
Densham, P. J., 217
Desrochers, M., 105
Dietrich, R., *133*
Doyle, T., *208*
Drezner, Z., 382
Dumas, Y., *105*

Eaton, D. J., 5, 16, 330
Ecker, J. G., 20,31
El-Tell, K., 319
Erkut, E., 9, 17, 363, 366, 372, 393
Erlenkotter, D., 14, 265, 269-270, 273-274, 302

Fatang, L., *390*
Fitzsimmons, J. A., 6
Floyd, R. W., 47
Flynn, J., 362
Fouska, N., 120
Francis, R. L., 11, 17
Friesz, T., *390*
Fulkerson, D. R., 54

Garcia-Diaz, A., 54, 61
Garey, M. R., 82-83, 87, 89-90, 94, 161, 203
Garfinkel, R. S., 9, 191
Geoffrion, A. M., 338
Ghosh, A., 15

491

Glover, F., 54, 374
Goldberg, D. E., 374
Goldberg, J., 133
Golden, B., 208
Goldman, A. J., 206
Graves, G. W., 338
Gray, R. G., 362

Haghani, A., 6, 122, 127
Hakimi, S. L., 6, 114, 161, 180, 202, 207-208
Handler, G. Y., 11, 183, 186, 190-191, 202
Harche, F., 15
Harrald, J., 5
Helgason, R. V., 35, 54
Hendrick, T. E., 4, 109
Hillier, F. S., 20, 31
Hillsman, E. L., 311
Ho, P.-K., 16, 200
Hoffman, A. J., 106
Hogan, K., 108,134
Homer, S., 90
Hooker, J. N., 31
Hopcroft, J. E., 35
Hopp, W. J., 14
Horowitz, E., 90
Hurter, A. P., 11

Jacobsen, S. K., 6
Jansma, G., 5
Jarvis, J. P., 4
Jayaraman, V., 92
Jensen, P. A., 35, 54, 59, 61
Johnson, D. S., 82-83, 87, 89-90, 94, 161, 203
Jones, P. C., 2, 76, 105-106, 387

Kanamori, A., 90
Kapur, K. C., 76
Kariv, O., 161, 180, 207-208
Karp, R. M., 88, 90
Karwan, M. H., 9
Kennington, J. L., 35, 54
Khasnabis, S., 76
Klimberg, R., 393
Klingman, D., 54
Kolen, A. W. J., 106
Kolesar, P., 14
Krarup, J., 10
Kuby, M. J., 9, *14*, 16, 307, 362, 366, 373, 390
Kupferschmid, M., 20, 31
Kutsal, M., 76

Laporte, G., 344
Larson, R. C., 6, 120, 133, 177, 275
Lawler, E., 47, 340

Lee, K., *14, 307*
Lenstra, J. K., 340
Lieberman, G., 20, 31
Liebman, J. C., 15
Lindner-Dutton, L., 9
List, G.F., 9, 371
Louveaux, F. V., 14
Love, R. F., 11, 389
Lowe, T. J., 2, 11, 105-106
Lustig, I., *31*

Magnanti, T. L., 35, 54, 88, 90, 303
Malandraki, C., 122, 127
Malini, E., 76
Maranzana, F. E., 213
Marianov, V., 134
Marks, D., 15
Marsten, R., 31
Martinich, J. S., 11, 191
Maze, T. H., 76
McCormick, S. T., 333
McGinnis, L. F., Jr., 17
Meadows, M., 94, 111, 134
Medina, B., 14
Megiddo, N., 114
Miller, H. J., 360
Minieka, E., 9, 54, 59, 191
Mirchandani, P. B., 6, 9, 11, 183, 186, 190, 200, 371
Mitwasi, M., *133*
Moon, I. D., 333
Moore, G. C., 319
Morris, J. G., 11
Mukundan, S., 200

Narula, S.C., 17, 319-320
Neebe, A. W., 191
Nemhauser, G. L., 42, 106, 274
Neuman, S., 9, 17, 363, 366, 372, *390*, 393
Nobert, Y., 344
Noble, B., 33

Odoni, A. R., 6, 177, 202
O'Kelly, M. E., 350, 352, 354-355, 357, 360, 363
Orlin, J. B., 35, 54, 88, 90
Osleeb, J. P., 14, 307, 390

Papadimitriou, C. H., 87, 90
Paz, L., 133
Perl, J., 16, 200, 344
Phillips, D. T., 54, 61
Phillips, N. V., 54
Plane, D. R., 4, 109

Author Index

Poola, M. S., *76*
Posner, B., *396*
Pruzan, P. M., 10

Qiang, R., *390*
Qingqi, S., *390*

Rao, M. R., 191
Ratick, S. J., 9, 14, 307, 362, 372, 390
ReVelle, C., *4*, 5, 9, 15, 108, 110, 134, 319, 371
Rinnooy Kan, A. H. G., 340
Rubin, D. S., 27, 40, 58
Rushton, G., 217

Sahni, S., 90
Sakarovitch, M., 106
Saltzman, M., *31*
Sasaki, M., 382
Schilling, D. A., 92
Schrage, L., 366
Shanno, D., *31*
Shenhuai, G., *390*
Shobrys, D., 9, 371
Shmoys, D. B., 340
Simchi-Levi, D., *396*
Simmons, D., *5*
Simons, G. R., 319
Soumis, F., *105*
Steiglitz, K., 87, 90
Stern, E., 4, 108
Stevenson, K. A., 4
Stewart, W., Jr., *208*
Subramanian, R., *31*
Swain, R., *4*
Sweeney, D. J., 14
Swoveland, C., 6
Szidarovszky, F., 133

Taillefer, S., 344
Tatham, R. L., 14
Teitz, M. B., 216
Tien, J. M., 319

Toregas, C., 4
Trudeau, P., *105*
Truscott, W. G., 14
Turnquist, M. A., 9

Ullman, J. D., 35
United States Department of Commerce, 479, 482
Uyeno, D., 6

Valenzuela, T., *133*
Van Roy, T. J., 14, 303
Vasudevan, J., 76
Verter, V., 366
Vertinsky, I., *6*
Vickson, R., *6*
Victor, D. J., 76

Wagner, H. M., 20, 27, 31, 40, 58
Walker, W. E., 4, 14
Wallace, W. A., *4*
Ward, J., *4*
Watanatada, T., 390
Weaver, J. R., 6
Wei, C., *390*
Wesolowsky, G. O., 11, 14
White, A. L., 9, 372
White, B. P., 382
White, J. A., 6, 17
Willemain, T. R., 4
Wolsey, L. A., 42, 106, 274
Wong, R. T., 303
Wyman, M. M., 9, 373

Xiaodong, Y., *390*
Xufei, S., *390*
Xusheng, W., *390*

Zemel, E., 114
Zhijun, X., *390*
Zionts, S., 31
Zografos, K., 9

Subject Index

Italicized citations of the form *n.m* are references to exercises at the end of the chapters, where *n* refers to the chapter number and *m* refers to the exercise number. All such references appear after standard page references.

0/1 knapsack problem, 89, 303
3-partition problem, 89

Application areas:
   airline crew assignment, 105-106
   ambulance bases, 1-7, 16-17, 339, 385-388
   automatic teller machines (ATM), 275, 318-319
   automobile production and distribution, 333-334, 336, 387
   assembly plants, 1, 15
   banking, 318-319
   car rental locations, *9.2*
   distribution centers or warehouses, 1, 17, 198, 309, 334-336, 386, *1.1*
   emergency highway vehicles, 1
   fast-food industry, 345
   fire stations, 1, 17, 391
   fortune tellers, *6.12*
   franchise outlets, 363, 373, *9.6*
   hazardous waste sites, 9, 15, 17, 363, 374, *8.18*
   health care facilities, 318-320, 327-333, 339, *8.4*
   hospitals, 9, 17, 318-320, 327-329, 331-333
   housing, 1, *1.3*
   landfills, 2, 9, 17, 309, 363, 371, 374
   libraries, 9, 237
   missile silos, 18, 363
   newspaper distribution, 345-348, 373
   nuclear reactors, 363, 374
   offices, 1
   prisons, 18, *1.3*, *9.4*
   post offices and postal services, 17, 318-319, 327-328, 333
   production plants, 1, 17
   retail outlets, 1, 3, 198
   schools, 9, 15, 17, 317-319, 327, 333
   tool selection, 2, 106
   vehicle emission testing stations, *1.4*, *9.3*

Binary search, 84, 173
Branch and bound, 20, 27, 99-101, 130, 132, 134-135, 274, 278-280, 282, 288-290, 292, 366, 679, 701, 702, *4.14*, *4.15*
   algorithm for solving integer programming problems, 102-103
   child node, 101
   fathoming nodes in, 101-102
   lower bound, 101
   parent node, 101
   root node, 101

Capacitated fixed charge location problem, 275, 277-278, 281, 283, 296-299, 302-303, *7.10*, *8.18*
   Bender's decomposition for, 292-302-303, 338, *7.11*
   continuous knapsack problem and, 285, 303
   linear programming relaxation of, 276
   Lagrangian relaxation algorithm for, 277-292
   mathematical formulation of the problem, 276, 294-297
COLORS.COL file, 404, 459-462
COLORSET program, 459-462
Column generation, 348
Complexity theory, 11, 80
Complimentary anticover model, 372
Cost per mile, 248-249, 252, 254-257
Coverage, 4, 92, 198, 309, 324, 347, 372, 393,

494

Subject Index

415, 453–454
dominated nodes, 95, 420
Covering problems, 198, 237, 309, 383, 386. *See also* Maximum covering; Maximum expected covering; Set covering

Decision problem, 83–85
Discrete location models, 10
Dispersion models, 363–364, 366, 373
DUALOC, 265

Exponential time algorithms, 82

Fixed charge facility location model, 13, 250, 302–303, 309, 338, 348, 383, *7.1. See also* Capacitated fixed charge location problem; Uncapacitated fixed charge location problem
  verbal statement of the problem, 9
Fixed costs of facility locations, 247–248, 250–257, 259, 278–279, 291, 298–299, 302, 336, 342, 347, 357, 372–373, 405–406, 436, *8.5*
Genetic algorithms, 374

Hamiltonian cycle problem, 88
Hierarchical facility location problems, 317–328, 330, 333, 373, *8.7, 8.8*
  budget constraint and, 326–327
  capacitated facilities and, 327–328
  coverage-based location models for, 324–326, 330
  fixed charge formulation and, 326
  globally inclusive service hierarchy, 320–321, 328
  locally inclusive service hierarchy, 320, 322
  maximum covering location problem, 325–326
  median-based location model for, 321–322
  successively exclusive facility hierarchy, 319–320, 324
  successively exclusive service hierarchy, 320–321
  successively inclusive facility hierarchy, 319–322, 324, 328
  successively inclusive service hierarchy, 320
Hub location problems, 349, 362, 373, *8.15, 8.16, 8.17*
  center model for, 360–361
  covering-based model for, 361–362
  fixed hub costs in, 357, 360
  fixed hub-spoke connection costs in, 359–360, 362

fully connected networks, 349–351
HEUR1 for, 355, 363
HEUR2 for, 355, 363
hub capacities and, 359–360
hub and spoke networks, 349–352
minimum spoke/hub flows in, 359–360, 362
multiple hubs and, 351
P-hub median model for, 353–358, 361–362
single-hub median model for, 352–353
single spoke assignment in, 357–360
sub-optimality of nearest assignment in, 354
Hypercube queuing model, 133

Integer programming, 18, 99–104
Interacting facilities, 328, 330, 333, 373
  coverage and, 330
  flows between facilities, 328, *8.9*
  proximity constraints and, 331

Lagrangian relaxation, 122, *4.14. See also* Capacitated fixed charge location problem; Maximum covering; Uncapacitated fixed charge location problem
  algorithm for, 122–123
  application to capacitated fixed charge location problem, 277–292
  application to maximum covering problem, 123–129
  application to P-median problem, 221–231
  application to uncapacitated fixed charge location problem, 262–265
  subgradient optimization and, 123, 126, 263, 280, 284, 290, 292
Linear programming, 18, 20, *2.1, 2.2, 2.3, 2.18*
  canonical form, 21
  complementary slackness conditions, 20–21, 24, 27, 29, 33–36, 99
  decision variables, 21
  dual problem, 20, 23–29
  feasible solution, 41
  inputs, 20
  interior point algorithm, 30–31
  primal problem, 20–21, 23, 25–29
  relaxation of integer programming problem, 100, 102, 201–202
  simplex algorithm, 27, 30–31
  slack variables, 24
  standard form, 21
  strong duality, 27, 37
  surplus variables, 21

Linear programming (*Continued*)
  unbounded solution, 25-27
  weak duality, 27
Location-routing models, 339, 342, 348, 372-373
l_p distance metric, 13

Manhattan distance metric, 12
Maxian model, verbal statement of the problem, 9
Maximum covering, 8, 13, 16, 92, 110, 135, 154-155, 198-200, 202, 215, 247, 311, 391, 401-402, 414-415, 419-420, 423, 433, 436, 441, 450, *4.10*, *4.11*, *4.14*, *4.15*, *4.16*, *4.21*, *4.22*, *7.3*, *8.2*, *8.3*, *8.5*
  greedy adding algorithm for, 113-115, 119, 122, 132, 135, 208, 416, 419-420
  greedy adding and substitution algorithm for, 116-119, 122, 129, 132, 135, 208, 215, 259, 374, 416, 419-420, 434
  Lagrangian relaxation algorithm for, 122, 125, 129, 135, 222, 419-424, 425, 441
  mathematical formulation of the problem, 110, 198, 312
  verbal statement of the problem, 4
Maximum expected covering, 130, 132-135, 198, 275, *4.24*
  mathematical formulation of, 131
  system-wide probability of being busy, 133
Maximum flow problem, 63, *2.17*
Maxisum model, verbal statement of the problem, 9
MENU-OKF program, 54, 64, 404, 459-463, 474, *2.8*, *2.19*
Minimum spanning tree, *2.16*
MOD-DIST program, 316, 406, 446, 450, 459-462, *4.17*, *7.2*, *8.2*
Multiobjective problems, 9, 309-311, 313, 313, 316-317, 331-332, 373, 390-391, *8.1*, *8.2*, *8.5*
  constraint method for, 316-317
  convex dominated solutions, 311, 317, 393
  dominated solutions and, 310
  duality gap solutions, 311, 317
  inferior solutions and , 310
  noninferior solutions and, 310, 313-317, 373, 397
  tradeoff diagram and, 310, 313, 316-317, 398
  undesirable facilities and, 372
  value path diagrams and, 393-396, 398, *9.1*
  weighting method for, 311, 313-314, 316
Multiproduct flows, 333

NET-SPEC program, 406, 446, 459-462, *4.10*, *7.2*
Network intersection points, 94, 111, 134
Network location models, 10-11
NIMBY (Not In My Back Yard) phenomenon, 363
Noxious facilities, 363
NP-complete problems, 86-88, 89-90, 96, 340, *7.9*, *7.10*
  strong sense, 89
NP-hard problems, 85, 88

Objective space, 390, 392-393, 396, 398
Obnoxious facilities, 363
Optimization problem, 83-84
Out-of-kilter flow problem, 54, *2.13*, *2.17*, *2.19*
  algorithm for, 59
  circulation flow, 54
  flow augmentation path, 60-61
  kilter diagram, 57-58, 60-63
  mathematical formulation of the problem, 54-55
  reduced cost of a link in, 57
  revising node labels in, 60-61

Partition problem, 89
P-center problem, 8, 13, 16, 105, 154-161, 198, 202-203, 237, 247, 309, 366, 383, 386, *4.12*, *4.13*, *4.18*
  absolute, 154, 157-161, 176, 180, 181-182, 184, 187-191
  absolute 1-center on a tree, 161
  absolute 1-center on an unweighted tree, 161-166, 191, *5.1*, *5.2*, *5.3*, *6.5*
  absolute 1-center on a weighted tree, 167-168, 172, 191, *5.4*, *5.5*, *5.6*, *6.5*
  absolute 2-center on a tree, 161, 191
  absolute 2-centers on an unweighted tree, 166-167, 191, *5.1*, *5.2*, *6.5*
  complexity of, 161-162
  mathematical formulation of the vertex center, 160
  verbal statement of the problem, 6, 154
  vertex, 154, 157, 160-161, 173-176, 178, 180, 184, 186, 191, *5.7*, *5.8*, *5.9*, *5.10*, *5.11*
  vertex 1-center on an unweighted tree, 165, 191, *5.1*, *5.2*
Planar location models, 10-11
Planning process, 383-384, 386, 388, 391-392
  actors in, 385-386, 388, 398
  analysis, 384, 386, 392-393, 395-398
  analysts in, 383, 386, 390-391, 395, 398
  communication and decision, 384, 392, 397-398

ns in, 385
constraints, 387-388
data validation in, 388, 390, 398
decision makers in, 383, 385-388, 390-393, 396, 398
goals in, 384-386, 388, 391, 398
implementation, 384, 397-398
model calibration in, 389-390
objectives in, 386-388, 391, 398
planners in, 383, 386-388, 390, 392, 396, 398
problem definition, 383-384, 386, 391, 397-398
P-median problem, 9, 13, 16, 200, 236-237, 247, 250-251, 258-259, 302, 309, 311-313, 321, 324, 330, 348, 352, 354-356, 360, 366-367, 371, 383, 386-387, 401, 414-415, 419, 423-424, 433, 436, 450, *6.6, 6.7, 8.2, 8.17*
1-median on a tree, 203, 206-207, 237, *6.1, 6.2, 6.3, 6.4, 6.5*
complexity of, 203, 208, 237
exchange algorithm for, 215-221, 232-238, 258-259, 710, *6.9*
Lagrangian relaxation algorithm for, 218, 221-222, 232-233, 227-236, 238, 262-263, 424-426, *6.8, 6.9, 6.10*
mathematical formulation of the problem, 201, 221
myopic algorithm for, 208-210, 212, 214, 219-221, 232-238, 699, 710
neighborhood in, 213-215
neighborhood search algorithm for, 213-215, 217, 220-221, 232-238, 260, 416, 424-425, *6.8*
node optimality in, 202-203, 208-209, 237
polynomial for fixed P, 203
substitution algorithm for, *see also* exchange algorithm for
verbal statement of the problem, 6, 200
Polynomial reducibility, 87
Polynomial time algorithm, 82, 85, 88, 90, 161, 203, 208, 237
Polynomial transformation, 87
PRINTER.CNS file, 404, 431, 469, 474-475
Problem:
decision, 83-85
instance of, 81
optimization, 83-84
size of, 81-82
Production/distribution model, 334-338, 348, 373, *8.10, 8.11, 8.12, 8.13, 8.14*
inventory costs and, 336-338
Pseudo-polynomial time algorithm, 82, 89

Queueing, 6, 275, 388

Satisfiability problem, 86
Set covering, 8, 13-14, 16, 92, 105, 107, 132, 134-135, 154-156, 173-175, 185-191, 202, 247, *4.1, 4.2, 4.3, 4.4, 4.5, 4.6, 4.7, 4.8, 4.9, 4.12, 4.13, 4.16, 4.17, 4.21, 4.22, 7.2, 7.4, 7.9*
backup coverage and, 108, 134
column reduction for, 95, 97-98, 134
complexity of, 94
existing facilities and, 109
mathematical formulation of the problem, 93
row reduction for, 96-98, 134
verbal statement of the problem, 4, 92
Set partitioning problem, 105
Shortest path problem, 18, 20, 41, 63-64, 92, *2.9, 2.10, 2.11, 2.12, 2.13, 2.14, 2.15, 2.16*
complexity of, 84-85, 87, 165, *3.1*
Dijkstra's algorithm for, 43-44, 53, 84
Floyd's algorithm for, 47-48
mathematical formulation of the problem, 42
verbal statement of the problem, 41
Simulation, 6
Solution space, 390, 392-393, 396, 398
Strongly polynomial algorithm, 88
SITATION program, 231-232, 238, 249, 264, 401, 446, 450, 459-462, 474, *4.10, 4.13, 4.14, 4.15, 4.16, 4.17, 5.10, 6.8, 6.9, 6.10, 6.12, 7.2, 7.4, 7.7, 8.2, 8.5*
command line options for, 439-441
computer requirements for, 402
demand data set, 412-413. *See also* input files
distance files, 407-408, 411, 456-458
input files, 405
maximum number of nodes in, 406
printing graphics screens with, 404, 429-430
quick guide to, 402, 409, 441-445
starting the program, 404
storing on a hard disk, 402-403

Tabu search, 374
Transportation problem, 18, 20, 63-64, 276-278, 281, 291, 293, 295-296, 298-299, 302-303, 336, *2.4, 2.5, 2.6, 2.7, 2.8, 2.20, 7.11*
algorithm for, 34-35
complementary slackness conditions for, 33-36
dual variables and sensitivity analysis, 36, 40-41, 301-302

Transportation problem (*Continued*)
mathematical formulation of the problem, 31
northwest corner method, 34, 36
verbal statement of the problem, 31
Traveling salesman problem, 81, 83, 84, 85, 88, 340, *3.3*
Trees, 11

Uncapacitated fixed charge location problem, 250-251, 261, 270, 276-278, 302-303, 348, 373, 401, 415, 419, 423, 427, 433, 436, 450, *7.5, 7.6, 7.9, 7.10, 8.5*
ADD algorithm for, 250-255, 257, 260-262, 264-265, 302, 401, 416, 427, *7.5, 7.7*
condensed dual formulation of, 266-268, 271, 273
DROP algorithm for, 250-257, 262, 265, 302, 401, 427, *7.5*
dual adjustment for, 272-274, 281, 283
dual ascent algorithm for, 261, 265-274, 277-278, 281-283, 302
dual of linear programming relaxation, 266-267
exchange heuristic for, 258-259, 262, 265, 302, 401, 418, 427
Lagrangian relaxation algorithm for, 261-265, 277-278, 302, 401, 428-429, *7.5*
linear programming relaxation of, 270, 273-274
mathematical formulation of the problem, 250, 265
neighborhood search heuristic for, 258, 260-261, 265, 302, 401, 416, 418, 428
verbal statement of the problem, 250
Undesirable facilities, 363, 374
equity and, 9, 372-373, *8.18*
maximin model for, 371
maxisum model for, 366-371
multiobjective models of, 371-373
routing and, 371

Vehicle routing problems, 340, 342

WILEY-INTERSCIENCE SERIES IN DISCRETE MATHEMATICS AND OPTIMIZATION

ADVISORY EDITORS

RONALD L. GRAHAM
AT & T Bell Laboratories, Murray Hill, New Jersey, U.S.A.

JAN KAREL LENSTRA
Centre for Mathematics and Computer Science, Amsterdam, The Netherlands
Erasmus University, Rotterdam, The Netherlands

ROBERT E. TARJAN
Princeton University, New Jersey, and
NEC Research Institute, Princeton, New Jersey, U.S.A.

AARTS AND KORST
Simulated Annealing and Boltzmann Machines: A Stochastic Approach to Combinatorial Optimization and Neural Computing

ALON, SPENCER, AND ERDÖS
The Probabilistic Method

ANDERSON AND NASH
Linear Programming in Infinite-Dimensional Spaces: Theory and Application

BARTHÉLEMY AND GUÉNOCHE
Trees and Proximity Representations

BAZARRA, JARVIS, AND SHERALI
Linear Programming and Network Flows

COFFMAN AND LUEKER
Probabilistic Analysis of Packing and Partitioning Algorithms

DASKIN
Network and Discrete Location: Models, Algorithms, and Applications

DINITZ AND STINSON
Contemporary Design Theory: A Collection of Surveys

GLOVER, KLINGHAM, AND PHILLIPS
Network Models in Optimization and Their Practical Problems

GONDRAN AND MINOUX
Graphs and Algorithms
(*Translated by S. Vajda*)

GRAHAM, ROTHSCHILD, AND SPENCER
Ramsey Theory
Second Edition

GROSS AND TUCKER
Topological Graph Theory

HALL
Combinatorial Theory
Second Edition

JENSEN AND TOFT
Graph Coloring Problems

LAWLER, LENSTRA, RINNOOY KAN, AND SHMOYS, EDITORS
The Traveling Salesman Problem: A Guided Tour
of Combinatorial Optimization

LEVITIN
Perturbation Theory in Mathematical Programming Applications

MAHMOUD
Evolution of Random Search Trees

MARTELLO AND TOTH
Knapsack Problems: Algorithms and Computer Implementations

MINC
Nonnegative Matrices

MINOUX
Mathematical Programming: Theory and Algorithms
(*Translated by S. Vajda*)

MIRCHANDANI AND FRANCIS, EDITORS
Discrete Location Theory

NEMHAUSER AND WOLSEY
Integer and Combinatorial Optimization

NEMIROVSKY AND YUDIN
Problem Complexity and Method Efficiency in Optimization
(*Translated by E. R. Dawson*)

PACH AND AGARWAL
Combinatorial Geometry

PLESS
Introduction to the Theory of Error-Correcting Codes
Second Edition

SCHRIJVER
Theory of Linear and Integer Programming

TOMESCU
Problems in Combinatorics and Graph Theory
(*Translated by R. A. Melter*)

TUCKER
Applied Combinatorics
Second Edition